The Origin of Chronic Inflammatory Systemic Diseases and Their Sequelae

The Origin of Chronic Inflammatory Systemic Diseases and Their Sequelae

Rainer H. Straub

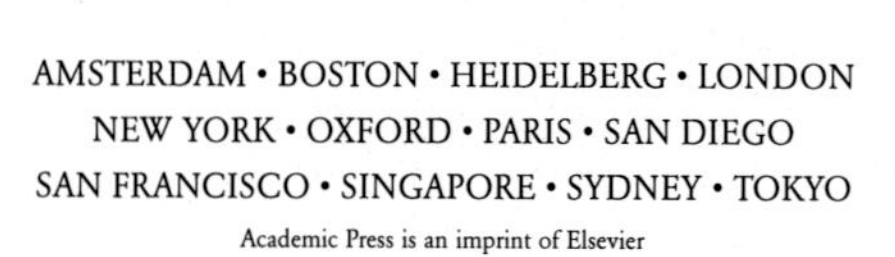

Academic Press is an imprint of Elsevier
225 Wyman Street, Waltham, MA 02451, USA
525 B Street, Suite 1800, San Diego, CA 92101–4495, USA
125 London Wall, London, EC2Y 5AS, UK
The Boulevard, Langford Lane, Kidlington, Oxford OX5 1GB, UK

Library of Congress Cataloging-in-Publication Data
A catalog record for this book is available from the Library of Congress

British Library Cataloguing in Publication Data
A catalogue record for this book is available from the British Library

ISBN: 978-0-12-803321-0

For information on all Academic Press publications
visit our website at http://store.elsevier.com/

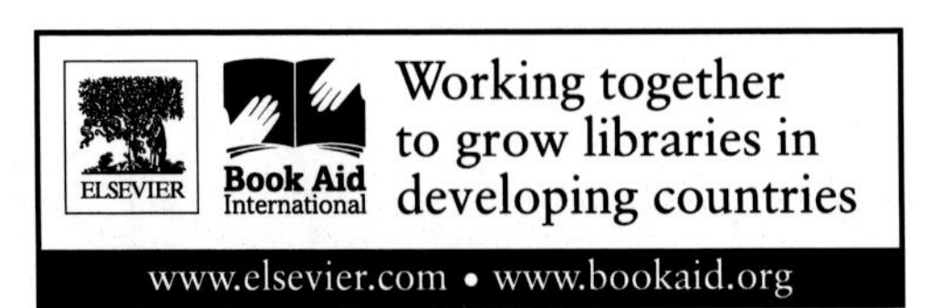

To all fellow researchers subject to persecution on the basis of religion, race, political reasons and nationality

Contents

Preface

One of my biggest disappointments in my beginnings in immunology was that it was fragmented in cells and and some molecules and we didn't have any way of considering a system that was regulated inside the body, the whole body.

Antonio Coutinho, immunologist. (Taken from Ref. 1.)

As an absolute outsider, I started to work in the field of immunology educated in basic neuroscience and clinical endocrinology. I was not trained in the hall of a successful immunology teacher, who published in excellent journals. I never applied for a position in an immunology laboratory because I wanted to become a real clinical doctor, a "country doctor." Even my postdoctoral time was spent outside immunology in the neuropharmacology laboratory at the University of Vienna. I did not spread the good gospel of a well-regarded immunologist, and I was never part of a distribution system in immunology that shared techniques and tools that multiply the success of collaborators. Because review boards consisted of peers that are subject to this influence, I did not receive grants until the age of 37. The bad part of this story is the fact that the things are gained late in life and that one has to be a pighead. The good part is that I have done it without the convincing power and the professional authority of an almighty teacher. This is a status that is typical to the newcomer.

Today, I know that new interpretations of nature often emerge when scientists are either young or newcomers.[2] In October 1995, I was young and a newcomer to immunology when I visited my first international meeting of the American College of Rheumatology in San Francisco. At the time, these meetings were not much influenced by pharmaceutical companies, and it was my impression that real immunology was presented with a clear link to the clinical problem related to the patient. This has changed much because nowadays, everything is clinical trial … clinical trial … clinical trial. There are some people who warn the community to think more of the etiologic basis of a given disease. I recall the excellent presidential speech into this direction of David Wofsy, San Francisco, but these statements usually go unheard.

Importantly, I met some outstanding colleagues with a similar interest not only in clinical immunology but also in endocrinology and neuroscience. This outstanding support group consisted of the clinically active rheumatologists Hans Bijlsma from Utrecht, Maurizio Cutolo from Genova, Ronald Wilder then from Bethesda, and Alfonse Masi from Peoria. Of the approximately 3000 presented abstracts, the 10 abstracts of the entire group including ours demonstrated research that linked immunology in chronic inflammatory systemic diseases to the then-esoteric subjects of endocrinology or neuroscience. This handful of people wanted to consider an immune system that was differentially regulated inside the whole body. In recent years, the number of professional supporters slowly increased. It is interesting that despite accepted research, not one of our more than 100 papers submitted to this meeting was ever presented as an abstract-based lecture. We outsiders were the splendid poster presenters. Better late than never, in 2012, I was pretty lucky to give the prestigious Philip S. Hench lecture of the American College of Rheumatology in San Diego, which was chaired by Eric L. Matteson.

In April 1996, I attended my first meeting of the Psychoneuroimmunology Research Society (PNIRS) in Santa Monica, which for me was the oasis in the desert. This was real interdisciplinary research presented by people trained in quite different disciplines. The important aspect of their research was the fact that they combined the thought styles[a] of at least two different research areas chosen from behavioral sciences, dermatology, endocrinology, gastroenterology, immunology, infectiology, neurosciences, pharmacology, psychiatry, psychology, rheumatology, tumor biology, and others. Attendees got infected by interdisciplinary ideas, were cross-pollinated, took up new information, and carried the interdisciplinary facts home. During these meetings, I met my first masters in an atmosphere of real interdisciplinary science.

Now, I realize that several scientific societies exist with a similar attitude toward interdisciplinary work in biomedicine such as the International Society for Neuroimmunomodulation (ISNIM), the International Society of Neuroimmunology (ISNI), the International Society of Psychoneuroendocrinology (ISPN), the Society on Neuroimmune Pharmacology (SNIP), and the German Brain Immune Network (GEBIN), to name the more important ones. In contrast to the classical thought schools of disciplinary research, members of the earlier-mentioned groups always try to bridge at least two different disciplines, and some people successfully bridged even three. These people were courageous and imperturbable, and I recognized in the mid-1990s that, although *in statu nascendi*, this was a big step forward toward a new interdisciplinary networking era. It is not the networking with people of different groups, which might also happen, but it is the idea to really understand the thought style of another discipline while already working with an own thought style.

As Kuhn pointed out, "acquisition of knowledge is discontinuous".[2] The philosophy of the previously mentioned societies consisted in the generation of an interdisciplinary atmosphere in order to shorten the interval in the discontinuous process of knowledge acquisition. Today, I might say that it was the strategy of these people to generate an outstanding positive environment

---

[a]Thought school or Denkschule, thought collective or Denkkollektiv, and thought style or Denkstil were expressions introduced by Fleck.[3]

for trainees in order to stimulate small "scientific revolutions," to use the metaphoric words of Kuhn.[2]

The disadvantage of our interdisciplinary research was the fact that we did not often develop new techniques or new molecular tools but rather used techniques and tools established in disciplinary research fields. If you study a network, the links—the lines between two nodes—are important and not so much the node itself. Necessarily, we do not find a node ourselves and completely rely on people who find nodes and, accordingly, genes. Ours is the *science of connections* or *interactions* (*Wissenschaft der Verbindung*).

During the past years, I sometimes thought that psycho-neuro-endocrino-immunology is so fragile that like a candle in the wind, it is endangered by being blown out by too many specialists. Most of my colleagues were subject to dishonorable attacks of disciplinary scientists in the 1980s and 1990s, which sometimes shattered their self-confidence, and some of them really stopped their career in the field because they lagged the necessary stamina, the supportive environment, or both. Nonetheless, the last two decades showed me many new and exciting aspects that recently started to interest journal editors, the scientific community, and funding agencies. Fortunately, the encouragement of funding agencies turned night into daybreak.

Many people outside the area possibly feel that numerous facts in our research are scattered and difficult to grasp, which is due to a general tendency of modern research toward the single molecule, the node. People often do not understand the network idea, which was quite different some five to six decades ago when everything was "system" or network.[4] On the other hand, it might well be that the network people have never tried to build up a strong paradigm that convinced the others. They only added to existing paradigms. They did not really show that network thinking leads to new paradigms, which might integrate the present node-directed molecular diversification. I am convinced that connecting the nodes leads to a wider understanding of nature's unsolved phenomena.

Kuhn pointed out:[2] "In the absence of a paradigm or some candidate of a paradigm, facts that could possibly pertain to the development of a given science are likely to seem equally relevant. This phase of early fact-gathering seems to be a nearly random activity." Personally, I think that the network people have overcome early random fact-gathering, and they are in the phase of confirming many experiments, which has given our interdisciplinary field the status of robust scientific knowledge.[1]

Looking back over the last three decades of active research, this is particularly true in the research field of psycho-neuroendocrine immunology of chronic inflammatory systemic diseases. These diseases are not only autoimmune diseases of systemic nature but also systemic immune system diseases induced by a chronic immune response against a harmless antigen (e.g., commensal flora as in Crohn's disease; allergic diseases are not included in the context of this book). The important point is the involvement of immune cells leading to chronic inflammation with a measurable systemic response that is transmitted to nearly all parts of the body. In contrast, only a small local inflammatory process is not a systemic disease and, thus, does not belong to the chronic inflammatory systemic diseases discussed in this book.

Concerning these diseases, I was part of the fact-gathering community, but I begin to realize that we need some conceptual framework. There are pressing and unsolved problems of chronic inflammatory systemic diseases, which can be attacked by these novel ideas in a way that the scientific community feels that significant progress can be made.

The book is written to better explain common sequelae of diseases, typical to chronic inflammatory systemic diseases, listed as follows: fatigue/depressive symptoms, anorexia, malnutrition, skeletal muscle wasting—cachexia, cachectic obesity, insulin resistance, dyslipidemia, alterations of steroid hormone axes, disturbances of the hypothalamic-pituitary-gonadal axis and fertility, elevated sympathetic tone, hypertension, volume expansion, decreased parasympathetic tone, inflammation-related anemia, bone loss, hypercoagulability, circadian rhythms of symptoms, disease exacerbation by stress, and sleep disturbances. These disease sequelae are typical for many chronic inflammatory systemic diseases such as rheumatoid arthritis, ankylosing spondylitis, systemic lupus erythematosus, psoriasis vulgaris, multiple sclerosis, inflammatory bowel diseases, and many more. Since similar phenomena appear during aging, many concepts in this book can also be applied to aged people (Chapter VI).

When one starts to think of pathophysiology of chronic inflammatory systemic diseases, attempts of explanation came mainly and still come from the field of immunology and closely related areas. This has important historical reasons that are demonstrated in Chapter I. For a long time, people were convinced to find the single magic reason for a given chronic inflammatory systemic disease, but many efforts were more or less unsuccessful. For most chronic inflammatory systemic diseases, there is no single magic bullet, which was verified by genome-wide association studies (GWAS). Chronic inflammatory systemic diseases depend on many small bullets from different areas. Part I of this book describes the important principles derived from basic immunology that are used to explain pathogenesis of chronic inflammatory systemic diseases.

In Chapter II, concepts of neuroendocrine immunology are introduced. This includes the influence of hormones of endocrine glands and neurotransmitters of nerve fibers on immune function with a special look on the situation in chronic inflammatory systemic diseases. This chapter also demonstrates the many abnormalities in the endocrine and nervous systems typically observed in many chronic inflammatory systemic diseases.

In Chapter III, the description of bioenergetics and energy regulation of the body explains common response pathways typical for systemic inflammation. Since energy regulation is tightly coupled to volume regulation, this chapter of the book also demonstrates the principles of water control. Energy and volume regulation are overarching principles that help to explain the uniform character of many systemic responses in chronic inflammatory diseases.

In Chapter IV, the theory of evolutionary medicine is introduced. Principles of evolutionary medicine clearly open a new page in pathophysiology. Evolutionary medicine demonstrates that disease-related genes, whether advantageous (by attenuating the disease) or disadvantageous (by boosting the disease), were not positively selected in the context of chronic inflammation (due to the negative selection pressure). Genes and their programs used in symptomatic chronic inflammatory systemic diseases were derived either from acute, highly energy-consuming physiological situations, which are terminated within 3-8 weeks (inflammatory episodes), or from programs that protect energy stores. Evolutionary medicine makes clear that patients with chronic inflammatory systemic diseases use adaptive programs typically applied to acute disease states but not useful in long-standing inflammation.

While Chapter III explains energy/water regulation in the healthy situation and energy demand of the immune system and Chapter IV describes the principles of evolutionary medicine, Chapter V combines the two columns of the theory (energy/water regulation and evolutionary medicine) to explain the basis of disease sequelae of chronic inflammatory systemic diseases. Since many aspects of the theory also apply to the aging process of normal people, Chapter VI demonstrates the link to aging research.

In Chapter VII, new hypothetical derivations are demonstrated that stem from the contents in Chapters I-VI. The concept of synchronization is introduced, which is a conceptual extension of circadian rhythms. While most of the book stands on solid ground after years of confirmatory fact-gathering, the last chapter has strong forward elements, which must stand the test of time.

This book is written to bring together previously controversial subjects of chronic inflammatory systemic diseases, and I think the new theory has two essential properties: the theory is sufficiently novel and open-ended to leave much room for further definition and precision to be gained by further conceptual thinking and practical experimentation.

Rainer H. Straub
Regensburg, 2015

Acknowledgments

I am deeply indebted to my wife Verena Straub and my children Isabella, Alexander, and Volker, who have built and still build a perfect smooth environment for a dedicated scientist and writer. They always encouraged me and they supported this lifelong dream and ongoing journey.

Special thanks go to Georg Pongratz, who once was my doctoral student before 2003, but now is a full-fledged scientist and a real spiritual brother. Many thanks go to Peter Härle and Reiner Wiest who followed with enthusiasm and dedication.

Thanks, for outstanding help and exactness over so many years, go to my technical assistants Angelika Gräber, Luise Rauch, Madlen Melzer, and Birgit Riepl.

Thanks go to more than 70 doctoral students and postdocs, particularly, Torsten Lowin, Zsuzsa Jenei-Lanzl, Poldi Miller, Christine Wolff, Silvia Capellino, Claudia Weidler, Markus Herrmann, Andreas Jeron, Thomas Frauenholz, Gebhard Berkmiller, Stefan Hrach, Martin Kees, Alexander Fassold, Stefanie Haas, Hubert Stangl, Susanne Klatt, Julia Kunath, Lars Jurzik, David Janele, Florian Grum, Barbara Ossyssek, Christain Günzler, Kristina Weber, Natalia Graf, Birgit Lehner, Selda Kizildere, and Markus Mayer.

Thanks go to Thomas J. Feuerstein, University of Freiburg, who showed me the first functioning basic research laboratory in 1987 and who stimulated me in 1990 to go into the direction of basic research. He was also instrumental in teaching the superfusion technique, which was part of my basic research methodology.

Thanks to outstanding friends and colleagues in Germany and all over the world who cooperated in multiple publications during the last 25 years. Particularly, thanks to my "amicone" Maurizio Cutolo and Frank Buttgereit, Martin Schmidt, Joachim Grifka (instrumental in the provision of human synovial tissue), David Jessop, Stephan von Hörsten, Hans Bijlsma, Alfonse Masi, Manfred Schedlowski, Piercarlo Sarzi-Puttini, Peter Angele, Franz Koeck, Marietta Leirisalo-Repo, Frieder Kees, Michael Meyer-Hermann, Marina Kreutz, Daniel J. Carr, Virginia M. Sanders, Markus Böhm, Hans-Georg Schaible, Susanne Grässel, Matthias Wahle, Christoph Baerwald, Stefan Reber, José Antonio Pereira da Silva, Marco Cosentino, Sven Anders, Jens Schaumburger, Volker Stefanski, Torbjørn Breivik, Sammy Bedoui, Thomas Schubert, Thomas Pollmächer, Hugo Besedovsky, Adriana del Rey, Norbert Sachser, Jochen

Kalden, Hanns-Martin Lorenz, Luigi Castagnetta, Firdaus Dhabhar, Fabiola Atzeni, Heike Nave, Heribert Schunkert, Frank Muders, Bernhard Krämer, Mathias Mack, Stephan Reinhold, Bernhard Banas, Michael Pfeifer, Klaus Stark, Leona Konecna, Daniela Männel, Jürgen Westermann, Ralf Paus, Jozef Rovensky, Inga Neumann, Christian Hafner, Karla Lehle, Sona Struharova, Hans-Jörg Linde, Hans-Peter Jüsten, Carsten Englert, Torsten Blunk, Bernhard Ugele, Georg Schett, Andreas Hess, Shaban Moussa, Christoph Stein, Rieke Alten, Paul-Peter Tak, Dominique Baeten, Roberto Pacifici, John Kirwan, Yrjö Konttinen, Vojtech Thon, Ernst Singer, and Norihiro Nishimoto.

Thanks to colleagues of the Department of Internal Medicine I of the University Hospital Regensburg, who until 2010 provided an integrative research atmosphere within an interdisciplinary department and cooperated in many publications, particularly, Tilo Andus, Cornelius Bollheimer, Hilke Brühl, Roland Büttner, Werner Falk, Martin Fleck, Thomas Glück, Volker Gross, Hans Herfarth, Bernhard Lang, Guntram Lock, Thomas Karrasch, Frank Klebl, Ulf Müller-Ladner, Elena Neumann, Florian Obermeier, Gerhard Rogler, Andreas Schäffler, Bernd Schnabl, Jürgen Schölmerich (the director of the department), Ulrike Strauch, Martin Zeuner, and Bettina Zietz.

Thanks go to Gary S. Firestein for inviting me to San Diego for a sabbatical, which was the starting point to write this book in 2010. Many thanks go to David Boyle being my coach in San Diego affairs during 2010.

In addition, I am thankful to colleagues from international scientific societies who shaped my path such as people from the PsychoNeuroImmunology Research Society like Mick Harbuz, Cobi Heijnen, Annemieke Kavelaars, Keith Kelley, Rodney Johnson, Robert Dantzer, Andy Miller, Nicolas Cohen, Kevin Tracey, Mike Irwin, Denise Bellinger, Dianne Lorton, Shamgar Ben-Eliyahu, Christopher L. Coe, Thom Connor, Moni Fleshner, Ron and Janice Glaser, Eric Smith, Linda Watkins, Jeff Woods, and Raz Yirmiya. There were important people from the American College of Rheumatology, who stimulated my work with their work, like David Pisetsky (who gave important editorial help on several papers of mine), Eric Matteson, Peter Lipsky, Ron Wilder, Leslie Crofford, Michael Lockshin, Bruce Cronstein, Steffen Gay, Harald Burkhardt, and Hans Carlsten.

Thanks go to the following organizations who funded my work: DFG (particularly, Research Unit FOR696 and many individual research grants), DAAD (German Academic Exchange Service), the Federal Ministry of Education and Research, the Volkswagen Foundation, and the University Hospital Regensburg (many thanks to the administration of the hospital for their very smooth handling of little things).

Finally, my thanks go to countless friends and colleagues not mentioned in the preceding text who shared thoughts and ideas over an educational and responsible drink.

Chapter 1

History of Immunology Research

Chapter Contents

In order to position the new theory presented in this book relative to present explanatory models in the field of immunology, an historic outline of important achievements in immunology and immunological paradigms relevant to the etiology of chronic inflammatory systemic diseases is given. It is not the idea to present a complete history of immunology or autoimmunity because this has already been done by merited scholars[5] and others in many reviews cited below. The achievements are summed up in Table 1 at the end of the collection in "Summary" section.

The Origin of Chronic Inflammatory Systemic Diseases and their Sequelae.
http://dx.doi.org/10.1016/B978-0-12-803321-0.00001-X

BEFORE 1945: THE EARLY DAYS

The history of modern immunology as a mature science started with the outstanding accomplishments of Louis Pasteur, Robert Koch, Emil von Behring and Shibasaburo Kitasato, Élie Metchnikoff, Charles Richet, Jules Bordet, and others in the field of bacteriology and immunology, a time often called *the golden age of immunology* (this is not to forget the variolation studies of Timoni, Sloane, or Jenner the century before). These golden age immunologists developed the concept of humoral immunity (antibodies and complement) and cellular immunity (phagocytes: macrophages and microphages (neutrophils)). While the German school favored the humoral aspect, the French/Russian school favored the cellular aspect, which appeared to be a significant conflict between the two groups (probably also stimulated by the war conflicts of the nineteenth century between France and Germany).

Although the cellular phagocyte theory of Metchnikoff was respected as a general biological phenomenon in the early twentieth century, most scientists started doing work on antibodies and complement rather than on cells.[5] Arthur Silverstein named this phenomenon "the growing humoral tide." For a period of almost 50 years, cells did not play a role in immunity due to the wrong dogma that antibodies can answer all problems of immunity and immunopathology.[5]

At the same time, microbiology concepts entered clinical medicine where physicians started to describe chronic inflammatory systemic diseases in more detail. For these diseases, microbes or toxins became important etiological factors because no other causative agents were at hand until the 1940s (reviewed, e.g., in Refs. 6,7). It was a time when people thought that inflammation was a deleterious reaction of no benefit to the host, and it was stated that microbes or toxins always trigger hostile inflammation. Today, we know that inflammation has most often positive regenerative or healing effects as long as it does not get chronic or destructive. Short-lived inflammation is often the necessary consequence of a combat with a foreign factor such as tissue alteration, microbes, or environmental toxins.

In clinical medicine, it was not really questioned how the body handles microbes, although phagocytes had been discovered by Metchnikoff and specific antibodies against microbes had been found by von Behring and Kitasato already in 1890 (followed by Paul Ehrlich's *Seitenkettentheorie* in 1897). Although Julius Donath and Karl Landsteiner, in 1904, described a type of hemolytic anemia caused by antibodies (the description of the first chronic inflammatory systemic disease against self: paroxysmal cold hemoglobinuria), chronic inflammatory systemic diseases—often linked to fever—were thought to be triggered by microbes.

In 1901, this was strongly supported by a famous statement of Paul Ehrlich who coined the term "horror autotoxicus," meaning that production of antibodies against self-antigens is principally impossible. As Yehuda Shoenfeld from the Sheba Medical Center in Tel Hashomer, Israel, puts it, "Ehrlich's statement prevented immunologists for decades from thinking of autoimmunity as a possible cause of chronic inflammatory systemic diseases." It was not only Ehrlich's declaration that stopped progress in immunology but also the move toward a more chemical direction, called immunochemistry, supported by the famous chemist Arrhenius.[8] Apart from Landsteiner's work (blood group antigens, hapten immunology, size of the immune repertoire, and cellular transfer of delayed-type hypersensitivity), this unproductive phase was called *the dark ages of immunology*.[5] In summary, with regard to chronic inflammatory systemic diseases in clinical medicine, microbes won the race for decades.

People thought that bacteria do not need to be in the affected tissue but effects might be indirect via toxins or allergic reactions (allergic reactions have been discovered by Arthus in 1903[9] and von Pirquet and Schick in 1905;[10] allergic mechanisms are not discussed in this book, but the interested reader is referred to a recent excellent summary[11]). William Hunter of London (in 1901), Frank Billings of Chicago (in 1912), and Edward Rosenow (in 1920) of the Mayo Clinic in Rochester supported the idea of "focal sepsis" and the surgical eradication of these foci.[12]

When extirpation of foci—for example, tonsillectomy or dental extraction—proved ineffective, doctors reinjected dead bacteria in an effort to boost resistance, which was called vaccine therapy.[6,7] This was applied to patients with rheumatoid arthritis, chronic iritis, myositis, psoriasis vulgaris, pemphigus, chronic thyroid disease, systemic lupus erythematosus, and some forms of vasculitis.[7] It was not any more applied to multiple sclerosis, Guillain-Barré syndrome, or type 1 diabetes mellitus because, beginning in the 1960s, vaccination was recognized as a possible deleterious trigger in these latter diseases and a new understanding of immunology was set in motion (see below).

In 1934, Fox and van Breemen, the founders of rheumatology, listed the causes of rheumatoid arthritis, and again, infectious foci were number one.[13] Although the concept of infectious etiology of rheumatic diseases was more and more discredited, bacteria such as *group B Streptococcus*, *Clostridium perfringens*, and *Mycoplasma* species and viruses played a major role in the etiology of these diseases (reviewed, e.g., in Refs. 6,7). Convincing results that infectious agents did not play a major role were provided by heroic transfer experiments in 1951, during which material of patients with rheumatoid arthritis was given to convict volunteers in Philadelphia. These experiments never elicited a chronic inflammatory systemic disease or an infection in the host.[14]

In the late 1940s and at the beginning of the 1950s, it was about time that things changed in etiologic thinking of chronic inflammatory systemic diseases. The next sections give a brief overview of outstanding achievements in basic immunology that lead to a much better understanding of immune system diseases. Etiologic aspects for chronic inflammatory systemic diseases taken from basic immunology are demonstrated thereafter.

1945-1960: IMMUNOLOGY REAWAKENS

World War II triggered research into various directions; among them are attempts to understand survival of skin and tissue transplants in burn and wound fatalities. This stimulated important studies to explain rejection of tissue grafts.[5]

An important step forward was the discovery of the histocompatibility gene locus in the mouse—the *H(istocompatibility)*-2 complex—by the work of Snell and Gorer to build the basis of molecular transplantation immunology.[15,16] This work was supported by the discovery, by Dausset in 1958, of a similar gene locus in humans.[17] People started to understand the large repertoire of the major histocompatibility complex (MHC). It remained for Benacerraf and colleagues and McDevitt to show that MHC genes also control active immune responses (see next section).[18,19]

Medawar's work in 1944 demonstrated that tissue homografts were routinely rejected, while autografts were not.[20] Rejection of foreign skin grafts followed the principles of specificity. Together with Billingham and Brent, in 1953, after having carried out most elegant studies in mice, also coined the term of "actively acquired specific tolerance": when mouse A received a graft of mouse B *in utero*, mouse A was tolerant to grafts of mouse B throughout life.[21] This clearly demonstrated that immunity does not always lead to the destructive process of tissue rejection. This is the first time that the expression "tolerance" was used, which is of outstanding importance today for the understanding of chronic inflammatory systemic diseases.

At approximately the same time, immunodeficiency diseases in the form of agammaglobulinemia were described by Bruton.[22] These diseases were important for the understanding of immune responses because they demonstrated that without immunoglobulins (and without immediate hypersensitivities of allergic nature), still, a delayed-type hypersensitivity immune response existed. Thus, antibodies were not the only Holy Grail of the immune response, but immune cells and secreted factors must also play a decisive role.

The second major advancement in the 1950s was the clonal selection theory of antibody formation developed over a decade by Sir Burnet, David Talmage, and Lederberg.[23,24] This theory said that an antigen triggers appropriate clonal precursors with specific receptors to induce cellular differentiation and proliferation leading to clonal expansion of daughter cells that produce specific antibodies toward the stimulating antigen (primary reaction). In a secondary reaction, antibody production is enhanced, faster, and closer-fitting. The absence of responses against self-antigens was explained by deletion of self-reacting clones early in ontogeny. The geneticist Lederberg added "that a unique primary amino acid sequence of an antibody is incorporated in a unique sequence of nucleotides in a gene for immunoglobulin synthesis." And in addition, antibody diversity is made possible by a high rate of spontaneous and random mutation of the DNA of the immunoglobulin gene in somatic cells.[24] Much later in 1976, by Hozumi and Tonegawa, the somatic mutation theory in contrast to germ line encoding was finally settled by the discovery of the immunoglobulin gene structure and the somatic rearrangement of immunoglobulin genes encoding for variable and constant regions.[25]

For chronic inflammatory systemic diseases of self-reactive (autoimmunity) and nonself-reactive characters, this theory was a solid basis to understand a part of chronicity, specificity, and diversity.

1960-TODAY: PURE IMMUNOLOGY

Several important discoveries further illuminated immunology. Several authors started to unravel the **subtypes of lymphocytes** (phagocytes were already known since Metchnikoff, antibody-producing plasma cells since the work of Fagraeus in 1948, and neutrophils/eosinophils/basophils since Paul Ehrlich). With the discovery of the antilymphocytic serum in 1950 by Woodruff and Forman,[26] the development of the hemolytic plaque assay by Jerne and Nordin,[27] and adoptive cell transfer experiments, important experiments were performed in the 1960s by several authors that differentiated between thymus-derived (T) cells and bone marrow-derived (B) lymphocytes (or bursa of Fabricius lymphocytes).[28–37] A first comprehensive review on B and T cells was published in 1973.[38]

This also led to the differentiation of T helper cells that are necessary to support antibody production from B lymphocytes.[34,39,40] A little later, T suppressor cells were added by Gershon et al.[41] and cytotoxic T cells by two groups of Cerottini and Janeway.[42–44]

The first T cell surface antigen-theta or Thy-1 (now CD90) was discovered by Raff, which made the study of these cells much easier.[45] T cells were detected by Thy-1 and other surface antigens of the Ly series (Ly1 = CD5, Ly2/Ly3 = CD8 α/β chain) and B cells by surface-bound immunoglobulins.[46]

At that time, it was thought that each immune function is carried out by a specialized subset of lymphocytes, which can be identified by its surface antigens. Subtype identification of different immune cells was largely boosted by the technique of monoclonal antibody generation by Köhler and Milstein in 1975[47] and by introduction of fluorescence-activated cell sorting cytometry by the Herzenberg laboratory in the early 1970s.[48]

Rodney Porter and Gerald Edelman exactly described the **immunoglobulin protein** and its different subtypes including the exact amino acid sequence (reviewed in Ref. 5). Class switch recombination, also known as isotype switching of immunoglobulins, was first demonstrated in the chicken by Kincade et al.[49] The hypervariable region of the immunoglobulin protein was discovered as the complementarity-determining residues (CDR) by Wu and Kabat.[50,51]

On the basis of the work of Benaceraff and colleagues and McDevitt, the role of the **MHC for immune responses** was discovered.[18,19] In 1973, Shevach and Rosenthal demonstrated that antigen is presented within the protein encoded by the MHC on macrophages and presented to T helper cells.[52] This was complemented by the work of Katz et al. and Kindred and Shreffler who showed the same presentation within the protein encoded by the MHC on B cells presented to helper T cells.[53,54] This was complemented by the demonstration of MHC class I restriction of cytotoxic T cell recognition of viral antigens of infected somatic cells by Zinkernagel and Doherty, which was the fundamental proof that cytotoxic T cells can distinguish self from altered self in the context of specific MHC class I-dependent presentation.[55]

These and other elegant studies defined the role of antigen presentation within the protein encoded by the MHC class I and class II toward cytotoxic T cells or T helper cells, respectively.[56–60] This was later complemented by refined studies of the MHC-encoded protein in the 1980s and 1990s, by the peptide in the niche, and by crystallographic studies that exactly defined the antigen niche in the class I and II molecules.[61–72] Probably, the work of Townsend and colleagues on the peptide in the niche was the most influential in this respect.[64] The work was complemented by the important role of the proteasome, an intracellular protein breakdown machinery,[73] that is responsible for peptide degradation necessary for antigen presentation in the MHC class I-encoded protein.[74,75] For presentation in MHC class II-encoded proteins, antigens are taken up into lysosomes, and peptides are recycled from lysosomes, transported to endosomes to bind MHC class II-encoded protein, and then expressed at the cell surface.[76]

Sometimes, phagocytized material of the lysosome pathway can be used for the presentation in the MHC class I-encoded protein, a mechanism called cross priming or cross presentation.[77,78] This is relevant in defense against many viruses and tumors, and it is important for vaccinations with protein antigens, which must be cross presented to activate cytotoxic T cells. Self-antigens are also cross presented, which typically results in deletion of autoreactive cytotoxic T cells, in a process named cross tolerance.[79]

Another important and far-reaching discovery in this period was the finding of **dendritic cells** by Steinman and Cohn in 1973.[80] Initially, these cells were described with "processes of varying length and width," which gave these cells their characteristic name. Besides macrophages and B cells, these cells turned out to be professional antigen-presenting cells using the MHC class II-encoded protein. The same group of people also characterized many functional roles of macrophages as recently summarized.[81]

In the 1980s, it was about time to discover the **structure and function of the T cell receptor**. This caused some trouble because it was thought that the T cell receptor might be, similar to the B cell receptor, an immunoglobulin-like molecule. The T cell receptor was initially characterized by monoclonal antibodies (reviewed in Ref. 82), and it was found to be largely different from the B cell receptor. On an mRNA level, Kronenberg et al. demonstrated that T cells do not possess genes typical for B cell receptor immunoglobulin.[83] In 1984, the golden year of the T cell receptor, several groups independently described genes of the different chains of the T cell receptor.[84–91] A first extensive review of the T cell receptor and partner molecules such as CD3 (on all T cells), CD4 (on T helper cells), and CD8 (on cytotoxic T cells) was given in 1987.[82]

During the same years, the nature of **natural killer** (NK) cells was defined by Lanier and colleagues.[92] Phenotypic, genetic, and functional studies supported that NK cells mediate non-MHC-restricted cytotoxicity against "NK-sensitive" targets.[92] Kärre and colleagues demonstrated that NK cells demonstrate a defense system that eliminates such cells that do not carry self-markers in the protein encoded by the MHC.[93] Importantly, NK cells have inhibitory receptors that block their killing activity and that bind to classical and nonclassical MHC class I-encoded proteins discovered by the group of Yokoyama.[94] Thus, the new principle of "missing self" was put forward. Today, we know that NK cells simultaneously express multiple receptors that work in concert to induce natural killing (stimulatory pathway) or to stop natural killing (inhibitory pathway) (table 1 in Ref. 95).

Further, important elements leading to an unprecedented huge increase in the variety of immune responses are secreted factors of different immune cells. These are **cytokines and growth factors**, and their natural inhibitors, the discovery of which started with interferons in 1957,[96,97] lymphokines such as macrophage inhibitory factor in 1966,[98,99] interleukin (IL)-2 in 1981,[100] IL-1 in 1984,[101,102] tumor necrosis factor in 1985,[103] IL-4 in 1986,[104,105] IL-6 in 1986,[106] and many many more. Those factors were made responsible for the fine-tuning of immune regulation leading to a hardly comprehensible diversity of regulatory switches. The ever-increasing list of these factors from the worlds of macrophages, T cells, B cells, other immune cells, and many other cell types forced researchers to categorize cytokines with respect to their major role in immune responses or elsewhere.

Let us make an example of the T helper cell world. In 1986, Mosmann and colleagues categorized responses of T helper type 1 and T helper type 2 cells according to secreted IL-2 and IFN-γ for T helper type 1 and IL-4 and IL-5 for T helper type 2.[107] Although this rather simple scheme is largely refined by T helper type 9, T helper type 17, regulatory T cells, and others, it was the most valuable starting point in order to understand grouping of T cell immune responses.

Sometimes, one might think that with all available soluble factors and costimulatory surface molecules, a T cell response should be more flexible than would be expected with 5-10 different T cell subtypes. Let us make a *gedankenexperiment* (thought experiment). Taking the vast number of cytokines and assuming that there is large redundancy, which would reduce the real number of possible influential factors to a number of 33,[a] a total of $2^{33} \approx 10^{10}$ different decisive cytokine cocktails would exist (with 50 influential factors, it is $2^{50} \approx 10^{15}$). In such a way, cytokine cocktails might be a crucial factor for the embodiment of the immune repertoire, and this has been nicely linked to immunogenetics of HLA class I and II inheritances.[108]

---

a. The number of 33 only reflects a 1-0 decision (the cytokine is present or absent or in high concentration and zero concentration), which in reality might be quite different because distinct concentrations can mean separate things. Thus, variety is probably much higher.

Since slight changes of the presented antigenic peptide can lead to changes in the cytokine cocktail,[109] why shouldn't it be possible that the cytokine cocktail itself defines the repertoire?[b] The entire field of cytokine research finally led to the development of anticytokine strategies for chronic inflammatory systemic diseases that entered clinical medicine in the 1990s. This was a major therapeutic breakthrough stimulated by basic immunology research (see below).

Entering the word **apoptosis** into PubMed in 2015 gives back more than 280,000 hits. Apoptosis was not known to the community before 1972. In 1972, Kerr, Wyllie, and Currie from the Department of Pathology of the University of Aberdeen published a paper describing "a new type of controlled cell deletion, which appears to play a complementary but opposite role to mitosis in the regulation of animal cell populations." Its morphological features suggest that it is an active, inherently programmed phenomenon, and it has been shown that it can be initiated or inhibited by a variety of environmental stimuli, both physiological and pathological.[110] Already in the original paper, all the morphological details are perfectly described.[c]

Then, in the late 1980s, only some 200 papers were published between 1972 and 1988; immunologists discovered apoptosis because it was demonstrated that TNF can induce apoptosis.[111] Many people recognized the potential of the new idea in 1988 and 1989, and with the publication of Peter Krammer's group in 1989 and a paper of Yonehara and colleagues, a first surface molecule was found that induced apoptosis, called Apo-1 or Fas antigen.[112,113] In 1991, the Fas antigen was characterized in more detail by a Japanese group.[114] This started an intensive search for signaling factors of apoptosis, and it turned out that apoptosis is a perfectly fine-tuned cascade of intracellular signaling events.[115,116] This concept also influenced pathophysiology thinking of chronic inflammatory systemic diseases because it is obvious that apoptosis is inhibited in many chronic inflammatory systemic diseases.

After the discovery of different types of immune cells in the 1960s, 1970s, and 1980s, researchers of the subject of leukocyte lineage defined the **checkpoints in leukocyte development** and their genetic control mechanisms. Let us make an example from the lymphocyte world. In 1976, Raff and Cooper, who identified precursor B cells in murine fetal liver and marrow expressing cytoplasmic μ heavy chains, suggested that these B cells were the precursors of newly formed B cells expressing cell surface immunoglobulin M.[117] The discovery of the recombination-activating genes 1/2 (RAG-1/2) in the late 1980s by Baltimore and colleagues was the ignition spark for a mechanistic explanation for DNA strand breakage in both B and T cell receptor rearrangements and, thus, an important help to understand the checkpoints.[118,119]

---

b. Thanks to the voluntary work of Horst Ibelgaufts, a homepage was generated that demonstrates the enormous diversity of the world of cytokines and other factors: http://www.copewithcytokines.de/.

c. In the original paper, they acknowledged a professor of the Department of Greek who gave them the idea to call it apoptosis (the second p is not spoken), because it means "falling off" of petals from flowers or leaves from trees.

During the 1990s and 2000s, this work was complemented by addition of further important genes and transcription factors that control precursors of B and T cells and, thus, their development and lineage commitment (reviewed in Refs. 120–123). In addition, these important studies defined a set of genetic factors that might be responsible for inherited immunodeficiency diseases or malignant diseases of hematopoietic cells, which do not belong to the chronic inflammatory systemic disease discussed in this book. It also defined subtypes of B cells, which play a role in chronic inflammatory systemic diseases against self (see below).

A similar hunt was not carried out with respect to monocytes, macrophages, and dendritic cells, but recent *in vivo* experimental approaches in the mouse have unveiled new aspects of the developmental and lineage relationships among these cell populations (reviewed in Ref. 124).[d]

Another important field of discovery in the 1980s with far-reaching influence until today was the demonstration of **cell adhesion molecules**, which were initially discovered as **costimulatory molecules** in the cross talk of T cells and antigen-presenting cells (and indeed, they can have both functions, adhesion to something and costimulation of another cell, reviewed in Ref. 125). The groups of Timothy Springer and Tadamitsu Kishimoto were instrumental in these discoveries.

Already in the 1970s and early 1980s, the antigens LFA-1, LFA-3, CD2, CD4, and CD8 were found to enhance antigen-specific functions by acting as cell adhesion or costimulatory molecules (reviewed in Ref. 125). The partner of LFA-3 was found in 1986 as CD2,[126] and the partner of LFA-1 was described to be the intercellular adhesion molecule-1 (ICAM-1).[127] Other important adhesion molecules are L-selectin and vascular cell adhesion molecule-1 described in 1989 and 1991, respectively.[128,129] The partner of the vascular cell adhesion molecule-1 is the integrin VLA-4 ($\alpha4\beta1$ integrin) found in 1990.[130] The number of costimulation factors and adhesion molecules tremendously increased, and in the year 2015, more than 4400 hits were found in PubMed for "costimulation" and more than 39,000 hits for "adhesion molecule."[e]

Directly linked to adhesion is the question of **vascular transmigration**. Although this subject is on the market since 1883 when Metchnikoff wrote "next to leukocytes, the vessels and the endothelial lining play the most important role in inflammation," the major advances were made in the last 25 years. Particularly, the refinements in the technique of intravital videomicroscopy

---

d. The special role of macrophage-like cells such as microglia, osteoclasts, and Kupffer cells is not mentioned in this history section.

e. Meanwhile, therapies exist that have been derived from the principles of adhesion or costimulation such as natalizumab for multiple sclerosis, which blocks T cell trafficking to the brain. Or in the therapy of rheumatoid arthritis, a blocking principle was introduced that inhibits the T cell-stimulatory communication of CD80/CD86 with CD28 based on a soluble form of CTLA-4. In addition, in the therapy of psoriasis, the interaction of CD2 with LFA-3 is now blocked by a soluble form of LFA-3.

together with the availability of large quantities of monoclonal antibodies against adhesion molecules led to a huge progress in this field (one of the first reviews was Ref. 131). It started with the demonstration that the three proteins, the selectins (E, L, and P), were important for attachment *in vivo*.[132] In addition, selectin ligands got into the focus of research. These ligands belong to a large group of surface molecules present on various cell types, called integrins according to their integral membrane function first discovered in the later 1980s.[133,134] The reader is referred to a recent comprehensive review.[135]

Along with attachment and transmigration, **chemoattractant cues** are important to guide cells to the battle place of inflammation. Leukocyte locomotion studies were performed already in the 1920s.[136] The first chemotactic peptide known to be released in the immune response was complement factor C5a.[137] Typically, chemotaxis studies in the 1970s have been performed not only with C5a, formyl-Met-Leu-Phe (fMLP), and platelet-activating factor but also with other factors such as arachidonic acid metabolites of the lipoxygenase (HETE) or cyclooxygenase (HHT) and with leukotriene B4.

The years 1987/1988 were essential for the discovery of many important new chemotactic proteins such as Gro,[138] RANTES,[139] IL-8,[140] KC,[138] platelet factor 4,[141] and many others. Between 1987 and 2002, a flood of new chemotactic proteins appeared that prompted several attempts to classify chemokines and their receptors, which finally led to the presently accepted classification installed by the IUIS/WHO Subcommittee on Chemokine Nomenclature.[142]

The chemokine field is a typical example how the rise in molecular biology in the second half of the 1980s led to a vast amount of new factors that needed a classification system. The speed of discovery was so immense that the individual researcher, who found a new protein, was among many other colleagues with similar ambitions. When in 1987, the cloning of a new cytokine or chemokine was published in a top journal like Science or Nature, by the end of the 1990s, these results only appeared in reputable standard journals. Molecular biology was made available for everybody leading to an unprecedented acceleration of discoveries. Today, the discovery speed is so high that important investigations defining the role of a new factor are often missing.

Connected to chemokines and adhesion molecules are **migration and homing of lymphocytes**. This field goes back to early observations in the 1930s as reviewed in 1959.[143] It was estimated in 1959 that "the combined output from the thoracic, cervical, subclavian, and right lymph ducts of the rabbit would replace all the lymphocytes in the blood about 11 times daily (24 times in the rat), which emphasize the short time spent by lymphocytes in the blood—about 2h in the rabbit and 1h in the rat. Although lymphocytes remain in the blood for only a few hours, they can survive for considerably longer periods, both in the tissues and when cultured *in vitro*."[143] It was recognized that the great majority of cells in the thoracic duct are cells that have recirculated from the blood and that lymphocytes repeatedly traverse this circuit during their lifetime. The route taken by the cells in passing from the blood into the lymph was not known in 1959.[143] In 1964, Gowans clarified that these cells leave the blood via high endothelial postcapillary venules[f] mainly in the lymph nodes.[144]

In 1976,[146] Schlesinger wrote that "lymphocyte subpopulations show a marked specificity in their distribution in various parts of the lymphoid system. It has been suggested that specific localization of lymphocytes depends on the presence of homing receptors on their cell surface." This demonstrates that the field moved to the question whether, or not, lymphocyte circulation is specific, meaning that there is homing of certain lymphocytes to certain tissues.

In the early 1980s, it was known that lymphocytes possess surface receptors that allow an easy attachment and penetration through high endothelial venules,[147,148] but the specific entrance of specific lymphocytes to a certain tissue was unproved. This changed in the year 1986 when Gallatin and colleagues demonstrated a review on specific homing.[149] Specific homing molecules were found that allow for homing of specific lymphocytes to distinct tissue (reviewed in Ref. 150). However, this theory was not unopposed because it was thought that not only entry into a certain tissue but also survival or resting time in the tissue would markedly influence the experimental investigation.[151]

Today, we know that many more markers beyond adhesion molecules such as chemotactic factors and expression of chemokine receptors and local stromal survival factors play an important role for selective homing responses.[152] Moreover, the idea of selective homing significantly contributed to the understanding of immunological memory in the T and B cell field (see below).[152,153]

During a typical immune response to an acute pathogen, antigen-specific T and B cells are activated, proliferate vigorously, and expand in a clonal fashion (Sir Frank Macfarlane Burnet, David Talmage, and Joshua Lederberg; see above). A large population of effector T and B cells is generated, and they will die in the subsequent contraction phase of the response. However, within the expansion phase, also, cells are generated that belong to the immunological memory, the **T and B cell memory** pools, which are primed and long-term available. The estimated half-life of T cell memory in humans is approximately 8-15 years[154] and of B cell memory up to 50 years.[155] The phenomenon of immunological memory is known since the first vaccination studies, but the underlying mechanisms are still unclear. This research field is in motion, a definite history is difficult to give, and thus, I rely on recent review articles.

For T cells, we are far away from understanding the exact memory mechanisms. Two broad groups of memory T cells have been separated by surface molecules as (A) effector-memory T cells, which express receptors for migration to inflamed tissues and display immediate effector function (CD62L− CCR7−), and (B) central-memory T cells, which express lymph node homing receptors

f. Notice that high endothelial venules can also exist in inflamed tissue as a secondary consequence of inflammation.[145] Thus, inflamed tissue can provide important means for transmigration of leukocytes to the local inflammatory site.

and lack immediate effector function (CD62L+ CCR7+).[156] Apart from this relatively simple picture, the situation is complex. It seems that several factors influence the decision of effector versus memory cells such as (1) the interaction of T cell receptor and MHC with the antigen (the strength and quality of the presentation; the stronger, the more "effector" and the weaker, the more "memory"), (2) the role of costimulatory molecules (supporting the strength of presentation), (3) the environmental situation including not only cytokines such as IL-7 but also other largely unknown tissue-specific factors, (4) the role of cellular metabolism (catabolic vs. anabolic and glycolytic vs. lipolytic), (5) the tissue location (skin vs. intestine vs. elsewhere), and (6) possibly asymmetrical division of activated T cells (reviewed in Ref. 157). The definition of memory cells is difficult because T memory cells may adopt tissue-specific signature phenotypes. It is still not clear whether, or not, an antigen needs to be present for long-term memory, but it seems that cross reactivity against one antigen can also protect from other antigens.[157] This field is sufficiently open and needs the big discoveries.

The situation is somewhat clearer in the B cell field. Two different cell types responsible for B cell memory were described: (1) the long-lived antibody-secreting plasma cells and (2) the long-lived memory B cells that can react immediately upon antigenic challenge (reviewed in Refs. 158,159). Both forms of B cell memory need T cell help,[160] but once plasma cell memory is established, T cells are not needed anymore to produce specific antibodies (reviewed in Ref. 159). Memory B cells survive in secondary lymphoid organs, while plasma cells survive in the "plasma cell survival niche" predominantly in the bone marrow and in secondary lymphoid organs.[161] For memory B memory cells, it is not exactly clear whether the continuous presence of antigen is a necessary requirement (might happen with and without presence of antigen). However, it is clear that long-lived plasma cells do not need antigen to survive.[159]

For both T cell memory and B cell memory, still, important mechanistic aspects are missing. Nevertheless, recent insight demonstrated that both types of memory can play an important role in chronic inflammatory systemic diseases (see below).

The adaptive immune system has an outstandingly **diverse immune repertoire** for antigen recognition by B and T cells generated on the basis of highly flexible gene rearrangements. The repertoire is so huge that approximately 10^{10} different α/β T cell receptors[162] and 10^{10} different B cell receptors exist.[163] The question whether the innate immune system has also some diversity in antigen recognition was unanswered for a long time. It has been proposed that there might be such a more primitive nonclonal, or innate, component of immunity that preceded in evolution the development of an adaptive immune system in vertebrates.[164] In 1996, by the group of Hoffmann, such a primitive immune defense has been demonstrated in *Drosophila* controlling fungal infections, called Toll with its ligand spätzle.[165] Soon after this discovery, in 1997, a human homologue of the *Drosophila* Toll protein was found by the group of Janeway.[166]

Today, we know many more similar proteins with an innate antigen recognition program not only on the surface of cells but also within the cell, and the name **pattern recognition receptors** is now applied. Antigenic factors of infectious agents bind to these pattern recognition receptors in order to initiate the innate and then the adaptive immune responses.[167] For example, lipopolysaccharides of Gram-negative bacterial walls bind to Toll-like receptor 4 (TLR4) as demonstrated by the group of Beutler.[168] The discovery of the pattern recognition receptors really linked innate immunity and adaptive immunity leading to a more complete understanding of the immune system. Importantly, some immunodeficiency diseases have their origin in genetic defects of these pattern recognition receptors.[167]

Central tolerance is directly linked to the function of the thymus whose role has started to be investigated by Jacques F.A.P. Miller at the beginning of the 1960s. He recognized that the thymus was instrumental for the immune response against microbes and skin grafts:[169,170] "Since the thymus is essential for the recovery of immunological function after total body irradiation, it seems likely that it is also responsible for the breakdown of tolerance under these circumstances. Further evidence for this comes from unpublished experiments which showed that tolerance breaks spontaneously in an intact animal but persists indefinitely after thymectomy. Is it possible, too, that the thymus is responsible for the breakdown of autotolerance, and hence for the development of autoimmune diseases?" What a prophetic idea!

A functional immune system requires the selection of T cells expressing receptors that are MHC-restricted but tolerant to self-antigens. This selection occurs predominantly in the thymus, where lymphocyte precursors first assemble a surface receptor. Although several studies in the 1970s and 1980s indicated the positive and negative selection process in the thymus, it was particularly in the years 1988 and 1989 when the experimental proof has been given by several independent groups.[171–179]

With the discovery in 1997 of the autoimmune regulator AIRE as a consequence of genetic analysis of patients with autoimmune polyglandular syndrome type I (APS 1, also called APECED) important mechanistic explanations were possible.[180,181] It turned out that the transcription factor AIRE is of outstanding importance for presentation of self-antigens by medullary epithelial cells of the thymus.[182,183] The basis to understand the negative selection and positive selection and, thus, central T cell tolerance was established. Moreover, as recent studies demonstrate, central tolerance is linked to peripheral tolerance since CD4+ CD25+ regulatory T cells represent a separate, thymus-derived lineage of CD4+ T cells (reviewed in Ref. 184). Regulatory T cells are discussed in the next paragraph.

Studies into the direction of central B cell tolerance are not yet in the same phase as those with T cell tolerance, but it seems clear that autoreactive B cells undergo stringent selection in either the bone marrow or peripheral circulation by deletion, induction of anergy, or receptor editing.[123,185] Recent studies point into the direction that B cell receptor editing is the most important part.[123,185]

Another important aspect has been the rediscovery of **regulatory T cells** in the 1990s. The first papers into this direction were published by the groups of Gershon, of Baker, and of Benacerraf and Katz already in the early 1970s.[186–188] This suppressor T cell concept was included into Jerne's network theory (discussed in Ref. 1) and as such abandoned when the network theory of anti-idiotypes was not further followed in the 1990s. The rise of the cytokine world and the large number of new costimulatory molecules and other surface receptors did not need these former suppressor concepts.[1]

Then, in the middle of 1990s, the new regulatory T cells returned on stage through the back door.[189] What seemed to be a reputable standard publication of Shimon Sakaguchi turned out to be a seminal paper for the second round of the T cell suppressor field. However, this time, the story was based on defined molecules such as CD4 and CD25 (the α-chain of the IL-2 receptor) and a suppressor cytokine cocktail with IL-4, IL-10, and TGF-β. Nevertheless, it took 5 years until Sakaguchi was able to show the new idea in a review article in Cell,[190] and 2000 was the breakthrough year of the new theory. Several important cofactors were soon discovered such as the glucocorticoid-induced tumor necrosis factor receptor-related protein, CTLA4, or the transcription factor Foxp3 by Sakaguchi's group.[191,192] Newer work demonstrated that epigenetic programming of regulatory T cells is an important way to maintain their functional role over time.[193]

Apart from the beneficial effects of regulatory T cells in several models of chronic inflammatory systemic disease (against self), these cells were successfully used in a transplantation model.[194] Most of the studies demonstrated that these regulatory T cells have suppressive activities when used in short-term applications. However, it remains open whether these cells, which appear in small numbers in the body, can keep the promise to treat chronic inflammatory systemic disease and transplantation-related graft-versus-host disease.[195–197] Nevertheless, it is increasingly clear that those cells largely contribute to peripheral immune tolerance.

Signal transduction is the process how an extracellular signal is translated into intracellular signals ultimately leading to regulation of gene transcription. Unraveling signaling cascades has a long history, but the first formulation of a theory was established in the 1930s to 1960s by Sutherland as the second messenger theory.[198] The prototypic second messenger was cyclic adenosine monophosphate (cAMP), which was a second messenger known in glucose metabolism and neuronal transmission.

In the early 1970s, a first study with human peripheral blood lymphocytes demonstrated the important role of cAMP for transcription in these cells.[199] Soon thereafter, dualistic effects of cAMP were reported, and cAMP and cGMP levels were measured upon stimulation of peripheral blood lymphocytes with phytohemagglutinin.[200] It was shown that ligands to adrenergic receptors increased intracellular cAMP levels leading to many intracellular changes in human leukocytes.[201] Throughout the 1970s until the midst of the 1980s, cAMP and cGMP were the main second messengers studied in immune cells (e.g., Ref. 202). It is interesting that further ideas from neurophysiology entered the field of immunology because it was demonstrated that cell membrane depolarization happened in B cells as a consequence of B cell receptor stimulation.[203,204] However, this neuronal aspect was not followed for a longer time.

Beginning in the midst of the 1980s, the field started to explode discovering a huge number of different signaling pathways. It would be beyond the scope of this book to demonstrate these pathways in detail. The reader is referred to important homepages that show different pathways in more detail http://www.genome.jp/kegg/.

The signaling cascades often demonstrate high redundancy so that many pathway factors can play a role in signaling of very different extracellular ligands. This is sometimes confusing, and it seems only logical that something is missing in our understanding. Many researchers interpret the cytoplasm as a space similar to a bucket without any physical separation. The soup in the bucket is recognized as a medium where everything can happen at the same time and at the same location. This is certainly not the case as new studies point out, which focus on compartmentalization of the cell (e.g., Refs. 205–208).

The latest fashion in immunology is the **field of epigenetics and of microRNAs**. Epigenetic means *in sensu stricto* modulation of the DNA by methylation or histone acetylation or similar secondary changes close to or at the DNA. Already in 1958, the term "epigenetic" has been used in order to separate genetic DNA-related aspects from nongenetic aspects, but mechanisms have not been mentioned.[209] Acetylation of histones and DNA methylation have already been described in the 1960s.[210,211] Respective enzymes were then described, but the link to the immune system came only in the midst of the 1980s.[212–217] Newer studies demonstrate that methylation and acetylation can have quite important modulatory functions in regulatory T cells[193,218–220] and in many other cells.

MicroRNAs are small noncoding RNAs transcript of a length of 21-23 nucleotides from specific microRNA genes located in intergenic regions with autonomous microRNA promoters.[221–223] These microRNAs have regulatory functions leading to inhibition or stimulation of gene expression. The existence of many microRNA genes in the human genome implicates a broad spectrum of functions and mechanisms. In recent years, it became clear that these microRNAs also influence important pathways in immunology. For example, the microRNA miR-181 has an influence on hematopoietic lineage differentiation to B cells[224] or granulocytes.[225]

As if it was not enough with all protein-encoding genes, the exact role of which needs to be determined for many of them, we now have an additional set of genes encoding microRNAs with important functions. Presently, the microRNA database "miRBase" lists 2588 microRNA genes for humans and 1915 for the mouse (http://www.mirbase.org/).

Summary

The abovementioned achievements are the highlights from the perspective of the author. I have provided citations for those seminal findings that helped me in organization of immunology with its many concepts (Table 1). In many instances, several authors contributed to a given field, and then, I have not mentioned an individual but presented several groups or an important review paper.

I was not able to touch all aspects of immunology and I have certainly forgotten some important things such as the details of the complement system, the anatomy of lymphoid tissues, the effector mechanisms of cell-mediated immunity, immunity to tumors, and allergy. These important aspects are mentioned in the subsequent chapters when they have got an important pathophysiological role in the context of chronic inflammatory systemic diseases. The above-given presentation mainly focused on the core elements of immunology, namely, hematopoietic cells.

TABLE 1 Major Achievements in Immunology

What	Who	When
Variolation	Timoni, Sloane, and Jenner	Eighteenth century
Bacteriology and link to immunology	Pasteur, Koch, and others	1880s
Phagocytes (monocytes, macrophages, neutrophils)	Metchnikoff	1880s
Antitoxic antibodies, passive transfer	Behring, Kitasato	1890s
Antibodies, B cell receptor, basophils, mast cells, eosinophils	Ehrlich	1897
Complement and fixation	Buchner, Bordet and Gengou	1880-1900
Allergic reactions, hypersensitivity	Richet, Bordet, Pirquet, Arthus	About 1900
First autoimmune disease	Donath and Landsteiner	1904
Blood group antigens	Landsteiner	1900
Immunochemistry	Arrhenius	1907
MNP blood group antigens, Rh blood group system	Landsteiner	1926, 1940
Histocompatibility complex	Snell and Gorer, Dausset	1945-1965
Histocompatibility complex has immune function	Benacerraf, McDevitt	1960s
Principles in transplantation and acquired specific tolerance	Medawar	1945-1955

TABLE 1 Major Achievements in Immunology—Cont'd

What	Who	When
Immunodeficiency disease	Bruton	1952
Clonal selection theory; self versus nonself	Burnet, Talmage, Lederberg	Late 1950s
Radioimmunoassay	Yalow and Berson	1960
Immunoglobulin structure	Porter, Edelman	1960s
Hypervariable complementarity-determining residues (CDR)	Wu and Kabat	1970
Immunoglobulin gene and somatic rearrangement of Ig genes	Hozumi, Tonegawa	1976
Different lymphocytes (B and T)	Many (Mitchell, Miller, Claman)	1960s
T helper cells	Many	1960s
T suppressor cells	Gershon, Benacerraf and Katz	Late 1960s
Cytotoxic cells	Cerottini, Janeway	Early 1970s
First T cell surface molecule (Thy-1)	Raff	1969
Class switch recombination of immunoglobulins	Kincade	Early 1970s
Antigen presentation within the MHC-encoded protein	Shevach, Rosenthal, Katz, Kindred, Shreffler	Early 1970s
MHC class I—restriction of cytotoxic T cell recognition, self versus altered self	Zinkernagel and Doherty	1970s
Fluorescence-activated cell sorting	Herzenberg	Early 1970s
Monoclonal antibodies	Köhler and Milstein	1976
MHC-encoded protein structure and function	Many	1980s
Immunogenetics, HLA as a risk factor	Many	1970-1990
Dendritic cells	Steinman and Cohn	1993
T cell receptor	Many	1984
Natural killer cell, missing self	Lanier, Kärre, Yokoyama	1986
Cytokines, growth factors	Innumerable many	1950+
Th1/Th2 paradigm	Mosmann	1986
Th17 paradigm	Many	Late 1990s

(Continued)

TABLE 1 Major Achievements in Immunology—Cont'd

What	Who	When
Apoptosis mechanisms	Kerr, Wyllie, and Currie; many	1972+
Lymphocyte lineage transcription factors	Many	Late 1980s
Recombination-activating genes 1/2 (RAG-1/2)	Baltimore	Late 1980s
Adhesion and costimulatory molecules	Many (Springer, Kishimoto)	1980s
Vascular transmigration	Many	1980s
Chemokines	Many	Late 1980s
Migration and homing of lymphocytes	Many (Gowans)	1920s+
The memory of the immune response	Many	1900+
Pattern recognition receptors, noninfectious self versus infectious nonself	Janeway, Beutler, Hoffmann; many	1990s
Danger theory	Matzinger	1990s
Central tolerance in the thymus	Miller, many others	1961, 1985+
Regulatory T cells (successors of T suppressor cells), CD25, CD4, Foxp3	Sakaguchi	1990s
Signal transduction in immune cells	Many	1960s+
Epigenetics	Many	1950s+
MicroRNA	Tuschl, Ambros, Bartel	2000s

PATHOGENIC EFFECTOR MECHANISMS OF CHRONIC INFLAMMATORY SYSTEMIC DISEASES

It cannot be the purpose of this book to describe the complete history of all chronic inflammatory systemic diseases with or without reactivity against self or foreign in all pathophysiological details. This has been done elsewhere in multivolume books.[226–229] It is more the idea to describe how principles from basic immunology were taken over to explain the pathophysiology of chronic inflammatory systemic diseases. This is done using examples that may guide the reader to understand the importance of the respective pathophysiological elements. The summary table at the end of this chapter demonstrates the most important pathogenic elements in chronic inflammatory systemic diseases.

Immunology reawakened after a dark period of immunology in the late 1940s. There were several important points for this renaissance, which stimulated investigators to think of autoimmunity:

1. Young scientist had few ties to old dogmas such as the "horror autotoxicus" of Paul Ehrlich.[5] The etiology of autoimmune diseases based on infectious agents was more and more discarded.
2. The Coombs test was important to detect autoantibodies.[230]
3. The introduction of immunofluorescent staining by Coons et al. in 1942[231] led to immediate staining of autoantibodies in the tissue.[232]
4. The introduction of Freund adjuvant[233] and development of animal models of autoimmunity revolutionized the field.
5. New paradigms of immunology started to fascinate clinically oriented researchers (e.g., Snell, Medawar, Jerne, Burnett, and numerous others).
6. Introduction of the Witebsky postulates[g] in the 1950s for autoantibodies and autoimmune diseases helped to shape the field.[234]
7. Transfer of antibodies and adoptive transfer experiments with cell types opened new avenues in the 1960s.

The above points and, particularly, animal models of disease were important elements to start to unravel human chronic inflammatory systemic diseases. For example, Rivers and his colleagues injected *Rhesus macaques* with normal brain extracts from rabbits and showed that nearly all monkeys developed acute central nervous system disease with immune cell infiltration and demyelinating lesions.[235] Rivers' group also noted that the disease-inducing capacity of the brain extracts paralleled their myelin content, providing the first hint that myelin was critically involved.[235] Thus, the experimental allergic (now "autoimmune") encephalomyelitis (EAE) model was born. In 1946, Kabat improved this hitherto cumbersome model by using the recently developed Freund adjuvant.[236] Similarly, already in the late 1930s, Smadel developed a model of chronic inflammatory nephritis by injecting anti-kidney serum into rats that was produced in rabbits immunized with rat kidneys.[237,238]

Other models based on Freund adjuvant[h] were established in different clinical fields such as experimental nephritis in 1949,[239] experimental arthritis in the midst of the 1950s,[240,241] experimental autoimmune thyroiditis in 1956,[242] or

---

g. Witebsky's postulates: (1) the direct demonstration of free, circulating antibodies that are active at body temperature or of cell-bound antibodies by indirect means; (2) the recognition of the specific antigen against which this antibody is directed; (3) the production of antibodies against the same antigen in experimental animals; and (4) the appearance of pathological changes in the corresponding tissues of an actively sensitized experimental animal that are basically similar to those in the human disease.

h. The meaning of Freund adjuvant is discussed later when the TLRs are mentioned.

experimental type 1 diabetes mellitus in 1966.[243] The idea was always the same: when using a strong immune stimulus such as Freund adjuvant and when using tissue of the target diseased organ, this should kick off the aggressive autoreactive disease against the target tissue.

These models[i] together with patient-derived material, for example, in autoimmune hepatitis,[244] were the solid ground to link basic and clinical immunology. It clearly demonstrated the autoaggressive aspect of the immune system leading to chronic systemic inflammation. Autoantibodies and autoreactive immune cells were found to play an important role. This was the renaissance of autoimmunity and infectious agents started to play a minor role.

Autoantibody, Immune Complex, and Complement

It took around 40-50 years between the first description of an autoimmune phenomenon by Donath and Landsteiner in 1904 and the rediscovery of autoantibodies against red blood cells by Coombs and colleagues in the 1940s.[230] Similarly, autoantibodies were found against solid organs such as the kidney,[245] liver,[244] and thyroid gland.[242,246,247] After publication of the clonal selection theory by Sir Frank Macfarlane Burnet, David Talmage, and Joshua Lederberg, the clonal cellular immune response with autoantibodies of a clonal specificity produced against an autoantigen had its solid footing.

During the same period, new autoantibodies were discovered against nucleoproteins, called antinuclear antibodies or anti-DNA antibodies in the laboratories of Miescher, Kunkel, and Ceppellini.[248–250] Similarly, the rheumatoid factor was described as a serum factor that binds to immunoglobulins.[251–254] The hunt for autoantibodies was stimulated leading to a vast panel of autoantibodies to different tissue antigens, called the "immunological homunculus."[255] [j]

It is important that the target antigen is closely linked to the target tissue. For example, an autoantibody in Goodpasture syndrome is directed toward glomerular basement membrane of the kidney and not against proteins of dermal epithelial cells such as desmoglein 1. This latter protein is reserved for the blistering autoimmune skin disease of pemphigus.[258] It was soon recognized that autoantibodies can build soluble immune complexes that can be deposited in the tissue leading to complement activation and an inflammatory cascade of events.[259,260] In 1967, this was called the "autologous immune-complex disease." It was obvious that autoantibody deposits were found in target tissues, where they can fix complement and

i. In addition, the introduction of specific gene modifications in mice in the late 1980s by the use of embryonic stem cells by Capecchi, Evans, and Smithies advanced the field of animal experimentation in the context of chronic inflammatory systemic disease.

j. A collection for rheumatoid arthritis is given in Ref. 255, a collection for multiple sclerosis is given in Ref. 256 and a collection for systemic lupus erythematosus is given in Ref. 257. Similarly, panels of autoantibodies have been described in other chronic inflammatory systemic diseases.

activate an inflammatory cascade.[261,262] Since these immune complexes can travel in circulation, the affected target organ can be diverse in a given disease such as systemic lupus erythematosus (brain, kidneys, serosa, and so forth).

Pathogenically relevant autoantibodies were found against the acetylcholine receptor in myasthenia gravis,[263,264] against the TSH receptor of the thyroid gland in Graves' disease,[265] against a myocardial nucleoprotein in fetal heart block children of lupus mothers, against membrane phospholipids in systemic lupus erythematosus, and against myeloperoxidase in order to activate neutrophils.[266–270] Many more good examples are now available that unambiguously demonstrate the pathogenic role of autoantibodies by binding to tissue targets. As a consequence, effector mechanisms are switched on such as antibody-dependent cell-mediated cytotoxicity; release of inflammatory mediators through stimulation of activating Fc receptors on NK cells, macrophages, or mast cells; opsonization of the antigen, which promotes phagocytosis by macrophages or inflammatory dendritic cells; or complement activation with subsequent assembly of the membrane attack complex, all of which activate a sequence of inflammatory events leading to inflammation in the target tissue.

Since not all antibodies can bind complement (only IgG and IgM), it was thought that antibody subtypes might play an important role. Indeed, there is a strong predominance of IgG antibodies as demonstrated, for example, in systemic lupus erythematosus,[271] in multiple sclerosis,[256] in pemphigus,[272] in autoimmune hepatitis,[273] in autoimmune Addison disease,[274] and in other chronic inflammatory systemic disease against self.[275] This is certainly an important feature that contributes to pathogenicity, and it also demonstrates an important shaping of the B cell response toward higher antibody affinity and memory.

While shortly after the discovery of "autologous immune-complex diseases," the pathogenic role of autoantibodies was not questioned, discussions about the pathogenic role of autoantibodies started in the 1990s.[276] The main reason for the skepticism was the fact that not all autoantibodies were pathogenically relevant. In 1993, it was summarized that "in autoimmune diseases, autoantibodies are the actual pathogenic agents of the disease [see examples above], the secondary consequences of tissue damage, or the harmless footprints of an etiologic agent."[277] This statement still holds true today, but the gaps in our knowledge become smaller and smaller.

In addition, we now recognize that patients diagnosed with a certain chronic inflammatory systemic disease may have quite different underlying immune pathologies although the clinical picture is very similar.[k] This has been demonstrated for multiple sclerosis or rheumatoid arthritis and is presently also *in*

---

k. This knowledge stems mainly from the quite different effects of new biologic therapies because some patients respond to B cell- or T cell-depleting therapies, whereas others do not.

vogue in other diseases.[278,279] In one patient, an autoantibody response may be outstandingly important, while in another patient of the same disease group, a similar autoantibody is an epiphenomenon. Thus, we have to be careful to reject the pathogenic role of an autoantibody on the basis of studies where patients with all types of pathologies were included in the analyses.

This section demonstrated the pathogenic role of autoantibodies, which today forms a solid paradigm to explain chronic inflammatory systemic diseases (including effector mechanisms). However, we are confronted with a huge variety of autoantibodies, which usually destroys our unitary consideration of a patient group. It clearly demonstrates that immunological heterogeneity is a common principle in chronic inflammatory systemic diseases.

B Cells and Plasma Cells

A series of checkpoints normally control B cell selection, both centrally in the bone marrow and in peripheral lymphoid tissues; in the latter, a second screening process for reactivity with peripheral self-antigens results in apoptosis, receptor editing, or anergy.[123,185] It is generally agreed that persistence of isotype-switched and somatically mutated autoreactive antibodies is a result of a break in B cell tolerance.[185]

It has been described that the control at the checkpoints can be disturbed in patients with systemic lupus erythematosus and rheumatoid arthritis.[280,281] These unfortunate control defects remain present in patients with systemic lupus erythematosus in remission.[282] These recent findings show that some of the chronic inflammatory systemic diseases against self are associated with alterations in B cell tolerance. B cells that do not undergo self-control at the checkpoints can participate in the germinal center reaction with subsequent expansion in class-switched B memory and plasma cell compartments.

Importantly, plasma cells can survive for long periods of time in the appropriate survival niches, and besides the normal B cell memory, they are an independent cellular component of immunological memory.[161] This mechanism is based on competition for survival niches between newly generated plasmablasts and older plasma cells.[161] Not only in the bone marrow but also, importantly, in inflamed tissue of patients with chronic inflammatory systemic diseases, plasma cells can survive for a very long time in the presence of chemokines and cytokines (CXCL12, IL-6, BAFF and/or APRIL, and TNF[161]).

We begin to realize that in the tissue of patients with chronic inflammatory lesions such as multiple sclerosis, rheumatoid arthritis, myasthenia gravis, psoriasis vulgaris, dermatomyositis, Sjögren syndrome, and even atherosclerotic lesions, the so-called follicle-like structures with B cell, T cells, dendritic cells, and plasma cells exist (tertiary follicle).[145,279,283–287] The role of these follicle-like structures is not completely understood, but they might be an important platform for local autoantibody production[285] and also survival of pathogenically relevant B cells, memory B cells, and long-lived plasma cells. In multiple sclerosis, a closer examination of B cell clones demonstrated that some B cell clones only appear in the intrathecal space but not in the periphery. In addition, these B cells demonstrate the signs of receptor editing that is typical for a coordinated shaping process within a follicle.[288–290]

Although the interest in B cells in chronic inflammatory systemic disease against self has historically focused on their autoantibody production (see above), we now recognize that B cells have multiple antibody-independent roles. B cells can efficiently present antigen and activate T cells,[291–295] they can augment T cell activation through costimulatory interactions,[296] and they can produce numerous cytokines that affect inflammation, angio- and lymphogenesis, and immune regulation. These nonantibody functions of B cells are most important in chronic inflammatory systemic diseases, and they start to build a paradigm on its own.[295]

As it has become more obvious that B cells contribute substantially to multiple human chronic inflammatory systemic diseases, highly targeted less toxic therapies focused on restoring normal B cell function and eliminating pathogenic autoantibodies are beginning to reduce the reliance on immunosuppressive drugs.[297] A monoclonal antibody against CD20 (rituximab) effectively reduces B cell numbers without significant toxicity,[297] and it works in rheumatoid arthritis, nonrenal systemic lupus erythematosus, immune thrombocytopenic purpura, Graves' disease, pemphigus, and others. Other B cell targets such as CD22, CD19, or BAFF are now in the focus with partly similar success. This clearly tells us that B cells play an outstanding role in some chronic inflammatory systemic diseases.

T Cells

In the 1970s, in experimental models, T cells were discovered to be stimulators of chronic inflammatory systemic diseases (against self).[298–304] Already in 1977, the defects in T suppressor cells were made responsible for the human disease of type 1 diabetes mellitus.[305] The concept of defective T cell suppressor cells remained present until the end of the 1980s when studies on central T cell tolerance, the new T helper type 1 and type 2 concept of CD4+ T cells, and cytokine regulation were pushed forward.

Defects in T cell tolerance were established on the basis of autoreactive T cells discovered in healthy subjects and in patients with rheumatoid arthritis.[306–308] Interestingly, one of the antigens for autoreactive T cells in rheumatoid arthritis was type II collagen,[307,308] which is the autoantigen in one experimental model of arthritis. In addition, patients with rheumatoid arthritis build antibodies against an epitope of collagen type II that is identical in the mouse model.[309] Moreover, collagen type II can be citrullinated leading to rheumatoid arthritis-relevant antigenic epitopes.[310] The significance of a defective central tolerance was finally established after discovery of the autoimmune regulator AIRE in 1997 in patients with APS 1 (also called APECED).[180,181] This demonstrated the key role of the thymus for central T cell tolerance.

In addition to the paradigm of a defective central tolerance, cytokines started to conquer immunology. On the basis of a polar cytokine response in animal models, the concept of T helper type 1 and T helper 2 was developed in the 1980s.[107] This construct was helpful in many ways because it provided the first explanation why different chronic inflammatory systemic diseases can be driven by either cellular/T helper type 1 (TNF, IFN-γ, IL-12) or humoral/T helper type 2 (IL-4, IL-5, IL-10, or IL-13) responses. It turned out that certain transcription factors were instrumental for T helper type 1 (T-bet) and T helper type 2 (GATA-3) polarization.[311,312] Historically, T helper type 1 organ-specific chronic inflammatory systemic diseases were separated from T helper type 2 humoral chronic inflammatory systemic diseases (Table 2).

In the 1990s, throughout all clinical disciplines, scientific conferences and journals were riddled with Th1/Th2 contributions. The paradigm survived until the later 1990s when regulatory T cells entered the stage,[189] and T helper type 17 cells complemented the picture at the beginning of the 2000s.[314] Indeed, T helper type 17 cells became a mainstay of the pathophysiological process in many diseases mentioned in Table 2.

TABLE 2 The Historical Th1/Th2 Paradigm Applied to Chronic Inflammatory Systemic Diseases

T Helper Type 1 (Th1)	T Helper Type 2 (Th2)
Rheumatoid arthritis	Systemic lupus erythematosus
Multiple sclerosis	Graves' disease
Crohn's disease	Myasthenia gravis
Hashimoto thyroiditis	Autoimmune hemolytic anemia
Type 1 diabetes mellitus	Antiphospholipid syndrome
Psoriasis vulgaris	Idiopathic thrombocytopenic purpura
Mesangioproliferative glomerulonephritis	Sjögren disease
Pemphigus	Ulcerative colitis
Primary biliary cirrhosis	
Graft-versus-host disease	

This separation still has some value, but it needs to be mentioned that within one patient, two different pathogenic pathways may exist in parallel, one being Th1 and the other being Th2. This was nicely demonstrated in an experimental model of chronic inflammatory systemic disease (against self) in mice.[313] Similarly, in a patient population with one chronic inflammatory systemic disease, one group of patients might demonstrate Th1- and the other Th2-dominated phenomena. In addition, the character of the disease might change so that at the beginning, another T helper-type response might prevail as in a later phase. Furthermore, some chronic inflammatory systemic diseases were never classified according to Th1/Th2 such as autoimmune hepatitis. In summary, in patients, the things are not as clear-cut as in animals.

The immune-controlling role of CD4+ CD25+ regulatory T cells was very clear in animal models of chronic inflammatory systemic diseases (reviewed in Ref. 190). With regulatory T cells, the important roles of TGF-β and the transcription factor Foxp3 were established.[192] Many studies indicated that patients with chronic inflammatory systemic diseases such as type 1 diabetes mellitus, systemic lupus erythematosus, Sjögren syndrome, and rheumatoid arthritis demonstrate alterations in the pool and function of regulatory T cells (reviewed in Ref. 197). Foxp3 received its important role when a Foxp3 loss-of-function mutation was discovered to be the cause of a chronic inflammatory systemic disease against self called "immune dysregulation, polyendocrinopathy, enteropathy, X-linked (IPEX) syndrome."[315,316]

However, the possibly favorable potential of regulatory T cells in humans is not clear although *ex vivo*-expanded regulatory T cells were adoptively transferred for the treatment of graft-versus-host disease. This therapeutic approach needs to stand the test of time because there are several critical issues that need to be considered (see Refs. 196,197).

But what about T helper type 17 cells? In important studies, Cua and coworkers demonstrated the role of IL-12 and IL-23 while studying the hitherto T helper type 1-related diseases of experimental autoimmune encephalitis and collagen type II arthritis.[317,318] It turned out that in both diseases, IL-23 and one subunit of its receptor IL-23p19 were much more important than IL-12 and its receptor. In previous studies, it had been found that IL-23 elicited production of the proinflammatory cytokine IL-17A (or short IL-17) from CD4 T cells of the effector and memory phenotype.[319]

In further studies, IL-17-deficient mice developed less severe experimental arthritis and encephalitis,[320,321] neutralization of IL-17 decreased disease severity,[322] and overexpression of IL-17 in the joints exacerbated disease.[323] The completion of these important studies was given when the decisive transcription factor RORγT was found to be responsible for T helper type 17 differentiation.[324] These studies started the T helper type 17 paradigm, which has now taken over to other diseases as well. More detailed reviews are given elsewhere.[314,325,326] A general scheme that is presently relevant for various differentiation pathways of T cells is given in Figure 1.

The importance of the T cell in human autoimmune disease is unambiguously demonstrated with T cell-directed therapies. A human fusion protein of CTLA4 coupled to the Fc part of immunoglobulin G has been launched in 2005 as a therapy for patients with rheumatoid arthritis (abatacept). CTLA4 on regulatory T cells or in its soluble form are important to block the interaction between CD28 and CD86 of CD4+ T effector cells and antigen-presenting cells. In addition, the T cell-depleting monoclonal antibody directed against CD52 (alemtuzumab) on T cells is a new therapy for patients with relapsing remitting multiple sclerosis. Similarly, a human fusion protein with LFA-3 coupled to the Fc part of immunoglobulin G, which blocks T cell activation, is an approved therapy in psoriasis vulgaris. There are more similar examples already approved

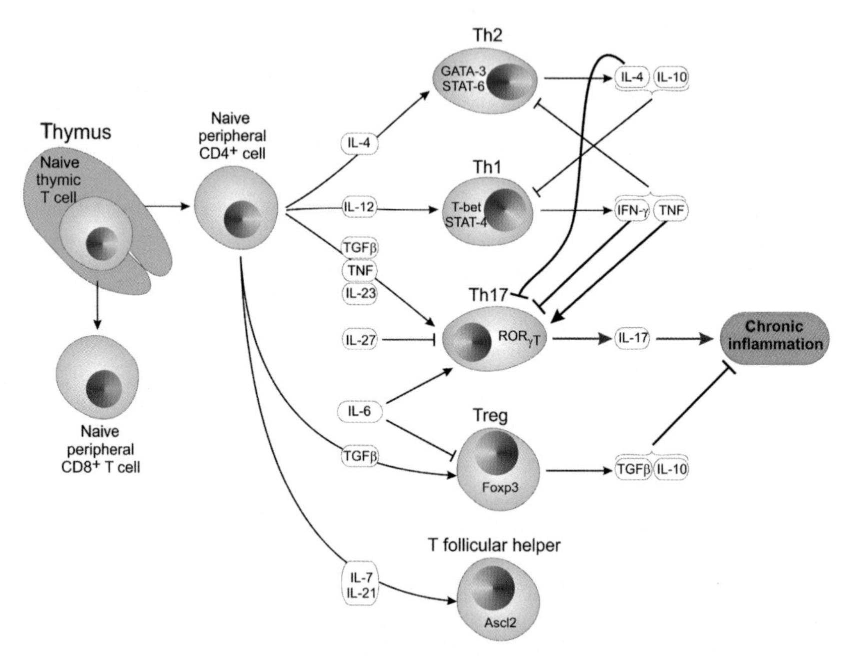

FIGURE 1 Differentiation of T cells. This diagram depicts the main T cell pathways (modified according to Ref. 314 and redrawn with concepts from Ref. 329). Lines with an arrow indicate a stimulatory effect, and lines with a bar at the end represent an inhibitory effect. The factors given in the cell nucleus are transcription factors relevant for development of T cells into the respective T cell subtype.

or in the pipeline that focus on T cell activation.[327] These studies ultimately demonstrated that autoreactive T cells play a particular role in chronic inflammatory systemic diseases.[1]

In addition, T cell memory is not well understood, and it has the potential to support a long-lived reaction against self or foreign antigens. The field of T cell memory in chronic inflammatory systemic disease awaits the big discoveries that will show how memory function perpetuates the chronic process.

MHC and Immunogenetics

In the 1960s and 1970s, it was discovered that the MHC has many different genetic loci that control the production of surface antigens on somatic cells. The number of polymorphisms at each of these loci is very numerous leading to a huge variety of different HLA combinations, which makes transplantation

1. The first T cell-directed therapies were performed with anti-CD4 antibodies.[328] Unfortunately, for the researchers, such a therapy depletes most probably the good and the bad ones, the regulatory and the effector T cells. The therapy was not effective.

so difficult. Similarly, it was the period when the link between HLA haplotypes and different autoimmune and other diseases was discovered.

The 1970s showed that certain HLA haplotypes appear to predispose to increased susceptibility to certain diseases. Some of these disease associations were observed with MHC class I-related antigens that involve HLA A, B, and C (e.g., the famous example of HLA-B27 for ankylosing spondylitis with an increased risk by factor 90), but most of the associations were found with MHC class II-related HLA D.[330] It is thought that this preponderance of class II versus class I, particularly, fostered research into the direction of CD4+ in contrast to CD8+ T cells.

The field of immunogenetics enriched clinical medicine particularly in all aspects of transplantation and transfusion medicine (graft-host matching). However, for a long time, MHC immunogenetics was not successful in explaining why patients with a certain haplotype develop more often a certain chronic inflammatory systemic disease. There were some early exceptions from this long-standing rule such as celiac disease where antigenic peptides are presented in a way that makes T cells more aggressive toward the respective antigen (reviewed in Ref. 331).

In recent years, we understood that the MHC class I and II haplotype diversity was important to overcome different types of infectious diseases.[108] For example, viral and intracellular bacterial infections are perfectly cleared in the presence of HLA-DQ6 (0601) with a strong IFN-γ response, while extracellular bacterial and fungal infections are better cleared in the presence of HLA-DQ8 (0302) and a strong IL-17 response.[108] The HLA-DR2/DR3/DR4 haplotypes have an intermediary role for intra- and extracellular infectious agents with a similarly strong IFN-γ and IL-17 response. While this diversity protected us from infectious agents, it also helped to present autoimmune peptides in a stronger and more perfect way as recently summarized.[108] As it was stated, HLA class I and II haplotypes "have survived bottlenecks of infectious episodes in human history because of their ability to present pathogenic peptides to T cells that secrete cytokines to clear infections. Unfortunately, they also present self-peptides to activate autoreactive T cells secreting the same proinflammatory cytokines…."[108]

Studies in the 1990s and 2000s redefined immunogenetics since genome-wide association studies presented some novel genes that encode an immune-active product considered an "immune response gene." The first respective study (in patients with psoriasis vulgaris) appeared in 1993 initiated by the Department of Pediatrics, University of Texas Southwestern Medical Center, Dallas.[332] A recent review summarizes the genome-wide association studies presently available and demonstrates a list of genes shared by chronic inflammatory systemic diseases (Table 3).[333] Some other aspects of genome-wide association studies are presented below in "Signaling Pathways."

Although geneticists are quite happy with these results, the individual risk is only marginally increased when having a respective polymorphism. For

TABLE 3 Shared Susceptibility Genes in Chronic Inflammatory Systemic Diseases[333]

Gene	Crohn	RA	T1D	Celiac	MS	SLE	Ps
PTPN22 (protein tyrosine phosphatase Lyp)	+	++	+++			+	
TNFAIP3 (TNF-induced protein 3)	+	++				++	++
IL-23R (interleukin 23 receptor)	+++	+					+
NRXN1 (neurexin 1 isoform β precursor)	+	+			+		
TRIM27 (tripartite motif-containing 27)	+	+	++				
C20orf42 (fermitin family homologue 1)			+	+	+		
TNKS (tankyrase)	+				+	+	
KIAA1109 (hypothetical protein)		+		++	+		
EPHA7 (ephrin receptor EphA7)		+	+	+			

MS, multiple sclerosis; Ps, psoriasis vulgaris; RA, rheumatoid arthritis; SLE, systemic lupus erythematosus; T1D, type 1 diabetes mellitus; TNF, tumor necrosis factor. +: *p*-value between 10^{-4} and 10^{-9}; ++: *p*-value 10^{-10} to 10^{-99}; +++: *p*-value above 10^{-100}.

example, the strongest association was found between type 1 diabetes mellitus and the PTPN22 polymorphism 1858C→T, rs2476601, with a *p*-value of 10^{-226}.[333] However, this increased the risk to develop type 1 diabetes mellitus in a person with the PTPN22 polymorphism only by factor 1.70-1.83, while the classical HLA association with DR3 and DR4 demonstrates an odds ratio of 36.0.[334] It is similar in rheumatoid arthritis: apart from the classical HLA class II association with DRB1 "shared epitope," the strongest factor in rheumatoid arthritis found in genome-wide association studies was the same PTPN22 polymorphism with an odds ratio of only 1.23-2.20.[335] Environmental factors operate in the same range: smoking or work that includes rock silica exposure (drilling or stone crushing) increased the chance to develop rheumatoid arthritis by factor 1.22-2.19 and 1.2-7.6, respectively.[335,336]

Under consideration of these small effects, geneticists think that combination of different SNPs and inclusion of environmental factors will provide the next level of knowledge, which should improve understanding of chronic inflammatory systemic diseases. This would, however, need a much larger number of affected patients and healthy controls.

Monocytes and Macrophages

Monocytes circulate in the blood, visit the bone marrow and the spleen, and do not proliferate like lymphocytes. The major role of these bone marrow-derived myeloid cells is immune surveillance because they can migrate into the tissue when respective cues are available. Migration to tissues and subsequent differentiation to other cell subtypes are dependent on microenvironmental factors that determine the inflammatory milieu.[124] Monocytes are the platform for the generation of not only macrophages in the tissues (alveolar, liver (Kupffer), bone (osteoclasts), lamina propria macrophages, and inflammatory lesions) but also inflammatory dendritic cells.[124,124]

In the 1960s, macrophages were thought to ingest and degrade antigens without many further functions. Still in 1970, the role of macrophages in autoimmune diseases was described by "metabolically degrading or sequestering antigen, [which] would remove the primary stimulus.[337]" During the 1970s, it was recognized that macrophages stimulate T cells by presenting the antigen in the MHC class II-encoded protein and secreting cytokines.[338] The cytotoxic role of macrophages in tumor biology was discovered.[339,340] Furthermore, the capacity of the macrophage to bind the Fc part of immunoglobulins with specialized receptors was demonstrated in the 1970s (reviewed in Ref. 341). Similarly, complement receptors, mannose receptor, and LDL scavenger receptor were found in the 1970s (reviewed in Ref. 342). This was later complemented by adhesion molecules, chemokine receptors, integrins, pattern recognition receptors (TLRs, C-type lectin receptors, NOD-like receptors), and others. At the beginning of the 1980s, TNF and IFN-γ were found to be the major stimulators of MHC class II-encoded protein expression on macrophages.[343–345]

The 1980s saw several studies that described how liposome-encapsulated bisphosphonates can specifically deplete macrophages *in vitro* and *in vivo*.[346] Using this important technique, many more *in vivo* functions of macrophages/dendritic cells were discovered in animal models of chronic inflammatory systemic diseases.[347–349] Then, macrophages were introduced in the pathophysiological thinking of clinical immunologists.[350–352] Initially, it was thought that autoantigen presentation is the major duty of classically activated macrophages in the context of autoimmune diseases, but then, many other functions were discovered.

In the 1980s and 1990s, macrophage cytokines were found such as IL-1, TNF, IL-6, IL-8, IL-10, IL-12, IL-18, IL-23, and also TGF-β. In addition, macrophages produce growth factors, chemokines, arachidonic acid metabolites, reactive oxygen species, nitrogen intermediates, complement components, proteolytic enzymes (e.g., matrix metalloproteinases), neuropeptides, hormones, and others. Macrophages are able to produce soluble factors that can exert many not only proinflammatory but also anti-inflammatory influences on the micro-environment (reviewed in Ref. 342,353–355). A macrophage-loving colleague of mine, Marina Kreutz, supported the all-rounder concept of macrophages: "Tell me about a factor that they cannot produce. There are none. The clue is to know how to stimulate them."

In recent years, macrophages were classified into (1) classically activated macrophages, (2) wound-healing macrophages, and (3) regulatory macrophages (with strong IL-10/TGF-β production) (reviewed in Ref. 355). In chronic inflammatory systemic diseases, mainly classically activated macrophages play a dominant role.[355] They can be activated by danger signals such as lipopolysaccharides, heat shock proteins, nuclear proteins (HMGB1, histones), DNA, RNA, and components of the extracellular matrix, where pattern recognition receptors play an important role.[355]

Upon stimulation, macrophages can polarize the T cell response by producing the T helper type 1-relevant IL-12. In addition, macrophages can support the recently described T helper type 17 cells because they produce TNF, TGF-β, IL-1, IL-6, and IL-23, all of which are decisive for T helper type 17 differentiation. However, they can also lead to an immune-regulatory situation by producing IL-10 and TGF-β, which would support the regulatory arm of T cells or B cells.

In different tissues, subtypes of macrophages fulfill specific functions, which are normally linked to tissue regeneration and immune surveillance. In the context of chronic inflammatory systemic diseases within inflamed tissue, specific macrophages can get activated and support a destructive process. For example, osteoclasts play a crucial role in subchondral bone destruction in rheumatoid arthritis. They can locally differentiate from monocytes/macrophages to osteoclasts with the help of RANKL and M-CSF.[356,357] This help is given by local T cells via RANKL and IL-17.[358–360]

The first anticytokine therapy was developed toward TNF because this cytokine of macrophages was thought to be the most important inflammatory factor in a sequence of inflammatory events demonstrated by the group of Feldman and Maini.[361–364] Similarly, IL-1 plays a central role and it can be neutralized with IL-1 receptor antagonist.[365] In addition, anti-IL-6 receptor therapy is now used in the treatment of patients with chronic inflammatory systemic diseases.[366] The successful neutralization of TNF, IL-1, and IL-6 indicates that macrophages (and inflammatory dendritic cells; see below), the major producers of these cytokines, play a proinflammatory role.[m]

The macrophage stands between the innate immune system and adaptive immune system since it can be stimulated by danger signals and can phagocytize

m. However, one has to be careful with any generalizations because anti-TNF therapy was not successful in all diseases (multiple sclerosis as an example). This indicates that compartments and the respective micromilieu have an important role. IL-17 is not produced by macrophages but its neutralization is also a fruitful strategy.[367,368]

foreign or self material, which can be presented in the form of antigenic fragments to CD4+ T cells. Thus, the macrophage plays a key role in the inflammatory process, which is not to forget that macrophages have also regulatory functions (type 2 macrophage). However, macrophages have no own memory so that an external stimulus from the environment or from B or T cells is necessary for the continuous activation in a chronic inflammatory systemic disease.

Dendritic Cells

Dendritic cells are closely related to monocytes and macrophages, and it is now established that they can differentiate from a common myeloid precursor of monocytes/macrophages/dendritic cells.[124] Three types of dendritic cells are presently separated.

Classical dendritic cells process and present antigen in secondary lymphoid organs (lymph nodes and spleen).[369,370] In the mature state, they have a high capacity to produce cytokines and to present antigens to CD4+ T cells. Classical dendritic cells are short-lived and replaced by circulating classical dendritic cell precursors.[371]

A second group of dendritic cells are long-lived plasmacytoid dendritic cells, which carry characteristics of lymphoid cells.[372] They differentiate from the abovementioned precursors or from a lymphoid precursor (not known), and they are present in the peripheral tissue and bone marrow. They react to viral infection by producing large amounts of type I interferons, but they can also present antigens to CD4+ T cells.[124]

A third group of dendritic cells are inflammatory dendritic cells that share many of the phenotypic features and functions of classical dendritic cells and macrophages. In contrast to classical and plasmacytoid dendritic cells, they differentiate from circulating and migrating monocytes under inflammatory conditions in the tissue.[124] These cells phagocytize antigen, migrate to secondary lymphoid organs, process and present antigen, and activate CD4+ T cells and B cells.[124] Inflammatory dendritic cells are mainly relevant in the control of chronic inflammatory systemic diseases within inflamed tissue and in lymphoid organs. The cytokine profile of these cells is detrimental for the T cell[373] and B cell response.[374]

Since it would be impossible to report about all different subsets of dendritic cells in all chronic inflammatory systemic diseases, an example for rheumatoid arthritis is given. In synovial tissue of patients with rheumatoid arthritis, inflammatory CD1c+ inflammatory dendritic cells were identified in close vicinity to T cells and expressing IL-12p70 and IL-23p19, which would give rise to proinflammatory T helper type 1 or T helper type 17 lymphocytes.[375] This is an example of the proinflammatory polarization of T cells. Another marker on inflammatory dendritic cells is CD1a, and cells positive for CD1a were found in synovial tissue of patients with rheumatoid arthritis and psoriasis arthritis.[376] Importantly, these CD1a+ inflammatory dendritic cells express RANK and RANKL, possibly contributing to osteoclast differentiation and T cell cross talk.[377]

Inflammatory dendritic cells derived from monocytes of patients with rheumatoid arthritis express higher amounts of proinflammatory cytokines such as IL-1, IL-6, TNF, and also IL-10 in comparison with cells from healthy subjects.[378] Importantly, monocyte-derived inflammatory dendritic cells can differentiate into osteoclasts in the presence of RANKL and M-CSF, which is supported by IL-1 and TNF in patients with rheumatoid arthritis.[379] In addition to inflammatory dendritic cells, also plasmacytoid dendritic cells exist in synovial tissue of patients with rheumatoid arthritis that are able to produce high amounts of type I interferons.[380] Type I interferons from plasmacytoid dendritic cells also play an important role in other chronic inflammatory systemic disease such as systemic lupus erythematosus, psoriasis vulgaris, type 1 diabetes mellitus, dermatomyositis, and Sjögren syndrome (reviewed in Ref. 381). An extensive review on dendritic cells in rheumatoid arthritis is given in Ref. 382.

These examples in the disease of rheumatoid arthritis demonstrate the presence of inflammatory dendritic cells in inflamed tissue and their mainly proinflammatory role with cytotoxic or humoral effector functions (stimulating T cells or B cells). This is similar in other chronic inflammatory systemic diseases. As such, dendritic cells have been implicated in the initiation and perpetuation of chronic inflammatory systemic diseases through the abolition of self-tolerance and the subsequent emergence of self-reactive T or B cells.

In contrast to inflammatory dendritic cells, there also exist tolerogenic dendritic cells that have been extensively characterized in graft-versus-host disease.[383] These cells might establish a regulatory milieu to overcome the continuous inflammatory process in chronic inflammatory systemic diseases. Presently, no routine therapies exist on the basis of tolerogenic dendritic cells that might lead to tolerance induction. However, an antigen-specific tolerogenic dendritic cell-based immunotherapy can be developed.[382,384] Recent preliminary clinical trials with tolerogenic dendritic cells used as autologous cellular therapy demonstrated beneficial effects in different chronic inflammatory systemic diseases (summarized in Ref. 385). This leads to the conclusion that dendritic cells are involved in chronic inflammatory systemic diseases as either proinflammatory stimulators or anti-inflammatory inhibitors.

NK Cells

The role of NK cells was long misunderstood because they were considered simple cytotoxic cells without any other function. Cytotoxicity was subdivided into (1) natural cytotoxic activity against tumor cells or virally infected cells (triggered via MHC class I) and (2) antibody-dependent cellular cytotoxicity directed against antibody-coated target cells (triggered via CD16, i.e., the Fc-γ receptor III). Now, it is realized that NK cells in addition to their cytotoxic activity can produce a panel of cytokines. The killer NK cells are $CD56^{dim}$ and CD16+, whereas the cytokine-producing NK cells are $CD56^{bright}$ and CD16-negative.[386]

The cytokine-producing NK cells are classified into NK1 cells (induced by IL-12/IL-18, and products are IFN-γ/TNF), NK2 cells (induced by IL-4, and products are IL-5/IL-13), regulatory NK cells (induced by IL-2, and products are IL-10/TGF-β), and NK22 cells in mucosa (induced by IL-23, and products are IL-22) (reviewed in Ref. 386–388). Somehow, they behave like brothers of T cells, which is expected because they are closely related to each other. While NK cells can assist dendritic cell maturation and T cell polarization, increasing evidence indicates that NK cells can also prevent and limit autoimmune responses via killing of autologous myeloid and lymphoid cells.[386,388]

In the healthy situation, NK cells are not activated because inhibitory messages to the NK cell dominate. In an activated situation, these inhibitory signals are switched off and activating signals are switched on so that NK cells start killing target cells and producing cytokines and cytotoxic factors to influence the micromilieu.

In recent years, NK cells have been studied in the context of atopic dermatitis, psoriasis vulgaris, pemphigus vulgaris, multiple scerosis, systemic lupus erythematosus, and juvenile idiopathic arthritis; and in animal models of type 1 diabetes mellitus and EAE (reviewed in Refs. 386–388). The pathophysiological concepts derived from these studies are far from clear, which depends on the many different functions of NK cells that cut surface at the moment (NK cells can be good and bad guys). This is also true for NK T cells with many functionally distinct subsets.[389]

The latest add-on in NK cell research is the finding of NK cell memory, which is long-lived, antigen-specific, and independent of B cells and T cells.[390–393] Such a memory function can be deleterious since it can lead to perpetuation of a chronic inflammatory process. Presently, no specific NK cell- or NKT cell-directed therapies are available. The NK cell field is widely open for many new discoveries.

Neutrophils

Neutrophils play critical effector and regulatory roles in a multitude of immune responses.[394] They eliminate unwanted pathogens and other components of their environment by phagocytosis. They produce toxic mediators and effector cytokines and chemokines, which in turn recruit and modulate the activity of other cells. However, the capacity for bacterial killing carries with it an implicit capacity for host tissue destruction as observed in inflammatory and autoimmune diseases; accordingly, neutrophil function must be tightly regulated.[394]

Neutrophils are activated and attracted to the site of inflammation by immediate chemotactic factors such as complement C5a, IL-8, fMLP, leukotriene B4, and similar ones.[394] Neutrophils phagocytize upon direct contact with danger factors (e.g., bacteria) or targets opsonized by antibodies and/or complement. Neutrophils possess Fc-γ receptor type I, II, and III and complement receptors.[394] This can lead not only to intracellular killing in phagosomes but also to release of neutrophil granule content such as reactive oxygen species, bactericidal proteins (defensins and others), proteinases (proteinase 3, elastase, matrix metalloproteinases), phospholipase A2, and others. It is important to mention that all these factors can be released in the context of chronic tissue inflammation.

The importance of neutrophils is best demonstrated in neutrophil deficiency syndromes such as leukocyte adhesion deficiency-I (CD18), chronic granulomatous diseases (NADPH oxidase defects), and Chediak-Higashi disease (*Lyst* defect and lysosome trafficking defect). These diseases are accompanied by severe immunodeficiency often leading to death within the first 10 years of life.

Neutrophil activity is mainly controlled by an early cell death with a half-life of approximately 8-10h in inflammatory and noninflammatory states.[395] Neutrophils play a proinflammatory role in several chronic inflammatory diseases such as crystal-induced arthritis, inflammatory bowel diseases, rheumatoid arthritis, ANCA-associated vasculitis (autoantibodies against neutrophil proteins such as LAMP-1,[396] myeloperoxidase, and proteinase 3), and systemic lupus erythematosus.[395]

Neutrophils as prototypic cells of innate immunity are independent of specific antigens, but nevertheless, they are highly effective in generating a proinflammatory and destructive milieu in chronic inflammatory systemic diseases.

Apoptosis

There exists an interesting mouse model called lpr/lpr where animals developed lymphadenopathy and a systemic lupus erythematosus-like autoimmune disease.[397] In addition, a human disease called Canale-Smith syndrome was similarly characterized by lymphadenopathy and autoimmunity in childhood.[398] Both diseases were directly linked to a defect in Fas signaling, a crucial pathway of apoptosis.[112,113,397,398] These findings have called the attention to apoptosis, which has been increasingly scrutinized during the last two decades.[399]

It has been delineated that clearance of apoptotic cells by macrophages or increased apoptosis is important to remove dangerous material.[400,401] In the normal situation, phosphatidylserine and other cellular membrane elements trigger efficient recognition and uptake of apoptotic material by phagocytes.[401–403] Recognition of these "eat me" signals by macrophages and dendritic cells leads to secretion of TGF-β and IL-10 and, generally, to an anti-inflammatory milieu.

Clearance problems can appear with defects in the complement system (C1q, CR1, CR2, CR3, and CR4) or the mannose-binding lectin pathway that predisposes to autoimmune diseases. This can lead to prolonged presence of dead material in the tissue. Defects in the clearance of apoptotic cells result in the occurrence of secondarily necrotic cells, which can lead to the release of danger signals and, thus, inflammatory responses. Uptake of necrotic cell material is mediated by Fc-γ receptor-related phagocytosis. In this context, nuclear autoantigens can be taken up by dendritic cells that present nuclear peptide fragments to T and B cells leading to autoimmunization. It is thought that this process is responsible for the generation of autoantibodies toward intracellular targets.[401,402] Although progress has been made in several areas, there is a lot more to learn.

With respect to autoimmune diseases, alterations in apoptotic pathways are made responsible for pathogenic aspects in multiple sclerosis,[404] pemphigus,[405] systemic lupus erythematosus,[400] rheumatoid arthritis,[406] Sjögren syndrome,[407] and others. There is either too much apoptosis, or too little; both change the tissue milieu in an unfavorable way. As long as defects in apoptotic pathways play a role, for example, with complement deficiency, the perpetuation of a chronic inflammatory systemic disease can be partly explained. A specific treatment of chronic inflammatory systemic diseases based on apoptosis pathway is presently not available.

Adhesion and Costimulatory Molecules, Vascular Transmigration, and Chemokines

It was obvious that adhesion and costimulatory molecules, vascular transmigration, and chemokines can be important targets for anti-inflammatory therapies because stopping lymphocytes, monocytes, dendritic cells, NK cells, neutrophils, and others from communicating or entering into tissue should soon lead to a reduction of proinflammatory cells at the site of inflammation. Every idea needs to stand the test of time and examination.

Blocking of the costimulatory interaction of CD80 and CD28 using CTLA4-Ig (abatacept) in rheumatoid arthritis was already discussed above. This successful therapy demonstrates that communication of T cells and antigen-presenting cells is critically important. Two biologics that target T cell adhesion and costimulation—an anti-LFA-1 antibody (efalizumab) and an LFA-3 fusion protein (alefacept)—have shown efficacy in psoriasis vulgaris. Another blocker of costimulation is the glatiramer acetate launched in 1997, which successfully inhibits the communication between T cell receptor and HLA DR on antigen-presenting cells, which is therapeutically successful in patients with relapsing/remitting multiple sclerosis.

In addition, trafficking of T cells can be inhibited by an antibody against α4β1 and α4β7 integrins (natalizumab), which was highly effective in patients with relapsing/remitting multiple sclerosis. The same antibody also worked in patients with Crohn's disease showing the role of vascular transmigration. Another humanized monoclonal antibody against α4β7 integrin (vedolizumab) proved to be successful in Crohn's disease.[n] Interestingly, no therapeutic principle exists

n. We have not yet talked about adverse events of all these therapies. All these therapies bear the risk of unwanted control of infection or the uncontrolled growth of tumors, which is best illustrated with natalizumab leading to progressive multifocal leukencephalopathy cause by the JC virus or TNF blockers that are linked to malignancies. However, the number of events relative to the number of successful treatments is quite favorable so that the drugs are used with prior thorough education of the patient.

on the basis of chemokines, which is probably very difficult because of the high redundancy of these molecules.

Although this is only a small list of current treatment possibilities with respect to adhesion and costimulatory molecules or vascular transmigration, it unambiguously demonstrates that these pathways are relevant in one or another disease. It is also obvious that not all therapies work in all disease, which demonstrates the typical immunological heterogeneity in pathophysiology. In addition, this is a field with a large number of failed treatment options although preclinical testing was very successful.[408]

Pattern Recognition Receptors

Sensing of infection is mediated by innate pattern recognition receptors,[164,166] which include TLRs, RIG-I-like receptors, NOD-like receptors, and C-type lectin receptors.[409] The intracellular signaling stimulated by these pattern recognition receptors lead to transcriptional expression of inflammatory mediators that activate many cells in a similar fashion, and NF-kappaB plays a central role.[409]

There are also numerous endogenous molecules that can activate pattern recognition receptors, which produced the theory of "danger signals."[410] Endogenous molecules are heat shock proteins, fibronectin, high-mobility group box chromosomal protein-1 (HMGB1), breakdown products of heparin sulfate and hyaluronic acid, GP96, uric acid crystals, and many others.[411,412] Since in the process of inflammation—for example, during cell death—also endogenous signals are released, pattern recognition receptors can play an important role in augmentation of chronic inflammatory systemic diseases against self.

In addition, pattern recognition receptors always play a role when infectious agents are somehow involved. This is most obviously the situation on body surfaces of the skin, the intestinal tract, the upper/lower respiratory tract, and the urogenital tract. Thus, the important link between NOD2, an important sensor of intracellular bacteria, and Crohn's disease[413,414] produced a paradigm shift in pathophysiological concepts of inflammatory bowel diseases and other chronic inflammatory systemic diseases such as graft-versus-host disease.[415] Since NOD2 became important in these diseases, the concept of pure autoimmunity or alloreactivity, respectively, needed to be revised. With the discovery of these pathophysiological triggers, the concept of chronic inflammatory systemic diseases stimulated by harmless foreign antigens was strongly supported.

It is time to think again of Freund adjuvant that was significant in many experimental autoimmune diseases at the beginning of this research (see above). Freund adjuvant is a solution emulsified in mineral oil with a mannide monooleate surfactant. Complete Freund adjuvant contains inactivated and dried mycobacteria, usually *Mycobacterium tuberculosis* (not part of incomplete Freund adjuvant). Exactly, these mycobacterial products are recognized by pattern recognition receptors.[416–419] Importantly, the ligands to different pattern recognition receptors can influence the subsequent direction of the immune response leading to either T helper type 1, T helper type 2, T helper type 17, or even regulatory T cell responses.[108,412] Thus, the fine-tuned synergism of different ligands to pattern recognition receptors became the Holy Grail of vaccine development.

In summary, the involvement of pattern recognition receptors in such diverse diseases as antiphospholipid antibody syndrome,[420] autoimmune hepatitis,[421] celiac disease,[422] type 1 diabetes mellitus,[423] inflammatory bowel disease,[421] multiple sclerosis,[424] rheumatoid arthritis,[425,426] systemic lupus erythematosus,[427] and others clearly demonstrates that the finding of these innate recognition receptors added to a paradigm shift in chronic inflammatory systemic diseases against self and against harmless foreign.

The Local Cellular Support System of Inflammation

The reader might have recognized that I have not mentioned nonhematopoietic cells such as fibroblasts, astrocytes, stromal cells, epithelial cells, keratinocytes, endothelial cells, mesangial cells, osteoblasts, chondrocytes, adipocytes, and myocytes. These cells define the immediate environment of hematopoietic immune cells, and they have their own important interactions with immune cells (e.g., Ref. 161). Today, we call them the "niche" of hematopoietic immune cells.

Usually, these cells define the special characteristics of a given tissue. They make the inflammatory response tissue-specific. For example, epithelial cells do not play a role in the inflamed joint, but they are instrumental in the inflamed gut. Thus, the importance of these cells for the pathophysiology of a given chronic inflammatory systemic disease depends on the tissue. Although several articles on these special cells already appeared early in the literature, their particular role in chronic inflammatory systemic diseases was installed in the last two decades.

Fibroblasts and Astrocytes

The fibroblast was initially discovered in wound repair responses and fibrosis reactions.[428–430] The fibroblast came to fame by the work of Gay in Zurich and Firestein in San Diego and their many international colleagues in the field of rheumatology.[431,432] Importantly, the fibroblast is one of the important destroyers of cartilage (matrix metalloproteinases), and these cells are much more mobile as initially thought.[433] The fibroblast plays its main role in joint diseases, but it also has an important role in other chronic inflammatory systemic diseases because it provides the extracellular matrix to induce overshooting fibrogenesis.

For example, astrocytes are responsible for the scarring reaction in the brain.[434–436] Astrocytes have hypertrophic characteristics in acute and chronic lesions of multiple sclerosis.[437] The centers of the lesions are hypocellular and contain naked axons embedded in a matrix of scarring fibrous astrocytes and lipid-laden macrophages and only few other cells.[437]

In addition, an exaggerated fibroblast response is a determining factor in systemic sclerosis.[438] Fibroblasts are responsible for strictures in Crohn's disease.[439] Many more similar examples can be given that demonstrate that fibroblasts have a decisive position in an inflammatory process. There are several attempts to influence these unfavorable actions of fibroblasts, but until now, no specific therapy exists that targets the fibroblast.

Stromal Cells of the Bone Marrow

These stromal cells obtained a significant role in the concept of the hematopoietic niche in the bone marrow. The hematopoietic niche in the bone marrow is important for memory functions of long-lived plasma cells.[161] Since memory is a crucial aspect in the continuation of chronic inflammatory systemic disease, these stromal cells can have a decisive role. In addition, stromal cells are able to present antigens to T cells, which would be an additional way to shape the immune response toward autoantigens.[440]

Endothelial Cells, Mesangial Cells, and Neovascularization

Endothelial cells can be direct targets in the inflammatory process such as in antiphospholipid antibody syndrome, glomerulonephritis, or vasculitis. The endothelial cell can be a critical player in atherosclerosis; and chronic inflammatory systemic diseases are often associated with atherosclerosis. Endothelial cells are responsible for transmigration of leukocytes into inflamed tissue leading to perivascular infiltrates (see above).

Endothelial cells can be activated to remodel vasculature. Proangiogenic factors like vascular endothelial growth factor, angiopoietin, TNF, and IL-8 are produced in the inflammatory process, which ends up in neovascularization. A balance of proangiogenic versus antiangiogenic factors determines the degree of neovascularization. Endothelial cells can have cross talk with other cells in the tissue such as dendritic cells with bidirectional effects.[441] In addition, nonendothelial cells can be transformed to become endothelial cells such as fibroblasts and dendritic cells.[441] Furthermore, endothelial cells of secondary lymphoid organs are able to present antigens to CD8 T cells.[440]

The pathophysiological role of endothelial cells has been demonstrated in numerous chronic inflammatory systemic diseases such as glomerulonephritis,[442] multiple sclerosis,[443] psoriasis vulgaris,[444] rheumatoid arthritis,[445–447] and systemic lupus erythematosus.[448] During the last decade, the endothelial cell and neovascularization were in the focus of several therapeutic approaches. However, many treatment attempts in patients were unsuccessful, which most probably demonstrates redundancy in this particular system. The successful therapies to influence transmigration of leukocytes have been demonstrated above.

Epithelial Cells

In chronic inflammatory systemic diseases, epithelial cells have been thought to be innocent bystanders of autoaggressive humoral and cellular reactions.

However, these cells play an important role in the inflammatory process. For example, epithelial cells of the inflamed salivary gland in Sjögren syndrome display high levels of immunoactive molecules such as IL-1, IL-6, IL-12, IL-18, TNF, and BAFF and also ICAM-1, VCAM-1, E-selectin, and many others that are known to support lymphoid cell homing, antigen presentation, and amplification of epithelial-immune cells interactions.[449] A wide panel of different chemotactic molecules are produced by epithelial cells.[449] Epithelial cells express proteins encoded by the MHC class II complex and costimulatory molecules such as ICAM-1 and B7, and it is thought that epithelial cells can present antigen.[449] Epithelial cells possess the TLRs that are relevant in upregulation of important proinflammatory molecules.[449]

Furthermore, epithelial cells play an important role in all chronic inflammatory skin diseases (see below keratinocytes), in different forms of thyroiditis (not well investigated), in autoimmune hepatitis and primary biliary cirrhosis,[450] in inflammatory bowel diseases,[451] in celiac disease, in glomerulonephritis, in type 1 diabetes mellitus, and in several other diseases with involvement of exocrine and endocrine glands. Most of the recently published studies demonstrate that epithelial cells are not innocent bystanders but very actively involved in these diseases. However, the field of epithelial cell research in the context of abovementioned chronic inflammatory systemic diseases is still in its infancy.

Keratinocytes

Keratinocytes can be the target of chronic inflammatory systemic diseases such as in the group of autoimmune intraepidermal blistering diseases of the skin, called pemphigus. The keratinocyte proteins desmoglein 1 and desmoglein 3 were found to play the key role for this group of diseases. While the autoimmune attack is directed against these important attachment proteins, keratinocytes are not innocent bystanders because they undergo apoptosis and stimulate several subsequent destructive events.[452]

The keratinocytes carry TLR molecules and are, thus, able to respond to many different endogenous or exogenous triggers in the local environment.[453] Binding of the autoantibodies to these desmosomal proteins induces a multitude of keratinocyte effects leading to keratin retraction and, thus, acantholysis—the basis of blistering diseases.[454] In addition, cytokine production of keratinocytes in response to autoantibody binding has been mentioned as a possible aggravating factor.[454]

A similar important role of keratinocytes is suggested in cutaneous lupus erythematosus because autoantigens targeted in lupus are clustered in two populations of surface structures on apoptotic keratinocytes.[455] Again, keratinocytes are not only targets but also active participants in the inflammatory process.

In psoriasis vulgaris, keratinocytes play an important role since they are responsible for hyperkeratotic lesions. In this disease, IL-23 is overproduced by dendritic cells and keratinocytes, and IL-23 stimulates T helper type 17 cells within the dermis to make IL-17 and IL-22. IL-22, in particular, drives keratinocyte hyperproliferation in psoriasis vulgaris.[456] In addition, keratinocytes produce chemokines to attract neutrophils, dendritic cells, and T cells such as IL-8, GROα, and CCL20. They produce defensins, heat shock proteins, S100 proteins (S100A12), and growth factors such as PDGF and VEGF, which influence the local microenvironment.[145]

The examples clearly demonstrate the complex role of these cells with functions quite different from pure bystander activity.

Oligodendrocytes

The target of the destructive process in multiple sclerosis are oligodendrocytes since most autoantigens such as myelin are produced by this cell type.[437] In addition, oligodendrocytes typically get lost in acute and chronic lesions, more so in the chronic than the acute form. However, oligodendrocytes are not the innocent bystanders because they can produce factors that directly influence the microenvironment like Nogo-receptor agonists such as Nogo, oligodendrocyte-myelin glycoprotein, and myelin-associated glycoprotein. These factors are important in axon repulsion and growth arrest of neurons.[437,457] Nogo itself can influence remyelination by oligodendrocytes.[458] Furthermore, pathologically described type III multiple sclerosis depends on a degradation process of oligodendrocytes that is possibly independent of immune cells.[437] Thus, oligodendrocytes are not innocent.

Adipocytes

The role of adipocytes in inflammatory processes has attracted much attention in recent years.[459] Since 1995, along with the discovery of leptin, it has been recognized that adipocytes secrete hormones and proteins such as adiponectin, visfatin, omentin, resistin, and cartonectin. Adipose tissue has been accepted as an endocrine organ.[460] Adipocytes produce TNF,[461] they express TLR, and they are responsive to lipopolysaccharides.[462] They can produce a large panel of cytokines, chemokines, complement factors, growth factors, and others.[463] In a situation where fat tissue is in close vicinity to an inflammatory process, an influence of adipocytes on immune cells is to be expected. Most probably, this can happen in creeping fat in Crohn's disease, ophthalmopathy in Graves' disease, synovial adipose tissue in rheumatoid arthritis, and mesenteric panniculitis to name a few.[463] But how can adipocytes influence the inflammatory process? Let us look on leptin.

Leptin has been demonstrated to influence T cell proliferation and T helper type 1 immune reactions.[464] Leptin stimulates reactive oxygen species production by stimulated polymorphonuclear neutrophils.[465] Leptin-deficient mice exhibit impaired host defense in Gram-negative pneumonia.[466] Furthermore, leptin is directly linked to the inflammation and immune response in chronic inflammatory systemic diseases.[467,468] These few examples demonstrate how an adipocyte factor can influence the inflammatory process indicating that adipocytes are certainly not innocent bystanders.

Myocytes

The myocyte is often the target of the autoimmune attacks because muscular surface, cytoplasmic, or nuclear proteins are recognized by autoaggressive immune cells. Examples for autoantigens are acetylcholine receptors (myasthenia gravis), muscle specific kinase (myasthenia), aminoacyl-tRNA synthetase (e.g., Jo-1 and many others in myositis), signal-recognition particle (myositis), and many more in dermatomyositis and polymyositis.[469] The autoantigenic process in myositis is described as partly T helper type 1-mediated with elevated levels of TNF and IFN-γ and a strong CD8+ cytotoxic T cell role in polymyositis, the role of which is presently unclear.[469,470]

The myocyte produces several factors that support the proinflammatory process once it is installed such as an NH2-terminal fragment of Jo-1 with chemotactic activity for T cells and monocytes.[471] Or cytokine-stimulated muscle fibers can secrete proinflammatory cytokines and chemokines, which facilitate the recruitment of activated T cells.[470,472,473] In addition, muscle fibers can express large numbers of not only MHC class I- and II-encoded proteins but also costimulatory molecules such as B7-H1 and ICOS ligand, which qualify them as facultative antigen-presenting cells.[473] The muscle fibers not only are targets but also generate a proinflammatory environment.

Eosinophils, Basophils, and Mast Cells

Although we know that these cells play a crucial role in allergic immune mechanisms (for review, see Ref. 11), their role in chronic inflammatory systemic diseases is not well defined. Of the mentioned cells, particularly, mast cells trigger proinflammatory effects not only by immediate release of proinflammatory cytokines (TNF, IL-6, and others) but also by factors such as tryptase, chymase, matrix metalloproteinases, chemotactic factors, and histamine.[474–484] Targeted silencing of mast cells protects against joint destruction and neoangiogenesis in experimental arthritis.[485] Similarly, mast cells influence severity in inflammatory bowel diseases,[486–491] experimental models of multiple sclerosis,[492–494] and several other diseases. However, recent discussions also see relevant anti-inflammatory roles of mast cells in autoimmune diseases.[495]

Signaling Pathways: Genome-Wide Association Studies

While cAMP and cGMP were the major signaling factors until the midst of the 1980, the molecular revolution of the 1980s added many additional signaling factors. While finding a specific disease-relevant signaling pathway was difficult during these early pioneering years, modern human genetics with the powerful tool of genome-wide association studies revolutionized the field. Today, we recognize many signaling proteins relevant in one or another chronic inflammatory systemic disease. Apart from the MHC genes, several newly discovered non-MHC genes are highly relevant in signaling cascades.

For rheumatoid arthritis, a recent study on more than 29,000 cases and more than 73,000 controls demonstrated a very large list of possible signaling factors related to this disease.[496] Highly relevant signaling factors were the following ones: TYK2 (tyrosine kinase 2; signaling of cytokine receptors), CD40 (cell surface molecule and receptor), IFNGR2 (interferon γ receptor 2), TNFAIP3 (negative regulator of the NF-kappaB pathway), PTPRC (CD45, cell surface molecule, receptor and protein tyrosine phosphatase), PTPN22 (protein tyrosine phosphatase, signaling of receptors), IL-6R (IL-6 receptor), FCGR2B (CD32, Fc fragment of IgG, low-affinity IIb receptor), IL-2 (cytokine), IL-2RA (IL-2 receptor α-chain), SH2B3 (member of the SH2B adaptor family of proteins, signaling of growth factor and cytokine receptors), and ICOSLG (inducible T cell co-stimulator ligand). There are many more relevant factors in rheumatoid arthritis as demonstrated in this genome-wide association study.[496]

It is highly interesting that the same signaling factors can play a role in different chronic inflammatory systemic diseases such as multiple sclerosis (*TYK2*, *CD40*, *TNFAIP3*, *PTPN22*, *IL-2*, *IL-2RA*, and others; Ref. 497), type 1 diabetes mellitus (*TNFAIP3*, *PTPN22*, *IL-2*, *IL-2RA*, *SH2B3*, and others; Refs. 498–500), systemic lupus erythematosus (*TYK2*, *TNFAIP3*, *PTPN22*, and *FCGR2B*; reviewed in Ref. 501), and Crohn's disease (*TYK2*, *PTPN22*, *IL-2RA*, *ICOSLG*, and others; Ref. 502). This tells us that shared immunostimulatory or immunoinhibitory signaling pathways are similarly used in the different diseases.

While there is important genetic overlap between phenotypically different chronic inflammatory systemic diseases, it is important to mention that these diseases have their own specific susceptibility genes relevant to signaling (more than shared ones). This is to be expected because tissue-specific cells, which I called the "local cellular support system," must play a decisive role. To make an example, when a chronic inflammatory systemic disease affects mucosa of the gut as in Crohn's disease, specific signaling factors relevant to epithelial cells but not to fibroblast-like synoviocytes should play a major role. Or if the disease affects pancreatic β-cells, signaling factors of this particular cell type must play an important role.

In conclusion, genome-wide association studies in different chronic inflammatory systemic diseases revealed important novel pathogenic signaling factors.

The reader might have recognized that neuronal and hormonal factors and respective cells were not included in the above description of pathogenic elements in chronic inflammatory systemic diseases. The field of neuroendocrine immune research resulted in many interesting pathogenic pathways, which are described in Chapter II of this book (Table 4).

TABLE 4 Major Paradigms of Pathophysiology of Chronic Inflammatory Systemic Diseases

Paradigm	Pathogenic Mechanisms
Neutrophils	Important innate immune cell with strong tissue-destructive properties (granules contain destroying molecules, Fc-γ receptors, and pattern recognition receptors)
Autoantigens are decisive for the disease	For example, myelin, TSH receptor, glomerular basement membrane, acetylcholine receptor, desmosomal proteins, collagen type II, DNA-histone, complex membrane phospholipids, and cyclic citrullinated peptides
Autoantibodies and immune complexes activate an inflammatory cascade	Antibody-dependent cell-mediated cytotoxicity, complement activation, Fc-γ receptor binding, opsonization, and antibody class switch to IgG
B cells	Checkpoint problems of B cell selection, expansion in class-switched B cell memory and plasma cell compartments
	Plasma cells are long-lived in bone marrow stromal niches and in follicle-like structures in inflamed tissue
	B cells present antigen via MHC class II; B cells produce numerous cytokines, chemokines, growth factors, etc.
T cells	T cells can be autoreactive toward defined autoantigens
	CD4+ helper T cells are classified into type 1 (TNF, IFN-γ, and IL-12; T-bet), type 2 (IL-4, IL-5, and IL-13; GATA-3), and type 17 (IL-17 upon stimulation with IL-23; RORγT)
	CD4+ regulatory helper T cells (TGF-β and IL-10; Foxp 3); regulatory T cells are thymus-borne; the thymus plays a central role in tolerance (AIRE)
	γ/δT cells have a role in inflammation
	CD8+ cytotoxic T cells are found in the target tissue
	Classification of diseases according to T cell response
	Costimulatory molecules are important and can be therapeutically targeted
MHC and immunogenetics	HLA associations were demonstrated and linked to specific cytokine sets
	HLA important in transfusion and transplantation medicine

(Continued)

TABLE 4 Major Paradigms of Pathophysiology of Chronic Inflammatory Systemic Diseases—Cont'd

Paradigm	Pathogenic Mechanisms
	Genome-wide association studies present novel "immune response genes" with small odds ratios (e.g., PTPN22)
Monocyte and macrophage type of cell ((1) classically activated macrophages, (2) wound-healing macrophages, and (3) regulatory macrophages with strong IL-10/TGF-β production)	Central cell in chronic inflammatory systemic diseases Phagocytosis of target material and inflammatory response (Fc receptor, scavenger receptor, pattern recognition receptors, and many more) and phagocytosis of tissue debris (complement receptor and others) Tissue destruction (e.g., osteoclasts and microglia) Antigen presentation to T cells and B cells Produce nearly all molecules of the body when adequately stimulated (cytokines, chemokines, growth factors, hormones, neuropeptides, etc.)
Dendritic cells	Antigen presentation, costimulation, chemotaxis, many proinflammatory local tissue responses But also tolerogenic function of dendritic cells, which might be used therapeutically
NK cells, NKT cells	Cytotoxicity relevant in chronic inflammatory systemic diseases (MHC class I, Fc-γ receptor, inhibitory and activating NK cell receptors) Cytokine production relevant in chronic inflammatory systemic diseases (NK type 1, NK type 2, NK type 22, regulatory NK)
Apoptosis and complement defects	Not only central to downregulate the clonal process of lymphocyte proliferation (contraction phase) but also relevant in somatic nonhematopoietic cells Defects in apoptosis lead to autoimmune diseases and lymphoproliferation due to accumulation of debris (e.g., link between systemic lupus erythematosus and complement defects)
Adhesion and costimulatory molecules, vascular transmigration, and chemokines	Central role in immune cell trafficking and immune cell activation Therapy with biologics demonstrates the important role in chronic inflammatory systemic diseases
Pattern recognition receptors	Central to proinflammatory stimulation with exogenous and endogenous ligands; the concept of chronic inflammatory systemic diseases stimulated by harmless foreign antigens is strongly supported; role of Freund adjuvant illuminated

TABLE 4 Major Paradigms of Pathophysiology of Chronic Inflammatory Systemic Diseases—Cont'd

Paradigm	Pathogenic Mechanisms
The support system of the tissue	Nonhematopoietic mechanisms play an important role
Fibroblasts, astrocytes, and myofibroblast	Proinflammatory; tissue-destructive and tissue-invasive phenotype; scarring fibrotic tissue response; autoantigen production
Stromal cells of the bone marrow	Support of hematopoiesis and immunological memory
Endothelial cells	Central to inflammatory neovascularization, central to transmigration of leukocytes, target of the autoimmune attack, dysregulated and leading to atherosclerosis, cross talk with other immune cells
Epithelial cells	Participate in the inflammatory process by producing cytokines, chemokines; can present antigens in MHC class II-encoded protein, can be activated via pattern recognition receptors
Keratinocytes	Autoimmune target (e.g., desmosomal proteins); produce a panel of cytokines, chemokines, growth factors
Oligodendrocyte	Autoantigenic target (e.g., myelin), produce repellent factors and lead to neuron growth arrest
Adipocytes	Produce a large panel of proinflammatory factors circulating in the body (e.g., leptin) or locally such as cytokines, growth factors, and chemokines; possess pattern recognition receptors; are in close proximity to inflamed tissue, supply of bioenergetic molecules
Myocytes	Autoimmune target; produce proinflammatory cytokines and chemokines; express pattern recognition receptors and MHC class I- and MHC class II-encoded proteins plus costimulatory molecules
Mast cells (eosinophils, basophils)	Proinflammatory mediators (histamine, TNF, IL-6, and many others)
Signaling pathways	
Multiple signaling factors	Factors of signaling pathways play a role. Some of them are common in different chronic inflammatory system diseases; some of them are disease-specific depending on the support system of the tissue

The table gives a sequence that reflects the historical development during the last century. This table is not complete (stem cells, hepatocyte, chondrocytes, osteoblasts, etc.)!

THE TRIGGER OF CHRONIC INFLAMMATORY SYSTEMIC DISEASES

Above, I described pathogenic effector mechanisms of chronic inflammatory systemic diseases. These effector mechanisms are quite well known in many instances. A huge number of diverse responses of cells with hematopoietic and nonhematopoietic origin are understood. This is not to say that we have discovered everything, but it should say that we have discovered quite a lot.

However, the difficult issue in pathogenesis research remains etiology. Etiology is the study about causation of a given disease. It was long thought that chronic inflammatory systemic diseases have one, and only one, important triggering factor that can explain etiology. This thought style originated in a time when sensational discoveries have been made, which delineated that only one factor is responsible for one disease. For example, the loss of insulin was responsible for high serum glucose concentration in type 1 diabetes mellitus,[503] or gliadin was responsible for celiac disease.[504,505] Still today, those marvelous discoveries are made when we recall the Canale-Smith syndrome (Fas)[398] or autoimmune polyglandular syndrome type I (AIRE).[180,181] Other monogenic examples from the field of autoinflammatory syndromes include hereditary periodic fevers and granulomatous and pyogenic disorders.[506] These syndromes sometimes mimic autoimmune diseases with similar signs and symptoms.

It is always much more satisfying and sensational when one responsible factor can be found, and the reputation of the discovering researcher(s) rises up to the sky. However, finding the one and only factor most often did not occur in almost all chronic inflammatory systemic diseases. Genome-wide and epidemiologic association studies were somewhat disappointing because they did not show the single magic bullet that we all expected. They demonstrated new targets that might be better called rubber projectiles than magic bullets. However, the rubber projectiles are often similar in different polygenic chronic inflammatory systemic diseases (Table 3).

In the light of current research, pathogenesis of most of the diseases is heterogeneous. It is not surprising that no single predominant mechanism for these diseases has emerged. As it always is the case when multiple pathogenic pathways are proposed, no leading paradigm evolves. This is a dilemma for the strict thinker because he wants to solve the pending problem. However, sometimes, problems cannot be solved due to financial, logistical, logical, complexity, and other problems. In such a situation, we are summing up the important etiologic concepts to make sure that we have not forgotten an aspect (Table 5). If we do not know which etiologic principle prevails, a principle of mediocrity should be used, which means that several different possibilities may exist in parallel.

The main focus of research during the last decades was directed toward points 1-4 of Table 5. Point 5 of the table is a very recent addition that I wish to highlight in this particular book (see Chapters II–VII). Historically, point 5 is the last add-on in the list originating in the fact that we are able to treat patients

TABLE 5 Etiologic Factors in Chronic Inflammatory Systemic Diseases[507]
1. Genetic susceptibility (gene polymorphisms; few monogenic, most polygenic)
2. Complex environmental priming (microbes, toxins, drugs, injuries, radiation, cultural background, geography, smoking, work-related factors such as silica dust)
3. Immune response (exaggerated and continuous immune response against self or harmless foreign)
4. Tissue destruction and rebuilding (continuous wound response without proper healing but fibrotic scarring)
5. Systemic response (support of the immune and wound response with desolate consequences for the rest of the body)

to a point where we start to focus on these systemic problems. The systemic problems become disease-relevant. Even with good therapies, point 5 of Table 5 becomes more and more important because our patients grow older and inflammation cannot be completely stopped. Since a mild proinflammatory condition is present even after a good therapy, long-term desolate consequences for the rest of the body materialize. We once called it the accelerated aging process.[508] But let us start to discuss points 1-4 of the list.

Twin studies, single genetic association studies, and genome-wide association studies have demonstrated that genes play a role (**point 1 of** Table 5). However, these studies also revealed that the individual effect of a susceptibility polymorphism was not as strong as we have expected (except those cases with monogenic autoinflammatory syndromes). Geneticists presently think that combinations of genetic factors in association with environmental factors will increase the significance of genetic/environmental associations. The environmental background was mainly supported by careful clinical studies from clinical epidemiologists adding to the growing list of environmental factors (**point 2 of** Table 5) (see section "Accumulation Theory" of Chapter IV).

I was always impressed by the influence of smoking. Before World War II, smoking was thought to be positive in order to strengthen the body. With increased life expectancy, the link between smoking and diseases in later life became obvious. With the first epidemiological study in the 1950s, smoking was clearly linked to lung cancer.[509] Today, we know that smoking is a serious factor in chronic inflammatory systemic diseases[510–512] (see section "Accumulation Theory" of Chapter IV).

When we think of environmental factors, we recall the microbe discussion at the beginning of the book. Although long-standing microbial infection was not the main factor in chronic inflammatory systemic diseases, the influence of infectious agents is still an important element in pathogenic considerations. It might well be that infectious agents transform cells so that continuous activation—not infection—plays a role. This has been described for endogenous retroviral elements that might be a trigger in fibroblast activation in rheumatoid arthritis.[513]

With respect to pattern recognition receptors and the link between the NOD2/CARD15 gene polymorphism and susceptibility to Crohn's disease, the microbe debate took a surprising twist. With these discoveries, the concept of chronic inflammatory systemic diseases stimulated by harmless foreign antigens was strongly supported. Defects in pattern recognition of microbes can lead to smoldering immune responses toward harmless commensal flora.

In addition, we know that microbes are important for molecular mimicry, a term coined by Damian in 1964.[514] The increasing list of molecular mimicry examples demonstrates that we are still a fair way off to understand these diseases (Table 6). In animal models, the mimicry phenomenon and epitope spreading are well established (recent review in Ref. 515).

With respect to **point 3 in** Table 5, the major present theme to explain the unwanted attack of self or harmless foreign (e.g., commensal flora) is breakage of tolerance and respective memory of the breakage. It is thought that either central or peripheral tolerance mechanisms have failed. On the basis of a huge and favorable diversity of our immune system, 94% of the population profit (during infections), while 6% suffer from chronic inflammatory systemic diseases (Figure 2). Evolutionarily positively selected immune systems—innate and adaptive—have served to overcome the infectious threat but have left us with the possibility of an unwanted immune response against self or harmless foreign.

TABLE 6 Examples of Molecular Mimicry in Chronic Inflammatory Systemic Diseases

Disease	Microbial Agent	References
Rheumatic fever	Group A *Streptococcus*	516
Guillain-Barré syndrome	*Campylobacter jejuni*	517
Psoriasis vulgaris	Group B *Streptococcus*	518
ANCA-associated vasculitis	Bacterial adhesin FimH	396
Autoimmune gastritis	*Helicobacter pylori*	519
Multiple sclerosis (some patients)	Epstein-Barr virus	520
Systemic lupus erythematosus (some patients)	Epstein-Barr virus	521

Some people believe that molecular mimicry is the most important causal factor. They say "that every patient can have its own mimicry factor. Studies in a patient population will not demonstrate the important mimicry factor that plays a role in an individual patient. If we would have enough time and money, we might find the individual mimicry factor of every single patient."

Innate global system	Adaptive immune system
Evolutionary very old	**Evolutionary young (400 m Year)**
Fixed in the genome	**Small rearrangeable gene segments (B cell)** **Prone to mutations and imperfection**
Recognition of a small number of patterns	**Recognition of a huge number of patterns (10^{10}, self and foreign)**
Discrimination of self and nonself is perfect	**Discrimination of self and nonself is imperfect**

FIGURE 2 The imperfect immune system. This figure presents modified contents presented by Charles Janeway, when he gave his honorary lecture for the Avery-Landsteiner Award for Immunology of the German Society of Immunology in 2002 in Marburg, Germany.

In addition, the adaptive immune system has learned to keep a memory of antigenic targets, which is an important prerequisite for the perpetuation of the inflammatory response. Memory is present with respect to T and B cells and has recently been discussed in NK and NKT cells (see section above). If we accept that an immune response against harmless self or harmless foreign is the trigger, then these memory mechanisms are of outstanding importance and should be a suitable target for future therapies (anti-B cell strategies already focus on this problem). Nevertheless, due to epitope spreading and more than one memory clone, a highly specific epitope-designed therapy will probably never exist. To conclude with Charles Janeway, "Some biological prize might be exacted for the benefits that the immune system endows upon the individual."

The last point in our discussion is tissue destruction/rebuilding as a trigger for chronic inflammatory systemic diseases (**point 4 in** Table 5). Tissue destruction not only might be the consequence of an unwanted immune response against harmless self or foreign but also can be a trigger on its own. Tissue destruction may lead to posttranslational modifications of extracellular proteins that can allow immune recognition of neo-self epitopes, which has been demonstrated for a key autoantigenic determinant in rheumatoid arthritis, cyclic citrullinated peptide.[522] Usually, extracellular matrix proteins are not cyclically citrullinated, but during an inflammatory response, this can transpire and a neo-self antigen can be borne.[310,523]

In addition, an important autoantigenic determinant of polymyositis is aminoacyl-tRNA synthetase, which can be liberated from damaged myocytes.[471] This intracellular protein never sees the light of extracellular space, but during damage, it may perpetuate the development of myositis by recruiting mononuclear cells that induce innate and adaptive immune responses. Since fragments of aminoacyl-tRNA synthetase have chemotactic properties, an aggravation of disease is to be expected upon degradation of the molecule.[471] In a similar way, intracellular autoantigens may be decisive after tissue destruction in other chronic inflammatory systemic diseases. Tissue destruction can be elicited by a harmless process such as infection, injury, or radiation, but the subsequent autoimmune process can lead to clonal expansion, autoagressive memory, and continuous self-attack.

An essential part can also be tissue destruction-related activation of nonhematopoietic cells such as fibroblasts. Not only in the situation of rheumatoid arthritis but also in many other diseases, these activated fibroblasts produce large amounts of extracellular matrix leading to the fibrous reaction known in most chronic inflammatory diseases. If a component of the extracellular matrix is the autoantigen (e.g., collagen type II), tissue destruction and activation of these cells are deleterious because more autoantigens are produced in a continuous way so that a positive reinforcing loop occurs.

A similar scenario exists with respect to oligodendrocytes, the producers of the myelin sheets in the central nervous system. Oligodendrocytes remain active in the border zone between multiple sclerosis plaques and the surrounding normal tissue. Activation of oligodendrocytes in the border area by factors in the plaques can lead to continuous activation of oligodendrocytes and to an increased production of the autoantigen (myelin) in the border zone. This can be significant for increasing plaque formation.

No matter how you look at it, here, the "danger theory" of Poly Matzinger has some additional implication. Tissue destruction leads to release of intracellular factors that can activate pattern recognition receptors. In a point-of-no-return situation, continuous stimulation of pattern recognition receptors might aggravate and perpetuate the disease (see also Chapter VII).

When we think of the five different etiologic causes of chronic inflammatory systemic diseases—genetic, environmental, immunological, tissue-destructive, and systemic—two different categories of influence exist: *a priori* and *a posteriori*. An *a priori* reason for a chronic inflammatory systemic disease is present long before the initiation of the disease. It is a prerequisite for a given disease. In contrast, an *a posteriori* reason for a chronic inflammatory systemic disease appears after the initiation or even the outbreak of the disease but significantly contributes to the perpetuation of the disease. For the five different etiologic factors, both *a priori* and *a posteriori* causes exist (Table 7).

For every etiologic factor of Table 7, *a priori* and *a posteriori* causes have been described. If we do not know which etiologic principle prevails, a principle of mediocrity should be used, which means that several different possibilities may exist for one phenotypically coherent disease.

Before we come to the heart of the book in the following Chapters II–VII (dealing with point 5 of Table 5), let us shortly think about the timescale of chronic inflammatory systemic diseases. Consideration of the timescale is important to understand many aspects of these diseases.

TABLE 7 ***A Priori* and *A Posteriori* Causes for Chronic Inflammatory Systemic Diseases**

Etiologic Factor	*A Priori*	*A Posteriori*
Genetic	Gender, gene polymorphisms (e.g., AIRE, Fas)	A gene becomes only important later in the disease course after the outbreak
Environmental	Geography, hygienic conditions, sun exposure, nutritional preferences, smoking, silica dust exposure	Change of environmental conditions due to the disease (e.g., social retreat aggravates major depression; reduction of sunlight exposure supports inflammation due to hypovitaminosis D)
Immunological	Primary autoimmunity (e.g., AIRE, Fas, complement deficiency)	Molecular mimicry as a consequence of infection (Table 6)
Tissue-destructive	Degeneration of cells such as in the case of oligodendrocytes in multiple sclerosis histopathologic type III	Damage leads to release of danger signals that trigger the perpetuation of the disease
Systemic	Primary defect of systemic regulatory hormonal axes such as the HPA axis[a]	Systemic disturbance induces changes of regulatory hormonal axis reactivity, which trigger the perpetuation of the disease (Chapter VII)

[a]*See the Fischer and the Lewis rat paradigm.*[524] *HPA, hypothalamic-pituitary-adrenal axis.*

THE TIMESCALE

The Asymptomatic Phase of Chronic Inflammatory Systemic Diseases

Since some years, it is known that immune phenomena are already present several years before manifestation of the chronic inflammatory systemic disease.[525,526] This tells us that an asymptomatic phase of a chronic inflammatory systemic disease exists (asymptomatic means without symptoms or preclinical). For example, antibodies against *cyclic citrullinated peptides* have been found to play an important role in patients with rheumatoid arthritis[527]; and citrullinated peptides play a role in multiple sclerosis.[528] These antibodies recognize self-proteins, which display the amino acid citrulline in a cyclic form on their protein surfaces. In two population-based studies,[525,526] it was demonstrated that these autoantibodies were present already many years before disease outbreak. It is further important that the levels of these autoantibodies sharply increase shortly before disease manifestation (Figure 3). This autoantibody rise reaches exponential dimensions 2 years before disease outbreak (Figure 3).

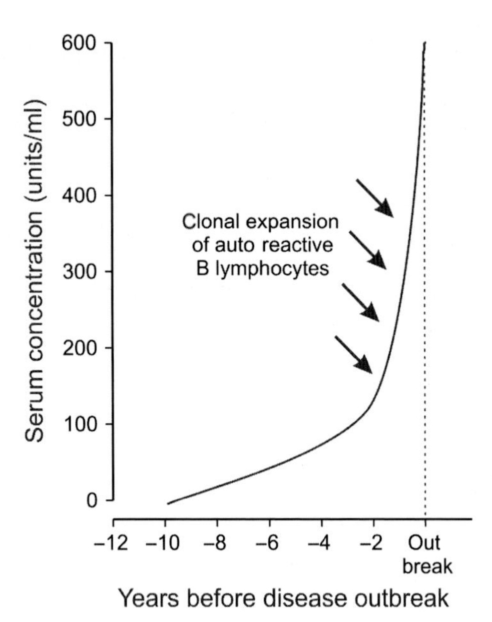

FIGURE 3 The increase of serum concentrations of autoantibodies against cyclic citrullinated peptides in patients with rheumatoid arthritis (schematic representation of information given in Ref. 526). The increase of the antibody concentration from year minus 2 toward the outbreak is indicative of a clonal expansion of autoreactive B cells, which enters the exponential phase 2 years before disease manifestation.

Since autoantibodies against cyclic citrullinated peptides are highly specific for rheumatoid arthritis and because the presence of these antibodies predicts the outbreak of the disease,[529,530] the autoimmune phenomenon *per se* is quasi-indicative of a causal role of these antibodies in the initiation of the disease. This is supported by findings with another autoantibody.

It was shown that antibodies to collagen type II were predictive for the development of rheumatoid arthritis later in life.[529] This is a significant finding because the most important humanlike animal model of rheumatoid arthritis is actuated by immunization against type II collagen (with Freund adjuvant; see above).

In addition, autoantibodies against type II collagen in mice and in humans with rheumatoid arthritis bind to the same epitope of the collagen molecule.[309] In the animal model, autoantibodies against collagen type II are pathogenic since transfer of these autoantibodies initiates arthritis in healthy animals:[309,531,532]

> *..., the arthritis transfer experiments ... of collagen type II epitope-specific human immunoglobulin G in the present investigation provide the first experimental evidence for the pathogenicity of autoantibody formation directed to an evolutionary positively selected conformational determinant that is located between amino acid 359–369 of the collagen type II triple helix. ... Moreover, the prevalence of immunoglobulin G autoantibodies binding to this particular conformational epitope in patients with rheumatoid arthritis clearly varies from other rheumatic*

conditions, suggesting disease-related differences in fine specificity of collagen type II - directed autoimmunity.

These findings position collagen type II and the respective autoimmune reaction at a central place in some patients with rheumatoid arthritis. Autoantibodies to other articular structures might play similar roles in other patients.[255] Since the advent of magnetic resonance imaging, we know that very similar asymptomatic phases exist in patients with multiple sclerosis.[437] In most cases, multiple sclerosis is subclinical early in its course. It has been demonstrated that occult disease activity, as detected by conventional magnetic resonance imaging, is 5-10 times as frequent as clinical active disease.[437]

The abovementioned autoimmune phenomena are described to demonstrate the concept of an asymptomatic phase of the disease. According to the presence of autoantibodies against cyclic citrullinated peptides and probably against other peptides as well, the asymptomatic phase in patients with rheumatoid arthritis can last up to 10 years.[525,526] This is a fascinating discovery because we immediately recognize that during the asymptomatic phase of the disease, several **endogenous factors** must be present, which counterbalance the outbreak of the disease (Figure 4). We have strong endogenous immunosuppressive factors. Interestingly, similar concepts also appear in other chronic inflammatory systemic diseases such as multiple sclerosis,[533–535] Crohn's disease and ulcerative colitis,[536] thyroiditis,[537] and type 1 diabetes mellitus.[530,538–542]

Sometimes, symptoms of a disease can be present for a very short period of time without any long-term symptomatic disease.[530] In rheumatology, a temporary form of arthritis is called palindromic rheumatism (the word "palindrome"

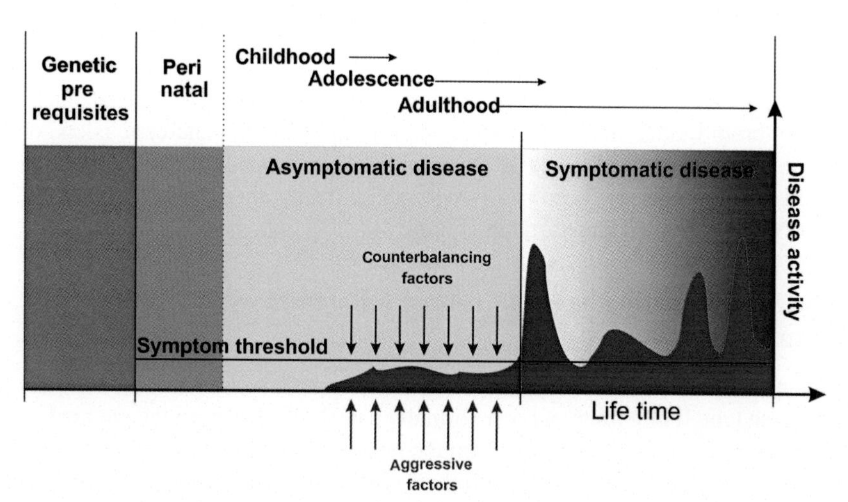

FIGURE 4 The asymptomatic and symptomatic phases of the chronic inflammatory systemic disease. In the asymptomatic phase of a chronic inflammatory systemic disease, some important stimulating phenomena might already be present. During the asymptomatic phase, counterbalancing endogenous factors outweigh aggressive factors. This behavior changes in the symptomatic phase.

has Greek roots of palin (πάλιν; "back") and dromos (δρόμος; "way, direction")). The disorder is defined by sudden short-lived attacks of arthritis (generally 1-3 days) often associated with swelling and redness, which go back spontaneously. About 47% of these patients develop a chronic inflammatory form of arthritis later in life.[530]

Interestingly, the same autoantibodies mentioned above can predict the development toward a chronic inflammatory systemic disease.[530] Similarly, in indeterminate colitis, antibodies against yeast predict the development of Crohn's disease, a chronic inflammatory systemic disease of the gut.[543] Additionally, in patients presenting with acute monosymptomatic brain disorders, production of immunoglobulins in the cerebrospinal fluid (the so-called oligoclonal bands) predicts progression to multiple sclerosis,[544,545] and patients with increased risk to develop type 1 diabetes present typical antibodies years before disease outbreak.[546] These are interesting examples because these patients might lead us to the abovementioned endogenous counterbalancing factors (Figure 4).

In the asymptomatic phase of a chronic inflammatory systemic disease, a limited number of cell types are involved. If we accept the etiologic concept of an antigen-driven disease (self, autoantigens or foreign, infectious agents and others), T cells, B cells, and antigen-presenting cells will play the major role in the asymptomatic phase (as easily recognized in animal models). During initiation of the disease, we recall that not many other cell types contribute to this process. This indicates that antigenicity is not instantaneously accompanied by a symptomatic disease, that is, the immune response is restricted to secret players. This "underground show" happens in lymph nodes and the spleen but not in the target organs, which are later affected in the symptomatic phase.

This scenario does not describe the situation in autoinflammatory syndromes, which are of monogenic origin and often immediately linked to a proinflammatory symptomatic disease.

The target structure of the inflammatory process, whether synovial membrane, brain tissue, intestinal mucosa, thyroid epithelium, pancreatic islets, or others, are not involved to a large extent in the asymptomatic phase. At the time point when the disease becomes symptomatic, many other local cell types get involved in inflamed target organs (Figure 4).

The Symptomatic Phase of Chronic Inflammatory Systemic Diseases

In a chronic inflammatory systemic disease, a target structure such as synovial collagen type II or a fragment of a normally harmless microbe in the gut are important stimuli to elicit the local immune response. If the number of aggressive and specific T cells or B cells with their outstanding capabilities to recognize these target structures outweighs immune-regulatory factors (endogenous immunosuppressive factors), the disease manifests in the respective compartment where the antigen is present. The circulating T cells and B cells (recall that they travel 10 times the entire body per day), which recognize target structures, start to produce proinflammatory cytokines and chemoattractant proteins.

Now, the local inflammatory process activates cell types, which usually reside in the tissue. For example, in the gut, these cells are epithelial cells, fibroblasts and myofibroblasts, endothelial cells, neuronal cells, myocytes, fat cells, mast cells, and others. In the joint, fibroblasts, osteoclasts, osteoblasts, chondrocytes, endothelial cells, adipocytes, nerve fibers, and others become involved. The type of cells involved depends on the local situation. It is clear that in a joint or in the gut, no cardiomyocytes exist. Thus, the response of cardiomyocytes to inflammation, which is specific to the heart, cannot happen in the joint or in the gut. This leads to a certain specificity of the inflammatory response in a given compartment due to local environmental conditions.

In addition to local cells, mobile cells are attracted such as neutrophils, monocytes, NK cells, basophils, eosinophils, and stem cells. As the role of these other cell types increases, the role of the initial secret players—the T cell, B cell, and antigen-presenting cells—probably decreases. Tissue destruction takes over with its own rules.

The characteristics of the disease gradually changes as the disease itself and the involved players grow older.[547] Necessarily, one has to ask the question whether, or not, therapeutic interventions must change over time. And exactly, this is the experience of a physician, who treats patients with chronic inflammatory systemic diseases, because formerly effective drugs might become ineffective during the course of the disease (e.g., therapy switch from anti-TNF antibodies to anti-CD20 B cell eradicating antibodies in rheumatoid arthritis).

The Plasticity Phase of Chronic Inflammatory Systemic Diseases in Early Life

Apart from the asymptomatic and the symptomatic phase, a third phase exists, which can have a strong influence on a later-manifesting chronic inflammatory systemic disease, the phase *in utero* and perinatally when plasticity of bodily systems is maximum.[548] The concept of "perinatal programming" is not new and has been investigated in the physiological neuroendocrine development of both the adrenal and gonadal axes.[549] Already at the beginning of the 1990s, it has been demonstrated that intrauterine growth and maternal nutrition are responsible for reorganization of endocrine and metabolic systems, which results in metabolic and cardiovascular pathologies in adult life.[550,551]

Recent important work demonstrated that treatment of pregnant female primates at mid to late pregnancy with adrenocorticotropic hormone to disrupt the HPA axis diminished secretion of IL-6 and febrile responses to a proinflammatory cytokine in a 2-year-old offspring.[552] Prenatal effects also appear to vary with the stage of pregnancy at which disruption occurs. Disruption during early pregnancy increases cellular immune responses, whereas prenatal stress exposure or glucocorticoid treatment during mid to late pregnancy can be immunosuppressive in the adult offspring.[552] The question appears whether, or not, these early stimuli *in utero* or postnatally can have an effect on the manifestation of a chronic inflammatory systemic disease.

Indeed, injection of endotoxin, a constituent of the bacterial cell walls, to pregnant rats changes the responsiveness of stress response systems in the offspring.[553] Injection of endotoxin increased the response of the HPA axis, which is demonstrated by increased secretion of glucocorticoids. In parallel, the offspring of endotoxin-injected mothers demonstrated less severe swelling during adjuvant-induced arthritis.[553] Similarly, it has been demonstrated that postnatal manipulations changed the severity of experimental allergic (or autoimmune) encephalitis later in life.[548]

Taken together, these data indicate that there is considerable plasticity in the developing systems. The presently available data consistently indicate that maternal factors influence the developing fetus and extend to perinatal lifelong programming of physiological responses.[548] In the fetal and perinatal phase of development of an individual, it seems that the orchestrated interplay of different systems such as the neuronal system, the endocrine system, and the immune system becomes imprinted for a lifetime. One may call this the "learned homeostasis" or the "learned self," whereby the learning phase is mostly restricted to a period between the status of embryo and the second postnatal year (Walter Bradford Cannon, 1871-1945, introduced the term homeostasis). In this vulnerable phase, "learned homeostasis" can be skewed from one extreme to the other so that a chronic inflammatory systemic disease can have a very different character though genetic prerequisites are identical.

Summary

This chapter introduced three phases of chronic inflammatory systemic diseases (in the sequence along the lifetime axis):

1. The plasticity phase without any indication of a chronic inflammatory systemic disease (no antigen-driven immune response present). The plasticity phase is constituted of the embryological, fetal, and perinatal phases. During these phases, the interplay of different systems in our body is programmed for a lifetime: the "learned homeostasis" or the "learned self." Respective research in humans is rare, although the importance of this phase has been recognized in animal research.
2. The asymptomatic phase of the disease without any symptoms but with clear indications of an antigen-driven immune response, which is decisive and causal for the later chronic inflammatory systemic disease. In this phase, the T cell, the B cell, and antigen-presenting cells have a dominant role. These cells play with the antigen in order to create a balance between an aggressive and a regulatory immune response. During this phase, these secret players push for a decision. If it is a decision for an aggressive combat, the

symptomatic disease starts because clonal expansion of aggressive lymphocytes outweighs clonal expansion of regulatory immunosuppressive lymphocytes. This phase is scientifically important because counterbalancing endogenous immunosuppressive mechanisms can be uncovered.

3. The symptomatic phase of the disease is characterized by the presence of typical symptoms of a chronic inflammatory systemic disease. Apart from T cells, B cells, and antigen-presenting cells, many other cell types of the local compartment get involved. This leads to a change of disease characteristics ("the disease ages").

Visualization of the three phases can be important in order to generate a broader research perspective. A chronic inflammatory systemic disease does not start when the first symptoms are present for 6 weeks (as it is sometimes recommended in diagnostic criteria). A chronic inflammatory systemic disease has a very long preceding period (Figure 5).

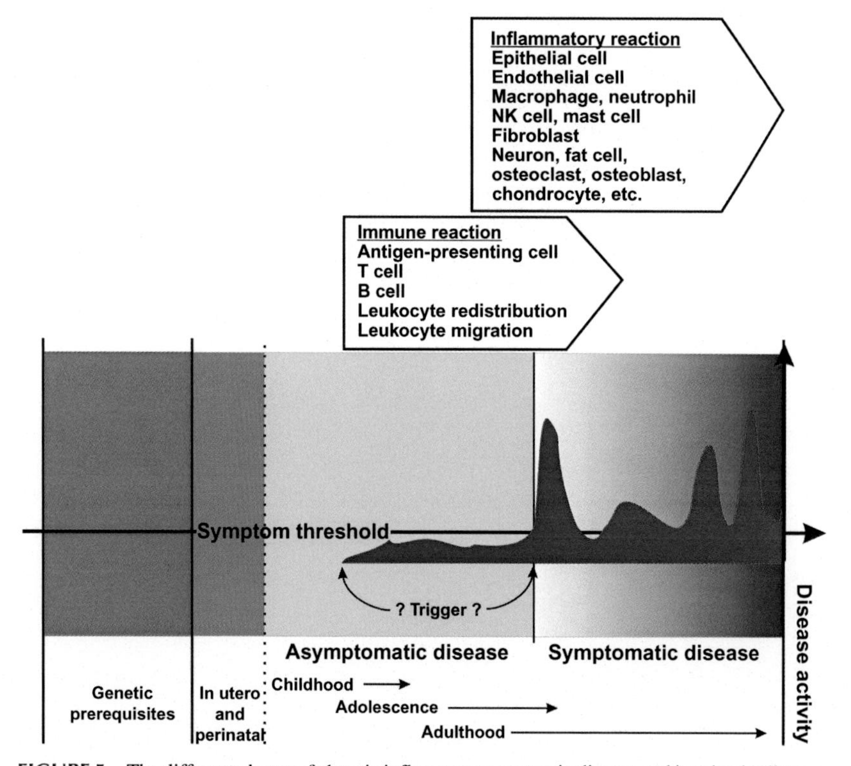

FIGURE 5 The different phases of chronic inflammatory systemic disease and involved cell types. When the disease grows older, the influence of antigen-presenting cells, T cells, and B cells decreases, whereas other cell types become more important during tissue destruction and local scar formation.

At this point, the question may appear how a chronic inflammatory systemic disease can go on forever. In Chapter VII, next to classical factors relevant for continuation of chronic inflammatory systemic diseases, some novel neuroendocrine immune mechanisms of disease perpetuation are discussed.

Chapter II

Pathogenesis and Neuroendocrine Immunology

To cross the threshold from where we are to where
we want to be, major conceptual shifts must take place ...
One such shift will be from studying elementary processes
to studying systems properties - mechanisms made up
of many proteins, complex systems of ... cells,
the functioning of whole organisms, and the interaction
of groups of organisms.

Kandel, Neurobiologist (taken from Ref. 554)

Chapter Contents

The Origin of Chronic Inflammatory Systemic Diseases and their Sequelae.
http://dx.doi.org/10.1016/B978-0-12-803321-0.00002-1

The cross talk of different systems such as the immune system and the neuroendocrine system can only be studied in an *in vivo* setting with all known difficulties. Although this is markedly more complex than studying single cells in culture or subcellular mechanisms, it is the only way to understand the mutual cross talk of the different bodily systems. To say it with the words of Mitchell (a complexity researcher), "linearity is a reductionist's dream, and nonlinearity [complexity] can sometimes be a reductionist's nightmare."[555]

The question in this chapter of the book is whether the neuronal and endocrine systems have any impact on the immune system and vice versa. If so, can neuronal and endocrine mechanisms explain some pathogenic aspects of chronic inflammatory systemic diseases?

HISTORICAL REMARKS

An influence of hormones on the immune system was first described when Hench successfully used purified endogenous glucocorticoids and adrenocorticotropic hormone (ACTH) in a patient with rheumatoid arthritis.[556] Soon thereafter, the anti-inflammatory effects of glucocorticoids were discovered in animals,[557] humans with other inflammatory diseases,[558] and laboratory assays with specific immune parameters such as antibody production and phagocytic capacity of neutrophils.[559] It was clear that ACTH has similar effects when used *in vivo* indicating the possibly indirect role of this hormone via an increase of adrenal steroidogenesis.[556]

In addition, the long known influence of pregnancy and the female-to-male preponderance in prevalence and incidence of several autoimmune diseases indicated a role of sex hormones such as 17β-estradiol or testosterone.[560,561] This women-to-men preponderance is particularly evident in reproductive years, when sex hormones are highest in women and men; for example, for systemic lupus erythematosus or systemic sclerosis, it is 9:1 (female:male). Furthermore, the induction of ovulation by gonadotropin or gonadotropin-releasing hormone analogs can result in flares, embryonic losses, or fetal deaths in patients with chronic inflammatory systemic diseases.[562]

Again, this indicates that, generally, an increase of female sex hormones can worsen a chronic inflammatory systemic disease. In sharp contrast, this seems to be the finding that some women develop an autoimmune disease such as rheumatoid arthritis after the menopause when serum sex hormone levels dramatically fall. This is not an unsolvable paradox compared to the above-mentioned role of sex hormones as will be demonstrated in the text below, because estrogens have very different effects on different types of immune cells.[329]

In addition, the clear connection between stressful live events on one side and exacerbations of chronic inflammatory systemic diseases on the other side demonstrates a role of stress hormones/neurotransmitters in these diseases (e.g., cortisol or noradrenaline).[563]

Similarly, the phenomenon that in different diseases hemiplegia spares the paretic side from inflammatory signs and symptoms supports an important propagating role of the nervous system (collected in Ref. 564; example in Figure 6).

Moreover, the circadian undulation of symptoms in chronic inflammatory systemic diseases links the rhythms of the central nervous and the neuroendocrine system to immunologically mediated phenomena (reviewed in Ref. 565). Indeed, this latter point is very important because it most clearly demonstrates the

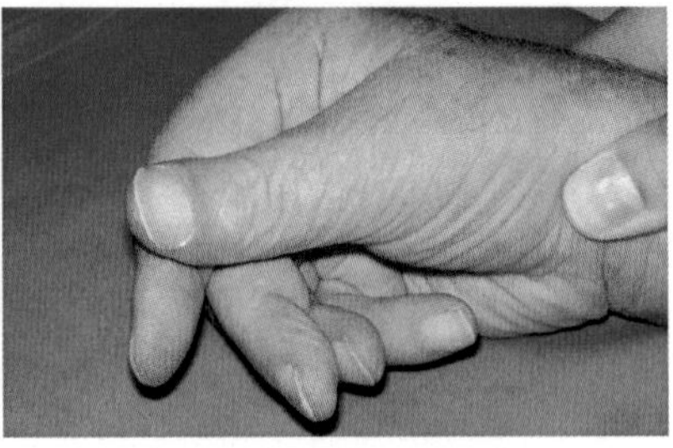

FIGURE 6 Hemiplegia spares the paretic right fingertips from ulcers in a patient with systemic sclerosis. The modified Rodnan skin score is a measure to estimate skin involvement. Photographs courtesy of Martin Fleck and Rotraud Meyringer.

link between the central nervous system, more specifically of the suprachiasmatic nucleus, and the peripheral immune system and the inflammatory machinery.

These clinical findings started a new research field and opened avenues for understanding the impact of the nervous and endocrine systems on the development and function of the immune system and, thus, on chronic inflammatory systemic diseases. Before the known pathogenic concepts from psychoneuroendocrine immunology will be demonstrated, some basic physiological aspects of systemic response systems are shortly recapitulated.

PHYSIOLOGICAL BASIS

A local immune/inflammatory process in the context of a chronic inflammatory systemic disease often induces a secondary systemic response. Two major divisions belong to the systemic response systems, the hormonal axes and the peripheral nervous systems. The response systems are delineated in the following list:

1. The hypothalamic-pituitary-adrenal (HPA) axis is the best-studied connection between the central nervous system and peripheral sites of immune response and inflammation. The HPA axis hormones comprise corticotropin-releasing hormone (CRH), ACTH, and several steroid hormones of the adrenal gland including cortisol, progesterone, and adrenal androgens such as androstenedione and dehydroepiandrosterone (DHEA) and, its biologically inactive degradation product, DHEA sulfate.
2. The hypothalamic-pituitary-gonadal axis includes gonadotropin-releasing hormone, luteinizing hormone/follicle-stimulating hormone, and the major bioactive steroid hormones of the gonadal glands, testosterone, and estrogens (plural because there is more than one estrogen).

3. The hypothalamic-pituitary-somatic axis includes growth hormone-releasing hormone, growth hormone, and insulin-like growth factors such as insulin-like growth factor 1 and respective binding proteins.
4. The hypothalamic-pituitary-prolactin axis. Prolactin has proinflammatory effects on the immune system.[566]
5. The melatonin pathway. Melatonin is produced in the pineal gland in the central nervous system, and melatonin circulates in the peripheral blood.
6. The fat tissue-hypothalamus axis with adipokines as major interconnecting factors such as leptin, resistin, omentin, and visfatin. Leptin activates the sympathetic nervous system.[567]
7. Hormones of the hypothalamic-pituitary-thyroid gland axis include thyrotropin-releasing hormone, thyroid-stimulating hormone (TSH or thyrotropin), and the two thyroid gland hormones thyroxin and triiodothyronine (T3) (the latter is the active hormone used in the periphery).
8. The glucose-regulating system includes mainly hypothalamic, pituitary, adrenal, and pancreatic pathways that serve gluconeogenesis, glycolysis, glycogen synthesis, and glycogenolysis in the liver (growth hormone, ACTH/cortisol, adrenaline, insulin, and glucagon).
9. The hypothalamic-autonomic nervous system axis with the two efferent pathways of the sympathetic nervous system (noradrenaline, neuropeptide Y, adenosine triphosphate (ATP), adenosine, enkephalins, and sometimes vasoactive intestinal peptide (VIP)) and the parasympathetic nervous system (acetylcholine).
10. The renin-angiotensin-aldosterone system (RAAS), which is directly related to the sympathetic nervous system, and arginine vasopressin as blood pressure-stimulating hormone systems.
11. The sensory nervous system with the afferent[a] function of transmitting pain signals and inflammation signals to the central nervous system and with the efferent[b] function of releasing neuropeptides such as substance P and calcitonin gene-related peptide (CGRP) into the vicinity of peripheral nerve terminals.

These interconnecting systems link the central nervous system and endocrine glands on one side to the immune system on the other side (and vice versa). The mutual cross talk is mediated by hormones, neurotransmitters/neuropeptides, and cytokines. In the context of this book, particularly in Chapter V, it will be demonstrated how inclusion of these neuroendocrine immune pathways in the pathophysiology of chronic inflammatory systemic diseases complements the major etiologic paradigms originating from immunology (point 5 of Table 5 in Chapter I). Some of the efferent response systems are illustrated in Figure 7.

---

a. "Afferent" means from the periphery to the brain.
b. "Efferent" means from the brain to the periphery.

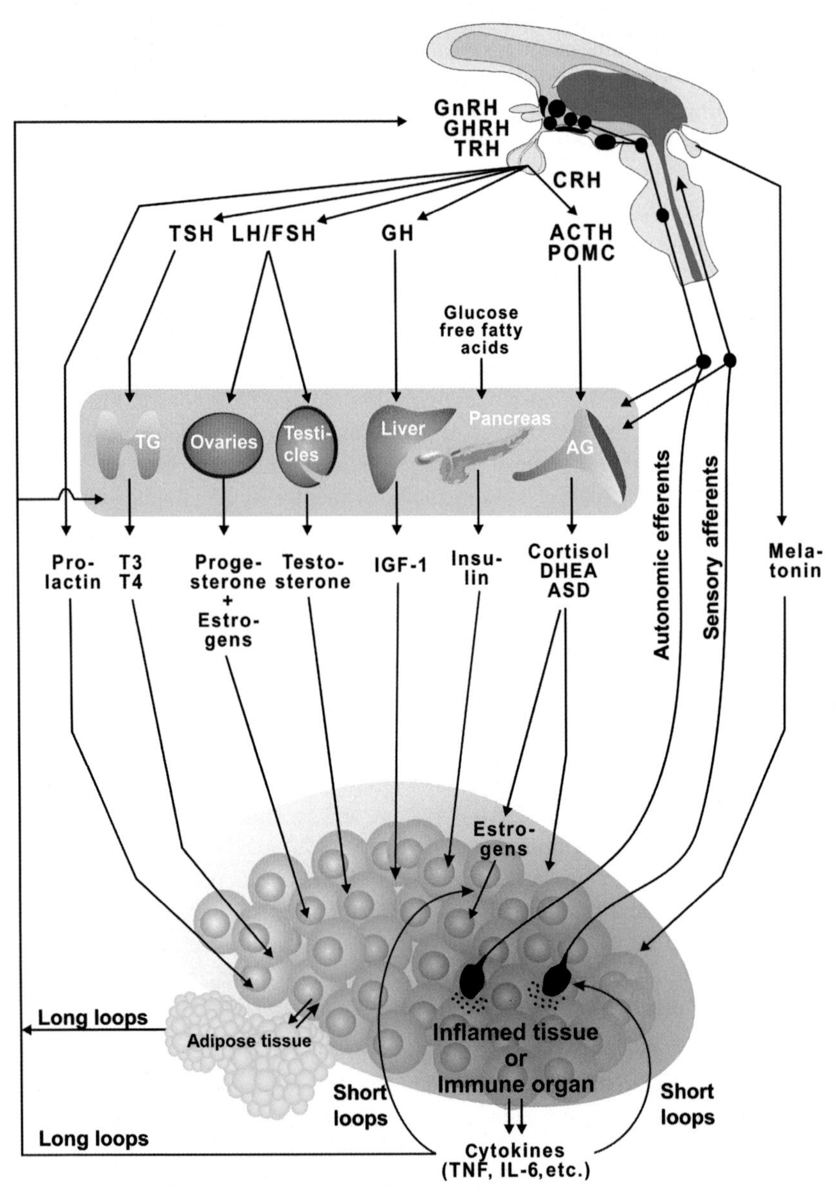

FIGURE 7 The basis of the mutual cross talk between the neuroendocrine system and the immune system (either in inflamed peripheral tissue or in immune organs). Over long and short loops, the systems influence each other. Hormones can be generated locally from hormone conversion (indicated for, e.g., estrogens). ACTH, adrenocorticotropic hormone; AG, adrenal gland; ASD, androstenedione; CRH, corticotropin-releasing hormone; DHEA, dehydroepiandrosterone; FSH, follicle-stimulating hormone; GH, growth hormone; GHRH, growth hormone-releasing hormone; GnRH, gonadotropin-releasing hormone; IGF-1, insulin-like growth factor 1; IL-6, interleukin-6; LH, luteinizing hormone; POMC, pro-opiomelanocortin; T3, triiodothyronine; T4, thyroxine; TG, thyroid gland; TNF, tumor necrosis factor; TRH, thyrotropin-releasing hormone; TSH, thyroid-stimulating hormone.

Receptors for Hormones and Neurotransmitters on Immune Cells

The last three decades have demonstrated that receptors for hormones and neurotransmitters are present on or in almost all cells of the immune system. It seems to be an exception when receptors for hormones and neurotransmitters are missing on or in a given type of immune cell. Indeed, there are some examples such as the β2-adrenergic receptor on T cells. After differentiation from naive T cells, only T helper type 1 cells carry β2-adrenergic receptors, while T helper type 2 cells lose this particular receptor.[568] Such a differential and time-dependent expression of receptors can shape the neuroimmune cross talk.

Sometimes, receptors are upregulated in the context of an inflammatory response such as the α1-adrenergic receptor.[569,570] A similar environment-dependent upregulation was observed for the estrogen receptor β, while the estrogen receptor α is downregulated during inflammation or hypoxia (reviewed in Ref. 329). The role of proinflammatory cytokines has been highlighted in inducing resistance to glucocorticoids,[571] growth hormone,[572] insulin-like growth factor 1,[573] and insulin.[574] Thus, receptor expression and signaling are very flexible and depend on environmental conditions.

In addition, intracellular signaling pathways of hormone and neurotransmitter receptors are dependent on environmental conditions, a fact that has been nicely demonstrated for G protein-coupled receptors.[575,576] Here, G protein-coupled receptor kinases play an important role in the downregulation of these receptors, and an inflammatory situation can decrease G protein-coupled receptor kinases, which can happen in certain immune cell subtypes.[576]

Nowadays, there exists a vast amount of literature with respect to hormone and neurotransmitter receptors on/in immune cells. It would certainly go beyond the scope of this paragraph to demonstrate a full picture. For more details on some receptors of hormones and neurotransmitters on immune cells, the reader is referred to the literature.[568,577,578]

Dual Roles of Hormones and Neurotransmitters/ Neuropeptides

For a pure immunologist, it is sometimes hard to understand bipolar roles of hormones or neurotransmitters because typical molecules in immunology are thought to have 0-1-effects.[c] This means that a molecule has either no influence

---

c. The 0-1-effect of a substance is an idealization in research. Most molecules have dual effects depending on concentrations and conditions. Newer studies present bell-shaped and U-shaped response curves that are much more realistic. Interestingly, I was often asked by paper reviewers why we observe such strange U-shaped and bell-shaped response curves. After having survived many reviews of my own papers, I must ask myself whether these reviewers have never worked in a laboratory. Natural systems often behave in ways that reach an optimum situation. An optimum situation is never achieved with an extreme 0-1-situation.

or a strong influence in a given test system. This is beautiful but it is not the same with respect to hormones and neurotransmitters. Thus, it is an important prerequisite to understand the function of hormones and neurotransmitters/neuropeptides by considering receptor-binding characteristics and other environmental conditions.

Let us take again the example for noradrenaline. This neurotransmitter binds to α- and β-adrenergic receptors, which exert opposite effects on intracellular signaling cascades (α1, increase of diacylglycerol and protein kinase C; α2, decrease of cAMP; and β, increase of cAMP). While noradrenaline binds to α-adrenergic receptors below the concentration of 10^{-7}-10^{-6} mol/l, it binds to β-adrenergic receptors at concentrations higher than 10^{-7}-10^{-8} mol/l (the typical tissue concentration of 10^{-7} mol/l is somewhere in the middle). Since α-adrenergic effects are markedly different from β-adrenergic effects, the concentration of noradrenaline in the environment of an immune cell is very important for adrenergic effects (Figure 8).

In the early experiments initiated in the 1980s, this has often been overlooked so that experiments were not easily interpretable. Importantly, such behavior is typical for many hormones and neurotransmitters/neuropeptides because often, more than one receptor can be a binding partner and different receptors have opposing intracellular signaling pathways. Further conditions that add to this bipolar receptor behavior are described below.

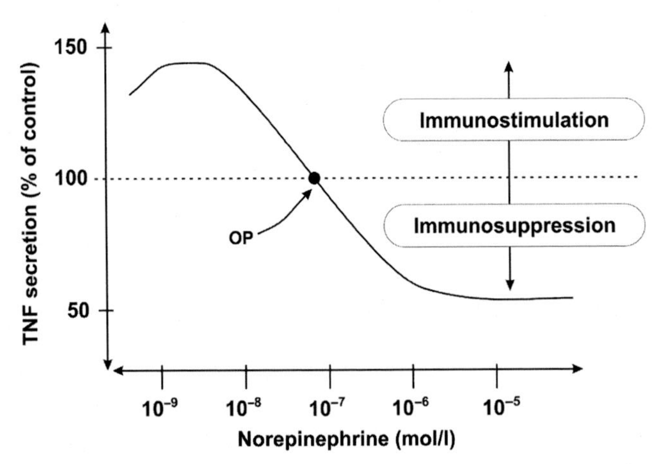

FIGURE 8 Effects of noradrenaline on the secretion of tumor necrosis factor (TNF). The effect is markedly dose-dependent (information merged from Refs. 579–581). OP is the optimum point at 10^{-7} mol/l, which is identical to the typical tissue concentration in the proximity of sympathetic nerve terminals under basal firing conditions. At high firing rates, this concentration can rise to levels of 10^{-5} mol/l in the proximity of sympathetic nerve terminals.[582,583] Plasma levels of noradrenaline are approximately 10^{-9} mol/l.

Availability of Hormones and Neurotransmitters/ Neuropeptides in the Tissue

Another important aspect is the availability of the hormone or the neurotransmitter/neuropeptide in the tissue in the proximity of immune cells. Availability is no problem for steroid hormones such as cortisol because they are often highly lipophilic and, thus, pass membrane and cell barriers nearly without restriction. It is quite difficult for hydrophilic molecules such as noradrenaline and many others. These molecules must be secreted in close proximity to an immune cell, which can only happen when the source of the molecule is nearby.

In the case of noradrenaline, sympathetic nerve fibers as source must be in the vicinity of immune cells. Necessarily, the tissue must be densely innervated by sympathetic nerve fibers.[d] Such close contacts between sympathetic nerve fibers and immune cells have been demonstrated in secondary lymphoid organs and elsewhere.[585–588] This is similar for most other neurotransmitters and neuropeptides. In contrast to steroid hormones, neurotransmitters/neuropeptides can specifically increase in a certain tissue block independent of a neighboring tissue block. Such a specific hardwiring can play an important role so that one organ receives a sympathetic input with a high firing rate, whereas a neighboring organ has no input at all.[589]

An additional important piece of evidence is the conversion of hormones and neurotransmitters/neuropeptides in the tissue. Progesterone, DHEA, and androstenedione are precursor hormones, which can be converted to downstream sex hormones. Importantly, sex steroid action can take place in the same cells where conversion takes place (intracrinology).[590] In postmenopausal women, nearly 100% of sex steroids are synthesized in the peripheral tissues from precursors of adrenal origin (in older men, peripheral conversion depends on age and can be up to 80%). Thus, after the menopause and during aging, the HPA axis can take over some functions of the hypothalamic-pituitary-gonadal axis and influence the immune/inflammatory reactivity. Peripheral conversion can be influenced by locally produced cytokines, growth factors, hormones, and neurotransmitters.[591]

Similarly, neurotransmitters can be locally converted to differentially active neurotransmitters. This has been nicely studied for ATP, known as a neurotransmitter of the sympathetic nervous system. ATP can be converted to ADP, AMP, and adenosine by enzymes located on the surface of immune cells and other

d. It is important to mention that most tissues in our body are perfectly equipped with blood vessels that provide necessary oxygen. Along the blood vessels, there exists always a perfect sympathetic innervation. In addition, sympathetic nerve fibers branch into the tissue fulfilling other roles than simple vasoregulation. For example, the synovial tissue is densely innervated by sympathetic nerve fibers.[584]

cells.[592–594] In addition, neuropeptides can be locally converted by peptidases leading to neuropeptide fragments, which are often antagonists at the specific neuropeptide receptor. A classic example is substance P, the major neuropeptide of the sensory nervous system, which can be converted to agonistic and antagonistic fragments.[595,596] Another example is given for neuropeptide Y whose fragments can be either agonistic, antagonistic, or inactive.[597,598] Both neuropeptides can be converted by, for example, dipeptidyl peptidase IV (CD26).[599]

There exists another way how hormones and neurotransmitters/neuropeptides appear locally in the inflamed tissue: they can be produced by activated cells in an inflammatory environment. The first hormone that was found to be produced by macrophages was ACTH, which has been demonstrated in the early 1980s by Blalock and Smith.[600] This intriguing finding was the trigger for others in studying local production of hormones and neurotransmitters/neuropeptides from activated immune cells (e.g., Refs. 601,602). An example of local cellular production of a neurotransmitter is given with noradrenaline, which can be produced by lymphocytes,[603,604] neutrophils,[605] macrophages,[606,607] or fibroblasts.[607]

Switching on the Systemic Response Systems

The concept of the activation of the central nervous system by localized immune responses goes back to the investigation of the fever response in the nineteenth century. In this research, it was recognized that the so-called—but unknown—pyrogens activate the central nervous system. This was vivified by important experiments of Besedovsky and colleagues in the 1970s.[608,609] Somewhat later, Blalock called the inflammatory activation of the immune system a sensory function or the "sixth sense," because it is a way how inflammatory signals can be forwarded to the conscious brain.[610] Today, we know that there are several ways how a localized inflammatory process or immune response can activate the entire system:

1. Cytokines with stable characteristics can travel in the circulation to remote organs such as the endocrine glands and the brain. The classical example is IL-6, which is a very stable and long-lived cytokine easily transported in the blood stream. Some researchers, including me, called IL-6 a hormone due to its many remote effects. These "traveling cytokines" activate endocrine glands and the central nervous system (via active transport across the blood-brain barrier by cytokine-specific carriers, via activation of endothelial cells of the blood-brain barrier and adjacent microglia, and via leaky circumventricular organs).[611]
2. Cytokines can locally activate afferent sensory nerve fibers, which has been demonstrated first with sensory fibers of the vagus nerve.[612,613] Later, it has been demonstrated for the glossopharyngeal nerve,[614] and in very elegant studies, it was shown for afferent fibers of sensory nerves supplying the knee joint.[615–617] Sensory nerve terminals in the periphery carry cytokine

receptors,[618] Toll-like receptors,[619] and receptors for mechanical, thermal, and chemical (pH) stimuli (see also Figure 21).

3. A third less respected mechanism is stimulation via locally activated and circulating immune cells. For example, it has been demonstrated that activated immune cells appear in the adrenal glands after colitis induction or endotoxin injection, which most probably modulates steroidogenesis.[620,621] A similar migration might happen toward other organs.

We recognize that this is an important aspect to switch on systemic responses of the body (point 5 of Table 5 in Chapter I), but what is the meaning of such an activation of remote organs? The classical view of immune activation of endocrine glands and the central nervous system is a feedback dampening of overshooting immune responses. This hypothesis has been put forward in the 1980s by distinguished scientists.[611,622,623] Much of the systemic response of the HPA axis and the sympathetic nervous system is, indeed, an immune inhibitory response where glucocorticoids and neurotransmitters play an important role.[624] However, this concept does not represent the full picture because an immunosuppressive activity of the central nervous system and endocrine glands is not meaningful in infectious diseases. In infectious disease, one wishes a strong immune response over a longer period of time to overcome the infectious threat. In addition, several hormones of the stress response system such as noradrenaline and also cortisol are not always immunosuppressive.[625,626]

In addition, as we will see later, the systemic response systems do not continuously slow down the immune system but would exert inhibitory activities only during a very short period of time. The activation of the central nervous system with, for example, injected IL-6 leads to a short-lasting activation of the HPA axis for a few days only.[627,628] Thus, a long-term immunosuppressive effect is not possible and, thus, the concept of an immune-inhibiting feedback dampening loop is too simple.

In Chapter III, a wider concept will be presented, which embeds the above-mentioned feedback dampening of overshooting immune responses into a more integral regulatory activity to manage the availability of energy-rich fuels and to control water content in the body. It will be demonstrated that energy regulation serves a superior goal and provides an integrative explanation for the activation of the general response systems. This will lead to comprehension of why systemic responses are elicited at all (point 5 of Table 5 in Chapter I).

Differential Activation and Inhibition of Response Systems

It was demonstrated above that a peripheral inflammatory/immune response can activate the HPA axis and the sympathetic nervous system. However, not all systemic response systems mentioned in Figure 6 are activated; some are even repressed. A classic example of a suppressed hormone axis is the hypothalamic-pituitary-gonadal axis, which is mediated by cytokines on a

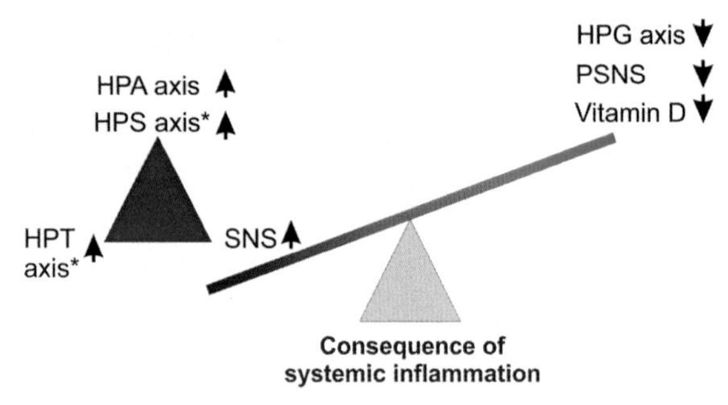

FIGURE 9 Consequences of systemic inflammation on hormonal and neuronal pathways. Upon systemic inflammation, the hypothalamic-pituitary-adrenal (HPA) axis, the sympathetic nervous system (SNS), and the hypothalamic-pituitary-thyroid gland axis (HPT) are activated. The HPT axis is soon blocked after initiation of inflammation (low-T3 syndrome). Some think that the HPT axis can be restored by TSH producing immune cells that enter the thyroid gland.[601] In longer-standing inflammation, the HPA axis is activated by circulating cytokines such as IL-6.[635] Acute systemic inflammation also activates the hypothalamic-pituitary-somatic (HPS) axis with growth hormone and insulin-like growth factor 1, while chronic systemic inflammation inhibits this axis (see children with chronic inflammatory systemic diseases and growth deficits). The asterisk indicates that these two axes are mainly active during the early acute inflammatory response. In contrast, the hypothalamic-pituitary-gonadal (HPG) axis, the parasympathetic nervous system (PSNS), and the vitamin D pathway are inhibited during early and chronic systemic inflammation. The red triangle illustrates the synchronization of activities of the HPA axis, the SNS, and the HPT axis.

hypothalamic, pituitary, or direct gonadal level as demonstrated by Catherine Rivier in San Diego.[629–633] Figure 9 summarizes the influence of a systemic inflammatory/immune response on hormonal and neuronal pathways.

It is important to understand that the HPA axis, the sympathetic nervous system, and the hypothalamic-pituitary-thyroid gland axis work together and stimulate each other.[634] This is most probably important for the synchronization of common effects that are necessary in energy regulation.

Influence of Hormones and Neurotransmitters/ Neuropeptides on Immune Function

Due to the existing immense information, a detailed description of the effects of individual hormones or neurotransmitters/neuropeptides on different immune functions would go beyond the scope of this book. This subject has been reviewed in many ways by distinguished authors. The interested reader is referred to the major reviews in this field (Table 8).

Moreover, some of the important effects of hormones and neurotransmitters/ neuropeptides on immune function are given in the following sections where some specific pathogenic concepts of psychoneuroendocrine immunology are demonstrated for chronic inflammatory systemic diseases.

TABLE 8 Reviews with Detailed Description of Hormonal and Neuronal Influences on Immune Function

Hormone or Neurotransmitter/Neuropeptide	References
CRH	636,637
ACTH	638
Cortisol	577,639–642
Androgens	643–646
Estrogens	329,647
Vitamin D	648–655
Growth hormone, insulin-like growth factor 1	602,656–660
Prolactin	602,657,660–663
Melatonin	664–668
Adipokines	459
Thyroid hormones (TSH, thyroxine, and triiodothyronine)	601,602,660,669
Insulin[a]	602,670–676
Sympathetic neurotransmitters (noradrenaline, neuropeptide Y, ATP, and adenosine)	677–685
Opioidergic peptides	686–693
Parasympathetic neurotransmitters (acetylcholine and nicotine)	694–699
Vasoactive intestinal peptide	700–703
Vasopressin	704–706
Substance P	707–711
Calcitonin gene-related peptide[b]	711–713

[a]For this hormone, no adequate review paper was found.
[b]For this neuropeptide, no specific review paper was found.

ENDOCRINE IMMUNE RELATIONS IN CHRONIC INFLAMMATORY SYSTEMIC DISEASES

Inadequate Secretion of Cortisol Relative to Inflammation

When Hench and colleagues successfully used purified endogenous glucocorticoids and ACTH in a patient with rheumatoid arthritis, he already recognized that the body of his patient was producing too little of this endogenous hormone to overcome inflammation.[556] It became very clear that glucocorticoids have strong anti-inflammatory effects.

The coming years witnessed that high therapeutic doses of glucocorticoids, which were needed to completely overcome inflammation, induced many unwanted side effects. This tells us that a long-standing increase of endogenous glucocorticoids should not be normal. Evolutionarily, long-standing high levels of glucocorticoids were not positively selected because it would lead to higher rates of infection and sepsis (see Chapter IV). Nevertheless, high glucocorticoids would help to overcome autoimmune inflammation.

Important evidence for anti-inflammatory effects of endogenous adrenal glucocorticoids in chronic inflammatory systemic diseases came from a heroic study in Italian patients with rheumatoid arthritis in 1986.[714] These authors tested the effect of metyrapone, an inhibitor of endogenous glucocorticoid production and secretion. Joint pain and tenderness increased significantly during metyrapone administration in these patients, who had never received corticosteroid therapy and who had no evidence of endocrine disease. They concluded that this observation implies that the normal circulating plasma cortisol exerts anti-inflammatory effects.[714] Nevertheless, the observed normal or somewhat increased circulating plasma cortisol levels were not high enough to overcome the chronic inflammatory systemic disease.

In the late 1980s, important animal models demonstrated a spontaneous defect of the HPA axis in obese strain chickens[715] and in rats of the Lewis strain.[524,716–718] While the chickens demonstrated spontaneous autoimmune thyroiditis, the Lewis rats were much more susceptible to antigen-induced inflammation such as arthritis, experimental autoimmune encephalomyelitis, and experimental colitis.[524,716–718] Both lines of evidence demonstrated that normal operation of the HPA axis is very important to dampen the chronic inflammatory systemic disease.

After these important findings, researchers tried to find the magic bullet of the HPA axis defect in human chronic inflammatory systemic diseases, which included studies on strong HPA axis activation with insulin hypoglycemia[719–721] and other stress tests.[722–724] However, only subtle stress tests showed some unresponsiveness of the HPA axis in patients with a chronic inflammatory systemic disease. The strong hypoglycemia stress elicited near normal results. The normal results are probably due to the strong endocrine response toward a life-threatening stimulus.

Furthermore, studies were included to find a genetic defect of the HPA axis, which were not really successful.[725,726] Similarly, in genome-wide association studies, no indication for a genetic effect with respect to the HPA axis appeared. I think that there is some work to be done because considering the combinations of genes and environmental factors might change the picture.[727] In addition, the so-called cofactors relevant for the function of enzymes of steroidogenesis were only recently investigated (see below). One can summarize that the HPA axis works normally, leading to normal or slightly increased levels of cortisol in patients with untreated chronic inflammatory systemic diseases. But what is normal? This question led to a long debate whether endogenous glucocorticoids are normal or whether they are inadequately low in relation to inflammation (e.g., Refs. 728,729).

Since the important work in 1977 of Besedovsky, del Rey, and colleagues,[608,609] and further refinements in 1986 by the same authors,[622] the stimulatory effect of immune system activation on the HPA axis is well known. Indeed, single cytokines such as IL-1β, TNF, IFN-γ, IFN-α, and IL-6 can activate the human HPA axis response.[627,730–734] In these situations, serum cortisol levels can increase by a factor of 7 as demonstrated in TNF-treated cancer patients[735] or by a factor of 5 in human volunteers treated with IL-6.[736] These studies clearly demonstrate activating effects of circulating cytokines on the HPA axis.

In healthy volunteers, Chrousos' group demonstrated a dose-response relationship between different doses of human recombinant IL-6 and cortisol or ACTH levels in a range of 3-300 pg IL-6/ml serum/plasma (Figure 10a and b).[736] Since patients with chronic inflammatory diseases range between 3 and 300 pg/ml of serum IL-6 (Figure 10a and b, *x*-axis), one would expect similar levels of serum cortisol or plasma ACTH in these patients as given in this study in healthy volunteers.

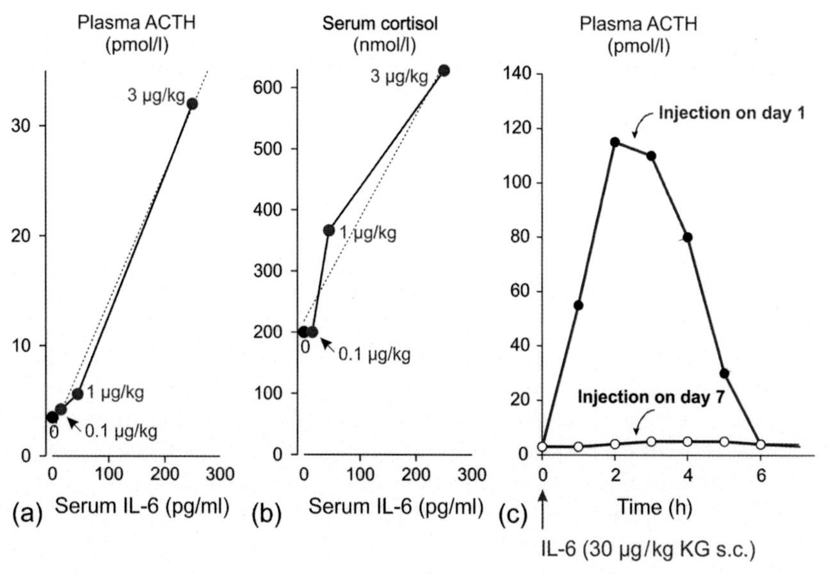

FIGURE 10 Adaptive downregulation of HPA axis activity upon repeated inflammatory stimuli. (a) Positive correlation between serum IL-6 levels and plasma ACTH after injection of indicated doses of IL-6 (red numbers) to healthy volunteers.[736] (b) Positive correlation between serum IL-6 levels and serum cortisol after injection of IL-6.[736] (c) Blunted ACTH response after repeated daily injection of IL-6 (red, first day; black, seventh day).[627] (c) The adaptive downregulation of HPA axis activity (here ACTH) during long-term elevation of serum IL-6.

The linearity of the relationship between cytokine level and hormone level in the mentioned range stimulated us to calculate the hormone/cytokine ratio in patients with chronic inflammatory diseases. If patients have cytokine levels in the abovementioned range, one would expect that a hormone/cytokine ratio should be similar in healthy controls compared to patients with chronic inflammatory systemic diseases. However, in patients with rheumatoid arthritis and reactive arthritis, hormone levels were much lower in relation to IL-6 or TNF.[737] A recent examination of data in patients with polymyalgia rheumatica showed the same results (not yet published). This was called inadequately low secretion of cortisol or ACTH in relation to stimulating cytokines such as IL-6 and TNF (disproportion principle).

The disproportion principle applies not only to patients with chronic inflammatory diseases but also to cancer patients who have repeatedly been treated with cytokines. Mastorakos and colleagues demonstrated that the ACTH response to repeated IL-6 injection was blunted after 7 days of daily therapy (Figure 10c).[627] The cortisol response was diminished after 7 days of treatment while this did not reach a significant level although area under the curve seems to be much lower.[627] In one representative patient, during 24 h of observation before and after IL-6 treatment, the cortisol response curve was much higher at baseline compared to a situation after 7 days of daily IL-6 treatment.[627]

A similar blunted response to IL-6 injection was observed for ACTH in a German study in cancer patients.[731] However, these authors did not demonstrate the loss of a cortisol response, which was interpreted as a possible direct stimulus of IL-6 on adrenal gland function. It is known that such a positive stimulus of IL-6 on adrenal gland-derived cortisol secretion can exist[635]; however, this most probably depends on the administered dose of IL-6.

With respect to another cytokine, an Austrian group demonstrated a blunted HPA axis response to repeated IFN-α injection over 3 weeks in patients with myeloproliferative disease.[734] The effect is really impressive because ACTH and cortisol response were nearly completely blunted after 3 weeks. In summary, continuously high levels of different cytokines can inhibit HPA axis function. How would the adrenal cortex respond when a cocktail of different proinflammatory cytokines would be injected repeatedly? This is the situation in an ongoing chronic inflammatory systemic disease.

From a biological standpoint, a continuously increased glucocorticoid level would be very critical in systemic severe infection because cortisol would inhibit important immunologic defense reactions of the body. Thus, it can be hypothesized that a short rise and fall of ACTH/cortisol was evolutionarily positively selected in order to support an early response of but not a long-term inhibitory influence on the immune system.[738] The question remains, why do the rise and fall happen? Several mechanisms have been suggested (Table 9).

Inadequate secretion of endogenous glucocorticoids can also be demonstrated by studying circadian rhythms of serum levels of cytokines and hormones.[739] The circadian rhythm of serum cortisol with respect to amplitude and period is very similar in healthy controls compared to patients with mild to

TABLE 9 Some Reasons for the Disappearing Responsiveness of the Hypothalamus-Pituitary-Adrenal (HPA) Axis During Extended Periods of Inflammation

The initially elevated cortisol levels will rapidly inhibit hypothalamic neurons stopping corticotropin-releasing hormone secretion (negative feedback regulation)
Prolonged increase of inflammatory stimuli with elevated TNF, IFN-α, or IL-6 serum levels will diminish the HPA axis responsiveness[627]
Longer-term elevated proinflammatory cytokines (IL-1β and TNF) influence the secretion of steroid hormones from the adrenal glands[740]
The age-related increase of serum IL-6 in healthy female and male subjects[741] may reduce the HPA axis responsiveness due to continuous stimulation[627]
The cytokine-stimulated increase of cortisol secretion from the liver (made from cortisone) enhances the negative feedback toward the hypothalamus, leading to low ACTH and adrenal androgens but normal to somewhat elevated serum cortisol[742]

moderate rheumatoid arthritis.[739] The serum concentration range of this endogenous glucocorticoid lies between 100 (nadir) and 300 nmol/l (acrophase). In contrast, serum levels of IL-6 are on average 10 times higher and the circadian rhythm of IL-6 is quite different in these patients compared to control. Particularly in the morning hours, the IL-6 curve is broadened and the decline of IL-6 is shifted toward later morning hours (reviewed in Ref. 565). Despite elevated levels of IL-6, the amplitude of the circadian rhythm of cortisol is not markedly different from healthy controls, which is indicative of inadequate cortisol secretion under these proinflammatory conditions.

One still might argue that all is adaptation. But what does adaptation mean when a proinflammatory cytokine such as TNF harms the HPA axis in many ways? Why doesn't adaptation lead to two or three times the normal cortisol levels? Why does a therapeutic dose of 2.5-5.0 mg prednisolone, which is approximately the amount a normal adrenal gland produces per day, is so helpful in many ways?[743–746] Isn't this an indication that endogenous production is too low in relation to inflammation? And isn't it an indication that successful low-dose glucocorticoid therapy is more of a substitution therapy of the adrenal glands similar as in some patients with relative adrenal insufficiency after emergency therapy on intensive care units (Figure 11).[747]

Let me conclude with the following citation of Drs. Jessop and Harbuz from Bristol: "the HPA axis has an inherent defect, which resided in the inability of patients to mount an appropriately enhanced glucocorticoid response to increased secretion of proinflammatory cytokines such as IL-1β, IL-6, [IFN-γ and others], and TNF. In other words, the HPA axis response is defective precisely because it is normal.[729]" In a chronic inflammatory systemic disease, such an inadequate response can play a disease-perpetuating role when cortisol secretion is small and, thus, anti-inflammatory activity of this hormone is low.

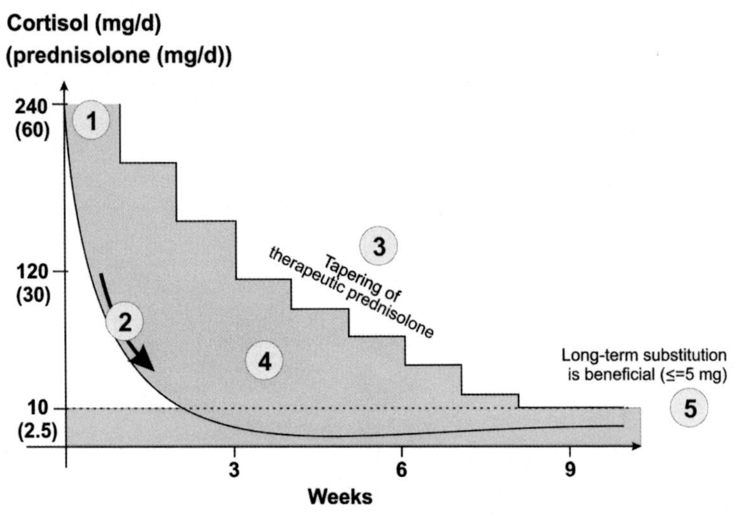

FIGURE 11 Inadequate cortisol secretion in relation to inflammation and the principle of supplemental therapy. The numbers in the figure are related to the following explanation: (1) During acute inflammation, the adrenal glands can produce large amounts of anti-inflammatory cortisol that can amount to 240 mg/day (which is equal in function to 60 mg of prednisolone; equivalent prednisolone doses are given in parentheses). (2) This high production rate cannot be maintained for a long time leading to a fast reduction of endogenous adrenal cortisol production. (3) The amount of cortisol needed would be much higher as depicted by the usual tapering dose of prednisolone as used for an acute inflammatory flare-up in a chronic inflammatory systemic disease. (4) The lilac area demonstrates the difference in cortisol needed and cortisol produced. There appears a deficit of produced endogenous cortisol, which is a mismatch between the need for anti-inflammatory cortisol not supplied by endogenous mechanisms and cortisol production. (5) After weeks of disease and treatment, endogenous cortisol production is lower as the normal mean basal daily cortisol production of 10 mg (equivalent to 2.5 mg prednisolone; the orange area). This situation is called "relative adrenal insufficiency" or inadequately low secretion of cortisol or ACTH in relation to stimulating cytokines, in which cortisol levels, although normal to moderately higher in absolute terms, are insufficient to control the inflammatory response. In such a situation with functional adrenal insufficiency, treatment with exogenous corticosteroids is a supplemental therapy.

The Link Between TNF and Hormone Axes

The discovery of TNF as an important proinflammatory molecule led to the successful development of therapeutic TNF-neutralizing strategies.[362–364] These studies have expanded our understanding about the possible mechanisms how TNF is involved in the pathogenesis of chronic inflammatory systemic diseases. Anti-TNF therapies are thought to act mainly on participating immune/inflammatory reactions. The effects are considered to be direct by neutralizing the TNF molecule or indirect by stimulating apoptosis in immune cells through a membrane-bound TNF molecule.

Beyond local effects on the immune system, anti-TNF treatment modulates systemic anti-inflammatory pathways such as hormonal and neuronal systems. It turned out that anti-TNF therapies over several weeks improve the HPA axis and the hypothalamic-pituitary-somatic axis.[748] This normalization of the "milieu intérieur" is an alternative mode of anti-inflammatory action.[748] In contrast, anti-TNF therapy over 12-16 weeks cannot normalize the hypothalamic-pituitary-gonadal axis, the increased activity of the sympathetic nervous system, and elevated serum levels of leptin.[748] It seems that these alterations are imprinted for a much longer period of time. Similarly, anti-IL-6 therapy can also improve some aspects of steroidogenesis in patients with chronic inflammatory systemic disease.[749]

In Table 7 of Chapter I, I discussed *a priori* and *a posteriori* causes of chronic inflammatory systemic diseases (see section "The Trigger of Chronic Inflammatory Systemic Diseases" of Chapter I). Using anti-TNF therapy in patients with rheumatoid arthritis, we were able to show that there exists a link between the TNF-induced *a posteriori* alteration of the HPA axis and subsequent worsening of the chronic inflammatory systemic disease (Figure 12).[750]

In this study, we measured at baseline serum levels of ACTH and cortisol as well as clinical improvement during anti-TNF antibody treatment, and serum cortisol at follow-up. It was demonstrated that those patients with good improvement and initially low serum cortisol levels demonstrated an increase of serum cortisol, which was opposite in patients with no or little improvement. This was the first study in a human chronic inflammatory systemic disease that demonstrates that inflammation-induced TNF interferes with HPA axis integrity, which is linked to the disease outcome (Figure 12).[750] TNF-dependent worsening of the HPA axis is a perfect example of an *a posteriori* cause mentioned in Table 7 of Chapter I.

The Possible Role of Adrenal Mitochondria in the Relative Decrease of Adrenal Steroid Secretion During Inflammation

The precursor of cortisol and adrenal androgens is circulating cholesterol from the liver in the form of HDL cholesterol, which is taken up in to adrenocortical cells via the scavenger receptor class B member 1. Then, cholesterol is stored in adrenal vesicles (cholesterol esters), which gives the adrenal cortex the classical, yellow-white, adipose tissue-like appearance. Stored cholesterol is responsible for the fast release of cortisol after challenge, because conversion from cholesterol to cortisol depends on only five enzymatic steps, some of which only happen in mitochondria (Figure 13). Thus, mitochondria are essential for steroidogenesis.

In recent studies, we demonstrated that mitochondria become markedly altered in the course of type II collagen arthritis in rats.[751] The number of altered swollen and cavitated mitochondria increased during the course of arthritis (maximum on day 55). The mitochondria looked as if they were treated with steroidogenesis blockers such as aminoglutethimide. Importantly, mitochondrial alterations were correlated to reduced breakdown of lipid droplets and increased lipid accumulation.[751] It is understandable that lipid breakdown must

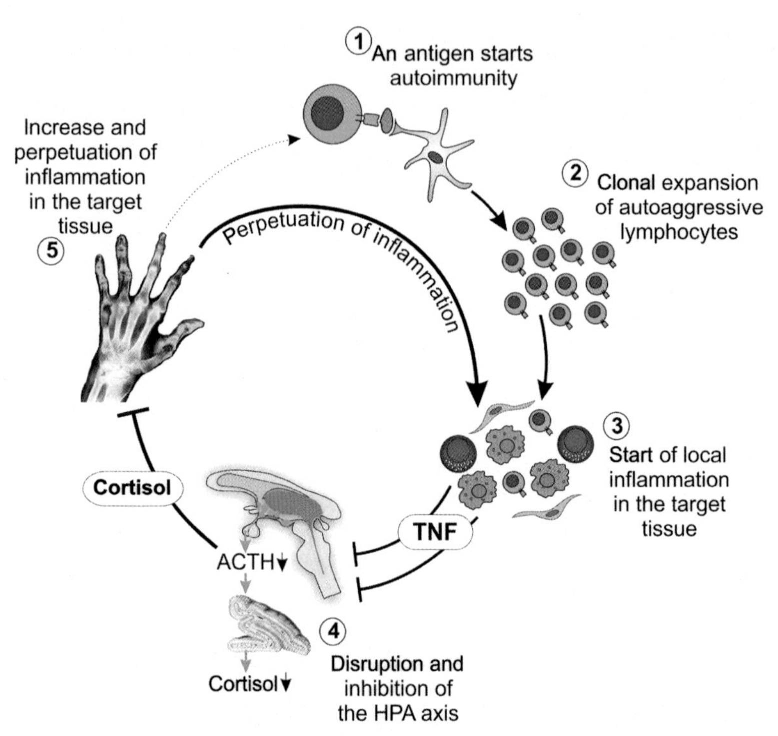

FIGURE 12 Position of the HPA axis in the pathophysiology of chronic inflammatory rheumatoid arthritis in relation to TNF and local joint inflammation. A double vicious circle is demonstrated: (1) The disease is most probably started by a shift of balance from the tolerant to the aggressive side of an autoimmune response against a harmless antigen. (2) The immune response leads to clonal expansion of T and/or B cells with an autoaggressive phenotype. (3) These autoaggressive cells start local tissue inflammation, which involves cell types such as macrophages, fibroblasts, and NK cells. Local inflammation leads to spillover of TNF into the systemic circulation. (4) A long-standing increase of circulating TNF inhibits the entire hypothalamic-pituitary-adrenal (HPA) axis on several organ levels. Serum cortisol levels are somewhat increased; however, the amount of cortisol is inadequate in relation to ongoing inflammation. (5) Endogenous cortisol would normally inhibit peripheral inflammation in the joints and elsewhere. Since cortisol levels are inadequately low, inhibition of inflammation is not sufficient. Insufficient suppression of inflammation perpetuates the disease process. Some patients with lower baseline serum levels of cortisol demonstrate good clinical response and elevation of serum cortisol under anti-TNF therapy. In this group of good responders, the TNF-induced break on the HPA axis seems to be particularly strong (according to Ref. 750).

be altered because mitochondria are the bottleneck of adrenal cholesterol breakdown and conversion to active steroids (Figure 13). If mitochondria become altered, lipid breakdown rate should be decreased. A key cytokine involved in decreased steroidogenesis is IL-1β.[751]

Although these findings were demonstrated in rats but not in humans, it is remarkable that IL-1β seems to play a highly important role in the human juvenile forms of arthritis. IL-1β neutralization was the most efficacious biologic in the juvenile form of systemic arthritis.[752] In addition, the neutralization

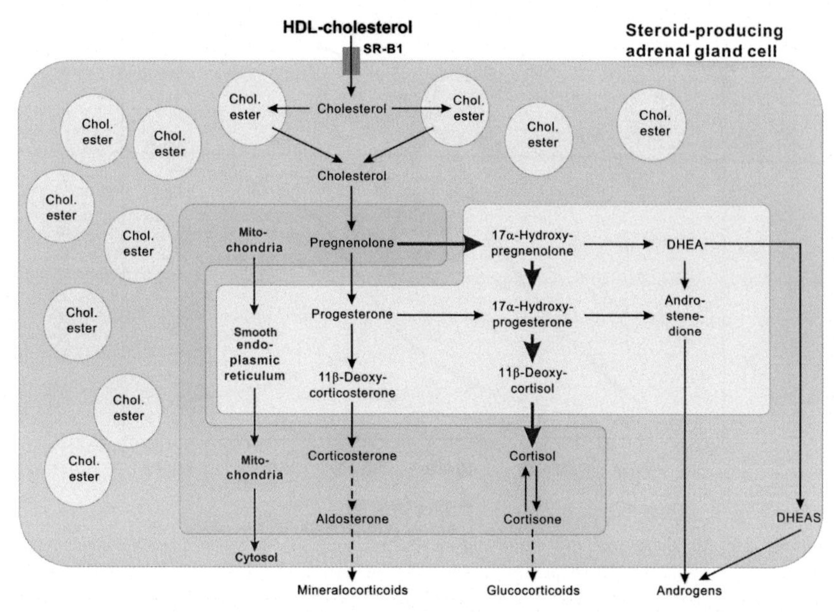

FIGURE 13 Steroidogenesis in adrenocortical cells in the cytoplasm (lilac), in the endoplasmic reticulum (light blue), and in mitochondria (orange). DHEA, dehydroepiandrosterone; DHEAS, DHEA sulfate; HDL, high-density lipoprotein; SR-B1, scavenger receptor class B member 1 that takes up HDL cholesterol.

of IL-1β is also a perfect strategy in autoinflammatory syndromes and type II diabetes mellitus.[753,754] Thus, IL-1β can be a key cytokine in human chronic inflammatory systemic diseases. Future studies need to address whether, or not, mitochondria are similarly altered in human diseases and whether neutralization of IL-1β increases endogenous glucocorticoids in the systemic form of juvenile idiopathic arthritis and autoinflammatory syndromes.

The Women-to-Men Preponderance in Rheumatic Diseases: The Role of Sex Hormones

The women-to-men preponderance in incidence and prevalence of chronic inflammatory systemic diseases is generally known. There are some arguments independent of sex hormones why women are more affected than men, but these are not discussed here.[755]

The women-to-men preponderance in incidence of, for example, rheumatoid arthritis is demonstrated in Figure 14. The incidence ratio for women/men decreases sharply after the menopause (yellow area in Figure 14), and this is similar in other chronic inflammatory systemic diseases. Thus, women are particularly more affected in the reproductive period, where serum levels of the major sex hormones 17β-estradiol and progesterone are approximately ten times higher in women than in men (opposite for testosterone).

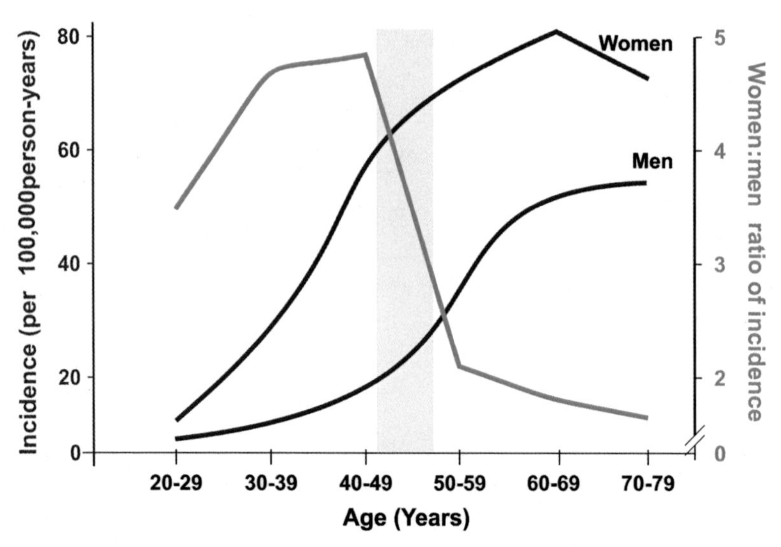

FIGURE 14 Incidence rate in women and men with rheumatoid arthritis. The two lilac curves delineate the separate incidence rates of women and men in rheumatoid arthritis.[756] The green curve delineates the ratio of rheumatoid arthritis incidence of women divided by the incidence ratio of men.[757] The yellow area depicts the time window of the menopause.

Thus, linking estrogens or progesterone to autoimmune diseases followed logical reasoning:

If estrogens or progesterone are high in the reproductive period, estrogens and progesterone are important driving factors of autoimmunity in women.

However, this idea was too simple because it does not explain why women with some autoimmune diseases demonstrated high incidence rates after the menopause, after which estrogens and progesterone dramatically fall (see example for rheumatoid arthritis in Figure 14). How can the same hormones, 17β-estradiol and progesterone, be stimulatory and inhibitory? To understand this apparent paradox, a quick side trip to therapy-induced remission with biologics is needed, because this obviously explains the concept of disease subtypes.

The remission rate of biologics is an important estimate to evaluate the pathophysiological role of cells/factors neutralized or depleted (Table 10). For example, when remission with a B-cell-depleting antibody appears in 25% of treated patients, it shows that B cells must be a main pathophysiological factor in 25% of rheumatoid arthritis patients. It also indicates that in 75% of rheumatoid arthritis patients, other important factors play pathophysiologically relevant roles. The question appears whether patients with remission compared to those with little response belong to different subtypes of rheumatoid arthritis.

Using the remission rate of biologics with targeted therapy effects, one can discriminate subtypes of chronic inflammatory systemic diseases with most

TABLE 10 Remission[a] with Different Biologics in Well-Controlled Randomized Clinical Trials

Treatment	Patients in Remission (%)
T-cell costimulation blocker (abatacept)	20
B-cell-depleting antibodies (rituximab)	25
Anti-IL-1β strategies in juvenile idiopathic arthritis	80
Anti-IL-6 receptor antibodies (tocilizumab)	25
Anti-TNF strategy (infliximab, etanercept, and adalimumab)	25-30

IL-1β, interleukin 1β; IL-6, interleukin-6; TNF, tumor necrosis factor.
[a]The remission rate depends on age and gender, the given dose, comedication, disease activity at start, previous DMARD failures, comorbidities, and the duration of treatment.

probably different underlying pathophysiologies. In other words, "disease is not disease!" such as "diabetes is not diabetes" (think of the two types of diabetes mellitus). The identification of distinct disease subtypes such as in rheumatoid arthritis and multiple sclerosis has also been suggested on histological grounds.[278,279] Histological studies delineated not only a B-cell subtype of rheumatoid arthritis and multiple sclerosis but also a T-cell subtype.[278,279]

From Table 10, one can extract that 25% of patients demonstrate a B-cell-driven disease (rituximab effect), whereas in 20%, disease is T-cell-driven (abatacept effect). Given the fact that macrophages are the main producers of TNF and IL-1 and IL-6 is produced mainly from fibroblasts, more than 50% of patients have a macrophage- or fibroblast-driven disease, or something else. These are rough estimations because T cells and B cells also produce TNF and IL-6. This short presentation shows that chronic inflammatory systemic diseases are probably driven by different leading cellular elements of the immune system such as B cells, T cells, macrophages, fibroblasts, or another cell. But what has it to do with the reproductive age and autoimmunity in women?

In order to understand this paradox, one needs to study the effects of 17β-estradiol and progesterone on these different immune cells as mentioned above. Figure 15 summarizes the effect of 17β-estradiol on B cells, T cells, macrophages, and fibroblasts as extensively reviewed recently.[329] It is obvious that 17β-estradiol at ovulatory to pregnancy levels stimulates many anti-inflammatory factors such as IL-4, IL-10, TGF-β, tissue inhibitors of matrix metalloproteinases, osteoprotegerin, and basic fibroblast growth factor, whereas proinflammatory factors are inhibited such as TNF, IL-1β, IL-6, matrix metalloproteinases 1 and 3, and MCP-1 (Figure 15). Importantly, only antibody

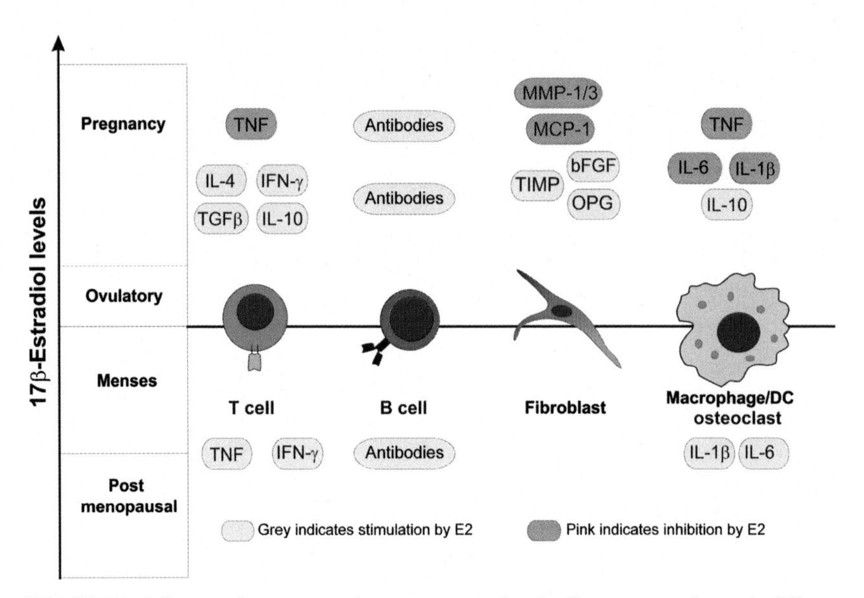

FIGURE 15 Influence of estrogens on important pro- and anti-inflammatory pathways in different cell types. On the *y*-axis, the concentration of estrogens is given. Depending on the concentration of estrogens, the factors in gray boxes are stimulated and factors in pink boxes are inhibited by estrogens. DC, dendritic cell; IFN-γ, interferon-γ; IL, interleukin; MCP-1, monocyte chemoattractant protein 1; MMP, matrix metalloproteinase; OPG, osteoprotegerin; TGF-β, transforming growth factor-β; TIMPs, tissue inhibitors of MMP; TNF, tumor necrosis factor.

production and B-cell immunology are stimulated at these higher levels of 17β-estradiol (also true for progesterone). It seems that B-cell immunology during pregnancy compensates for the limitations of T-cell immunology.

In contrast, the fall of 17β-estradiol to postmenopausal levels is a stimulus for proinflammatory cytokine production by T cells and macrophages, and B cells are also stimulated (Figure 15). One can summarize that 17β-estradiol and progesterone at concentrations, which appear in the reproductive phase of a woman, mainly activate B cells but inhibit T cells, macrophages, and fibroblasts.

In men, testosterone inhibits all types of immune cells so that the increase of chronic inflammatory systemic disease incidence in older men is explained by the loss of protective testosterone. Now, let us have a look at Figure 16.

If the premenopausal women-to-men preponderance depends on the difference in sex hormones, one can hypothesize that the activating effect of 17β-estradiol and progesterone on B cells makes the difference. This does not mean that these hormones cause the disease in an exclusive manner, but they provide the adequate milieu for the outbreak of the disease. Here, I wish to highlight the work of Diamond's group, who exactly demonstrated that the estrogenic milieu is very important for the initiation of a chronic inflammatory systemic disease.[758]

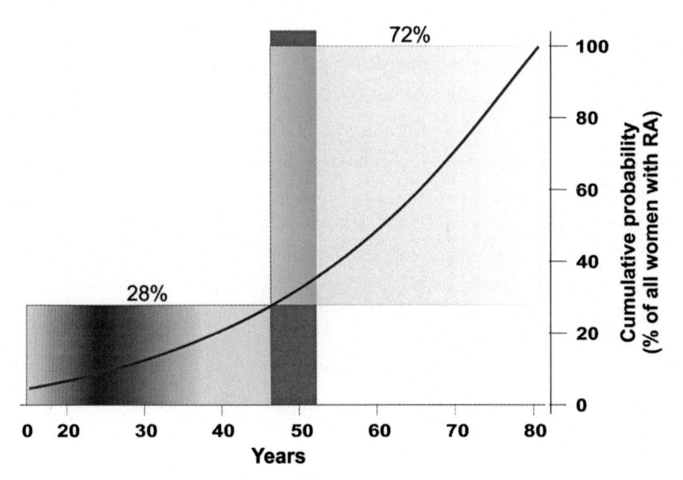

FIGURE 16 Cumulative probability to develop the disease in women over time. The example of rheumatoid arthritis is given. Approximately 28% of women develop rheumatoid arthritis before the menopause, whereas 72% of women develop rheumatoid arthritis during or after the menopause. Thus, the well-known women-to-men preponderance in the ages between 20 and 45 is generated by only 28% of women. Necessarily, the disparity in sex hormones may play a role in only 28% of women. The different red colors in the panel indicate the levels of 17β-estradiol, which are highest (deep purple) between 20 and 25 years of age, but get very low later on (light yellow).

It is hypothesized that the B-cell-driven subtype of chronic inflammatory systemic disease establishes in the reproductive age in the presence of high levels of 17β-estradiol and progesterone. Importantly, the remission rate of B-cell-depleting therapies in rheumatoid arthritis is similar to the percentage of rheumatoid arthritis women, who develop the disease in the reproductive years (Table 10 and Figure 16). In contrast, it is hypothesized that those subtypes establishing in later ages (>45 years) are dependent on T cells, macrophages, fibroblasts, or other estrogen-inhibited cell types.

The first experimental evidence comes from multiple sclerosis research: early-onset patients with secondary progressive multiple sclerosis demonstrated a B-cell subtype of the disease.[759] A similar finding was demonstrated in rheumatoid arthritis where patients with early-onset disease had more autoimmune-relevant CD5+ B cells as compared with late-onset patients.[760] Those patients with an onset after the age of 65 had 0% of the autoimmunity-relevant CD5+ B cells.[760]

In addition, in systemic lupus erythematosus, thyroiditis, autoimmune hepatitis, and many other autoimmune diseases of young ages, which are mainly B-cell-driven, no similar increase of incidence appears after the menopause. There exists no similar T-cell-/macrophage-/fibroblast-driven subtype of these seemingly B-cell-driven diseases.

In conclusion, we start to recognize that B-cell-driven diseases may reach the maximum incidence rate in the reproductive years (Figure 17). In contrast, it

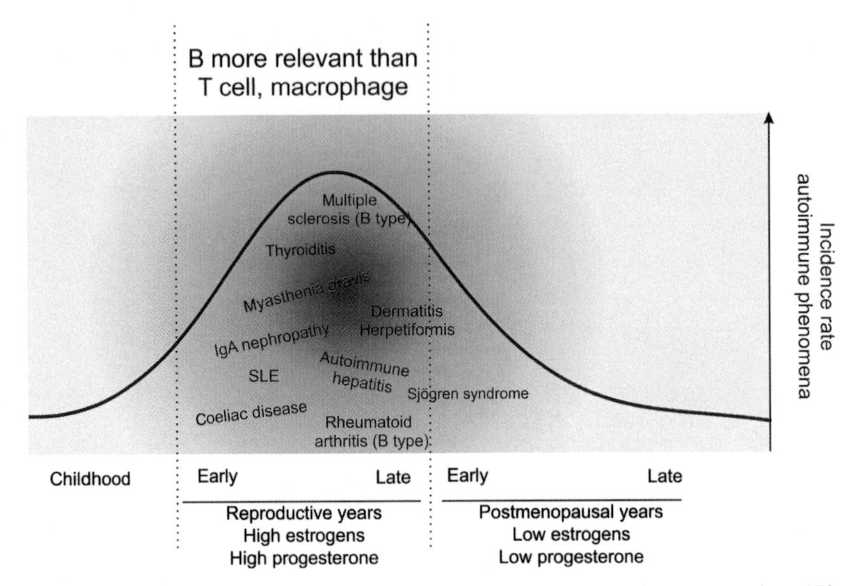

FIGURE 17 B-cell-driven diseases appear in reproductive years. Sex hormones such as 17β-estradiol and progesterone stimulate the B cell but inhibit T cells, macrophages, and fibroblasts. The concentrations necessary for this differential effect appear in the ovulatory phase and during pregnancy (→reproductive period).

is hypothesized that diseases that appear after the menopause are mainly driven by T cells, macrophages, and/or fibroblasts (Figure 18). Thus, defining subtypes of diseases would be fundamental for adjustment of biologic therapies.

Loss of Adrenal and Gonadal Androgens

It is well known that activation of the HPA axis by proinflammatory stimuli leads to a parallel decrease of the HPG axis activity (Figure 9 and Refs. 630–632). This can be substantiated by decreased levels of follicle-stimulating hormone and luteinizing hormone, and it is even more evident by looking on serum testosterone and serum DHEAS in chronic inflammatory systemic diseases. Particularly, serum testosterone and serum DHEAS become very low.[761–781] This can lead to clinically relevant problems in these patients such as lower mean testicular volume, lower median total sperm count, and lower sperm motility.[782]

Since testosterone has anti-inflammatory properties, the decline of this hormone supports the proinflammatory process. In the adrenal and gonadal glands, the loss of DHEA and DHEAS is—among other reasons (Table 9)—attributed to a blockade of the second step of the P450c17 enzyme by factors such as IL-1β and TNF (Figure 19, upper right: the step from 17α-hydroxypregnenolone to DHEA). In addition, it has been demonstrated that the conversion of the biologically inactive DHEAS (which is already low in the serum) to the active

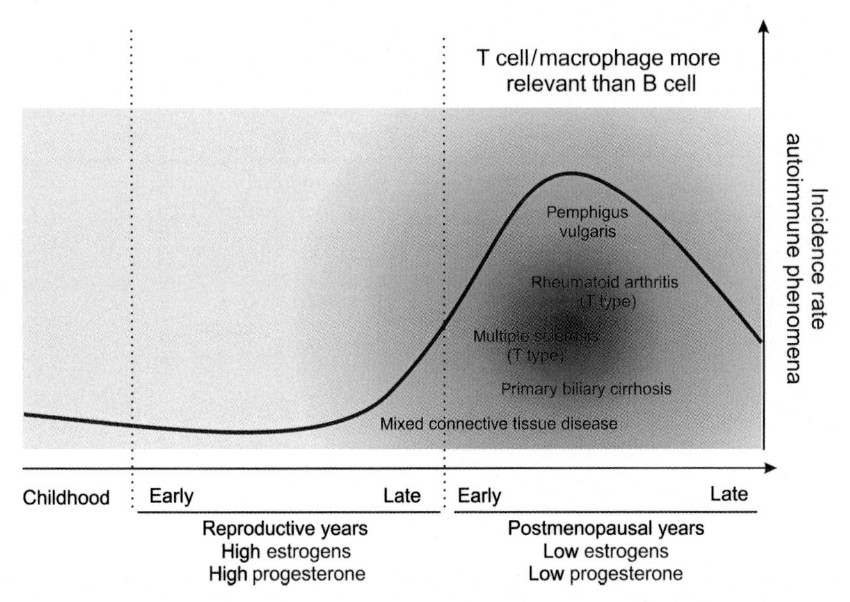

FIGURE 18 Non-B-cell-driven diseases appear outside of the reproductive period (before menarche or after the menopause). The figure demonstrates diseases that establish after the menopause. Since sex hormones such as 17β-estradiol and progesterone inhibit T cells, macrophages, and fibroblasts at ovulatory to pregnancy levels, their decrease after the menopause is a stimulus for immune reactions involving these cells. Since the same disease appears in men to a similar extent, the loss of testosterone and other androgens is similarly conducive.

DHEA is inhibited by TNF in cells of inflamed tissue (Figure 19, bottom).[783] This supports the further loss of anti-inflammatory androgens in the tissue at the locally inflamed site. In addition to the marked reduction of local androgens, the remaining DHEA is converted into the proinflammatory 7β-hydroxy-DHEA and into proproliferative 16α-hydroxyestrone and 16α-hydroxyestradiol, which is stimulated by TNF (Figure 19, bottom).[784,785] The role of serum sex hormone-binding globulin has not been studied in detail but might have an impact on androgen results because it is the binding protein of androgens.[786]

In contrast, adrenal androgens can be relatively increased in very acute forms of human chronic inflammatory systemic diseases.[737] It was hypothesized that there can be a transition from initially high adrenal androgen levels acutely to low levels in human chronic inflammatory systemic diseases in the later phase. Mechanisms of transition from high to low production rates of anti-inflammatory androgens need to be investigated in longitudinal studies. An important target is the second lyase step of the P450c17 enzyme because this enzyme is responsible for the conversion of prehormones such as 17α-hydroxypregnenolone to active androgens such as DHEA (Figure 19).

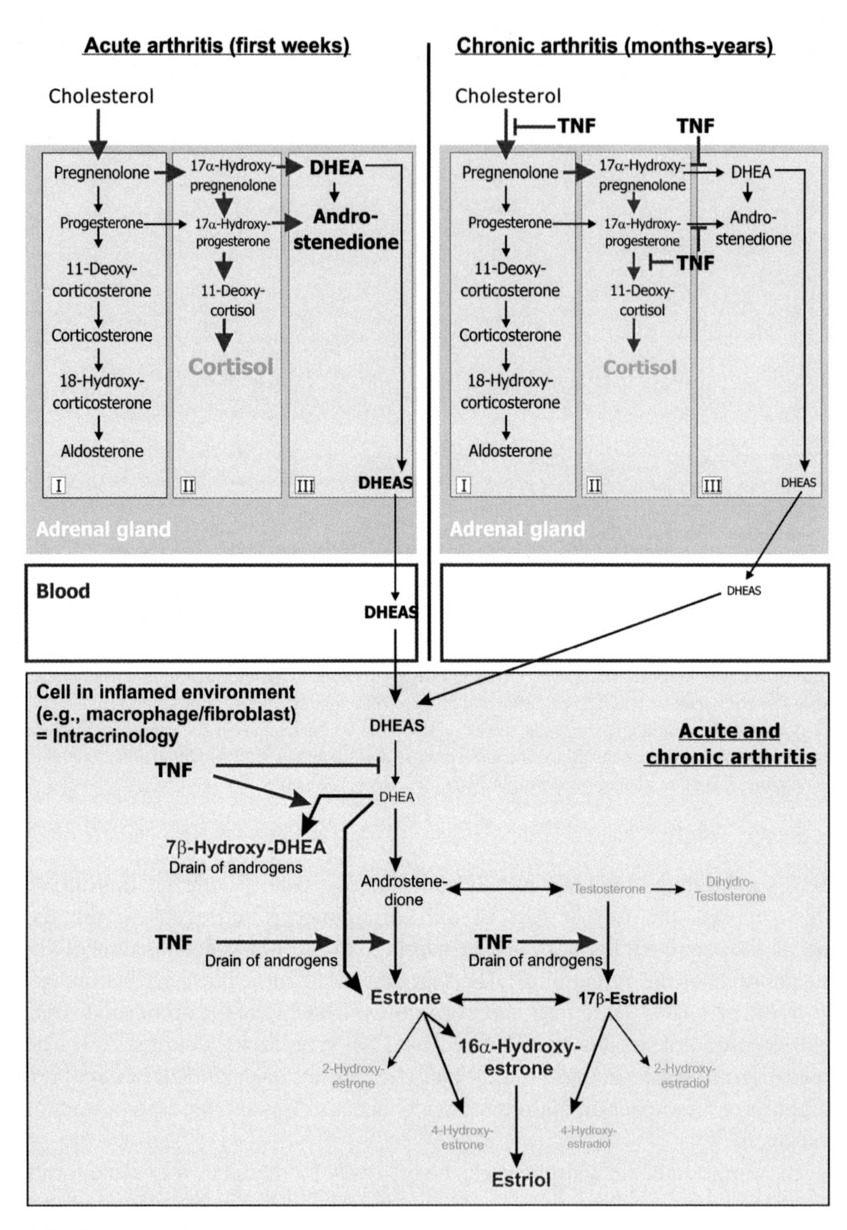

FIGURE 19 See figure Legend on Opposite page.

New information demonstrates that there exists in patients with rheumatoid arthritis an important polymorphism in the gene of *CYB5A* on chromosome 18, which determines the amount of androgens converted from androgenic precursors.[787] Cytochrome B5 type A is the important supportive cofactor of the second lyase step of the P450c17 enzyme (Figure 19, from 17α-hydroxypregnenolone to DHEA). Both large genome-wide association studies (Wellcome Trust Case Control Consortium and North American Rheumatoid Arthritis Consortium) revealed rheumatoid arthritis-associated polymorphisms in the CYB5A gene: rs1790834 ($p=0.0073$, OR=0.83) and rs1790858 ($p=0.0095$, OR=0.44), respectively. This was replicated in a Slovak rheumatoid arthritis sample with 521 rheumatoid arthritis cases and 321 healthy controls.[787] For rs1790834 localized in intron 1 of the CYB5A gene, a favorable association was found with an odds ratio of 0.69 (0.45-0.86), that is, a protective effect of the rare allele. Functional tests clearly demonstrated that rheumatoid arthritis carriers of the favorable allele produced more androgens from the precursor pregnenolone when synovial fibroblasts were used as test system.[787]

In conclusion, adrenal and gonadal androgens are significantly decreased, which supports a proinflammatory milieu in general and, especially, in inflamed tissue.

Increased Conversion of Androgens to Estrogens

Several studies now demonstrated that androgens such as testosterone, androstenedione, and DHEA are rapidly converted to estrogens in inflamed tissue (Figure 19, bottom).[788,789] Particularly, macrophages and fibroblasts are able to promote estrogen generation from precursor androgens. This is an additional cause why the levels of anti-inflammatory androgens are decreased in inflamed tissue. On the other hand, measurable estrogen serum levels are, thus, relatively normal because of the spillover from the inflamed tissue into the circulating blood.

FIGURE 19 Inadequate availability of anti-inflammatory steroid hormones in chronic inflammation. The size of the font indicates availability, production, and concentration of the respective factor (small font, little; big font, much). Hormones in red color are proinflammatory while hormones in green color are anti-inflammatory. In the left upper part (blue box), the reaction of the adrenal gland during an acute inflammatory episode is demonstrated: the major pathways (in red) to cortisol (glucocorticoid pathway, II) and DHEA (androgen pathway, III) are stimulated in the acute situation. In the right upper part (blue box), the reaction of the adrenal gland during chronic inflammation is delineated. The major pathway to cortisol is still activated leading to relatively normal cortisol serum levels albeit increased cytokine levels. This pathway predominance to cortisol relative to DHEA and androstenedione leads to markedly lower serum levels of DHEAS, the main precursor of androgens in peripheral cells (gray box at the bottom). In the bottom part (gray), the conversion of steroid hormones in peripheral cells such as the macrophage or fibroblast is depicted. DHEAS is converted to downstream androgens and estrogens. The proinflammatory TNF interferes with several hormonal conversion steps (a line with a bar at the end indicates inhibition, whereas an arrow indicates stimulation or conversion to another hormone). DHEA, dehydroepiandrosterone; DHEAS, DHEA sulfate; TNF, tumor necrosis factor.

It is now accepted that cytokines such as TNF, IL-6, and IL-1β activate the aromatase complex in nongonadal peripheral cells such as macrophages (Figure 19, bottom). In addition, a positive correlation was documented between erythrocyte sedimentation rate and serum levels of 17β-estradiol.[781] The increased aromatization of androgens has been demonstrated in inflamed tissue of patients with chronic inflammatory systemic disease.[633,789] Importantly, these estrogens are further converted to proproliferative estrogens such as the 16-hydroxy forms of estrone and 17β-estradiol (Figure 19, bottom).[785] In addition, the proproliferative 16-hydroxyestrogens appear in significantly higher amounts in the tissue (Figure 19, bottom).[785] In contrast, the 2-hydroxylated forms, which have many anti-inflammatory activities, are not detectable in supernatants from inflammatory cells.[785] Furthermore, the molar ratio of urinary 16-hydroxyestrogens divided by 2-hydroxyestrogens is highly increased in women and men with chronic inflammatory systemic diseases when compared to healthy subjects.[790] This shows that the conversion in the tissue is independent of biological sex and that it has most probably a proinflammatory consequence.

DHEA and the Risk of Developing a Chronic Inflammatory Systemic Disease

The adrenal hormone DHEA and androstenedione are important precursors of peripheral anti-inflammatory androgens and low levels have been considered a susceptibility biomarker in studies of rheumatoid arthritis in premenopausal-onset women.[791,792] In these studies, prerheumatoid arthritis subjects demonstrated significantly lower DHEAS and androstenedione serum levels 12 years before onset of disease as compared to respective controls. This was supported by a recent study in patients with seronegative rheumatoid arthritis because, now, serum testosterone levels were lower in patients that later developed seronegative rheumatoid arthritis.[793]

Two other investigations in patients with rheumatoid arthritis did not confirm the earlier findings, but used different assay methods.[794,795] Indeed, in these two studies, the serum levels of DHEAS were inadequately low or high, respectively, which led to criticism of the assay techniques used. The production capacity of the adrenal glands might be better studied using functional tests (CRH test, ACTH test, or circadian rhythms of hormones). The theory of the "small adrenal engine" that is not sufficiently compensatory for the development of experimental and human chronic inflammatory systemic diseases is still valid, because it was so decisive in animal models.[716] In addition, the new polymorphism found in the *CYPB5A* gene clearly indicates that an androgen production defect might be present in women long before the outbreak of the disease.[787]

CRH in Local Inflammation

The proinflammatory role of local CRH and its presence in synovial fluid of patients with rheumatoid arthritis are known.[796] In 2001, CRH mRNA was found upregulated in the tissue of patients with chronic inflammatory systemic diseases, but not in normal tissue.[797] Inflammatory cytokines enhance the transcriptional activity of the human CRH promoter in primary synoviocytes. In addition, the CRH antagonist antalarmin ameliorated an experimental form of a chronic inflammatory systemic disease.[798] In conclusion, although central CRH is involved in the anti-inflammatory activity of the HPA axis, peripheral CRH is proinflammatory in local inflammation.

Elevated Levels of Melatonin and Prolactin

Melatonin and prolactin have been linked to chronic inflammatory systemic diseases since both hormones at normal to slightly elevated concentrations stimulate many aspects of the immune system.[668,799] The dose-response curve of prolactin for immune stimulatory effects follows a bell-shaped curve with a maximum at two times the serum concentration.[800] Importantly, both hormones are indeed somewhat elevated in patients with chronic inflammatory systemic diseases (two times the normal level).[801,802] At these concentrations, both hormones exert proinflammatory effect on several immunologic/inflammatory pathways. The immune-supportive role of the two hormones has also been delineated in the context of circadian rhythms of immune function, where both hormones are suspected of supporting the nightly activation of the immune system.[803] The increase of serum levels at the beginning of sleep was linked to the increase of serum cytokines in the middle of the night (see below). Others have demonstrated elevated levels of serum melatonin in patients from northern Europe compared to southern Europe,[804] which was linked to the elevated incidence and prevalence of chronic inflammatory systemic diseases in northern Europe.

In conclusion, many hormonal pathways are shifted toward a more proinflammatory situation, a fact that certainly supports the perpetuation of these diseases (Figure 20).

Vitamin D Hypovitaminosis

Vitamin D, besides having well-known control functions of calcium and phosphorus metabolism, bone formation, and mineralization, also has a role in the maintenance of immune homeostasis. Vitamin D was included in etiology thinking because of the observed geographic distribution of chronic inflammatory systemic diseases. Areas with high sunlight exposure, the principal inducer of vitamin D synthesis, have a relatively low prevalence of chronic inflammatory systemic diseases and vice versa.

Importantly, low levels of circulating vitamin D are associated with a higher incidence of chronic inflammatory systemic diseases.[648,653,655,805] Three major effects on the immune system have been assigned to vitamin D: (1) enhancement of innate immunity through increased recruitment of macrophage and macrophage differentiation, (2) prevention of autoimmunity through inhibition of the ability of dendritic cells to induce T helper type 1 and T helper type 17 immune responses, and (3) promotion of self-tolerance through permissive effects on the generation of T helper type 2 or T regulatory cells.[648,649,651,652,655]

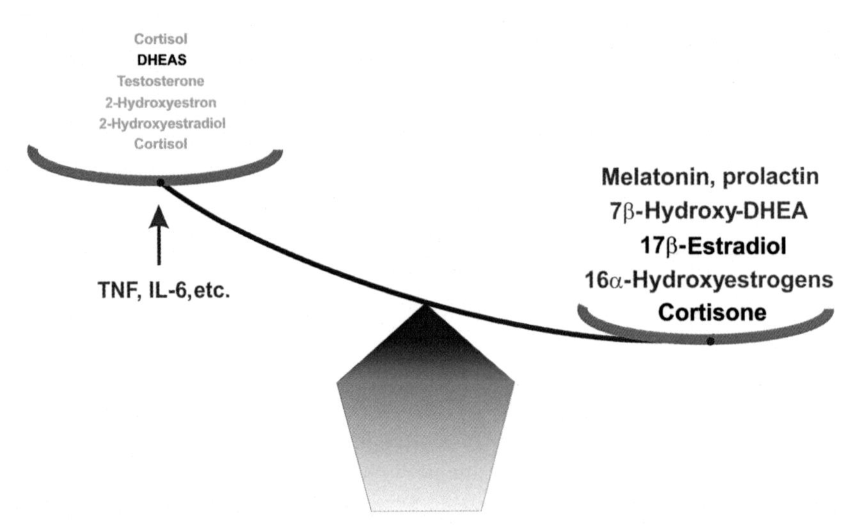

FIGURE 20 The hormone pendulum swings to the proinflammatory site. Hormonal pathways are shifted toward a more proinflammatory situation. The red (green) color indicates a proinflammatory (anti-inflammatory) influence. These are many good reasons to support the perpetuation of chronic inflammation. DHEA, dehydroepiandrosterone; DHEAS, DHEA sulfate; IL-6, interleukin-6; TNF, tumor necrosis factor.

Considering these effects of vitamin D on the immune system, it is understandable that hypovitaminosis D can be a negative element in chronic inflammatory systemic diseases. Hypovitaminosis D is most probably a consequence of disease-related anorexia and reduction of sunlight exposure. This is a typical *a posteriori* cause of chronic inflammatory systemic diseases (see Table 7 and section "The Trigger of Chronic Inflammatory Systemic Diseases" of Chapter I). As a consequence of detected hypovitaminosis D, vitamin D3 substitution is recommended to many patients with chronic inflammatory systemic diseases.[648,650,652,653,655] However, no randomized, double-blind, placebo-controlled trial exists that would support the beneficial effects of vitamin D in chronic inflammatory systemic diseases.

The Hypothalamic-Pituitary-Thyroid Gland Axis

Thyroid hormones induce oxygen radical production in neutrophils;[806] IFN-γ-stimulated expression of MHC class II-encoded proteins;[806] IL-6, IL-8, and IL-12 secretion from different cell types;[806] lymphocyte proliferation;[806] IFN-γ-stimulated natural killer cell activity;[806] and superoxide anion production in human alveolar neutrophils and macrophages.[807] Thyroid hormones are required for normal B-cell production in the bone marrow.[808] These genomic effects are complemented by nongenomic effects of thyroid hormones.[806] Thyroid hormones can bind to the integrin αvβ3 to switch on a cascade of signaling events through either MAP kinase, phospholipase C, protein kinase Cα, ERK1/2, or hypoxia-inducible factor -1 leading to enhanced cytokine and growth factor action and angiogenesis.[806]

Apart from classical actions of thyroxine (T4) and T3, the TSH has many supportive effects on the immune system.[602] Although thyroid hormones have also some anti-inflammatory actions, the usual concentrations of hormones of the hypothalamic-pituitary-thyroid gland axis exert many stimulatory effects on the immune system and inflammation. The question remains whether thyroid hormones like the biologically active T3 are changed during acute and chronic inflammation.

Elevation of systemic inflammation such as during injury, inflammation, or starvation leads to the nonthyroidal illness syndrome (NITS) with the following features:[809] (i) downregulation of hypothalamic TRH; (ii) lowered secretion of TSH, free T4, and free T3; (iii) decreased levels of circulating free T3 due to decreased peripheral T4→T3 conversion; and (iv) increased metabolism of biologically active T3 to inactive reverse T3 (rT3). All mechanisms lead to the deterioration of the hypothalamic-pituitary-thyroid gland axis. Cytokines play an important role in this sequence as demonstrated for IL-6. After injection of IL-6 into healthy volunteers, T4 and free T4 increased after 4 h, but T3 levels were reduced 24 h later, which indicates a rapid downregulation of this biologically active hormone by IL-6.[810]

Although not many studies exist in chronic inflammatory systemic diseases, it seems likely that the levels of thyroid hormones are low as demonstrated in systemic lupus erythematosus, Kawasaki disease,[811,812] and other diseases (Table 11).

TABLE 11 Thyroid Function in Chronic Inflammatory Systemic Diseases

Disease and Finding	References
Celiac disease: low T3	813,814
Crohn's disease: low T3	815,816
Multiple sclerosis: hypothyroidism[a] and low T3	817
Multiple sclerosis: normal function	818
Polymyalgia rheumatica: hypothyroidism	819
Psoriasis vulgaris: normal function	820
Rheumatoid arthritis: normal function	821,822
Sjögren's syndrome: normal function	823
Systemic sclerosis: hypothyroidism[a]	824,825
Systemic lupus erythematosus: hypothyroidism[a] and low T3	811,826
Type 1 diabetes mellitus: hypothyroidism[a] and low T3	827,828

[a]*Hypothyroidism is often linked to concomitant thyroid autoimmunity. No studies were found for pemphigus and ankylosing spondylitis.*

However, it remains unclear whether this is a direct effect of increased circulating cytokines (similar to NITS) or secondary antithyroid autoimmunity with functional defects. A recent investigation in rheumatoid arthritis patients found a decrease in TSH levels during TNF-neutralizing therapy.[829] This might be interpreted as a consequence of TNF-induced reduction of peripheral thyroid hormones and, consequently, upregulation of TSH, but exact mechanisms remain enigmatic.

Importantly, recent experimental studies have shown that downregulation of the central part of this axis observed during acute and chronic inflammation does not necessarily induce decreased thyroid hormone levels in key metabolic organs such as the liver, skeletal muscle, and adipose tissue (summarized in Ref. 830). The differential regulation of local thyroid hormone availability mainly depends on the expression of activating deiodinases 1 and 2 (D1 and D2) and inactivating deiodinase 3 (D3).[830]

For example, during acute inflammation, in the muscle, the hormone-activating D2 increases, whereas the hormone-inactivating D3 decreases, which would lead to higher muscular T3 levels.[830] This is different in chronic inflammation where D2 and D3 are elevated, and, as it was described, the net effect on T3 is a reduction in active T3.[830] A similar concept exists in the liver but the exact pathways are far from clear.[830]

Local deiodinase expression and function were tested in locally inflamed tissue in the chronic inflammatory systemic disease of rheumatoid arthritis (Pörings AS, Lowin T, Straub RH, unpublished data). In this compartment, rT3 levels (degraded T3) in superfusate of synovial tissue were higher in rheumatoid arthritis compared to control osteoarthritis. Staining of synovial tissue revealed the expression of thyroid-converting enzymes D1, D2, and D3, thyroid hormone transporter MOT-8, and nuclear receptors TR-α and TR-β. The addition of TNF or IL-1β to synovial fibroblast cultures increased protein levels of D1, D3, and TR-α. In rheumatoid arthritis, serum levels of rT3, the degradation product of T3, are higher in synovial fluid than in plasma, which is opposite for T3 and T4. Serum rT3 relative to free T3 was higher in rheumatoid arthritis than control osteoarthritis. These data demonstrated that thyroid hormones are metabolized locally in the inflamed joint. This local inactivation is more pronounced in rheumatoid arthritis than osteoarthritis (higher rT3 levels). Since cytokines alter the expression of D1-D3 convertases and thyroid receptors, the thyroid hormone system might become dysregulated in the joint during chronic inflammation (Pörings AS, Lowin T, Straub RH, unpublished data).

In neutrophils, the inactivating D3 is highly expressed.[830] In these cells, the release of inorganic iodide was related to improved killing of bacteria.[830] This is a very attractive concept because it might well explain the stimulating effects of thyroid hormones on phagocytosis and killing independent of genomic effects via thyroid hormone receptors.

In conclusion, the hypothetical sequence of events during inflammation might be as follows: (i) There is a rapid increase of thyroid hormones for the first 4h (cooperation with HPA axis and sympathetic nervous system);

(ii) Then, a rapid downregulation of the hypothalamic-pituitary-thyroid gland axis is established (as observed as low T3, NITS); (iii) This is accompanied by differential expression of deiodinases with relatively normal local T3 in metabolic organs, which might serve the activated immune system by the breakdown of energy-rich substrates (known for skeletal muscle); (iv) Activated neutrophils would be nourished by these circulating substrates and, in parallel, increased D3 expression to provide inorganic iodide necessary to kill bacteria; and (v) Increased metabolization of thyroid hormones is observed in rheumatoid arthritis supporting the active D3 mechanism. However, many parts of this story are still unknown for most chronic inflammatory systemic diseases.

Insulin and Insulin-Like Growth Factor 1 are Proinflammatory; Insulin Resistance is a Proinflammatory Signal

In earlier years, the direct effect of insulin on immune cells was demonstrated to be proinflammatory, mainly, by supporting proliferation of immune and other cells.[831] In recent years, insulin was categorized as an anti-inflammatory hormone because it is able to remove energy-rich fuels such as glucose and free fatty acids from circulation that would nourish the immune system.[832,833] Although this latter concept seems reasonable with intact insulin signaling, it will not work under insulin resistance of the liver, skeletal muscle, and fat tissue, and this is exactly what happens in chronic inflammatory systemic diseases. Thus, the balance between systemic anti-inflammatory insulin effects and local proinflammatory insulin effects will influence the role of this hormone in a given situation. It is proposed that insulin resistance is the critical determinant of the pro- or anti-inflammatory effect of insulin, because only the liver, skeletal muscle, and fat tissue become insulin-resistant but not leukocytes (see also section "Insulin Resistance" of Chapter V).[834]

Insulin-like growth factor 1 was demonstrated to have mainly proinflammatory effects.[835,836] The aspects of insulin-like growth factor 1 are stimulation of hematopoiesis, T and B lymphopoiesis, increase of natural killer cell activity, priming of macrophages and neutrophils for radical production, increase of TNF production from macrophages, sensitization for mitogen stimulation, and enhanced primary antibody responses *in vivo*.[835]

In chronic inflammatory systemic diseases such as rheumatoid arthritis and systemic lupus erythematosus, hyperinsulinemia and insulin resistance were described (see also section "Insulin Resistance" of Chapter V).[837–839] In addition, insulin-like growth factor 1 resistance was described in patients with rheumatoid arthritis,[840] and insulin-like growth factor 1 levels are typically decreased in chronic inflammation.[841–843] Thus, both pathways through insulin and insulin-like growth factor 1 receptors are not intact in chronic inflammatory systemic diseases. The high insulin levels and insulin resistance would support the proinflammatory process in chronic inflammatory systemic diseases (direct, insulin activation of immune cells; indirect, provision of energy-rich fuels; see also section "Insulin Resistance" of Chapter V).

Growth Hormone: Dual Roles in Adults and Children

The growth hormone accelerates the recovery of the immune system following transplantation of various cell types, and it replenished the severely affected T-cell compartment in HIV patients.[657] Treatment with drugs that stimulate growth hormone secretion or treatment with growth hormone can all have positive effects in restoring aspects of the aged immune system.[657] Immune cells carry the growth hormone receptor, and growth hormone signaling involves JAK2-STAT-Ras-MAP kinase pathways that are shared by proinflammatory cytokine-signaling pathways (summarized in Ref. 657).

Growth hormone given to healthy volunteers slightly but significantly increased serum TNF and serum IL-6.[844] Growth hormone was made responsible for the chronic smoldering inflammation during aging, which can be studied in growth hormone pathway-deficient mice that demonstrate less inflammation and increased longevity.[845] Furthermore, growth hormone primes neutrophils for the production of lysosomal enzymes and superoxide anions, supports survival of memory T cells, increases immunoglobulin secretion of B cells, and stimulates thymulin secretion by thymic epithelial cells, natural killer cell activity, phagocytosis, oxidative burst, and killing capacity of neutrophils or macrophages.[602] Transgenic mice overexpressing growth hormone or its receptor exhibit overgrowth of the thymus and spleen and display increases in mitogenic responses to concanavalin A.[602]

Growth hormone serum levels were lower, normal, or slightly elevated in patients with rheumatoid arthritis and systemic lupus erythematosus (summarized in Ref. 846). Thus, in patients with chronic inflammatory systemic diseases, growth hormone serum levels behave in a similar way than cortisol serum levels. In other words, there is no clear increase or decrease of hormone levels in serum. One might argue that these hormones are not much involved because of unchanged serum levels. However, the anti-inflammatory role of endogenous cortisol was visualized in chronic inflammatory systemic diseases by blocking endogenous cortisol production with metyrapone.[714] We learned that the growth hormone has important immunostimulating effects, and the question appears whether inhibition of growth hormone release by somatostatin also demonstrates the effects of the endogenous hormone in chronic inflammatory systemic diseases, opposite in effect compared to endogenous cortisol when blocked with metyrapone.[714]

Open therapies with the growth hormone inhibitor somatostatin in small studies demonstrated anti-inflammatory effects such as the reduction of synovial membrane thickness[847] and improved clinical symptoms such as morning stiffness and other *American College of Rheumatology* criteria in rheumatoid arthritis.[848,849] Although knowing that somatostatin has direct suppressive effects on immune cells and nociceptive nerve fibers,[850] somatostatin might also block growth hormone release on a systemic level in the pituitary gland. Thus, one can hypothesize that anti-inflammatory effects of somatostatin or other growth hormone blockers are expected on the basis of inhibition of energy expenditure and inhibition of immunostimulation, two functions of growth hormone.

This sounds pretty reasonable, but experimental proof in this complex growth hormone-insulin-like growth factor 1 system needs to be demonstrated.

During growth in children and adolescents, the situation might be quite different because energy-rich fuels like glucose and amino acids and expedients like calcium/phosphorus are stored in the growing skeletal muscle and bone (they are not provided to the active immune system). Growth hormone might be judged in a different way in juvenile forms of chronic inflammatory systemic diseases due to anabolic effects on skeletal muscle and bone growth.[851] Indeed, in juvenile forms of chronic inflammatory systemic diseases, growth hormone can have favorable growth-promoting but no anti-inflammatory or proinflammatory effects (summarized in Ref. 851). From this point of view, one can hypothesize that growth hormone effects depend on a balance between storage and expenditure of energy-rich fuels. Growth hormone might shift this balance toward energy storage in children and toward energy expenditure in adults.

Angiotensin II is Proinflammatory

While the systemic effects of angiotensin II are related to hemodynamic and metabolic functions, local RAAS pathways support proinflammatory, proliferative, and profibrotic activities via angiotensin type 1 (AT1) receptors that couple to G protein subunits Gq and Gαi (recently summarized in Ref. 852). AT2 receptors also couple to Gαi proteins, a G protein that supports proinflammatory pathways. In different organs such as the kidney, heart, and vasculature, angiotensin II induces an inflammatory response by fostering the expression of proinflammatory chemokines, responsible for tissue accumulation of immunocompetent cells.[852] Angiotensin II via AT1 receptors is also a proinflammatory factor in a lupus mouse model.[853] Angiotensin-converting enzyme inhibitors of different types reduce the severity of collagen type II-induced arthritis.[854,855]

Acute infectious disease leads to the upregulation of the RAAS in a mouse model of cytomegalovirus infection.[856] Injection of lipopolysaccharide into rats increased activity of the RAAS.[857] Patients with sepsis demonstrate increased activity of the RAAS.[858] Angiotensin-converting enzyme is upregulated in synovial tissue of patients with rheumatoid arthritis leading to higher availability of angiotensin II in inflamed joints.[859] While the proinflammatory role of angiotensin II is well established, there are very little studies that addressed serum levels of hormones of the RAAS in humans. Two Russian studies identified increased levels of angiotensin II and aldosterone in patients with rheumatoid arthritis and systemic lupus erythematosus, but this awaits further confirmation.[860,861]

In conclusion, all these findings indicate that the RAAS is activated in acute and chronic inflammation. Since the RAAS exerts proinflammatory effects in addition to its function as an energy expenditure hormonal system, it is perfectly able to support the reallocation of energy-rich fuels to the activated immune system. In addition, water retention with these hormones leads to volume overload (see Chapters III and V).

Adipokines are Proinflammatory

Adipokines, such as leptin, IL-6, TNF, visfatin, adiponectin, resistin, omentin, and others are a group of cytokines secreted by the adipose tissue. Per definition, adipokines are produced by adipocytes, which are closely related to fibroblasts. In chronic inflammatory systemic diseases, most interest was devoted to leptin.[862]

Leptin from the adipose tissue is stimulated by proinflammatory cytokines such as TNF and IL-1β.[863–867] These cytokines seem to stimulate short-term release of stored leptin although its production may even be inhibited during long-term *in vitro* stimulation with TNF.[868] Thus, the acute cytokine-driven rise in leptin may support the initial proinflammatory response. Human leptin stimulates the proliferation and activation of human circulating monocytes via the leptin receptor, which is a member of the IL-6 receptor-related cytokine receptors.[869] Leptin is a fundamental factor for human T-cell proliferation, and it induces T helper type 1 cells and T helper type 17 immune reactions.[464,870] Moreover, fat mass in humans is directly related to white blood cell count, which is closely related to leptin serum levels,[871] and leptin stimulates oxidative species production by stimulated polymorphonuclear neutrophils.[465] Thus, leptin is an "acute-phase protein of the fat tissue," which supports the immune system during a short-term inflammatory disease.

Indeed, leptin-deficient mice exhibit impaired host defense in Gram-negative pneumonia,[466] and starvation with low serum leptin levels leads to immunosuppression.[872] Interestingly, serum leptin levels are increased in sepsis survivors as compared with nonsurvivors.[873] All these factors indicate that leptin has been evolutionarily positively selected for an acute inflammatory response. During some chronic inflammatory disease such as rheumatoid arthritis, the serum levels of leptin are also increased.[874–876] Similar to the acute infectious situation, most probably the proinflammatory load increases serum leptin, which depends on fat mass. Moreover, leptin inhibits the important second step of the adrenal P450c17 reaction (Figure 19, upper boxes, 17α-hydroxypregnenolone to DHEA), which adds to the loss of anti-inflammatory androgens.[877] This leptin-induced hypoandrogenicity is a proinflammatory signal.

In conclusion, leptin is a systemic indicator of the available energy stores and, thus, may be necessary for the fine-tuning of the energy-consuming immune response (see Chapters III and V). In this respect, it supports the proinflammatory pathways of the immune system. Many, not all, adipokines have similar immune system-supporting effects. The interested reader is referred to further literature.[459,463,878–880]

Hormonal Influence on T Helper Type 17 Cells and T Regulatory Cells

At the beginning of endocrine immune research, the concept of CD4-positive T helper type 1 and T helper type 2 was the focus. During the last decade, the important IL-17A-dependent T helper type 17 pathway and T regulatory cells (earlier called suppressor T cells) were added. The effects of hormonal influence on these cells are demonstrated in Table 12. A general inhibitory influence on T helper type 17 cells appeared for glucocorticoids, 17β-estradiol, progesterone, and vitamin D3 (Table 12).

With respect to T regulatory cell pathways, insulin, α-MSH, 17β-estradiol, progesterone, testosterone, glucocorticoids, and vitamin D3 increase T regulatory cell function. Only leptin was demonstrated to inhibit T regulatory cell function (Table 12).

In conclusion, there exists a natural set of endocrine factors that is responsible for inhibiting T helper type 17 cells or pushing T regulatory cells.

TABLE 12 Hormonal Immune Modulation of T Helper Type 17 (Th17) Cells and T Regulatory Cells

Described Effect	References
T helper type 17 cells	
Leptin increases Th17 cell generation	870
Aldosterone promotes Th17 immunity	881
Growth hormone-releasing hormone stimulates IL-17 production	882
Glucocorticoids inhibit Th17 functions	883–885
17β-Estradiol inhibits Th17 immune responses	886–888
2-Methoxyestradiol inhibits IL-17A in arthritis	889
Ovariectomy increases Th17 functions (reversed by 17β-estradiol)	890
Progesterone inhibits Th17 functions	891,892
Vitamin D3 inhibits IL-17 production	893–899
T regulatory cells	
Leptin inhibits T regulatory cell function	900
Insulin upregulates T regulatory cell function	901
α-MSH increases T regulatory cell function	902,903
17β-Estradiol increases T regulatory cell function	886,904–907
Progesterone supports T regulatory cell functions	892,908,909
Testosterone supports T regulatory cell pathways	910
Glucocorticoids increase T regulatory cells and functions	911–916
Vitamin D3 induces T regulatory cell responses	894,917–920

NEUROIMMUNOLOGY IN CHRONIC INFLAMMATORY SYSTEMIC DISEASES

Introduction[e]

For over 2000 years (since Celsus and Galen), clinicians recognized that cardinal features of neurogenic responses, such as redness, warmth, swelling, and pain, are rapid sequelae of inflammation. Neurogenic vasodilatation reported in 1876 by Stricker and 1901 by Bayliss[922,923]; the inflammatory axon reflex with erythema observed in the 1910s by Bruce and by Breslauer[924,925]; the flare response reported by Lewis around 1930 with erythema, hyperalgesia, and edema[926]; rediscovery of the antidromic vasodilatory flare response and dorsal root reflex by Chapman et al.[927]; and Kelly's and Jancsó's more extended concept of neurogenic inflammation in the 1960s[928,929] were all expressions of the same principle: the influence of sensory afferent nerve fibers on acute inflammation and on cardinal clinical signs of inflammation. In the last three decades, our view has expanded to include the sympathetic and parasympathetic efferent nervous systems in inflammatory/immune control.

The concept of neuronal regulation of inflammation is supported by reports of patients with hemiplegia and chronic inflammatory systemic diseases, where the paralytic side is protected from inflammation (Table 13). Cases have been reported in whom hemiplegia manifested long after outbreak of chronic inflammatory disease or long before, leading to protection independent of disease onset (Table 13).

Neuronal regulation of inflammation is dependent on a robust innervation of lymphoid organs and the direct influence of neurotransmitters/neuropeptides on immune cells. While sympathetic nerve fibers usually follow arteries (some fibers also branch into vessel-free regions), sensory nerve fibers have their own routes along vessels or independent of the vasculature. In addition, nerve fibers of the parasympathetic nervous system innervate many tissues in the head, neck, and trunk of the body while upper and lower limbs are excluded. Most support for a neuroimmune contact comes from the innervation of lymphoid organs, where nerve fibers are responsible for neuronal regulation of immune responses.[588,930,931]

Receptors for neurotransmitters are present on almost all immunocompetent cells. There are some exceptions, such as the absence of β2-adrenergic receptor on T helper type 2 cells.[568] The differential and time-dependent expression of receptors can shape the neuroimmune cross talk. Sometimes, receptor expression is increased or decreased in the context of an inflammatory response.[569,570,932,933] The time-dependent involvement of different immune cells and receptor expression in the course of a given disease is probably important for neuronal regulation of inflammation.

In addition, intracellular signaling pathways of neurotransmitter receptors are dependent on environmental conditions, which has been demonstrated for

---

e. Parts of this chapter are taken from Ref. 921.

TABLE 13 Role of Neuronal Innervation for the Development of Chronic Inflammatory Systemic Diseases[921]

Situation	Modulation of Disease Symptoms	References
Poliomyelitis paralysis	RA only on the nonparalyzed side	934
Hemiplegia	RA only on the nonparalyzed side	935–947
Hemiplegia	RA vasculitis only on the nonparalyzed side	948
Hemiplegia	Gouty arthritis only on the nonparalyzed side	949
Hemiplegia	Skin changes in pSS only on the nonparalyzed side	950
Hemiplegia	Skin changes in pSS only on the nonparalyzed side	Figure 6
Hemiplegia	Psoriatic arthritis only on the nonparalyzed side	951
Sensory denervation	Respective finger is spared from psoriatic arthritis	952
Brachial plexus lesion	Shoulder inflammation in a PMR patient only on intact side	953
Hemiplegia	DTH skin lesions more marked on the nonparalyzed side	954
Hemiplegia	Hemochromatosis arthritis only on the nonparalyzed side	955

DTH, delayed-type hypersensitivity; PMR, polymyalgia rheumatica; pSS, progressive systemic sclerosis; RA, rheumatoid arthritis.

G protein-coupled receptors.[575,576] Another control process of these receptors involves regulators of G protein signaling, and TNF can lead to increased desensitization of Gα protein-coupled receptors.[956] For more details on some receptors of neurotransmitters on immune cells, the reader is referred to the literature (e.g., reviewed in Refs. 568,578).

The role of neuronal influence can be explored by examining three key phases of the inflammatory process: (A) phase 1 includes first inflammatory actions within the first 12h, (B) phase 2 describes inflammation from several hours to several days until resolution of inflammation (the normal wound-healing process), and (C) phase 3 starts with the onset of chronic inflammatory systemic disease that does not properly resolve. Thus, it is meaningful to start with an acute transient inflammatory episode such as a wound response after injection of foreign material into the skin (by the way, this is how animals are usually immunized with an autoantigen together with Freund adjuvant to develop a chronic inflammatory systemic disease).

Acute Inflammation (Phase 1: The First 12 h)[f]

Recognition of Foreign or Pathogenic Material—Immune and Pain Pathways

After the injection of foreign material into the skin, there are two categories of recognition: (1) recognition by local cells and (2) systemic recognition. These two forms of recognition are interwoven and the strength of the local response accounts for the magnitude of systemic involvement (see section "The Neuroendocrine Systemic Response"). Local recognition happens via surface receptors on local cells (e.g., pattern recognition receptors) and by secreted factors.

Systemic recognition of foreign material occurs in highly specialized nerve endings of sensory afferent, nociceptive nerve fibers (they might be called neuronal "laesioceptors"). Nerve endings of sensory afferent nerve fibers possess an impressive array of receptors that are responsible for instant signaling in the nerve fiber (Figure 21).[957,958] Upon the introduction of foreign material, infectious agents can pose a threat, which elicits a neuronal response via pattern recognition receptors on polymodal nociceptors, for example, the Toll-like receptors (Figure 21).

In addition, factors such as bradykinin, prostaglandins, and cytokines from activated mast cells and other cells can stimulate respective receptors on sensory nerve terminals (Figure 21). Considering these mechanisms, peripheral recognition of foreign material by nociceptors is part of the innate immune response. Moreover, mechanical irritation, noxious cold/heat, and low pH concentration stimulate the sensory afferent nerve fiber (Figure 21). Altogether, this leads to an orthodromic action potential that stimulates neurons in the dorsal root ganglion and releases, for example, substance P into the wounded peripheral tissue (efferent function of sensory afferents). The spreading reaction is attributed to the axon reflex and the dorsal root reflex, which contribute to antidromic activation of neighboring sensory afferents and local expansion of the immediate flare response.[957,959,960]

Substance P is one of the strongest chemotactic and vasodilatory factors, which provokes instant plasma extravasation and accumulation of neutrophils, monocytes, and other cells.[961–963] Substance P and other neuropeptides increase vascular leakage and the interstitial fluid volume in connective tissue capsules, tendons, and skeletal muscles leading to stiffness. In addition, substance P immediately stimulates activities of mast cells, monocytes, macrophages, dendritic cells, and neutrophils to reflexively increase local proinflammatory responses. Parallel to substance P, also CGRP is released that has strong vasodilatory and chemotactic activities. The third sensory neurotransmitter is the excitatory amino acid glutamate whose proinflammatory effects have been described.[964]

---

f. Parts of this chapter are taken from Ref. 921.

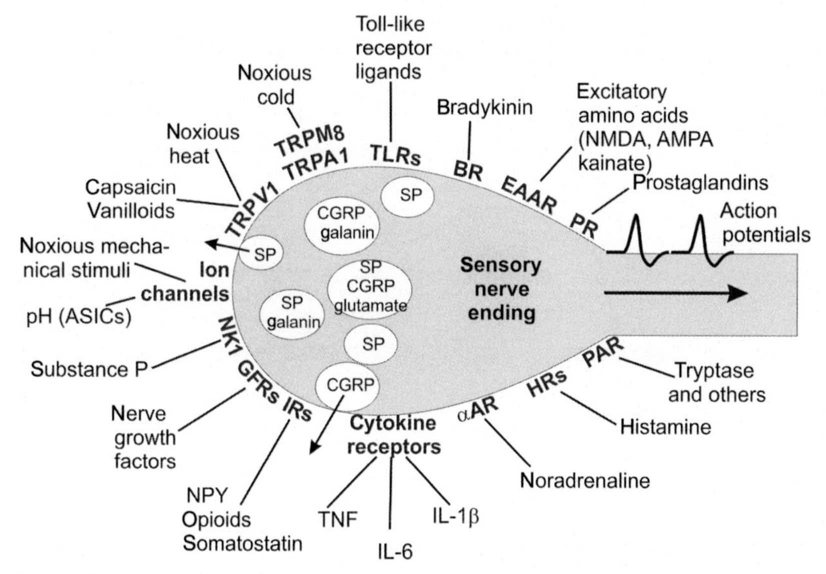

FIGURE 21 The mechanisms of a polymodal nociceptor. "Laesioceptor" might be another name because not only pain signals are recognized. The figure schematically depicts receptors on and neuropeptides of nerve fiber endings of sensory afferent nerve fibers. The list of receptors is not complete. α-AR, α-adrenergic receptors; AMPA, α-amino-3-hydroxy-5-methyl-4-isoxazolepropionic acid; ASICs, acid-sensing ion channels; BR, bradykinin receptors; CGRP, calcitonin gene-related peptide; EAARs, excitatory amino acid receptors; GFRs, growth factor receptors; HRs, histamine receptors; IL, interleukin; IRs, inhibitory receptors; NK1, neurokinin 1; NMDA, *N*-methyl-D-aspartate; PARs, proteinase-activated receptors; PRs, prostaglandin receptors; SP, substance P; TLRs; Toll-like receptors; TNF, tumor necrosis factor; TRPA1, transient receptor potential ankyrin 1; TRPV1, transient receptor potential cation channel V1.

The fourth neurotransmitter of sensory afferent nerve fibers is galanin, which possibly has dual proinflammatory and anti-inflammatory roles depending on receptor subtypes (but data are limited with respect to effects on immune cells).[965–967] All these neurotransmitters/neuropeptides are locally secreted into the vicinity of the peripheral nerve terminal.

In addition to local effects of these neurotransmitters/neuropeptides, pain signals are transmitted to the brain and elicit a systemic response (Figure 22). The pathways ascend through sensory nerve fibers (Aδ or C fibers), the neurons in the dorsal root ganglion, the neurons in the spinal medulla, and the contralateral spinothalamic tract in order to reach the medial and lateral thalamus, cortical areas S1/S2, the hippocampus, and other brain regions responsible for affective components of pain (anterior cingulate cortex, the insula, and the prefrontal cortex; Ref. 968) (Figure 22). Nearly all parts of the pain pathway can be sensitized under the influence of inflammatory stimuli. Sensitization means stabilization and amplification of nociceptive stimuli.

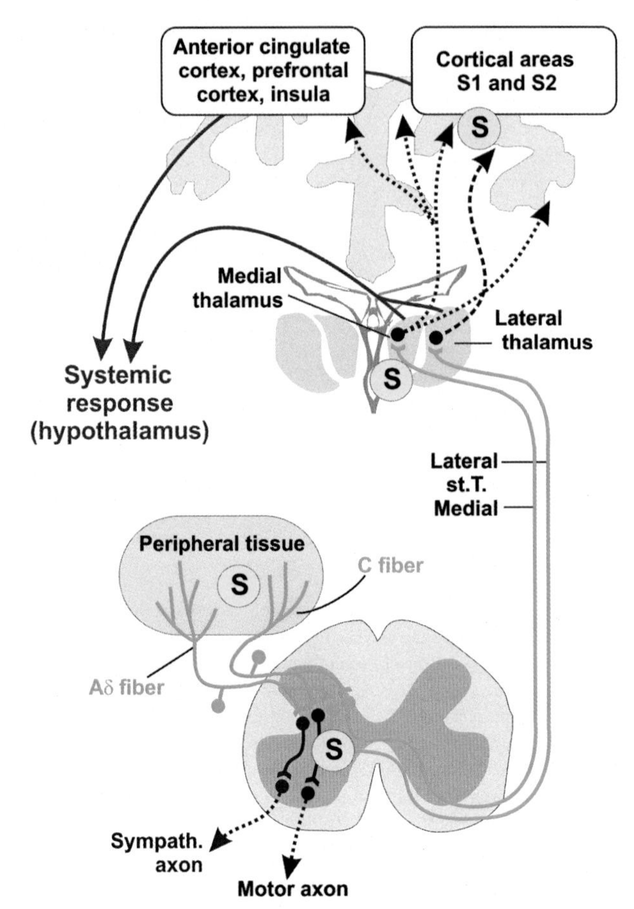

FIGURE 22 Pain pathways in the human body. Upon activation of Aδ and C fibers in the peripheral tissue, sensory afferents transmit the signals to the dorsal root ganglia and, finally, to the spinal medulla. The signal is transmitted to the thalamus and cortex via the spinothalamic tract. On all levels, sensitization of the input can happen, leading to stabilization and amplification of the pathway ("S" in yellow circular area). Interneurons transmit the signal to sympathetic efferents and motoneurons in order to induce immediate responses. This latter connection leads to a compartmentalization of the response because only site-specific and segmental sympathetic nerve fibers and somatomotor nerve fibers are involved. S, sensitization; st.T., spinothalamic tract; sympath., sympathetic.

Peripheral Sensitization

Sensitization appears during the earliest phase of inflammation as demonstrated in the kaolin/carrageenan or similar instant chemical models. Nevertheless, sensitization is a dynamic process changing over time as demonstrated by inflammation-induced induction of transient receptor potential vanilloid 1 (TRPV1) receptors on dorsal root ganglion neurons, gradual infiltration of macrophages into the dorsal root ganglion, or bilateral long-term upregulation of bradykinin receptor B2 in the dorsal root ganglion and dorsal horn.[969–973] Thus, sensitization plays a role throughout all inflammatory phases but the underlying mechanisms might change over time.

In normal tissue, nociceptors have high thresholds. However, during inflammation, these thresholds are lowered and nociceptors are sensitized.[958,974] Lowering of the nociceptor threshold is a consequence of converging stimulatory inputs into the nerve terminal via different receptor pathways (Figure 21). These high-threshold units, defined as nociceptors by their high mechanical threshold, become sensitized and start to respond to light pressure and movements in the working range of the joint (Figure 23).

Most of these units are thin myelinated (Aδ) fibers or unmyelinated (C fibers) fibers. Furthermore, mechanoinsensitive and thermoinsensitive "silent" nociceptors are sensitized in the inflamed tissue, and they start to respond to mechanical and thermal stimuli during inflammation.[958,974] This class of receptors is characterized by long-standing responses to algogenic factors, and they are important in neurogenic inflammation.[958,977] These mechanisms are summarized under the heading of peripheral sensitization (the "S" in a yellow circular area in Figure 22).

Importantly, the injection of proinflammatory cytokines such as IL-6, TNF, and IL-17A into the joint leads to a huge increase in the number of action potentials recorded from afferent fibers supplying the joint. These cytokines have the potential to sensitize afferent nerve fibers in the joint to mechanical stimulation contributing to mechanical hypersensitivity.[615,617,978] These effects can be blocked by anticytokine therapy with biologic agents,[615,978] and it is expected that this also inhibits the release of proinflammatory substance P and other neuropeptides. This is another way to stop the inflammatory and nociceptive process.

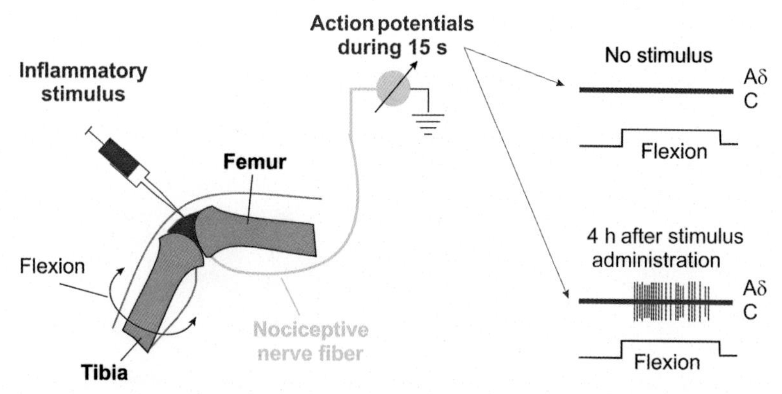

FIGURE 23 Mechanisms of peripheral sensitization. Injection of an inflammatory stimulus leads to an increased number of action potentials as recorded from afferent fibers supplying the joint.[975,976] Peripheral sensitization is mediated by a plethora of heterogeneous receptors on afferent fibers (see Figure 21).

Central Sensitization

In the dorsal root ganglion and spinal cord, peripheral inflammation makes neurons hyperexcitable and more susceptible to input from sensory nerve fibers (the "S" in Figure 22). This amplifies the response by the additional activation of adjacent and even remote spinal neurons far away from the inflamed region leading to the expansion of the receptive field.[958,974] The peripheral inflammatory response increases the expression of substance P, CGRP, and bradykinin with their respective receptors in the dorsal root ganglion and dorsal horn.[975,976] In the dorsal horn, substance P potentiates the release of factors such as glutamate and aspartate.[979] The ipsilateral response can lead to contralateral coactivation of the dorsal root ganglion and sensory afferents,[969] which might contribute to symmetrical manifestations of inflammation.[980,981] Bilateral upregulation of, for example, neurokinin 1 and bradykinin 2 receptors has been demonstrated whereby this phenomenon was strictly segmental and not general (Figure 24).[969]

Sometimes, spinal sensitization persists beyond the peripheral nociceptive or inflammatory process, and the character of pain changes from an inflammatory to a neuropathic form.[982] In experimental arthritis, such a shift from inflammatory to neuropathic features of sensitization has been demonstrated by the increased expression of a typical marker of neuropathic pain, ATF3, and by the favorable effects of gabapentin treatment in a postinflammatory phase of hypersensitivity.[982]

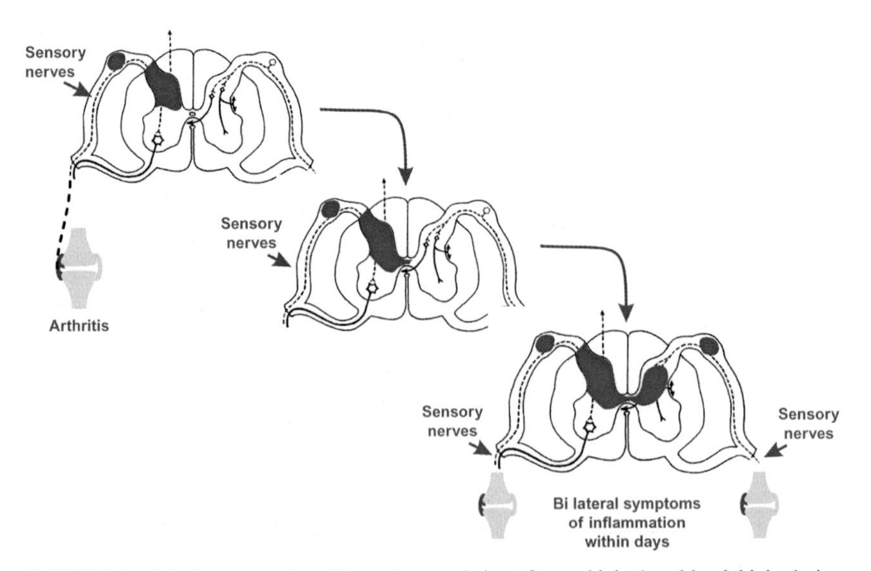

FIGURE 24 Spinal cross section. Bilateral upregulation of neurokinin 1 and bradykinin during ipsilateral inflammation might lead to contralateral sensitization. This might also constitute a risk factor for symmetry in chronic inflammatory joint diseases.

Spinal sensitization is often a consequence of an increased release of excitatory amino acids (glutamate and aspartate), substance P, CGRP, neurokinin A, and galanin from nociceptor neurons and upregulation of respective receptors in the spinal medulla. Enhanced release can be induced by peripheral inflammation.[957,958,974,983] Non–NMDA receptors but also NMDA glutamate receptors are relevant in chronic peripheral inflammation (Figure 25).[984,985]

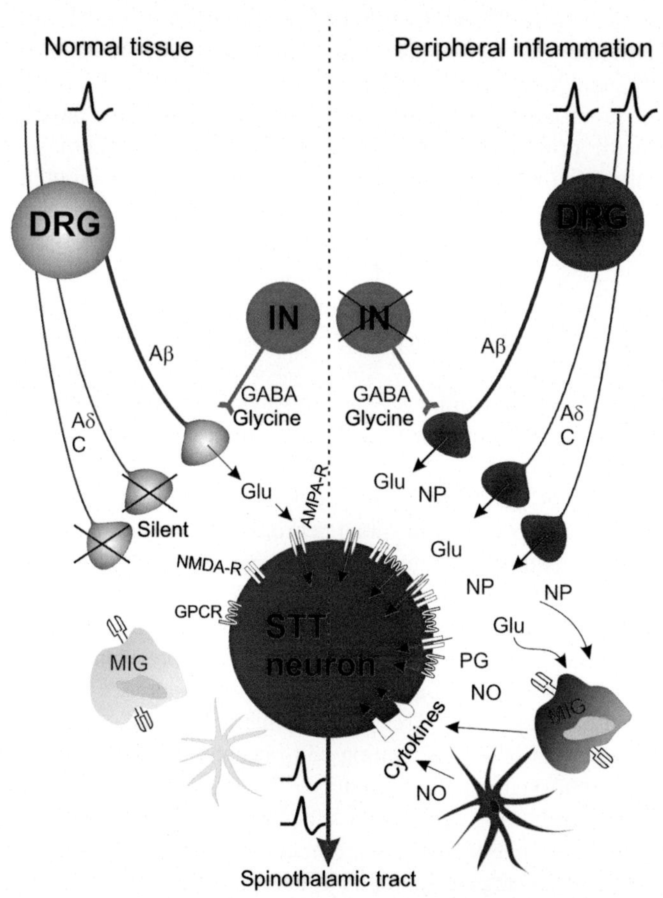

FIGURE 25 Spinal central sensitization. Left side: In the normal situation, only Aβ fibers are activated (upon mechanical stimuli), which are low-threshold nonnociceptor fibers that release glutamate (Glu). On the postsynaptic neuron, only AMPA receptors (AMPARs) are activated and opened.[975,976] Right: In the inflammatory situation, previously high-threshold Aδ and C fibers are activated by pressure and motion leading to the release of glutamate and neuropeptides (NPs) such as substance P and calcitonin gene-related peptide (CGRP). This leads to the activation of the postsynaptic membrane via AMPA receptors (AMPARs) and NMDA receptors (NMDARs), neuropeptide receptors, prostaglandin receptors, and cytokine receptors (particularly, IL-1β, IL-6, and TNF). These changes lead to long-standing hypersensitivity. AMPA, α-amino-3-hydroxy-5-methyl-4-isoxazolepropionic acid; DRG, dorsal root ganglion; GPCR, G protein-coupled receptor; IN, interneuron; MIG, microglia; NMDA, *N*-methyl-D-aspartate; NO, nitric oxide; PG, prostaglandin; STT, spinothalamic tract. The star-formed cell is an astrocyte.

Sensitization can be mimicked experimentally by intrathecal administration of substance P or NMDA via an increase of prostaglandins or cyclooxygenase-2.[986] In addition to the activating pathway, there are inhibitory pathways via, for example, γ-aminobutyric acid (GABA) or glycine.[987] Second, spinal sensitization is dependent on microglial cells and astrocytes that can aggravate pathological pain states, where cytokines and chemotactic factors play an important priming and perpetuating role (Figure 25).[988,989] Cytokines such as IL-1β, IL-6, and TNF are dominant in cytokine-induced hypersensitivity,[988,989] and these cytokines are induced in the spinal cord during experimental arthritis.[990]

Several proinflammatory intracellular signaling pathways have been implicated in the priming of microglia and pain-processing neurons. It is not easy to distinguish whether signaling cascades in neurons, microglia, or other cells are important because experimental studies using microdialysis or intrathecal administration of pathway inhibitors do not target a specific cell type. Nevertheless, these studies clearly demonstrated the importance of factors such as nuclear factor kappa-light-chain-enhancer of activated B cells (NF-κB),[991] protein kinase A,[992] protein kinase C,[993,994] c-Jun N-terminal kinase,[995] JAK/STAT3 signaling pathway,[996] p38 mitogen-activated protein kinase,[997–999] Src family kinase,[1000] and arachidonic acid pathways[1001]. These pathways typically lead to intraneuronal calcium and sodium accumulation, which is the major excitatory signal.[957,974]

For example, the p38 pathway was demonstrated to be an important proinflammatory signaling cascade in spinal neurons and microglial cells in experimental arthritis.[997] Phosphorylated p38 is increased in microglial and neuronal cells during the course of experimental arthritis. The intrathecal administration of a specific p38 inhibitor led to not only decreased synovial inflammation but also suppressed articular cytokine and protease expression as well as joint destruction measured by radiographic and histology scores.[997] This effect was dependent on the presence of TNF in the spinal cord. TNF can be a signaling element upstream of p38 by activating p38-phosphorylating kinases or downstream of phosphorylated p38 that induces TNF secretion. Importantly, it was demonstrated that intrathecal, but not subcutaneous, TNF neutralization with etanercept inhibited p38 phosphorylation and peripheral inflammation.[997] The positive effect of intrathecal spinal TNF neutralization was confirmed in another model of inflammation.[616] In this model, not only was peripheral inflammation favorably influenced, but also pain-related behavior was drastically reduced.[616]

It is interesting that all mentioned pathways belong to proinflammatory cascades initially described in peripheral immune cells. In contrast, anti-inflammatory pathways such as the adenosine A1, the β-adrenergic, and δ-/μ-opioidergic receptor pathways are inhibitory in microglia activation paradigms.[1002–1006] For instance, spinal administration of an A1 adenosine receptor agonist markedly reduced inflammation and bone and cartilage destruction in an experimental arthritis model.[1002,1003] The administration of the A1 adenosine receptor agonist also decreased nuclear c-Fos expression in the superficial and deep dorsal horns of the spinal medulla. In addition, the A1 adenosine receptor agonist decreased the density of astrocytes in these areas.[1003] This indicates that next to neurons and microglial cells, also astrocytes are involved in sensitization.

Finally, it is an important question why peripheral and central sensitization was positively selected during evolution. Sensitization of pain has a protective role because it warns about potential danger, enables us to remove noxious stimuli (with attached microbes), and stimulates wound management. Furthermore, avoidance of painful situations in the future would be desirable. This is nicely indicated by the fact that peripheral inflammatory stimulation of sensory neurons can induce central IL-1β release in the hippocampus,[1007] a cytokine that is instrumental in hippocampal learning phenomena.[1008] Sensitization is an amplification factor, which should last as long as painful/noxious stimuli are present (or even a little longer to stimulate wound management). Thus, sensitization is a supportive factor of innate immunity. It is hypothesized that it has been positively selected as a learning phenomenon, which will not stop until inflammation is terminated (the stimulus is removed).

The Neuroendocrine Systemic Response

Some aspects have already been discussed in section "Physiological Basis". Parallel to the local inflammatory reaction, hormonal, and neuronal systemic responses are engaged. The hormonal response system is mainly the HPA axis, which is stimulated through the activation of sensory pathways, through circulating cytokines, or through circulating cells.[611] Activation of the HPA axis can happen on adrenal, pituitary, and hypothalamic levels.[611] Neuronal efferent response systems are the sympathetic nervous system and the parasympathetic nervous system. Other endocrine response systems are described in the sections above.

In the course of inflammation, the sympathetic nervous system is activated via direct spinal interneurons that link sensory inputs to sympathetic output (Figure 22, blue lines). This has the advantage that the input defines the location of the output, which leads to confinement of the response to the affected area. In addition to the central coupling of sensory and sympathetic pathways, sympathetic nerve fibers communicate with sensory nerve terminals by way of α2-adrenergic and prostaglandin cross signaling at the level of the peripheral nerve terminal.[1009,1010] This inflammation-induced cross signaling leads to higher activity of sensory afferents. Systemically, the sympathetic nervous system like the HPA axis is activated through the stimulation of sensory pathways or through circulating cytokines (e.g., IL-6).

While systemically relevant inflammation is coupled to an increased sympathetic nervous tone and an increased activity of the HPA axis (albeit inadequately low in relation to inflammation = disproportion principle; Figure 10) (Figure 7), the activity of the parasympathetic nervous system and the HPG axis is inhibited.[631,1011] This leads to the well-known dissociation of sympathetic versus parasympathetic and HPA axis versus HPG axis activity (androgens are low), respectively.

The activation of the HPA axis and the sympathetic nervous system at the onset of inflammation prepares the immune system for most naturally occurring immune challenges (reviewed in Ref. 1012). Activation of the HPA axis mobilizes immune cells leading to redistribution of neutrophils, monocytes, and NK cells (reviewed in Ref. 1012). The sympathetic nervous system can support the very acute inflammatory process in phase 1 because of six main mechanisms: (1) mobilization of immune cells from systemic stores (similar to the HPA axis),[1012] (2) support of plasma extravasation,[1013] (3) remodeling of tissue by inducing matrix metalloproteinases,[1014,1015] (4) stimulation of nociceptors via α2-adrenergic and prostaglandin cross signaling,[1009,1010] (5) chemoattractant activity of sympathetic neurotransmitters (e.g., Ref. 1016), and (6) liberation of free fatty acids and glucose necessary for the activated immune system (requirement of energy-rich fuels; see Chapter III). In summary, during the first hours of inflammation, the HPA axis and the sympathetic nervous system are mainly proinflammatory (in phase 1).

Vagal afferents from the intestine and liver play an important role in modulating a systemic milieu that increases or decreases the magnitude of very acute inflammatory hyperalgesia, which depends on the agent, the stimulus strength, and the adrenaline secretion from the adrenal medulla (reviewed in Refs. 1017,1018). The vagal nervous tone determines the overall reflex modulation of very acute inflammatory processes, and vagal afferents are important in perception of inflammatory conditions in the abdomen.[612,613] Reports in the last decade demonstrated that a lipopolysaccharide-induced inflammation can be inhibited by electrical vagus stimulation of the distal end of the dissected vagus nerve.[1019] These very acute vagal effects were dependent on the sympathetic innervation of the spleen.[1020] In addition, carrageenan-induced leukocyte recruitment into a preformed subcutaneous air pouch was inhibited by vagus nerve stimulation of the intact vagus nerve.[1021] This was done without dissection of the vagus nerve so that afferent and efferent vagus nerve effects cannot be separated.[1021]

Noteworthy, the administration of an intrathecal p38 MAP kinase inhibitor (mentioned above), which has favorable effects in experimental arthritis, largely increases vagal activity.[1022] Since spinal application of p38 MAP kinase inhibitors blocks aspects of central sensitization,[997] one would expect also blockade of segmental sympathetic outflow as demonstrated in Figure 22 (blue lines). A decrease of central pain signaling, thus decreased hypothalamic activation of HPA axis and sympathetic nervous system, and diminution of segmental sympathetic outflow most probably increase parasympathetic reflex activity. Particularly, in very early inflammation, this should be a favorable anti-inflammatory feature. These acute vagus experiments were complemented by experiments in longstanding chronic inflammation models (see below phase 3).

Intermediate Inflammation (Phase 2: Between 12h and Several Days/Few Weeks)[g]

Introduction

Almost all experimental systems dissecting neuroinflammatory pathways have been performed under very acute inflammation conditions (within minutes to 12h reflecting an experimental working day; phase 1). Much less information is available after 12h until termination of uncomplicated inflammation with a normal wound healing. Within the mentioned time span, many additional immune/inflammatory responses appear such as increase of local cell accumulation, antigen transport to secondary lymphoid organs, antigen processing, clonal expansion of lymphocytes, release of lymphocytes from secondary lymphoid organs, and access of antigen-specific cells into the target tissue.

In this phase of inflammation, mechanisms discussed in the very early phases can still apply. However, they might change over time because tissue innervation is altered and many more players are involved. Immune/inflammatory effector responses relevant in this phase are modulated by neurotransmitters, and these responses can differ from very acute inflammation.

Local Cell Accumulation in Inflamed Tissue

Local cells accumulate in inflamed tissue as a consequence of cell mobilization and chemotaxis. The major neurotransmitters/neuropeptides of sensory afferents (substance P) and of sympathetic efferents (noradrenaline) are potent chemotactic factors for innate immune cells such as neutrophils, monocytes, and eosinophils. The direct chemotactic effect of substance P has been demonstrated by injecting substance P into the skin, which leads to the upregulation of the endothelial adhesion molecule E-selectin (CD62E) and, for example, attraction of eosinophils to the injection site.[1023] Similarly for the sympathetic nervous system, the lack of catecholamine production in animals with a deletion of the dopamine β-hydroxylase gene leads to a strong reduction of leukocyte accumulation in the adventitia/periadventitia of vessels.[1024]

Substance P and noradrenaline also have strong chemotactic effects *in vitro*.[1016,1025] Similarly, the sympathetic cotransmitter neuropeptide Y and the sensory cotransmitter CGRP have chemotactic effects, too.[1016,1026] The effects of substance P and noradrenaline can be amplified by increasing secretion of potent chemotactic factors such as IL-8.[1027,1028] In addition, noradrenaline and substance P can upregulate matrix metalloproteinases to soften the tissue.[1014,1015,1029] Thus, the major neurotransmitters of the two neuronal systems induce an immediate response that supports innate immunity and chemotaxis of leukocytes.

---

g. Parts of this chapter are taken from Ref. 921.

The Immediate Change of Neuronal Innervation

Upon entering of monocytes and neutrophils into the tissue, these cells become activated and can engulf pathogens or foreign material. Activation of these cells is mediated by pattern recognition receptors and other proinflammatory mediators and inflammasome-derived products leading to activation of neutrophils and differentiation of entering monocytes into macrophages or inflammatory dendritic cells.[124]

In striking contrast, noradrenaline and its potent cotransmitter adenosine (made from sympathetically released ATP) inhibit many proinflammatory effects of activated innate immune cells such as monocytes, macrophages, NK cells, and neutrophils via β-adrenergic and A2 adenosine receptor signaling (summarized in Ref. 1030). In this context, it is important to mention that ectonucleotidases (CD39 and CD73), which convert purine precursor neurotransmitters such as ATP to adenosine, are increased in inflammation.[1031–1033] A classic example of A2-adenosine-mediated or β-adrenergically induced inhibition of cells is the strong negative effect on neutrophil or monocyte/macrophage phagocytosis and on the function of dendritic cells. The important role of adenosine as an inducer of T regulatory cells has been carefully documented.[1034] Since, under healthy conditions, there is a balanced innervation of the tissue with similar densities of sensory and sympathetic nerve fibers, a too strong influence of sympathetic neurotransmitters would be counterproductive for innate immunity. The question appears whether sympathetic influence can be specifically decreased while substance P influence can be increased and whether it is accomplished through innervation density of nerve fibers.

Activated macrophages and stimulated tissue fibroblasts start to produce nerve growth factor (NGF), which supports the outgrowth of sensory and sympathetic nerve fibers equally well. In other words, NGF is not specific for sensory or sympathetic nerve fibers. Indeed, inflammatory tissue releases large amounts of NGF as, for example, substantiated in rheumatoid arthritis or in experimental arthritis.[1035,1036] And one might hypothesize that the licking response to wounding, during which large amounts of salivary NGF enters the wound area, is a neurotrophin signal for nerve fiber growth (recall the work of the Nobel Laureate Rita Levi-Montalcini).

Other guiding factors are nerve fiber-repellent factors. Activated macrophages and fibroblasts also produce nerve repellent factors such as semaphorin 3C and semaphorin 3F.[1037,1038] These two factors specifically repel sympathetic nerve fibers and have no effect on sensory nerve fibers.[1037,1038] In contrast to sympathetic nerve fibers, sensory nerve fibers sprout under the influence of NGF into inflamed tissue leading to a preponderance of substance P over sympathetic neurotransmitters.[1039] Such a sensory hyperinnervation is also observed in skin wounds, while sympathetic nerve fibers are absent.[1040] The loss of sympathetic nerve fibers is a rapid process that is observed soon after initiation of experimental inflammation.[1041–1043] It can also be observed *in vitro* using repellent factors in neurite outgrowth assays (within few hours).[1038]

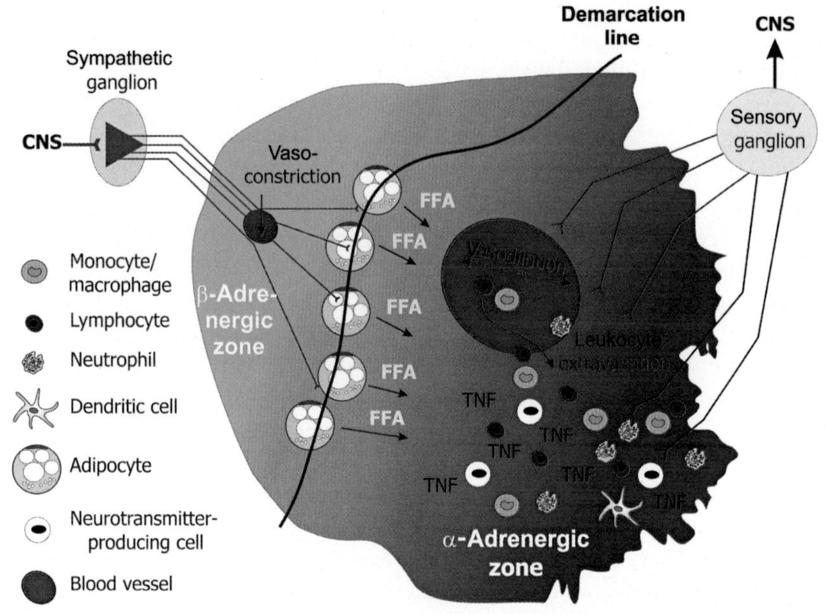

FIGURE 26 Loss of sympathetic nerve fibers and sprouting of sensory nerve fibers into inflamed tissue. Hypothetical model: loss of sympathetic nerve fibers leads to the generation of two quite distinct noradrenergic zones. In a zone with low concentrations of neurotransmitters (the red α-adrenergic zone), only α-adrenergic effects are possible due to the affinity of noradrenaline to the two receptor subtypes (high for α and low for β). However, in the vicinity of sympathetic nerve terminals, α- and β-adrenergic effects can be expected (green β-adrenergic zone). Sympathetic nerve fibers on the healthy site of the demarcation line support β-adrenergic mechanisms such as the release of free fatty acids from adjoining fat tissue, whereas on the other site of the demarcation line, noradrenaline supports proinflammatory α-adrenergic signaling and pain induction via α2-adrenergic receptors on nerve terminals of nociceptive neurons. In parallel, sensory nerve fibers sprout into inflamed tissue leading to a dissociation of innervation between sympathetic and sensory nerve fibers. The proinflammatory milieu becomes stabilized (this creates zones of "permitted local inflammation"). The dissociation is a consequence of specific nerve fiber repulsion of sympathetic but not sensory nerve fibers. In the symptomatic phase of the disease, neurotransmitter-producing cells appear, whose anti-inflammatory capacities are too small to overcome inflammation. CNS, central nervous system; FFA, free fatty acids; TNF, tumor necrosis factor.

Repulsion of sympathetic nerve fibers and sprouting of sensory nerve fibers can be crucial in creating a proinflammatory environment in the intermediate phase of inflammation. As shown in Figure 26, the appearance of two noradrenergic zones (β, normal/healthy; α, inflamed tissue) can be a consequence of this process.

It is important to mention that loss of sympathetic nerve fibers is observed not only in inflamed tissue but also in the spleen[1041,1042,1044] and in the lymph nodes. In the former, the loss of sympathetic nerve fibers appears in the white pulp (T-cell proliferation area), and, similarly, these fibers are not observed in B-cell follicles.[585,1041] In the same animals, sympathetic nerve fibers sprout into the hilus area of the spleen and do not reach the distal white pulp or follicles. Thus, a proinflammatory milieu is established in secondary lymphoid organs similar to peripherally inflamed tissue.

The Role of Neurotransmitters in Antigen Transport to Secondary Lymphoid Organs and Immune Response

After antigen capture, a further important aspect of inflammation is the transport of processed antigenic material to secondary lymphoid organs. Transport to lymphoid organs is mediated by lymphatic vessels, whose pumping efficiency is decreased by β-adrenergic pathways and stimulated by α-adrenergic signaling.[1045,1046] In addition, migration of antigen-loaded dendritic cells is stimulated via α1-adrenergic mechanisms.[1047] Immature dendritic cells migrate upon α1-adrenergic influence, but CD40-stimulated mature dendritic cells do not (those that arrived in secondary lymphoid organs and encountered a T-cell contact via CD40-CD40 ligand).[1047] Thus, the rapid establishment of an α-adrenergic zone in the peripheral tissue probably is a prerequisite for migration of dendritic cells. In addition, substance P supports dendritic cell maturation and activity.[1048,1049]

Catecholamines and its cotransmitter adenosine influence the direction of the immune response, whether T helper type 1, T helper type 2, or others. At the beginning of neuroimmune research, the concept of CD4-positive T helper type 1 and T helper type 2 was in the focus. Detailed *in vitro* experiments demonstrated that noradrenaline via β-adrenergic pathways inhibits T helper type 1 cell priming by inhibiting IL-12 and IFN-γ and stimulating IL-10 of dendritic cells.[578,1050] Many *in vitro* and *in vivo* studies indicate that noradrenaline supports the T helper type 2 cell immune response and the B-cell response, while it suppresses the T helper type 1 cell response (reviewed in Ref. 682). Substance P, on the other hand, does not demonstrate such a dichotomy. The last decade added the important IL-17A-dependent T helper type 17 pathway and T regulatory cells (earlier called suppressor T cells). The effects of neurotransmitter/neuropeptide influence on these cells are demonstrated in Table 14.

The picture for Th17 modulation is not always clear-cut, which might depend on the way of IL-17 induction. Not only is IL-17 induced by IL-6 and TGF-β, but also it can be stimulated by TNF and IL-1β. Neuroendocrine signals that increase cyclic AMP also increase IL-6 but, particularly, TGF-β,[1051] but the same pathways inhibit TNF. Thus, the presence of TNF is very critical in determining an inhibitory or stimulatory effect via cyclic AMP pathways as demonstrated in bacterial infections.[1052] A general stimulatory picture appeared for substance P and CGRP.[1053,1054] An ambivalent situation was found for typical modulators of cyclic AMP such as β2-adrenergic agents, adrenaline and noradrenaline, adenosine, and dopamine (Table 14). Thus, the concept of the β- to α-adrenergic shift and sensory hyperinnervation described in Figure 26 may largely influence the direction of the local T-cell immune response. Recently, it was demonstrated that the removal of the sympathetic nervous system reduces the IL-17 response in draining lymph nodes.[1055] Removal of the α-adrenergic zone described in Figure 26 may be involved in IL-17 reduction in this model of inflammation.

TABLE 14 Neuroimmune Modulation of T Helper Type 17 Cells and T Regulatory Cells

Described Effect	References
T helper type 17 cells	
Substance P stimulates the Th17 phenotype	1056,1057
Calcitonin gene-related peptide increases IL-17 production[a]	1054
A β2-adrenergic agonist increases Th17 relative to Th1[a]	1058
Dopamine via D5 receptors potentiates Th17-mediated immunity[a]	1059
Dopamine via D1 receptors stimulates IL-6 dependent IL-17 production[a]	1060
Adrenaline primes dendritic cells to facilitate not only IL-17A but also IL-4 production[a]	1061
Removal of the sympathetic nervous system reduces IL-17A	1055,1062
Adrenaline and noradrenaline inhibit LPS-stimulated IL-17 at high concentrations (β-adrenergic)[a] (LPS induces TNF)	1063
Adenosine via A_{2A} receptors inhibits Th17 lymphocyte generation[a]	1064
Dopamine via D1 receptors inhibits Th17 functions[a]	1065
T regulatory cells	
Tolerogenic effects of adenosine have been described	1034
Dopamine through D1 dopamine receptors inhibits T regulatory cell function	1066
The sympathetic nervous system has no influence on the numbers of T regulatory cells	1062
The sympathetic nervous system increases T regulatory cells	1067

D, dopamine; IL, interleukin; LPS, lipopolysaccharide; TGF-β, transforming growth factor-β; Th17, T helper type 17; TNF, tumor necrosis factor.
[a]*An increase of cyclic AMP is expected with this neuroimmune factor. This can increase IL-6 and TGF-β, two stimulators of IL-17A. However, the same neuroendocrine factors can inhibit TNF via cyclic AMP, which in turn can lead to inhibition of TNF-induced IL-6, TGF-β, and IL-17A. Thus, the presence of TNF is decisive for up- or downregulation of IL-17 pathways.*

With respect to T regulatory cell pathways, dopamine via D1 dopamine receptors were demonstrated to inhibit T regulatory cell function (Table 14). The exact role of the sympathetic nervous system on T regulatory cells is presently not known because of controversial findings (Table 14).

In addition to reactions on T cells, noradrenaline inhibits antigen presentation by epidermal Langerhans cells, which is β-adrenergically mediated.[1068] Already in the late 1980s, it was demonstrated that surface expression of the antigen-presenting molecule HLA class II was inhibited by β-adrenergic signaling.[1069,1070]

In summary, the sympathetic nervous system has many inhibitory roles via β2-adrenergic and A2-adenosine receptors when T helper type 1 cell priming participates (e.g., in arthritis). The opposite occurs in a situation with T helper type 2 conditions since noradrenaline stimulates IL-4 and IL-10, along with many stimulating effects on B cells and antibody production (e.g., in systemic lupus erythematosus).[578] Since noradrenaline and adenosine have strong inhibitory effects on secretion of TNF, IFN-γ, and IL-2 via β-adrenergic and A2 adenosine pathways, the general inhibitory aspect of these neurotransmitters at high concentrations, along with VIP and CGRP, is well established.[701,1071]

At low concentrations of noradrenaline, when α1/2-adrenergic signaling is dominant, there are even stimulating effects on TNF and IL-17.[579,1072] When a β-adrenergic influence in the peripheral tissue and in secondary lymphoid organs is reduced due to a loss of sympathetic nerve fibers (Figure 26), a proinflammatory T helper type 1 or type 17 cell priming may prevail. Such a dichotomy does not exist for proinflammatory substance P.

Clonal Expansion of T and B Cells

The antiproliferative effect on T cells of noradrenaline via β-adrenergic receptors has been documented by many investigators.[578,682] The proliferative response of CD8+ T cells is inhibited to a greater extent than CD4+ T cells, presumably because CD8+ T cells have a higher number of β-adrenergic receptors, and this effect is mediated via the inhibition of IL-2 secretion (reviewed in Ref. 682). A proliferative effect of noradrenaline via β-adrenergic receptors is known for B cells.[578,1073,1074] Similarly, the proliferative effect of substance P on T and B cells is common knowledge. The supportive effect of noradrenaline on antibody production has been demonstrated many times.[578,1073,1075] These dichotomous effects of noradrenaline shape the immune response induced by T helper priming antigens or autoantigens. Catecholamine effects depend on the stage of T- or B-cell activation because naive cells are influenced in different ways as compared with mature antigen-selected cells.[1073] Thus, the timing of the neurotransmitter influence is mandatory.

Resolution of Inflammation and Tissue Repair

Upon clearance of a pathogen, the resolution of inflammation or the reconstitution of normal tissue is the final step. Inflammation often leads to a preponderance of sensory nerve fibers (sensory hyperinnervation) over sympathetic nerve fibers that are reduced in inflamed areas. In acute wounds, both nerve fibers disappear, but the reappearance of sensory nerve fibers seems to start earlier than reinnervation with sympathetic nerves (Table 15).[964,1040,1041,1053,1076–1086]

In general, reinstallation of sympathetic nerve fibers is a very long process as substantiated in transplanted organs (>4 weeks), after tibial nerve crush (8-12 weeks), after chemical sympathectomy in the spleen (3-8 weeks), and after monophasic arthritis (4-8 weeks).[1087–1090]

TABLE 15 Behavior of Nerve Fibers and Their Neurotransmitters/Neuropeptides in Wound Reactions

Nerve Fiber Type	Change During Wound Reactions	References
Sensory nerve fibers	Sensory nerve fibers are lost after 2 days but reappear after approximately 7-14 days	1093–1095
	Substance P and CGRP promote wound healing	1096–1100
Sympathetic nerve fibers	Fast loss of sympathetic nerve fibers and reappearance after approximately 14 days	1101
	Catecholamines block wound repair via β-adrenergic receptors	1102,1103
	Noradrenaline inhibits wound macrophages and neutrophils	1104,1105
	Catecholamines support later reepithelialization	1106–1109
	Adenosine supports the wound-healing response via A2 receptors (mediated through increase of fibrosis)	1110,1111

Note: Experiments with 6-hydroxydopamine, the sympathetic nerve fiber toxic substance, are not included because this substance affects not only sympathetic nerve fibers.
CGRP, calcitonin gene-related peptide.

In a typical wound reaction, substance P promotes wound-healing responses while catecholamines have negative and positive effects on wound healing such as the inhibition of wound macrophages/neutrophils but support of later reepithelialization (Table 15). Moreover, stressful events that release sympathetic neurotransmitters and glucocorticoids lead to wound-healing problems.[1091,1092] From this point of view, a preponderance of substance P-positive nerve fibers over sympathetic nerve fibers would be favorable for wound healing. Sensory hyperinnervation is probably supportive.

Chronic Inflammatory Systemic Disease (Phase 3)[h]

Neuronal Influences on Chronic Inflammatory Disease in Animal Models

Chronic inflammatory disease occurs when inflammation fails to resolve and tissue repair is inadequate. The neuronal elements can contribute to this process.

h. Parts of this chapter are taken from Ref. 921.

Most studies investigated the role of the sympathetic nervous system in the adjuvant arthritis model in Lewis rats.

The proinflammatory role of substance P and sensory nerve fibers in this model was demonstrated early in the 1980s.[1053] Additionally, substance P is proinflammatory for human synoviocytes and monocytes.[1112] Nociceptive fibers in the draining dorsal lymph nodes must play a critical role during this induction phase because local capsaicin treatment of these lymph nodes markedly decreased disease severity.[1113] Although the proinflammatory effects of substance P and other tachykinins are widely known, substance P antagonist therapies were not effective, which is probably due to the redundancy in the tachykinin system.

In experiments to study the role of the sympathetic nervous system in adjuvant induced arthritis, most studies focused on a time period of 14-40 days. From these studies, it is evident that the overall peripheral sympathectomy or blockade of adrenergic receptors (particularly via β-adrenergic receptors) prior to or at injection of Freund adjuvant diminishes the severity of joint inflammation during the entire observation period of 40 days (e.g., Refs. 1114,1115). Similarly, the sympathetic nervous system plays an aggravating role in the collagen-induced arthritis (CIA) model, which is an autoantigen-driven chronic inflammatory disease. This might result from increased CD4+ CD25+ FoxP3-negative T cells as recently demonstrated.[1062] When the sympathetic nervous system was destroyed before immunization and up to day 18 after immunization, sympathectomy markedly reduced the severity of arthritis.[1062,1116] However, when the sympathetic nervous system was destroyed after outbreak of the disease, sympathectomy strongly aggravated arthritis.[1116] In the monoarthritis model, sympathectomy prior to antigen injection also ameliorated the disease.[1055]

The surprising dual role of the sympathetic nervous system might be explained as follows (hypothesis). In the very early induction phase after injection of adjuvant/antigen, mobilization, migration, and chemotaxis of proinflammatory cells such as neutrophils, NK cells, and monocytes directed to the site of adjuvant/antigen injection play a decisive role. In addition, targeted repulsion of sympathetic nerves in secondary lymphoid organs supports antigen presentation and the switch to an aggressive effector immune response via α-adrenergic pathways.[1115] Under conditions with sympathectomy in secondary lymphoid organs, due to a dominant α-adrenergic signaling, arthritis gets more severe due to improved antigen presentation, stronger T helper type 1 and type 17 immune reactions (aggressive phenotype for tissue-specific autoantigens), and probably downregulation of several regulatory elements such as IL-4 and IL-10 (see intermediate inflammation above).[1115] Since the effect of prior sympathectomy is long-lasting, these initial events are important for the later inflammatory disease, which has also been demonstrated in atopic dermatitis and experimental colitis.[1117,1118] However, in the chronic phase of the disease, the influence of the sympathetic nervous system largely changes.

One of the main changes is the loss of sympathetic nerve fibers in inflamed tissue and in secondary lymphoid organs as already discussed above (Figure 26). Nerve fiber loss, starting with the onset of disease,[1041,1042,1119] turns the essentially anti-inflammatory influence of sympathetic neurotransmitters at high concentrations into a proinflammatory one at low concentrations (Figure 26). In addition, recently described tyrosine hydroxylase-positive cells with anti-inflammatory capacities appear in lymphoid organs and arthritic tissue.[606,607,1120] In chronic inflammatory disease, the number of these cells in secondary lymphoid organs increases over time, and they appear shortly after disease outbreak in the inflamed joint.[607,1121] Since these cells can be eliminated by 6-hydroxydopamine treatment (the classical sympathectomy technique),[1121,1122] the anti-inflammatory influence of these cells is soon lost after experimental chemical sympathectomy leading to a proinflammatory aggravation.[1116] The loss of tyrosine hydroxylase-positive cells probably leads to an overall proinflammatory situation because these cells might have tolerogenic activities.[604] The quite different effects of this sympathectomy tool are now explained by an early destruction of sympathetic nerve fibers, which are proinflammatory (acute inflammation), and a later destruction of anti-inflammatory catecholamine-producing cells in the chronic inflammatory disease, which are anti-inflammatory.[1122]

Finally, the influence of the parasympathetic nervous system has attracted increasing interest. The α7 subunit of the nicotinic acetylcholine receptor is especially relevant in the regulation of inflammatory responses,[1123] which led to the concept of the cholinergic anti-inflammatory reflex. Further experiments with agonists of the α7 subunit of the nicotinic acetylcholine receptor demonstrated the anti-inflammatory importance of this cellular pathway in animal experiments and human cells.[1124–1128] This particular nicotinic receptor is highly expressed on macrophages and fibroblasts of patients with rheumatoid arthritis.[1127,1129]

At the moment, it is unclear how the vagus nerve influences the synovial inflammatory disease. Four different possibilities exist how favorable cholinergic effects on joint inflammation can be explained: (1) the cholinergic influence supports sympathetic inhibition of splenic proinflammatory immune responses, (2) the cholinergic influence directly affects cells in draining lymph nodes of the trunk, (3) the cholinergic influence affects cells in the gut (which play an important role as substantiated in the HLA-B27 rat model of arthritis and colitis), and (4) nonneuronal acetylcholine release appears within inflamed synovial tissue.[698,1042,1130]

Neuronal Influence on Endothelial Cells and Neoangiogenesis

Angiogenic effects of sympathetic neurotransmitters have been demonstrated in tumor models and in tumor cells. Since tumor cells often demonstrate quite different signaling pathways, generalizability of these findings for different forms of inflammation might be critical. Nevertheless, studies in this field are the only source of information. Noradrenaline has been implicated in angiogenesis by inducing vascular endothelial growth factor expression in tumor cells via β-adrenergic effects while direct trophic effects on endothelial cells were demonstrated via α1-adrenergic effects (reviewed in Ref. 1131). These effects were potentiated by hypoxia. In addition, the sympathetic cotransmitter neuropeptide Y (neuropeptide Y) has angiogenic activities shown in animals deficient of the neuropeptide Y receptor type 2. Dopamine via D2 receptors has universal inhibiting effects on angiogenesis (reviewed in Ref. 1131).

As long as sympathetic nerve fibers are present, α- and β-adrenergic and neuropeptide Y effects are possible. When nerve fibers are lost, only in the border areas along the demarcation line in Figure 26 with still existing normal innervation, these typical effects of sympathetic neurotransmitters/neuropeptides can be expected, because tissue concentration would be high enough. In an area of lost sympathetic nerve fibers and replacement by catecholamine-producing cells, mainly α-adrenergic and dopaminergic effects predominate because the concentration of noradrenaline is low and dopamine is an important neurotransmitter produced.[607,1132,1133] Thus, angiogenesis might be supported by sympathetic neurotransmitters in the healthy border zone of inflammation but not in the middle of the inflammatory zone. Nevertheless, in the sympathetic α-adrenergic zone, sensory hyperinnervation exists so that higher levels of substance P can be expected.

Substance P was demonstrated to support capillary growth *in vivo* in a rabbit cornea model and in a rat sponge assay via NK1 receptors. In addition, substance P stimulates the proliferation of different endothelial cell types (reviewed in Ref. 1134). *In vivo* experiments showed that endogenous substance P could be implicated in neoangiogenesis connected with inflammation (reviewed in Ref. 1134). Thus, the two neurotransmitter systems of catecholamines/neuropeptide Y and substance P probably influence angiogenesis in inflammation.

Neuronal Influences on Fibroblasts and Adipocytes

Sympathetic neurotransmitters modulate the function of fibroblasts by inducing proliferation, collagen gene and protein expression, and fibroblast migration via α1-adrenergic receptors.[1135–1138] In contrast, fibroblast undergo increased apoptosis and autophagy via β-adrenergic signaling.[1139,1140] In addition, noradrenaline induces secretion of IL-6, IL-8, and matrix metalloproteinase 2 from fibroblasts via β-adrenergic receptors.[1141–1145] Under conditions with an α-adrenergic zone due to nerve fiber loss (Figure 26), one can expect proliferative responses of sympathetic nerve fibers on fibroblasts. This can support the fibrotic process in chronic inflammatory lesions.

The proliferative effects of substance P on fibroblasts are well documented. Substance P supports the growth-promoting effect of IL-1 in cultured fibroblasts.[1146] These substance P effects were mediated through NK1 receptors.[1147] In addition, substance P induces migration of human fibroblasts *in vitro*.[1148] While substance P demonstrates the proliferative effect on fibroblasts, CGRP has no similar role but it stimulates fibroblast IL-6 secretion.[1149,1150] In conclusion, the establishment of an α-adrenergic zone together with sensory hyperinnervation induces a proliferative effect that can lead to scarring reactions typical in chronic inflammation.

In recent years, adipocytes in the proximity of inflammatory lesions have gained enormous interest due to their proinflammatory activities.[463,1151] These fat cells might be important targets of neuronal influences in order to support the inflammatory process. Indeed, it is well known that the sympathetic nervous system via β-adrenergic pathways is instrumental to release energy-rich free fatty acids that are used by different immune cells as energetic substrates (recently reviewed in the context of chronic inflammation in Ref. 1152). Fat tissue in the proximity of inflamed lesions is perfectly innervated by sympathetic nerve fibers. Thus, noradrenaline is present in adequate amounts to stimulate lipolysis via β-adrenergic receptors (Figure 26). Importantly, α2-adrenergic stimulation leads to an inhibition of lipolysis (reviewed in Ref. 1153). Thus, the balance between the lipolysis-promoting β-adrenergic receptor activation and lipolysis-inhibiting α-adrenergic receptor activation dictates the degree of lipolytic activity and provision of energy-rich free fatty acids to consumers. In addition, the sensory neuropeptide substance P supports proliferation and anti-apoptotic pathways in fat depots, which might contribute to the development of inflammatory fat accumulation demonstrated *in vitro* and *in vivo*.[1154,1155]

Neuronal Influences on Osteoclasts and Osteoblasts

Neurotransmitters of the sympathetic nervous system (catecholamines) and neuropeptides (VIP, substance P, and CGRP) play a role in normal bone homeostasis. We have to distinguish the different phases of osteoblast and osteoclast differentiation because the influence of neurotransmitters/neuropeptides depends on the differentiation stage. Catecholamines inhibit bone formation via β2-adrenergic receptors by stimulating osteoclast differentiation and by inhibiting osteoblast function.[1156–1158] This can be opposed by agonists of α-adrenergic receptors.[1159,1160] Since noradrenaline has a higher binding affinity for α-adrenergic receptors than β-adrenergic receptors, the local concentration of this neurotransmitter most probably determines the effect (low concentrations can support bone formation; high concentrations inhibit bone formation).

In addition, VIP and CGRP support bone formation by inhibiting osteoclastogenesis and by stimulating osteoblasts (reviewed in Refs. 1161,1162), which is different for substance P that preferentially stimulates osteoclasts and, thus, bone loss.[1163,1164] All these mechanisms have been detected in cells of healthy subjects or normal animals and/or in cell lines. No such studies have been carried out in primary osteoblasts or osteoclasts from inflamed animals or patients with chronic inflammatory systemic diseases. This must be subject of future studies because in chronic inflammation, the situation might be quite different.

Changes of the Nervous System in Patients with Chronic Inflammatory Systemic Diseases: Increased Activity of the Sympathetic Nervous System

Several studies demonstrated that patients with chronic inflammatory systemic diseases have an elevated sympathetic nervous system tone.[1165–1169] Increased sympathetic activity could be related to the increased risk of cardiovascular events as observed in rheumatoid arthritis patients.[1170] Such an increased sympathetic tone may be a consequence of relatively decreased serum levels of cortisol in relation to inflammation since there exists cooperativity of cortisol and noradrenaline on a molecular level via the β-adrenergic receptor signaling cascade.[1171,1172] Functional loss of one factor probably upregulates the other factor in order to maintain functions such as blood glucose homeostasis, regulation of the bronchial lumen, and blood pressure control. Since TNF is relevant for the observed alterations of the HPA axis leading to inadequately low cortisol secretion (see above), its neutralization may change the increased sympathetic tone.

A recent study confirmed the increased sympathetic tone in patients with rheumatoid arthritis and also in systemic lupus erythematosus, which was accompanied by a relatively normal tone of the HPA axis (called "desynchronization of the HPA axis and the sympathetic nervous system" during chronic inflammation because an increase of the tone of both axes can be expected during acute inflammation).[1592] Interestingly, 12 weeks of anti-TNF therapy only slightly decreased the sympathetic activity as measured by plasma neuropeptide Y levels.[1592] Thus, desynchronization persists after longer treatment, and it appears that TNF is not the sole factor responsible for this phenomenon. A similar increase of sympathetic activity has been demonstrated in patients with Crohn's disease and ulcerative colitis.[1173]

It should be mentioned that an elevated activity of the sympathetic nervous system might not cause increased local sympathetic neurotransmitters in the inflamed joint because sympathetic nerve fibers are lost.[606] In addition to the increased sympathetic nervous tone, one observes a decrease of the parasympathetic outflow.[1011] Such a decrease of the parasympathetic system is probably an unfavorable signal because this will impede on the anti-inflammatory role of the vagus nerve.[1174]

Loss of Sympathetic Nerve Fibers and Sprouting of Sensory Nerve Fibers

The loss of sympathetic nerve fibers and sprouting of sensory nerve fibers have already been described in Figure 26. Loss of sympathetic nerve fibers has been described in synovial tissue of patients with rheumatoid arthritis,[606,1175–1177] in the oral mucosa of patients with lichenoid reactions,[1178] in the inflamed Charcot foot,[1179] in inflammatory endometriosis,[1180,1181] in chronic pruritus and prurigo nodularis,[1182] in Crohn's disease,[1118] and others. Sprouting of substance P-positive sensory nerve fibers has been observed in gastritis,[1183] in esophagitis,[1184] in psoriatic skin disease,[1185] in synovial tissue of patients with rheumatoid arthritis compared to osteoarthritis,[1177] in Charcot foot,[1179] in chronic pruritus and prurigo nodularis,[1182] in endometriosis,[1181] in Crohn's disease,[1118] and others. There is also a marked preponderance of substance P-positive nerve fibers over CGRP-positive nerve fibers in rheumatoid arthritis synovial tissue, which supports inflammation since CGRP would have some anti-inflammatory properties.[1186] In general, all these findings support the concept of a nonspecific reaction favoring inflammation.

This concept is supported by additional results that in later stages of inflammatory joint disease, α1- and α2-adrenergic receptors seem to gain a more prominent role. In patients with juvenile idiopathic arthritis, functional α1-adrenergic receptors on leukocytes are induced, whereas the receptors are absent on leukocytes of normal donors.[570] These effects seem to be cell type-specific since vascular α-adrenergic receptor-mediated vasoconstriction in inflamed joints is downregulated leading to elevated perfusion.[1187,1188] Suppression of α-adrenergic receptors on vasoconstrictors and parallel induction on leukocytes would be a proinflammatory stimulus because this leads to vasodilatation (vessels), proliferation of lymphocytes, and secretion of proinflammatory cytokines such as IL-6 and TNF (leukocytes).

Interestingly, AP-1 and NF-κB binding sites are present in the regulatory region of the α1-adrenergic receptor promoter that may be responsible for the induction of α1-adrenergic receptor expression in monocytes by IL-1β and TNF.[1189] In another study, synovial fibroblasts of patients with rheumatoid arthritis could be activated to proliferate by α2-adrenergic agonists, which is mediated via phospholipase C, protein kinase C β II, and MAP kinase.[1190] *Ex vivo* studies with peripheral blood mononuclear cells from rheumatoid arthritis patients revealed that only in patients with high disease activity, catecholamines mediate their effects via α-adrenergic receptors.[1191] In addition, peripheral blood mononuclear cells of patients with rheumatoid arthritis demonstrate lower numbers of β-adrenergic receptors supporting the overall α-adrenergic preponderance.[932] Thus, there seems to be a receptor defect, recently also confirmed in a model of chronic inflammatory systemic disease.[933]

From these data, it seems as if α-adrenergic signaling becomes more relevant in the later stage of this chronic inflammatory disease [1192]. This is referred to as the "β-to-α-adrenergic shift" during disease progression (Figure 26).

Cells Positive for Neurotransmitters/Neuropeptides Appear in the Tissue

In recent years, several reports appeared that described the presence of immune cells in inflamed tissue with the capacity to produce neurotransmitters or neuropeptides (already touched above). These studies demonstrated the production of substance P from rheumatoid arthritis and osteoarthritis synovial fibroblasts.[1193,1194] The production of this neuropeptide was linked to a proinflammatory situation.

Rheumatoid arthritis synovial fibroblasts and macrophages possess the enzyme machinery for catecholamines.[606,607,1120] The production of catecholamines seems to exert anti-inflammatory effects in synovial cells of patients with rheumatoid arthritis and experimental arthritis.[607,1121,1122] Catecholamine production was linked to T regulatory cells in patients with multiple sclerosis.[604,1195] Cells in inflamed lesions can also secrete endogenous opioids with anti-inflammatory activities (reviewed in Ref. 1196). Even the production machinery for acetylcholine exists in cells of the inflamed tissue, which might also be an anti-inflammatory signal.[1130]

The role of these cells is not clear, but it seems that the anti-inflammatory activities of catecholamine-, opioid-, and acetylcholine-producing cells are insufficient to overcome the proinflammatory domination. It needs to be investigated whether a change of neurotransmitter/neuropeptide secretion can be achieved by therapeutic drugs leading to higher production of anti-inflammatory neurotransmitters/neuropeptides.

Summary

Various neuronal systems exhibit both pro- and anti-inflammatory roles depending on many influences: (1) time point in relation to the start of the inflammatory process, (2) antigenic stimulus shifting the immune response, (3) tissue location and environment, (4) involved cell types and time-dependent receptor expression, (5) changing tissue innervation, and (6) appearance of neurotransmitter-producing cells. Thus, the general statement of "the neuronal anti-inflammatory reflex" is meaningless outside this context.

Three decades of experimentation and clinical investigation demonstrated the proinflammatory role of the nervous systems in chronic inflammatory disease (sensory vasoregulation, peripheral and central sensitization, neurotransmitter-induced chemotaxis and cell mobilization, sympathetic nerve fiber loss, sensory hyperinnervation, β- to α-adrenergic shift, etc.). Due to adaptive programs, the net effect is unfavorable in chronic inflammatory systemic diseases with immunity against self-antigens or harmless foreign antigens. Unfortunately, many anti-inflammatory activities of the different nervous systems are switched off (β-adrenergic, A2-adenosinergic, opioidergic, cholinergic, and others). Therapeutic targets that capitalize on the anti-inflammatory effects of the nervous system or neurotransmitters have considerable potential to modulate chronic inflammatory systemic diseases.

Cooperation of the Sympathetic Nervous System and HPA Axis

A characteristic endocrine dilemma of many chronic inflammatory systemic diseases is cortisol deficiency in relation to the level of inflammation measured in the systemic circulation (see above). In addition, a higher degree of tissue inflammation was associated with a reduction of sympathetic innervation (see above). Both factors probably contribute to an overall proinflammatory situation in inflamed tissue due to a lack of anti-inflammatory cooperativity of cortisol and noradrenaline.

In asthma therapy, parallel topical (or systemic) treatment with a β-adrenergic agonist and glucocorticoids has an additive antiobstructive effect.[1197] Cooperative effects of cortisol and noradrenaline lead to an increase of glucocorticoid receptors, β-adrenergic receptors, intracellular cyclic AMP, protein kinase A, and cAMP response element-binding protein, a sequence of events that has been demonstrated in various cell types.[1171,1172,1198–1203] Increase of these intracellular mediators is accompanied by a strong anti-inflammatory response in various immune cells. Furthermore, cortisol supports the production of noradrenaline and adrenaline from sympathetic nerve terminals and adrenal medulla by inducing the synthesizing enzymes. Thus, one would expect that a parallel increase of cortisol and noradrenaline with local physiological concentrations of 10^{-8}-10^{-6} M would be more anti-inflammatory than each substance alone. Presumably, a dissociation of these two factors may be deleterious in chronic inflammatory systemic diseases. In addition, the circadian rhythm of the activity of the HPA axis and the sympathetic nervous system are identical in order to provide cooperative effects (Figure 27).

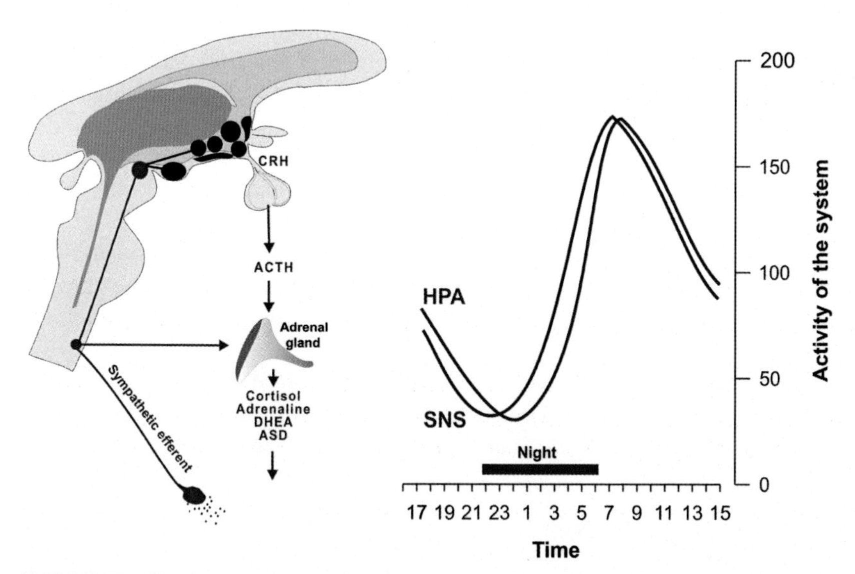

FIGURE 27 Simultaneous rise and fall of activity of the hypothalamic-pituitary-adrenal (HPA) axis and the sympathetic nervous system (SNS) under normal conditions. Both systems reach their maximum in the early morning hours. This would lead to cooperativity of the HPA axis and the sympathetic nervous system. ACTH, adrenocorticotropic hormone; ASD, androstenedione (adrenal androgen); CRH, corticotropin-releasing hormone; DHEA, dehydroepiandrosterone (adrenal androgen).

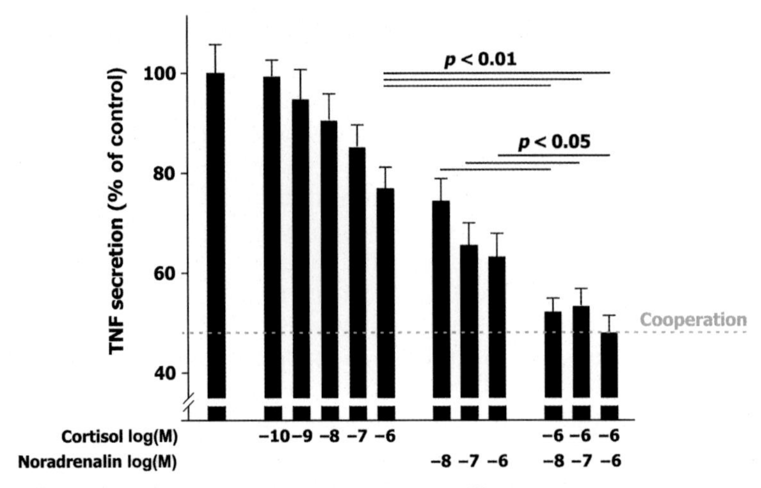

FIGURE 28 Cooperativity of cortisol and noradrenaline.[581] Mixed synovial cells are incubated with cortisol, noradrenaline, or both substances together. TNF is measured 12h after the start of incubation. The cooperative effect is demonstrated as a stronger inhibition of TNF as compared to each substance alone. Further reading in Ref. 581.

Indeed, in mixed synovial cells of patients with rheumatoid arthritis, the concomitant administration of cortisol and noradrenaline has a stronger inhibitory effect on TNF secretion as compared to each substance alone (Figure 28).[581]

Furthermore, rheumatoid arthritis patients with prior prednisolone treatment and presence of sympathetic synovial innervation had significantly lower levels of synovial inflammation as compared to patients without prior prednisolone treatment and presence of sympathetic synovial innervation or as compared to patients with prior prednisolone treatment but without sympathetic synovial innervation (Figure 29).[581] This indicates that cooperation of glucocorticoids and noradrenaline leads to a more anti-inflammatory situation in the tissue microenvironment. Since both factors are diminished in the chronic disease, the usually expected cooperation is absent leading to a proinflammatory milieu.

In conclusion, cooperation of neuronal and hormonal pathways can be important to establish an anti-inflammatory environment. Disruption of cooperation might be a significant risk factor of inflammation (see also Chapter VII).

The Opioidergic Pathway

Sympathetic nerves release endogenous opioids such as β-endorphin, methionine enkephalin, and leucine enkephalin into the vicinity of nerve terminals. Importantly, cells in inflamed tissue also produce the same opioids.[1204] Since the excitability of sensory nerve fibers is blocked by endogenous opioids, it is known that these opioids have analgesic properties. In addition, endogenous

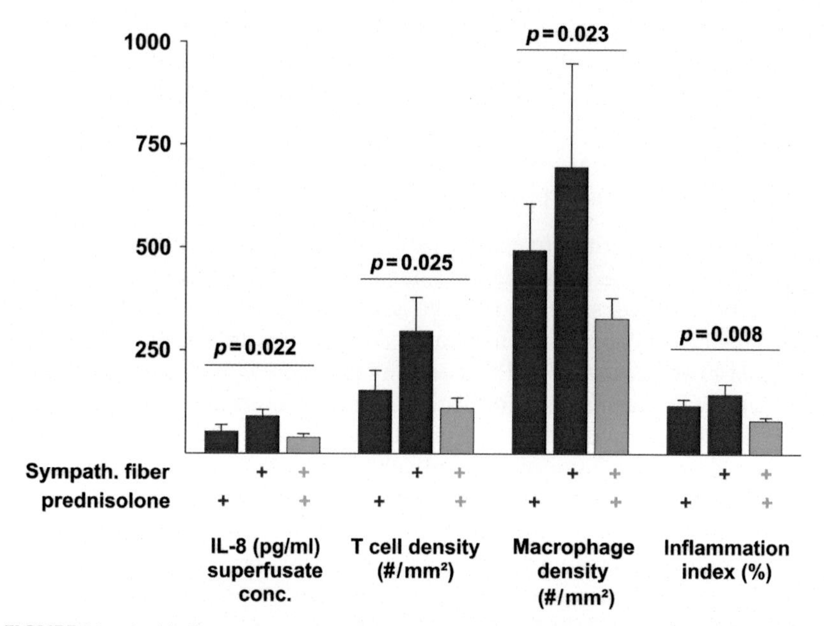

FIGURE 29 Anti-inflammatory cooperativity of corticosteroids and the sympathetic nervous system in patients with rheumatoid arthritis.[581] Histological and functional parameters of inflammation in patients with rheumatoid arthritis in relation to prior prednisolone therapy and histological presence of tyrosine hydroxylase-positive sympathetic nerve fiber. Units are given in parentheses. The plus sign indicates the presence of the respective factors. The *p*-value indicates the significant difference in the comparison of all subgroups. With respect to all markers of inflammation, the green group demonstrates the lowest levels of the respective inflammatory factor. Data are given as means ± SEM. IL-8, interleukin-8 from synovial tissue. Further reading in Ref. 581.

opioids serve as anti-inflammatory factors, particularly via ligation of the μ-opioid receptor on immune cells and by inhibition of substance P release from sensory nerve endings.

In chronic arthritis, μ-opioid receptor expression is downregulated on sensory nerves.[1205] This leads to a concomitant loss of analgesic effects of the most potent endogenous μ-opioid receptor agonist endomorphin-1.[1205] In parallel, the two other opioid receptors (kappa and delta) were found to be decreased on synovial fibroblasts in patients with rheumatoid arthritis.[1206] It was further shown that TNF and IL-1β downregulate kappa and delta opioid receptors.[1206] These changes of opioid receptor in arthritis most probably establish a proinflammatory environment. However, the exogenous administration of endomorphins has still an anti-inflammatory effect in patients with rheumatoid arthritis and osteoarthritis and in animals with experimental arthritis.[1207]

In conclusion, the loss of sympathetic nerve fibers with endogenous opioids is accompanied by the appearance of opioid-producing cells.[1196,1204,1208] However, due to inflammation-induced downregulation of opioid receptors, a proinflammatory environment is established, and local opioids are not powerful enough to overcome proinflammatory domination. Nevertheless, treatment with exogenous opioids can exert beneficial analgesic and anti-inflammatory effects.[1209] Whether opioids can be included into standard therapy in chronic inflammation is a long-standing open question to be answered.

Vasoactive Intestinal Peptide

VIP has many similarities to noradrenaline (via the β2-adrenergic receptor and cyclic AMP), because it switches on similar intracellular cascades.[702] Although VIP has been known since the late 1960s, its strong anti-inflammatory role in animal models of chronic inflammatory systemic diseases was discovered during the last two decades.[702] VIP has anti-inflammatory effects in synovial cells of patients with rheumatoid arthritis and in the experimental model of arthritis.[1210–1212] The levels of VIP are increased in the synovial fluid of patients with rheumatoid arthritis, but most probably, this anti-inflammatory compensatory mechanism is not sufficient to block the proinflammatory process.

VIP receptor type 1 was found to be expressed on synovial fibroblasts of patients with rheumatoid arthritis and osteoarthritis, but the levels were significantly lower in rheumatoid arthritis fibroblasts.[1213] TNF can downregulate the expression of the VIP receptor type 1, which is most probably a deleterious effect. Since VIP was found in nonadrenergic neurons of sympathetic origin,[1214,1215] the loss of sympathetic nerve fibers can be a critical step in losing this important anti-inflammatory neuropeptide. No VIP-related therapies are currently on the market, but nonpeptide agonists to VIP receptor types 1 and 2 are being developed for therapy, which are tested in experimental and human chronic inflammatory systemic diseases.

The Anti-Inflammatory Cholinergic Pathway

The neural cholinergic anti-inflammatory pathway has been described in the context of acute sepsis.[1019] It was demonstrated that the nicotinic acetylcholine receptor α7 subunit on macrophages is an essential regulator of inflammation.[1123] Electrical stimulation of the vagus nerve markedly decreased the production of proinflammatory cytokines.[1174] When compared with macrophages, monocytes are refractory to the cytokine-inhibiting effects of acetylcholine, which might explain the relatively little effects of acetylcholine on human peripheral blood mononuclear cells.[1216] These interesting new findings point toward a strong anti-inflammatory effect of the parasympathetic nervous system via acetylcholine.

In addition, the vagus nerve has a similar anti-inflammatory effect in an experimental model of arthritis and in rheumatoid arthritis.[1126,1127,1217–1219] However, an extensive case control study of patients undergoing vagotomy did not demonstrate an increased risk of rheumatoid arthritis in vagotomized compared with nonvagotomized people.[1220] This indicates that susceptibility to rheumatoid arthritis is not increased by vagotomy. However, severity and the course of rheumatoid arthritis might be influenced, but this has not been tested.[1220]

The findings of an anti-inflammatory cholinergic reflex need some critical considerations: (1) Cholinergic nerve fibers do not innervate the joints. Thus, a possible cholinergic effect must be indirect via the modulation of secondary lymphoid organs in the belly or gut (not in the spleen because cholinergic innervation is missing) or via local cellular production of acetylcholine. (2) Acetylcholine is produced by immune cells, but its production has not been demonstrated in the cells of patients with chronic inflammatory systemic diseases. (3) The isolated consideration of only one subunit (α7) of the nicotinic acetylcholine receptor is critical because of the complexity of nicotinic and muscarinic acetylcholine receptors. (4) Presently, the importance of the nicotinic acetylcholine receptor α7 subunit in relation to the course of chronic inflammation is not known. It might well be that the cholinergic system similarly to the sympathetic nervous system has dual roles depending on the timing of intervention. In conclusion, the new findings of a "cholinergic anti-inflammatory reflex" are interesting and probably important, but its role in chronic inflammatory systemic diseases needs further investigation.

Summary

The following list summarizes the major concepts from psychoneuroendocrine immunology. These findings have markedly enriched the understanding of many symptoms and signs in chronic inflammatory systemic diseases. Personally, I recognize that my colleagues were helpful in establishing this new understanding. Due to space constraints, not all important aspects are listed.

- Genetic prerequisites for chronic inflammatory systemic diseases were found in the neuroendocrine system. Animal studies demonstrated that genetic defects of the HPA axis are linked to more severe chronic inflammatory systemic diseases, but similar findings were not reported in humans.
- Glucocorticoids such as cortisol are anti-inflammatory at high concentrations ($\geq 10^{-7}$ mol/l). Patients with chronic inflammatory systemic diseases show inadequate serum cortisol levels relative to inflammation (disproportion principle); TNF, interferons, and IL-6 play important roles.
- Low-dose glucocorticoid therapy can be viewed as adrenal gland substitution therapy in the chronic phase of disease.
- In an animal model of chronic inflammatory systemic disease, adrenal mitochondria become severely altered and adrenal lipid depots are not degraded, which can be an important reason for inadequate glucocorticoid secretion. IL-1β is an important mediator of alteration in a rat model.
- Estrogens play an important role for the women-to-men preponderance in the incidence of many chronic inflammatory systemic diseases. A likely explanation is the estrogen/progesterone-dependent influence on different

immune cell subtypes relevant to chronic inflammatory systemic diseases during reproductive years.

- Patients with chronic inflammatory systemic diseases demonstrate an absolute loss of anti-inflammatory androgens. Loss of androgens depends on cytokine-mediated deficits in glandular production and increased conversion from androgens to estrogens in inflamed tissue (androgen deficit). All must be viewed as a strong proinflammatory factor.
- Low levels of serum androgens such as DHEAS prior to disease manifestation predispose to rheumatoid arthritis (not known for other chronic inflammatory systemic diseases).
- CRH is a proinflammatory hormone in inflamed tissue, while it has an anti-inflammatory role in the context of the HPA axis activation and secretion of glucocorticoids. In the chronic phase in experimental models of inflammatory systemic diseases, CRH is downregulated in the hypothalamus while vasopressin is upregulated (activation of the sympathetic nervous system).
- Serum levels of prolactin and melatonin, two proinflammatory hormones, are increased in chronic inflammatory systemic diseases.
- Hypovitaminosis D is observed in many chronic inflammatory systemic diseases. Since vitamin D has anti-inflammatory capacities toward the adaptive immune response, the loss of this hormone is thought to aggravate chronic inflammatory systemic diseases.
- Thyroid hormones such as T3 become decreased similarly as in the NITS. Presently, it is unclear whether low T3 levels also appear in inflamed tissue and in activated local immune cells. Low T3 levels in inflamed tissue would reduce the inflammatory capacity of immune cells.
- Insulin resistance and insulin-like growth factor 1 resistance are observed in chronic inflammatory systemic diseases (with hyperinsulinemia and insulin-like growth factor 1 loss). Hyperinsulinemia in this context is most probably a proinflammatory signal because insulin stimulates many aspects of immunity (immune cells do not become insulin-resistant).
- The RAAS is activated in chronic inflammatory systemic diseases, and angiotensin II has many proinflammatory activities.
- Similarly, serum levels of adipokines such as leptin are increased in chronic inflammatory systemic diseases. Adipokines exert many proinflammatory activities.
- Sensory nerve fibers are dominant in inflamed tissue, which supports pain perception and sensitization. Sensory nerve fibers, substance P, and CGRP are instrumental in neurogenic inflammation. Sensitization of pain pathways is also installed on a spinal level.
- Cytokines such as TNF, IL-1β, IL-6, and IL-17 play an important role in peripheral and spinal sensitization.
- The sympathetic nervous system has a proinflammatory role in the early phase of experimental arthritis but an anti-inflammatory role in the chronic phase. The influence in the chronic phase might depend on anti-

inflammatory catecholamine-producing cells. Cooperation with the HPA axis (cortisol) is disturbed due to a loss of sympathetic nerve fibers in inflamed tissue and inadequate cortisol secretion and function.

- Recently described neurotransmitter-producing cells (catecholamines, opioids, and acetylcholine) might have ant-inflammatory roles, but they are not capable of inhibiting ongoing inflammation.
- Anti-inflammatory opioidergic pathways are altered due to a loss of sympathetic nerve fibers and downregulation of opioid receptors on immune cells.
- An increased sympathetic activity is observed in chronic inflammatory systemic diseases, which stimulates an atherogenic milieu and the proinflammatory RAAS (hypertension).
- VIP is an anti-inflammatory factor, but local production of VIP does not overcome the proinflammatory dominance.
- The cholinergic pathway might play an anti-inflammatory role in chronic inflammatory systemic diseases but its exact role is not established.

Chapter III

Energy and Volume Regulation

Queequeg caught a terrible chill which lapsed into a fever; and at last, after some days' suffering, laid him in his hammock, close to the very sill of the door of death. How he wasted and wasted away in those few long–lingering days, till there seemed but little left of him but his frame and tattooing.

Herman Melville, American novelist (Taken from Moby Dick, Ref. 1221)

Chapter Contents

The interplay of the nervous, endocrine, and immune systems is recognized to be involved in the pathophysiology of chronic inflammatory systemic diseases (Chapter II). This was the result of a long and winding cognitive and experimental process in medicine. However, the individual aspects of endocrine immune and neuroimmune connections and the related problems described in Chapter II are considered to be more or less independent of each other. Typically, investigators in this research field study on one or two of these individual aspects. Likewise, physicians usually accept individual symptoms as mere independent "accidents of inflammation."[a]

On the other hand, we ask why do chronic inflammatory systemic diseases affect the whole body and produce myriad debilitating and disabling symptoms that make people sick. The question appears whether there is a common denominator, which would explain the multitude of systemic disease sequelae, immune aberrations, endocrine alterations, and neuronal changes.

a. The WHY behind the accident is not important, the HOW TO TREAT is mainstream.

The Origin of Chronic Inflammatory Systemic Diseases and their Sequelae.
http://dx.doi.org/10.1016/B978-0-12-803321-0.00003-3

In the winter of 2008-2009, through 10 weeks of passionate work, I included systemic energy regulation into the pathophysiology of chronic inflammatory systemic diseases. This field was prepared by scholars in the fields of surgery, emergency medicine, and nutrition research, who focused on acute stressful events such as surgery, injury, hemorrhage, and sepsis knowing that these episodes are highly energy-demanding. It turned out that energy regulation on the systemic level is an important addition to pathogenetic concepts of chronic inflammatory systemic diseases. Particularly, the combined study of energy regulation and evolutionary medicine (the next Chapter IV) brought new insight. This chapter describes the basis of systemic energy and volume regulation.

ESTIMATION OF ENERGY DEMANDS

Usually, we need 7000-8000 kJ (1672-1910 kcal)/d to cover basal metabolic activities.[1222] With a moderate workload, roughly 10,000 kJ (2388 kcal)/d), is needed (approximately the standard metabolic rate). A bicyclist of the *Tour de France* requires 30,000 kJ/d (7164 kcal/d) and a male ironman participant 42,000 kJ (10,029 kcal) in 8 h (which can be maintained only for a short period of time).[1222, 1223] The majority of an individual's caloric intake is lost as heat and only a small proportion of 10-15% is used for muscular work.[1224, 1225] More data for comparison are provided in Table 16.

In the human body, the distribution of energy-rich fuels is key to survival and is tightly regulated, involving several organs. Uptake of external energy occurs in the intestinal tract (external energy-rich fuels: glucose, amino acids, and fatty acids; maximum uptake per day 20,000 kJ (4776 kcal)). Storage happens in the fat tissue (12 kg of triglycerides in the body = 500,000 kJ (119,400 kcal)), in the liver (150 g glycogen = 2500 kJ (597 kcal)), and in the skeletal muscles (300 g glycogen = 5000 kJ (1194 kcal); 6-7 kg protein = 50,000 kJ (11,940 kcal)). The glycogen store in the skeletal muscle of 300 g (=5000 kJ (1194 kcal)) is not available for the entire body because skeletal muscle glycogen is only locally used.[1222] One can separate two groups of neuroendocrine factors that either store energy in the adipose tissue, skeletal muscle, and liver or are responsible for the release of energy-rich fuels from stores to consumers (Table 17). Since calcium and phosphorus are important expedients to many cellular reactions of energy expenditure, the bone as the storage organ of these two ions is also included in Table 17. The data presented in Table 17 were recently published in a review article.[1231]

During short-lasting activity, systemically available energy-rich fuels are supplied by the liver (breakdown of glycogen and gluconeogenesis), but when activity lasts for several hours, fat stores start to break down (triglyceride lipolysis). Here, the liver can be seen as a switchboard to produce glucose from other energy-rich substrates. The provision of energy-rich fuels to the entire body in the form of glucose and fatty acids is mainly mediated by mediator substances of the sympathetic nervous system (adrenaline and noradrenaline mainly via β2-adrenergic receptors), of the hypothalamic-pituitary-adrenal

TABLE 16 Energy Expenditure of Systems and Organs Under Various Conditions[1152, 1222, 1227]

System/Organ	Energy Expenditure per Day, kJ/d (kcal/d)
Total body basal MR	8000 (1910)
Total body MR with usual activity (sedentary)	10,000 (2388)
Total body MR of a *Tour de France* bicyclist	30,000[a] (7164)
Total body MR of a marathon runner (projected per day)	140,000[a] (33,432)
Total body MR during minor surgery	11,000 (2627)
Total body MR with multiple bone fractures	up to 13,000 (3104)
Total body MR with sepsis	15,000 (3582)
Total body MR with extensive burns	20,000 (4776)
Total body daily uptake (absorptive capacity in the gut)	20,000 (4776)
Immune system MR in a quiescent state	1600[c] (382)
Immune system MR moderately activated	2100[c] (501)
Central nervous system MR	2000 (478)
Muscle MR at rest	2500 (597)
Muscle MR activated	2000++ (478++)
Liver[b] MR (including immune cell activity)	1600 (382)
Kidneys MR	600 (143)
Gastrointestinal tract[b] MR (including gut immune system, without liver, kidney, and spleen)	1000 (239)
Spleen (erythrocytes plus leukocytes; 90% anaerobic)	480 (115)
Abdominal organs (together)[b] MR	3000-3700 (716-884)
Lung[b] MR (including lung immune system)	400 (96)
Skin[b] (including skin immune system)	100 (24)
Heart MR	1100 (263) (and more when activated)
Thoracic organs (together)[b] MR	1600-2400 (382-573)

Abbreviations: kJ, kilojoule (10,000 kJ = 2388 kcal); MR, metabolic rate.
[a]*Such a high energy expenditure can be maintained only for a very short period of time.*
[b]*Energy need is difficult to estimate independent of the immune system in these organs.*
[c]*See derivation of energy need in the following text. Leukocytes use all types of fuels, but the main source is glucose and glutamine making roughly 70% of the fuels needed.*[1228–1230]

TABLE 17 Summary of Energy Storage and Energy Expenditure Hormones

	Fat Tissue	Muscle	Liver	Bone
Energy storage hormones				
Insulin	Uptake[a]	Uptake[a]	Uptake[a]	
Insulin-like growth factor 1[b]	Growth	Growth	Uptake[a]	Growth
Androgens[b]		Growth		Growth
Estrogens[b]	Gynoid fat distribution	Glucose uptake[a]		Growth
Vitamin D		Growth		Growth
Osteocalcin[b,c]		Glucose uptake[a]	Glucose uptake[a]	Growth
Vagus nerve	Uptake[a]		Uptake[a]	
Energy expenditure hormones				
Cortisol[d]	Release[a]	Release[a]	Release[a]	Release[a]
Sympathetic nervous system (noradrenaline/adrenaline)[d]	Release[a]	Release[a]	Release[a]	Release[a]
Growth hormone[d]	Release[a]	Release[a]	Release[a]	Growth via IGF-1
Thyroid hormones (T3)[d]	Release[a]	Release[a]	Release[a]	Release[a]
RAAS[d]			Release[a]	Release[a]

Abbreviation: RAAS, renin-angiotensin-aldosterone system.
[a]*Uptake/release of energy-rich fuels into/from the respective tissues.*
[b]*Increase of insulin sensitivity.*
[c]*Support of androgens.*
[d]*Decrease of insulin sensitivity (induction of insulin resistance).*
Summarized in Ref. 1231.

(HPA) axis (mainly cortisol and growth hormone), of the pancreas such as glucagon, and through the renin-angiotensin-aldosterone system (angiotensin II) (Table 17). During starvation, energy-rich fuels are primarily supplied by protein breakdown (during the first 2-3 days, exploitable protein store 6-7 kg, which is roughly 50,000 kJ (11,940 kcal)) and adipose tissue (starting from day 3 onward, triglyceride breakdown and fatty acid conversion to ketone bodies in the liver).[1222] This latter process prevents protein breakdown from skeletal muscles.

CALCULATION OF ENERGY REQUIREMENTS BY THE IMMUNE SYSTEM

As demonstrated in Figure 30, an estimate of energy expenditure of leukocytes (lymphocytes, granulocytes, monocytes/macrophages, and others) demonstrates that approximately 1600 kJ/d (382 kcal/d) is needed when these cells are not activated (in the basal metabolic state, excluding cellular movement), which can rise to 1.750-2.080 kJ/d (418-497 kcal/d) when mildly to moderately activated (the original information for all mentioned calculations is derived from Refs. 1232–1236).

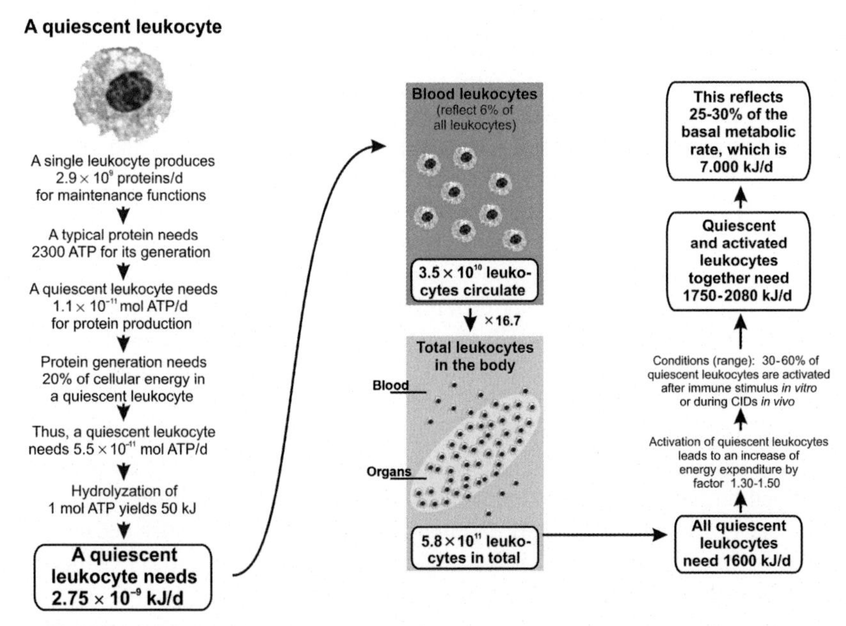

FIGURE 30 An estimate of energy expenditure by leukocytes per day. In the original studies, leukocytes were activated with concanavalin A, or leukocytes were taken from patients with rheumatic or infectious diseases. This yielded an activation factor of 1.30-1.50. Leukocytes include all cells involved in immune reactions. The original information for all calculations is derived from Refs. 1232–1236. Notice that 1600 kJ corresponds to 382 kcal, 1750-2080 kJ correspond to 418-497 kcal, and 7000 kJ corresponds to 1672 kcal. Abbreviation: CIDs, chronic inflammatory systemic diseases.

This calculation of the energy demand of the immune system has been published.[1152] With the number of 5.8×10^{11} leukocytes in the body (Figure 30), one can also estimate the volume and the weight of the entire immune system (volume will be used in section "Bone Loss" of Chapter V). Assumed an immune cell has a diameter of 8-20 µm and the classical formula for a perfect sphere is used, the volume of one cell is 2.7×10^{-13}-4.2×10^{-12} m^3 (268-4188 fl; 10^{-15} l). Multiplied with the total number of leukocytes, we obtain a volume of 0.16-2.43 l. The upper number of 2.43 l might be exaggerated due to the fact that larger immune cells such as macrophages and dendritic cells demonstrate a large diameter but are not spherical, while smaller lymphocytes and neutrophils are spherical. When we assume that a specific density of leukocytes is approximately 1.05 kg/l, all leukocytes weigh 0.17-2.55 kg. According to a tabular data collection,[1237] the weight of all lymphocytes is 1.5 kg, which is a very similar number than the one calculated above.

From the calculations in Figure 30, an inflammation-induced increase of the metabolic rate of approximately 150-480 kJ (36-115 kcal) appears to be a normal figure with mild to moderate inflammation, but it can go up to 5000 kJ (1194 kcal) in sepsis or 10,000 kJ (2388 kcal) with extensive burn wounds.[1225] Leukocytes use all types of fuels, but the main source is glucose and glutamine making roughly 70% of fuels needed.[1228, 1229] It is a general picture that activated cells such as endothelial cells, fibroblasts, T cells, and M1 macrophages mainly use the glycolytic pathway (reviewed in Ref. 1238). A minor part is added by glutamine and very little by free fatty acids.[1238]

This is interesting because the glycolytic pathway terminates at lactate, and this does only yield two molecules of ATP per one molecule of glucose, while full metabolism of one glucose molecule through the Krebs cycle yields 38 molecules of ATP.[1239] Thus, complete metabolism would be much more economical. However, the glycolytic pathway is much faster than complete metabolism through mitochondrial oxidative phosphorylation, so that glycolysis is more economical per time.

It is important that variously differentiated immune cells such as T effector cells versus T regulatory cells or M1 macrophages versus regulatory M2 macrophages use different energy-rich fuels. It turns out that regulatory or anti-inflammatory leukocytes mainly use free fatty acids while the activated effector cells use glucose.[1238] When the immune system is activated, the glycolytic pathway is the most important one. Necessarily, the removal of lactate from an inflamed tissue (the end product of the glycolytic pathway) is essential. Thus, the tissue needs to be adequately vascularized, probably not so much because of oxygen provision but more so because of lactate removal. Lactate is reactivated to glucose in the liver in a special Cori cycle between an inflamed tissue and the liver, which needs four ATP and two GDP molecules in the process of gluconeogenesis (inflammatory Cory cycle in contrast to the muscular Cory cycle of lactate reprocessing). Under consideration of these numbers, the question arises as to how energy-rich fuels can be provided to the immune system when energy is limited (recall that there is the maximum absorption rate in the gut of 20,000 kJ/d (4776 kcal/d), and the usual amount of food intake is 10,000 kJ/d (2388 kcal/d)).

CIRCADIAN ALLOCATION OF ENERGY-RICH FUELS IN A HEALTHY SUBJECT

Fuel allocation to the immune system is daytime-dependent as demonstrated in Figure 31. In a fasting subject, fuel provision to the body is started by an activation of the HPA axis and the sympathetic nervous system in the morning hours (Figure 31). These hormones including glucagon support fuel provision mainly for the brain and skeletal muscles by stimulating triglyceride lipolysis, β-oxidation of fatty acids, glycogenolysis, gluconeogenesis, and little protein breakdown.

In addition, these hormones inhibit many aspects of the immune system except the secretion of natural polyclonal antibodies (supported by the sympathetic nervous system) and leukocyte traffic in the blood (supported by the sympathetic nervous system and HPA axis; e.g., i.v. injection of cortisol or noradrenaline in the case of clinical emergency leads to leukocyte increase in the blood).[625, 803, 1240, 1241] Leukocyte traffic in circulation is needed for immunosurveillance of the tissue, and this is mainly a daytime job.

Starting in the afternoon and reaching the minimum at midnight, low hormone levels prevent energy-rich fuel provision to the brain and skeletal muscles. In parallel, hormonal inhibition of the immune system is largely decreased (Figure 31, blue half). In contrast, during the night, energy-rich fuels are mainly allocated to the immune system, repair systems, and—in case of children—growth of the body (Figure 31, the yellow half). An important hormone for glucose allocation to the immune system shortly after sleep onset is the growth hormone, which stimulates gluconeogenesis. Since immune cells use mainly glucose, growth hormone-associated provision of glucose is important for nightly immune activation (glucose is upregulated after sleep; Ref. 1242). In addition, serum levels of ketone bodies, free fatty acids, and glycerol rise from 8 p.m. until midnight by a factor of 3-7,[1243] which represents another important fuel source for the immune cells.

From this point of view, the circadian rhythms of the neuroendocrine and immune systems belong to an important program necessary for the allocation of energy-rich fuels to daytime and nighttime consumers. Since the brain and skeletal muscles need a lot of energy-rich fuels during the day, the major activities of the immune system are located during sleep times. Sleep protects energy stores by approximately 25-30% as demonstrated in excellent studies with indirect calorimetry in humans.[1244, 1245] Importantly, during sleep, it is mainly the brain that reduces glucose consumption by 25-30%.[1246]

In conclusion, energy regulation during the period of 1 day is important to understand the general principles of fuel storage and supply by neuroendocrine pathways.

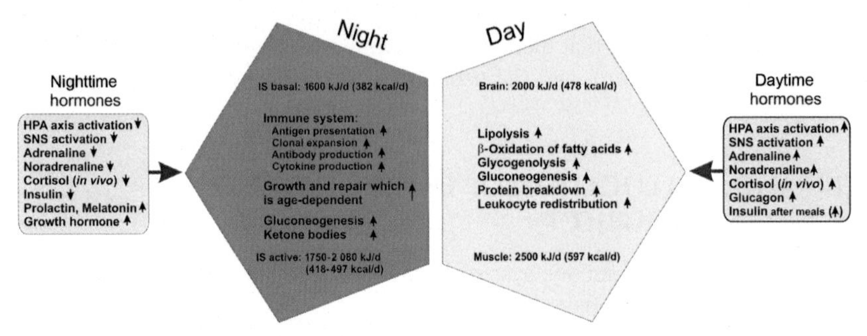

FIGURE 31 Circadian use of energy-rich substrates by different consumers. During the day (yellow pentagon), mainly the skeletal muscles and brain use energetic substrates. The hypothalamic-pituitary-adrenal (HPA) axis and the sympathetic nervous system (SNS) are instrumental in providing these fuels to the brain and skeletal muscles. Respective metabolic pathways are switched on. During the night (blue pentagon), mainly the immune system (IS) and growth-related mechanisms are activated. Immune-activating hormones such as prolactin, melatonin, and growth hormone stimulate the immune system in a milieu with a small influence of the HPA axis and the SNS. Growth hormone-related gluconeogenesis and the provision of ketone bodies are important for the immune system and growth-related mechanisms.

ENERGY REGULATION IN LOCAL INFLAMMATION AND SPILLOVER INFLAMMATION

As the immune system needs a lot of energy-rich fuels,[1247] local inflammation somewhere in the tissue must be supported by fuel provision from local or systemic stores. If local inflammation is confined to a small space (e.g., a thorn of a rose with attached bacteria in the skin), local fuel stores and circulating glucose are preferentially used (note that the immune system does not store energy-rich fuels). I hypothesize that local stores are made available by extracellular protein breakdown by matrix metalloproteinases yielding substrates such as proline, hydroxyproline, glycine, glucuronic acids, and others. These substrates can be used in ATP-generating pathways (Table 18).

It was shown that cells of starved animals demonstrate inhibited transcription of inflammation-related genes but strongly increased transcription of genes involved in matrix degradation.[1248] In addition, local lipolysis leads to the release of free fatty acids, which can be used as energy-rich substrates by immune cells. An important element in local inflammation is adequate supply of calcium and phosphorus, which can be provided by increased local bone turnover when inflammation is in the proximity of the bones. Proinflammatory cytokines (TNF, IL-6, IL-17A, and IL-1β) and parathyroid hormone-related peptide are instrumental.[1249]

When inflammation is strong, a spillover of cytokines (e.g., IL-6), circulating activated immune cells, and stimulation of sensory nerves signal inflammation to the rest of the body (Figure 32). The deeper meaning of these messages is not inhibition of the immune system as often stated (the classical view with the negative feedback), but it is much more an appeal for energy-rich fuels: it is an "energy appeal reaction." This has been recently demonstrated after injection

TABLE 18 Components of Extracellular Matrix and Breakdown

Extracellular Matrix	ATP via
Collagen → proline, hydroxyproline	Citric acid cycle
Glycosaminoglycans → *N*-acetylglucosamine[a]	Glycolysis
Glycosaminoglycans → glucuronic acid[a]	Glycolysis
Collagen → glycine	Citric acid cycle

[a]*Hyaluronic acid, chondroitin sulfates, dermatan sulfate, heparin, heparan sulfate, and keratin sulfate are glycosaminoglycans.*

of IL-1β that led to strong hypoglycemia as a fast consequence of fuel allocation to the immune system and other IL-1β-activated pathways.[1250]

Spillover inflammation activates the HPA axis (cortisol) and the sympathetic nervous system (adrenaline and noradrenaline), it induces sickness behavior (cessation of the skeletal muscle, brain, and gut activity), and it reduces sexual activity and reproduction (downregulation of the hypothalamic-pituitary-gonadal axis). All this is part of a program to allocate energy-rich fuels to the immune system by mobilizing glucose from the liver, amino acids from distant skeletal muscles, lipids from distant fat stores, ketone bodies from the liver, and calcium from distant bones (Figure 32).[567] I hypothesize that spillover inflammation involves the entire body to divert fuels to the activated immune system.

At this point, the question appears, which factors can stimulate the energy reallocation program during acute inflammation? A seminal study demonstrated the interrelation between the dose of subcutaneously injected recombinant human IL-6 (rhIL-6), serum levels of IL-6, and increase of energy expenditure in healthy volunteers.[736] It was demonstrated that injection of 0.1 μg rhIL-6/kg b.w. increased serum levels of IL-6 to approximately 10-15 pg/ml, 1.0 μg rhIL-6 led to 45 pg/ml, 3.0 μg stimulated a serum level of 250 pg/ml, and 10 μg was accompanied by an IL-6 serum concentration of more than 1000 pg/ml. In parallel, the maximal increase of metabolic rate in percent of basal metabolic rate was 4%, 7.5%, 18%, and 25%, respectively.[736] This means that a visible influence on energy regulation was observed already at a serum level of 10-15 pg/ml, but the effect was small in these healthy volunteers. In contrast, serum levels of 45 pg/ml were related to an increase in metabolic rate of 7.5%, which would amount to approximately 750 kJ/d (179 kcal/d) in a normal-sized healthy subject (standard metabolic rate: 10,000 kJ/d (2388 kcal/d)). Thus, an increase of serum IL-6 from 1-2 pg/ml as in healthy subjects[741] to 45 pg/ml already induces a marked energy expenditure program.

Chronic inflammatory systemic diseases like rheumatoid arthritis are accompanied by markedly elevated serum levels of IL-6 ranging from 40.0 pg/ml before anti-TNF therapy to 8.0 pg/ml after anti-TNF therapy.[737] Thus, the levels are much higher as compared to healthy subjects (1-2 pg/ml; Ref. 741). Untreated patients with rheumatoid arthritis should increase their daily energy expenditure

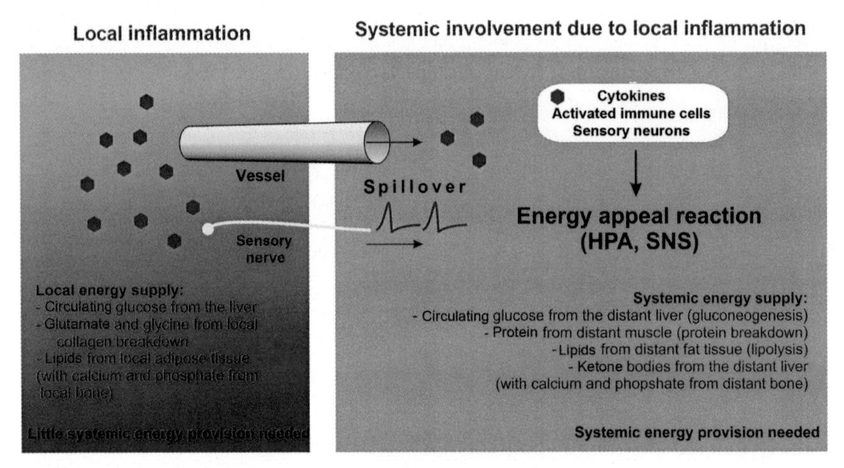

FIGURE 32 Local and spillover inflammation. Local inflammation does not switch on a program of systemic provision of energy-rich fuels. Inflammatory cells use local supplies of energy-rich fuels. When inflammation reaches a certain threshold, spillover of cytokines/activated immune cells and stimulation of sensory nerve fibers announce inflammation to the rest of the body leading to an "energy appeal reaction." At the beginning, the "energy appeal reaction" is mainly driven by the sympathetic nervous system (SNS) and the hypothalamic-pituitary-adrenal (HPA) axis. These neuroendocrine systems provide fuels to the inflammatory process.

by 750 kJ/d (179 kcal/d) (standard metabolic rate: 10,000 kJ/d (2388 kcal/d)). Now, we understand that not only cytokines in the circulation (Figure 7) but also cytokines and other stimulating factors at sensory nerve endings (Figure 21) are critical to induce a systemic response. The systemic response starts a program to redirect energy-rich fuels from stores to activated immune cells. I called it "energy appeal reaction"; others called it "energy demand reaction."[1251]

TOTAL CONSUMPTION TIME

One important aspect of acute systemic inflammation is accompanying sickness behavior,[1252] anorexia, and cachexia, which will be discussed in more detail in Chapter V of the book (see the introducing citation at the beginning of this chapter). Acute self-limiting infectious diseases can be very energy-consuming although, in the presence of disease-induced sickness behavior and related anorexia,[1152, 1253] the intake of energy-rich substrates can be significantly inhibited.[1254]

In considering energy stores available during these acute responses, it is important to note that storage occurs primarily in the fat tissue (12 kg of triglycerides in the body of a contemporary person: 500,000 kJ (119,400 kcal)) and in the skeletal muscles (6-7 kg skeletal muscle protein: 50,000 kJ (11,940 kcal)).[1224, 1225] Under conditions of sickness behavior and anorexia without uptake of energy-rich substrates but an increased sickness-related metabolic rate, the total amount of stored energy would only last for 19-43 days in females and 28-41 days in males (Table 19).

TABLE 19 Total Consumption Time in Human Evolution

Species	Date Range	Body Mass (kg)	Sickness-Related Metabolic Rate[a]	Stored Energy (kJ)	Total Consumption Time (day)
Females					
Australopithecus afarensis	3.9-3.0 Ma	29	7925	152,448	19.2
Australopithecus africanus	3.0-2.4 Ma	30	8061	163,943	20.3
Paranthropus boisei	2.3-1.4 Ma	34	8581	195,237	22.8
Paranthropus robustus	1.9-1.4 Ma	32	8325	177,921	21.4
Homo habilis	1.9-1.6 Ma	32	8325	177,921	21.4
Homo ergaster	1.9-1.7 Ma	52	10,612	404,052	38.1
Homo erectus	1.8 Ma-200 ka	52	10,612	404,052	38.1
Homo neanderthalensis	250 ka-30 ka	52	10,612	372,884	35.1
Homo sapiens	100 ka-1900	50	10,406	294,034	28.3
Homo sapiens	Today (the United States)	74	12,660	545,052	43.1
Males					
Australopithecus afarensis	3.9-3.0 Ma	45	9872	275,502	27.9
Australopithecus africanus	3.0-2.4 Ma	41	9423	238,194	25.3
Paranthropus boisei	2.3-1.4 Ma	49	10,302	303,277	29.4

(Continued)

TABLE 19 Total Consumption Time in Human Evolution—Cont'd

Species	Date Range	Body Mass (kg)	Sickness-Related Metabolic Rate[a]	Stored Energy (kJ)	Total Consumption Time (day)
Paranthropus robustus	1.9-1.4 Ma	40	9308	214,241	23.0
Homo habilis	1.9-1.6 Ma	37	8952	198,282	22.2
Homo ergaster	1.9-1.7 Ma	66	11,956	485,500	40.6
Homo erectus	1.8 Ma-200 ka	66	11,956	485,500	40.6
Homo neanderthalensis	250-30 ka	70	12,313	509,846	41.4
Homo sapiens	100 ka-1900	65	11,865	377,130	31.8
Homo sapiens	Today (the United States)	86	13,648	558,908	41.0
Animals					
Domestic pig (adult)	65 Ma distance[b]	100	13,009	754,611	58.0
Domestic fowl (adult)	300 Ma distance[b]	3.7	1159	21,177	18.3

Ma, million years ago; ka, thousand years ago.
[a]*The sickness-related metabolic rate is given as the basal metabolic rate multiplied by a factor of 1.5 (a 150% increase), which was demonstrated to be a good energy expenditure measure for moderate activation of the immune system.*[1152, 1247]
[b]*Distance means time in evolution to the most recent common ancestor (see Chapter IV). A 10,000 kJ corresponds to 2388 kcal. To obtain the numbers in kcal, divide it by factor 4.2 kJ/kcal.*
The information was taken from a recent publication.[1255]

The number is relatively similar for domestic fowl that have an evolutionary distance to *Homo sapiens* of 300 million years. In other words, an acute consuming infectious disease that uses all energy stores can only last until the stores are empty, say 19-43 days. This necessitates that, for example, an adaptive immune response with antigen uptake, antigen transport, antigen presentation, clonal expansion of T and B cells, lymphocyte egress from secondary lymphoid organs, circulation of lymphocytes, transmigration through vessel walls, appearance in inflamed tissue, eradication of the infectious invader, and shrinkage of clonally expanded lymphocytes must fit into this prespecified time frame of 19-43 days. Indeed, the full picture of this immune response beautifully fits into this time frame. This outstanding example demonstrates that an ill individual (the host of the microbe) and the microbe (the invader) experienced long-standing coevolution.

Due to the physical restrictions of energy storage under natural Paleolithic conditions, an acute infectious disease may not last much longer than 3-6 weeks. It can be hypothesized that energy considerations can help explain the relatively constant time course of an adaptive immune response in the context of acute infection. If an acute inflammatory response lasted longer than the time point at which energy stores were exhausted, the affected person would probably have died due to inanition and starvation, preventing transfer of favorable genes to any offspring. If, however, an immune response lasted for a shorter period, the affected person could survive and be able to transfer favorable genes to offspring (coevolution over eons made it possible).

We perceive a chronic disease when it lasts for more than approximately 6 weeks. This time span is relevant because either energy or, similarly, calcium is lost from stores in the context of sickness behavior/anorexia. This time span is also visible when looking on germinal center reaction of B cells,[1256] maximum of calcium loss during immobilization, healing time of uncomplicated fractures, typical wound responses, and empirical definition criteria of chronic inflammatory diseases such as rheumatoid arthritis and juvenile idiopathic arthritis (Table 20).

These examples demonstrate a time frame from 21 to 60 days (3-8 weeks), during which an acute inflammatory disease or a repair process is usually terminated. Any longer-standing process of inflammatory character might represent a starting chronic inflammatory systemic disease.

VOLUME REGULATION AND INFLAMMATION

Acute inflammatory episodes require abundant energy and are often accompanied by local and systemic water loss (Table 21). Local water loss is particularly important when an inflamed tissue has an exposed surface area to the skin or to inner surfaces of the gastrointestinal, respiratory, or urogenital tract.

For example, water loss through skin wounds depends on the surface area and can amount to $0.35\,ml/cm^2/d$ (area: $30 \times 30\,cm^2 = 315\,ml/d$).[1258] Another example

TABLE 20 Time Frame of Typical Acute Inflammation-Related Phenomena

Acute Inflammation-Related Phenomena	Time Span in Days
Total consumption time (time of death under complete anorexia; see Table 19)	19-43
Calcium consumption time[a]	50
Germinal center reaction of B cells (antibody affinity maturation and termination)	21-28
Maximum of immobilization-induced bone loss	Until 42
Typical wound response after a scratch	21-28
Definition criterion of RA[b]	Longer than 42
Definition criterion of juvenile idiopathic arthritis[b]	Longer than 42

[a] *A theoretical time to depletion of all calcium from nonskeletal tissues in the presence of normal renal calcium excretion and without supply from the bone (calculations in Chapter V).*
[b] *As used in classification criteria in rheumatoid arthritis and juvenile idiopathic arthritis.*[1257]

TABLE 21 Water Loss During Transient Inflammatory Episodes[1255]

Systemic loss of water
By triglyceride and glycogen breakdown from fat tissue and liver
By gluconeogenesis from lactate (Cori cycle between liver and inflamed tissue)
By skeletal muscle breakdown in the context of cachexia
By complete amino acid breakdown in the urea cycle in the liver
By sweating during fever
By perspiration during fever
By cell proliferation (leukocyte proliferation in lymphoid tissue)
Local loss of water from inflamed tissue
Loss into inflamed upper and lower airways (rhinitis, perspiration, and expectoration)
Loss from wounds by evaporation
Loss from inflamed tissue into the gastrointestinal tract (diarrhea and vomiting)
Third-space fluid shifts into pleural cavity, peritoneal cavity, and similar cavities

of water loss relates to insensible perspiration. Under normal conditions in adults, insensible perspiration through the skin and respiratory tract can reach 0.5 ml/kg/h (e.g., 80 kg and 24 h = 960 ml/d).[1259] During surgical intervention, which is a model of mild acute inflammatory activation, water loss by insensible perspiration can amount to 1 ml/kg/h (e.g., 80 kg and 24 h = 1820 ml/d).[1259] These numbers show that loss of water can be quite high during acute inflammatory episodes.

In addition to physical loss of water via outer and inner surfaces, many metabolic reactions that degrade storage forms of energy-rich substrates need water for hydrolysis (Figure 33).[1222, 1239, 1260] For example, the breakdown of one molecule of glucose from glycogen needs two molecules of water (Figure 33, pink area in the left upper corner); triglyceride breakdown needs three molecules of water for three molecules of free fatty acids (Figure 33, pink and yellow areas on the left side); and degradation of skeletal muscle protein needs particularly many molecules of water (Figure 33, brown area in right upper corner).[1239] If amino acids are completely degraded within the urea cycle in the liver, one molecule of water is needed for one molecule of amino acid (Figure 33, pink area in the left upper corner).

Furthermore, the generation of the special Cori cycle between the lactate-producing inflamed tissue and the glucose-producing liver,[1261] involving gluconeogenesis, needs five molecules of water to reprocess one molecule of lactate (Figure 33, red lines with arrows).[1239] Finally, the operation of a proper adaptive immune response with proliferation of B cells and T cells and proliferation of innate immune cells (neutrophils and monocytes) in primary and secondary lymphoid organs needs enormous amounts of water (Figure 33, green area in the left lower corner).[1239] Water loss described above is water loss in the system (the liver, skeletal muscle, fat tissue, lymphoid organs, and others). It is not local water loss in the inflamed tissue because, here, in contrast, water is produced.

Water is formed when glucose, amino acids, and free fatty acids undergo degradation in the Krebs cycle and during oxidative phosphorylation; these processes occur in the mitochondria of activated cells in inflamed tissue (Figure 33, orange area). Thus, important differences exist for water loss from the system and the generation of water from locally inflamed tissue. The amount of locally formed water can be large (Figure 33, orange area), but usually, water is lost rapidly in the environment of inflamed tissue via outer and inner surfaces (Table 21).

To overcome systemic water loss during acute inflammatory episodes and events such as injury, hemorrhage, and burns, a water retention system is activated. The key elements of this system are the sympathetic nervous system that activates the renin-angiotensin-aldosterone system[1262] and the HPA axis with ACTH, aldosterone, and cortisol. ACTH can stimulate aldosterone as well as cortisol, which also has mineralocorticoid activity.[1263] It is not a simple coincidence that the sympathetic nervous system and the HPA axis can redirect energy-rich substrates from energy stores to the activated immune system and serve as major water retention systems. In this context, it is important to recall that the acute adaptive response described by Selye in the 1940s was considered a

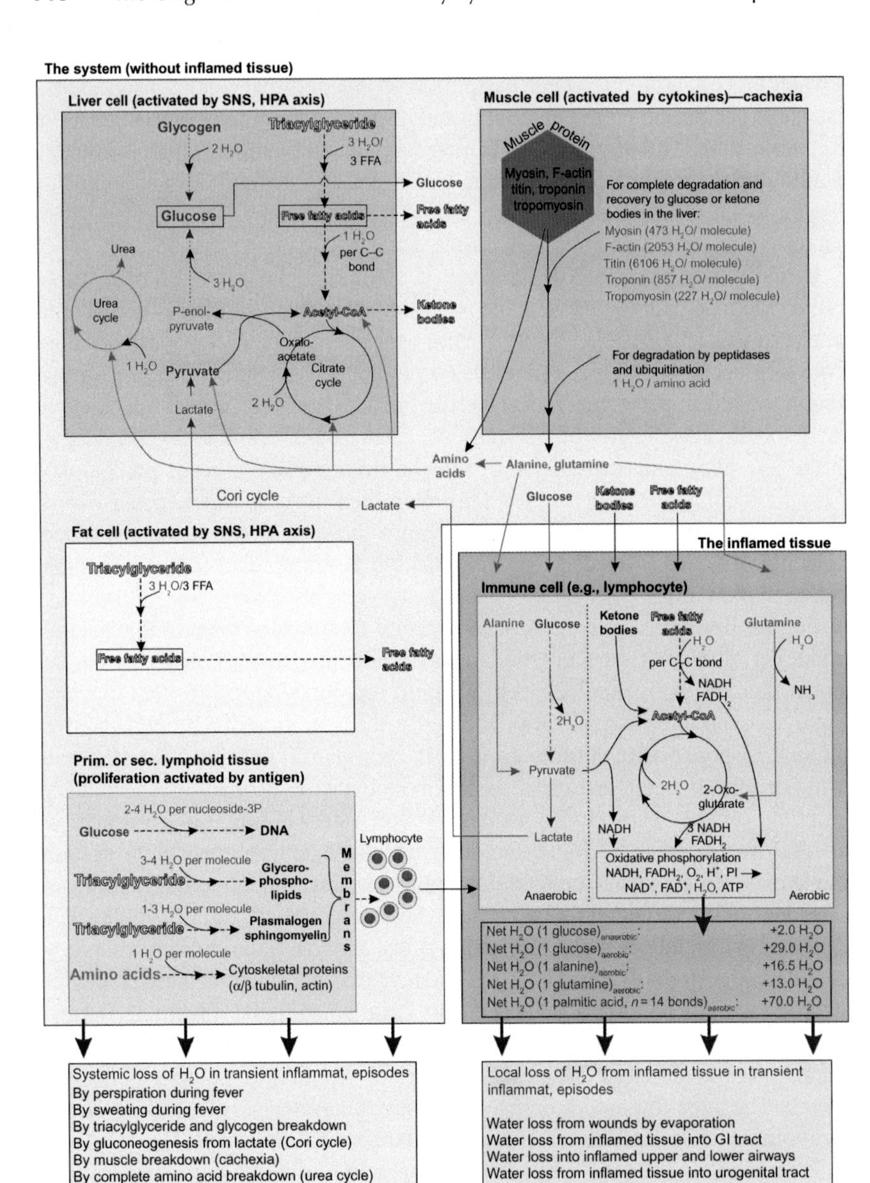

FIGURE 33 See figure Legend on Opposite page.

"nonspecific" alarm reaction.[1264] Today, we can say that the Selye alarm reaction was not "nonspecific" but rather had important physiological functions as both a highly specific energy appeal reaction and a specific water retention reaction.

The sympathetic nervous system and the HPA axis are supported by other alarm hormones such as vasopressin (which is lipolytic and water-retentive), growth hormone (glucogenic, lipolytic, and water-retentive), insulin (lipogenic and water-retentive), and others. Table 22 adds to Table 17. During transient inflammatory episodes, due to sickness behavior and anorexia, one may think that lipogenic insulin may not be a big player because insulin secretion activated by the presence of carbohydrates would be downregulated. However, as will be described in Chapter V of the book, insulin resistance and hyperinsulinemia are hallmarks of inflammation.

These ideas can explain other activities of physiological mediators and provide insights in findings that, for example, water retention hormones such as angiotensin II have proinflammatory activities (see also Chapter II).[1265] Angiotensin II uses the important proinflammatory signaling cascade of NF-κB activation similar as TNF.[1266] The inhibition of the renin-angiotensin-aldosterone system by angiotensin-converting enzyme inhibitors or angiotensin II receptor antagonists exerts anti-inflammatory effects as summarized elsewhere.[1267] In the setting of acute inflammatory episodes such as infectious disease, it is important that the energy appeal reaction and the water retention reaction support the proinflammatory process to eliminate the infectious agent. It can be hypothesized that the water retention hormone angiotensin II is immunostimulatory in this respect.

In conclusion, the water retention system shows important similarities to the energy provision system. This possibility is supported by the proinflammatory activity of some water retention hormones. It is therefore important to consider the role of the water retention system in settings of chronic inflammation

---

FIGURE 33 Water fluxes in the system (blue box) and inflamed tissue (orange box). Blue box: It is demonstrated that the liver cells (lilac box) need water for degradation of glycogen, triglycerides, and amino acids in the urea cycle and for gluconeogenesis in the context of the Cori cycle (red arrows in the blue box). Similarly, degradation of skeletal muscle proteins in the skeletal muscle (brown box) and triglycerides in fat tissue (yellow box) withdraws water from the system. An enormous amount of water is needed for the proliferation of immune cells in primary and secondary lymphoid organs (green box) due to the generation of RNA/DNA, membranes, cytoskeletal proteins, and many others. Orange box: The inflamed tissue is separated from the system. In inflamed tissue, activated cells, mainly immune cells, use provided energy-rich substrates such as glucose, amino acids such as glutamine and alanine, ketone bodies, and free fatty acids to overcome the inflammatory state. In the process of complete degradation of energy-rich substrates to CO_2 and H_2O, enormous amounts of water are generated (black-edged box in orange box). In an acute inflammatory situation, both the system and the inflamed tissue lose water by several pathways (given as boxes below the blue and orange boxes). Abbreviations: C–C bond, carbon–carbon bond; FFA, free fatty acids; GI tract, gastrointestinal tract; HPA axis, hypothalamic-pituitary-adrenal axis; Pi, inorganic phosphate; SNS, sympathetic nervous system. The calculations of water fluxes are derived from biochemistry literature.[1222, 1239, 1260]

TABLE 22 Summary on How Energy Storage and Energy Expenditure Hormones Influence the Kidneys

	Kidneys
Energy storage hormones	
Insulin	Water retention
Insulin-like growth factor 1	
Androgens	
Estrogens	
Vitamin D	Calcium/phosphorus retention
Osteocalcin	
Vagus nerve	
Energy expenditure hormones	
Cortisol	Water/sodium retention
Sympathetic nervous system (noradrenaline/adrenaline)	Water/sodium retention
Growth hormone	Water/sodium retention
Thyroid hormones (T3)	Water/sodium retention
RAAS	Water/sodium retention

Abbreviation: RAAS, renin-angiotensin-aldosterone system.
Parts of this table have been published in Ref. 1231.

such as may occur in aging and in rheumatic and autoimmune diseases undergoing medical treatment (after entering the chronic phase) (see Chapter V, "Hypertension").

SUMMARY

The following list summarizes the major concepts of this chapter:

- Usually, we need 10,000 kJ/d (2388 kcal/d) in our sedentary way of life. Important consumer organs are the skeletal muscles (2500 kJ/d (597 kcal/d)), brain (2000 kJ/d (478 kcal/d)), and immune system (1600 kJ/d (382 kcal/d)). Under mild to moderate acute inflammation, the immune system needs an additional 400-500 kJ/d (96-119 kcal/d), which can rise to 5000 kJ/d (1194 kcal/d) in sepsis.
- Hormones and neuronal systems are important in order to regulate energy storage and energy expenditure (Table 17). The same hormones play important roles in water retention and volume regulation (Table 22).

- Energy is regulated in a circadian fashion. Nighttime consumers are the immune system and repair/growth systems.
- Local inflammation can spillover so that a systemic reallocation program of energy-rich fuels is started. Thus, the Selye alarm reaction is not unspecific but rather specific in order to reallocate energy-rich fuels and to induce water retention.
- The total consumption time is the time evolved until all stored reserves are consumed. It amounts to 19-43 days in humans and hominoids. Other typical repair processes are usually terminated between 21 and 60 days. The time point of 42 days is used as a critical threshold in the definition of rheumatoid arthritis, a chronic inflammatory systemic disease (see Table 20).
- Acute inflammation is often accompanied by severe water loss, which necessitates that inflammatory pathways induce not only an energy reallocation program but also a water retention program. Similar hormonal systems (sympathetic nervous system, HPA axis, and renin-angiotensin-aldosterone system) are active.
- This chapter demonstrated that neuroendocrine regulation of energy allocation and water retention are strongly coupled.

Chapter IV

Evolutionary Medicine

Acting like a sponge, chronic illness soaks up personal and social significance from the world of the sick person.

Arthur Kleinman, medical anthropologist (taken from Ref. 1268)

Chapter Contents

INTRODUCTION

Evolutionary medicine combines information from different disciplines such as evolutionary biology, anthropology, and zoology with human medicine in order to create new paradigms for investigating and understanding human disease.[1269] Sometimes, people argue that evolutionary medicine is teleological in nature, but exactly the opposite is true. Teleology looks for final causes in nature. That is, analogous to human purposes, teleology says, "nature inherently tends toward definite ends and purposes." A form of teleology is Lamarckism with the famous example of giraffes stretching their necks to reach leaves high in trees (the action), which generate offspring with slightly longer necks (the purpose). However, evolutionary medicine explains nature in an opposite way because it is based on the principles of Darwinian evolution. Evolutionary medicine is completely in accord with Ernst Mayr's statement that "adaptedness… is an a posteriori result rather than an a priori goal-seeking."[1270]

The Origin of Chronic Inflammatory Systemic Diseases and their Sequelae.
http://dx.doi.org/10.1016/B978-0-12-803321-0.00004-5

Infectious and noninfectious hazards were present throughout human evolution and evolution of our nonhuman and nonmammalian ancestors. With rats and mice, we share many similar inflammatory mechanisms, although the last common ancestor of humans and rodents lived 65 million years ago.[1271] Humans and chickens had their last common ancestor 310 million years ago, but both species use immunoglobulin gene rearrangement and somatic hypermutation to shape the antibody response toward infectious agents.[1271] Even the shark, 420 million years away, uses a very similar immune system like ours with T cell receptors and immunoglobulin diversity based on V(D)J recombination and the RAG1/2 system.[1272]

It is different for lampreys and hagfish, called jawless fish, which use a more primitive immune system that already consists of variable lymphocyte receptors.[1272] Thus, when we talk about activation of the immune system in the context of the most prevalent danger of infection, humans share many similarities with nonhuman ancestors back in evolutionary time. Table 23 gives an overview of distance to the last common ancestors of humans.

We recognize that acute inflammation must have been a common and old problem during evolution. This is the reason why sharks, birds, rodents, and

TABLE 23 Distance to the Last Common Ancestors of Humans

Species	Years BCE (Mio Years)
Gorillas, chimpanzees, orangutans	6.5
Rats, mice, rabbits	65
Pigs, cows, goats, horses	65
Dogs, cats	65
Extinct dinosaurs	310
Chickens, snakes, crocodiles, turtles	310
Frogs, salamanders	360
Coelacanth	400
Sharks, rays	420
Lampreys, hagfish	460
Sea cucumbers, sea urchins, sea squirts	515
Squids, octopus, snails, lobsters, crabs	530
Insects	530
Sponges	540
Plants, fungi, bacteria	>600

humans share similar immunologic pathways because infectious threats are at least 420 million years old. Of course, infections exist since bacteria exist, our earliest ancestors some billion years ago.

In a similar way, birds, rodents, and humans share many neuroendocrine immune pathways responsible for bodily homeostasis such as the HPA axis and the sympathetic nervous system (Walter Bradford Cannon, 1871-1945, introduced the term homeostasis). Many of the abovementioned hormones and neurotransmitters or at least functional orthologs are present in our ancestors. For example, glucocorticoids of the HPA axis are known in vertebrate animals such as birds, jawed fish, and even in lampreys.[1273, 1274] Functional analogs of corticosteroids—the brassinosteroids—exist in plants in order to regulate immunity, for example, in rice.[1275]

The bilaterally organized sympathetic nervous system is present in jawed fish like sharks and rays and in all vertebrates, but it is absent in lampreys and hagfish,[1276] but lampreys produce catecholamines in monoamine-containing neurons. Even bacteria possess functional orthologs of tyrosine hydroxylase, which converts L-thyroxine to L-DOPA, the first catecholamine in synthesis toward noradrenaline and adrenaline.[1277] Thus, catecholamine production exists for more than a billion years, and it is not surprising that bacteria can use catecholamines as growth factors.[1278]

In addition, our nonhuman ancestors, as studied in birds and rodents, show similar behavior during inflammation such as sickness behavior[1279] or anorexia as compared to humans.[1254] Rainbow trouts can become anorectic under stressful conditions.[1280] Even omnivorous crocodiles develop anorexia and lethargy during infection.[1281]

When we observe cross talk between the central nervous system, the hormonal system, and the immune system, many similarities exist between humans, vertebrates, invertebrates, and even plants. One can summarize that the cross talk between systems has been positively selected throughout evolutionary history to maintain bodily homeostasis.

However, if homeostasis is disturbed by stressful events such as infection (with inflammation), the different systems need to cooperate in order to overcome the threat. The cross talk of these systems has coevolved with infectious agents. This reaction to inflammatory stress is relatively uniform in many species. When stress events such as infection and inflammation are short-lived, there is a good chance to overcome the threat. In Chapter III, I described that energy reserves in the form of energy-rich substrates define the duration of controllable short-lived stressful events (total consumption time; Table 19). The question appears whether chronic long-standing inflammation demanding a significant amount of extra energy does exist in the wild.

It is hard to imagine that fish, birds, or rodents with energy-consuming chronic inflammatory systemic diseases do exist in the wild. Who cares for the affected individual bringing food, keeping a safe place, providing assistance, and so on? Similarly, handicapped humans can be immediate victims to dangerous threats as demonstrated by Jared Diamond.[1282] He described the killing of a traditional New Guinean Dani boy in a small war between tribal Dani people because he could not run fast enough due to an injured leg. What was the chance of a human or nonhuman ancestor to survive in the presence of chronic inflammatory systemic disease that requires a lot of extra energy?

From this point of view, there is a highly critical point in the course of chronic inflammatory systemic diseases, which separates the asymptomatic phase from the symptomatic phase of the disease. I call it the asymptomatic-to-symptomatic threshold or a-s threshold (Figure 34). Genes can be transferred only in the asymptomatic phase of a chronic inflammatory systemic disease because reproduction is not impeded. These transferred genes may confer an increased or decreased risk for a chronic inflammatory systemic disease. After initiation of chronic inflammation, however, reproduction is inhibited. Thus, specific genes, which may have attenuated or boosted chronic inflammation in the symptomatic phase, are not positively selected. One can conclude that genes can be transferred before the a-s threshold, meaning before onset of chronic inflammatory systemic diseases, and these genes end up inside descendants, but the genes relevant in the symptomatic phase end up in the stomach of predators and maggots (Figure 34).

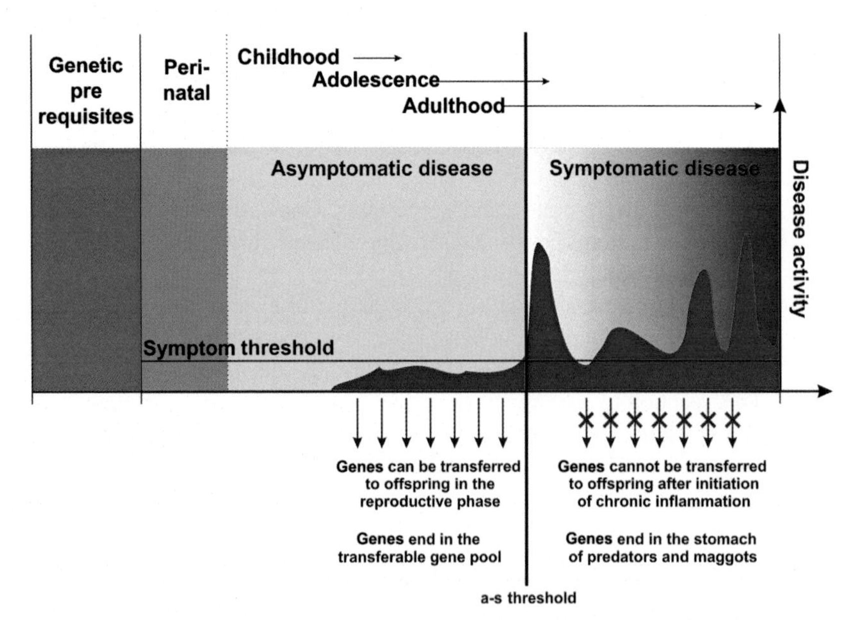

FIGURE 34 The critical point between asymptomatic disease and outbreak of a chronic inflammatory systemic disease, the asymptomatic-to-symptomatic threshold or—in short—a-s threshold. The a-s threshold is given by the bold vertical line.

ARE DISEASE-RELATED GENES POSITIVELY SELECTED DURING EVOLUTION?

Disease-related genes, whether advantageous (by attenuating the disease) or disadvantageous (by boosting the disease), can only be transferred to the progeny if the transfer is not blocked. Blocking in this particular sense means prevention of reproduction, which can happen in several ways (Box).

1. High Negative Selection Pressure

Chronic inflammatory systemic diseases lead to loss of reproducibility because affected individuals are excluded from competition fights for nutrients, positions in the group, and sexual partners. Sometimes, they die in competition fights or, more often, in fights with predators due to their handicap. They also lose their reproducibility because a chronic inflammatory systemic disease is accompanied by severe loss of sex hormones and sexual function, because the reproduction system is switched off during inflammation and severe psychological or somatic stress (this is more elaborately explained in the subsequent part of this chapter; see below). Furthermore, the disease itself leads to early death because no appropriate therapies were available. All conditions imply a *high negative selection pressure.* Under these circumstances, even a young individual in the best years will be a desperate case under the influence of a chronic inflammatory systemic disease.

2. No Selection Pressure At All

Many chronic inflammatory systemic diseases of today manifest in higher ages. Most of the mentioned diseases become apparent after the age of 25 years (Table 24). In the Stone Age, life expectancy was approximately 25 years.[1283, 1284] It was 28 years in Classical Greece and Rome, 33 years in medieval Britain, and 37 years by the end of the nineteenth century in Western Europe. It is still very low at 45 years in underprivileged areas such as Sub-Saharan Africa according to the World Resources Institute. Thus, our human ancestors did not suffer from the chronic inflammatory systemic diseases that we know today. This is even more obvious in our nonhuman ancestors, which use many similar homeostatic pathways as described above but do not reach a similar age than primates.

Prolongation of life as a consequence of hygienic and nutritive practices and modern medicine was not predicted by biological evolution. It is clear that genes were transferred to offspring before the a-s threshold (Figure 34). However, due to the late onset of the disease, a possibly favorable or unfavorable effect of such a gene on the chronic symptomatic disease course has not been subject to selection pressure at all.

3. There was no Time for Natural Selection

Albeit the diagnostic criteria of most chronic inflammatory systemic diseases have not been established before 1960, some publications are available that demonstrate that various chronic inflammatory systemic diseases did not exist some 100-200 years ago. This has been claimed for rheumatoid arthritis and inflammatory bowel diseases. If this is correct—and some indications really exist—then the time span for natural selection was probably too brief.

TABLE 24 Age at Disease Onset in Different Chronic Inflammatory Systemic Diseases

Disease	Mean Age at Onset (Years)	References
Ankylosing spondylitis	26	1285, 1286
Multiple sclerosis	27	1287
Systemic lupus erythematosus	35	1288
Ulcerative colitis	35	1289
Dermatitis herpetiformis	38	1290
Adult rheumatoid arthritis	48	1291
Pemphigus	48	1292, 1293
Primary biliary cirrhosis	49	1294
Adult Sjögren's syndrome	60	1295

All mentioned factors strongly speak against the transfer of genes, which might have influenced the symptomatic chronic inflammatory systemic disease (i.e., after the a-s threshold). Although some diseases such as arthritis, inflammatory bowel disease, psoriasis vulgaris, multiple sclerosis, and ankylosing spondylitis also appear in a juvenile form, the abovementioned items #1 and #3 still apply to the juvenile forms. With respect to item #1, the loss of reproducibility is a severe problem in a young individual with a chronic inflammatory systemic disease.

In chronic inflammatory systemic diseases, sexual function and sex hormones are blocked, which can be derived from several studies. The group of George Chrousos published an important study with human IL-6. IL-6 is a pleiotropic cell hormone (a cytokine) that is released during inflammatory episodes leading to serum levels of 100-1000 pg/ml and more. They found that a single injection of IL-6 blocked the release of testosterone, by approximately 50%, and it needs 7 days until the restoration of normal serum levels (Figure 35). In the acute phase of inflammatory diseases, IL-6 is always increased and, thus, affects sexual function and sex hormone release for a long time. During a chronic inflammatory systemic disease, this cytokine remains elevated if no appropriate treatment is applied. This is a stunning example how an inflammation-related cytokine can heavily influence sex hormone secretion.

There is another clear evidence that proinflammatory cytokines such as IL-6 and TNF, another important proinflammatory cytokine, directly inhibit the gonadal cells responsible for testosterone secretion.[1296] The injection of another inflammatory cytokine, IL-1β, into the brain of experimental animals decreased

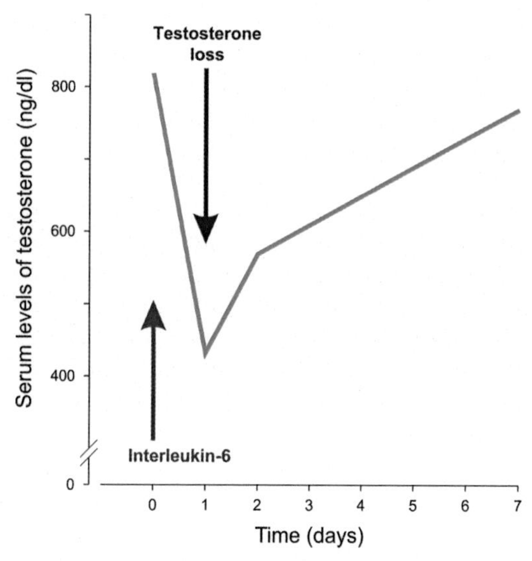

FIGURE 35 Schematically demonstrated influence of IL-6 on serum levels of the male sex hormone testosterone. IL-6 was injected on day 0 (red arrow), and serum levels of testosterone were measured in the subsequent 7 days. The most dramatic fall of testosterone is already observed on day 1 after injection (blue arrow).[1307]

testosterone production for a long time.[632] The same cytokines influence sperm motility resulting in reduced mucosal penetration properties and decreased male fertility.[1297] Similarly, the female reproduction is influenced by inflammatory cytokines such as IL-1 and TNF,[1298, 1299] and infectious episodes reduce female fertility.[1300] Others demonstrated the deleterious effect of inflammation on the menstrual cycle leading to amenorrhea.[1301] Several reports described sexual dysfunction in juvenile forms of chronic inflammatory systemic diseases,[1302–1305] and even with optimum treatment nowadays, fertility is reduced in girls with chronic inflammatory systemic diseases.[1306]

George Chrousos and his group concluded in the before-mentioned publication:[1307] "This may be adaptive for limited periods of time, as it would direct energy to defense rather than reproduction." These authors discussed their findings in the context of a short-lived inflammatory episode (an infectious episode), but it also applies to chronic inflammatory systemic diseases. In chronic inflammatory systemic diseases, we not only know that it is a short-lived episode of inflammation but also, under natural conditions without adequate therapy, observe progression of the disease until death. Such a linkage between blunted reproductive functions and proinflammatory cytokines has not allowed the evolutionary acquisition of well-adjusted responses to impede or modulate the chronic development of a disease.

In conclusion, a disease-related gene, attenuating or boosting inflammation in the symptomatic phase of the disease, has not escaped the predator's or the maggot's stomach. Nevertheless, those genes transferred to the progeny early in life might play a decisive role.

ACCUMULATION THEORY OF CHRONIC INFLAMMATORY SYSTEMIC DISEASES

Under natural conditions, a chronic inflammatory systemic disease would be hardly expressed for a long time in the form of a chronic disease because of the need to cope with the environmental threats and to compete for nutrients. Natural selection has limited the opportunity to exert a direct influence over the process of being chronically ill (see above). However, an individual that would suffer from a chronic inflammatory systemic disease later in life can transmit to the progeny, before the a-s threshold, the genetic background that predisposes to the disease.

Some related genes have long been recognized in chronic inflammatory systemic diseases. The best known associations between chronic inflammatory systemic diseases and genes have been established for the abovementioned HLA system (immunogenetics as described in Chapter I), which is an important element of antigen presentation (Table 25).

Sometimes, the risk to develop a disease in the presence of a certain subtype of the HLA system can be high (Table 25). However, patients without suspect HLA proteins develop the mentioned diseases, too. And 10-25% of the population carry risk HLA genotypes, but the prevalence of any of the chronic inflammatory systemic diseases is 0.1-1%,[1312] and thus far below the expected 10-25%. There have to be other reasons because a certain HLA protein is not necessary and not sufficient to elicit a certain chronic inflammatory systemic disease.

Today, we label such an alteration a genetic disease-promoting risk factor. If the alteration of an HLA protein is not the single causative factor, we expect that other genetic factors are also relevant. Indeed, chronic inflammatory systemic diseases have a multifactorial genetic background as demonstrated, for example, in rheumatoid arthritis (for other diseases, see Chapter I).[496] Therefore, HLA genes were retained in the offspring because in most cases the disease did not develop due to the lack of additional factors. In rheumatoid arthritis, many other genetic prerequisites have been described (Figure 36). The genetic prerequisites for a chronic inflammatory systemic disease can be retained over generations and generations via individuals that never express the disease because relevant cofactors either may not occur or are expressed following very irregular patterns. We also know that the accumulation of several genetic risk factors increases the risk to develop a chronic inflammatory systemic disease. This is the accumulation theory (Figure 36).

There is another important aspect that needs to be mentioned. From monozygotic twin studies, we know that genes contribute to the appearance of chronic inflammatory systemic diseases. In these twin studies, the concordance rate is

TABLE 25 Human Leukocyte Antigen (HLA) Association of Diseases

Disease	HLA Association	Relative Risk	References
Narcolepsy	Various HLA-DR and DQ alleles (DQB1*06:02)	>90	1308
Ankylosing spondylitis	HLA-B27	90	1285, 1286
Type 1 diabetes mellitus	HLA-DR3, 4	20	1309
Dermatitis herpetiformis	HLA-DR3	15	1290
Adult rheumatoid arthritis	HLA-DR3, 4	10	1291
Systemic lupus erythematosus	HLA-DR3	6	1288
IgA nephropathy	HLA-DR4	4	1310
Multiple sclerosis	HLA-DR2	4	1287
Hashimoto's thyroiditis	HLA-DR4	3	1311

The list is ordered according to the relative risk. Only some examples are given for those diseases that also appear during adulthood.[330]

determined as the quantitative statistical expression for the concordance of a given genetic trait. Some examples are given in Table 26.

The concordance rate is somewhere between 0% and 50%, and on average for the mentioned diseases in Table 26, it is 25%. Thus, one needs to consider gene-environment interactions, because more than three quarters of influential factors seem to depend on the environment. From this point of view, the full accumulation theory also needs to include the environmental aspect (Figure 37). This is often forgotten although the environmental influence can be as high as 100% as given in a somewhat surprising Danish study in patients with rheumatoid arthritis (Table 26).[1313]

At this point, the environmental example of smoking and rheumatoid arthritis is presented (was already discussed in section "The Trigger of Chronic Inflammatory Systemic Diseases" of Chapter I). Already in the 1970s, a link was demonstrated between smoking and increased serum levels of autoantibodies.[1319] The first epidemiological studies on the link between smoking and the risk to develop rheumatoid arthritis appeared in the early 1990s.[1320–1322] One study showed that smoking increased the risk to develop rheumatoid arthritis by

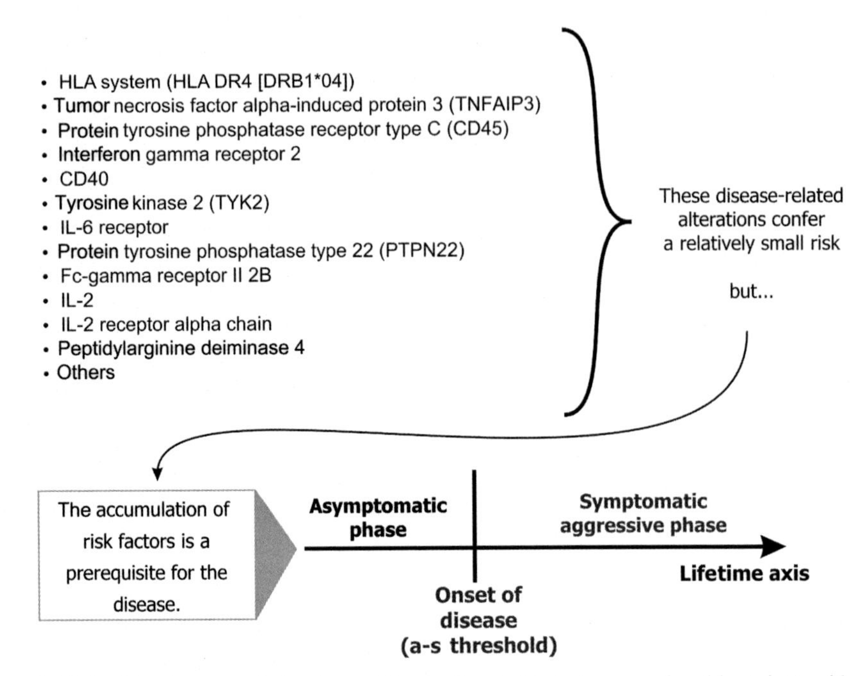

FIGURE 36 The accumulation theory. The mentioned risk factors were found in patients with rheumatoid arthritis (in different populations).[496] The risk factors confer only a relatively small risk to develop the disease, but they are important prerequisites. Prerequisites represent the basis of a chronic inflammatory systemic disease, and they are already present after conception, which means that on the lifetime axis, they are around at the very beginning of life. Abbreviations: HLA, human leukocyte antigen; IL, interleukin; TNF, tumor necrosis factor.

TABLE 26 Concordance Rate in Twin Studies in Chronic Inflammatory Systemic Diseases

Disease	Monozygotic	References
Rheumatoid arthritis	0-21	1312–1314
Hashimoto's thyroiditis	17	1315
Systemic lupus erythematosus	11-24	1312, 1316
Multiple sclerosis	6-31	1312
Graves' disease	22	1312
Type 1 diabetes mellitus	13-38	1312
Psoriasis	35	1317
Ankylosing spondylitis	50	1318

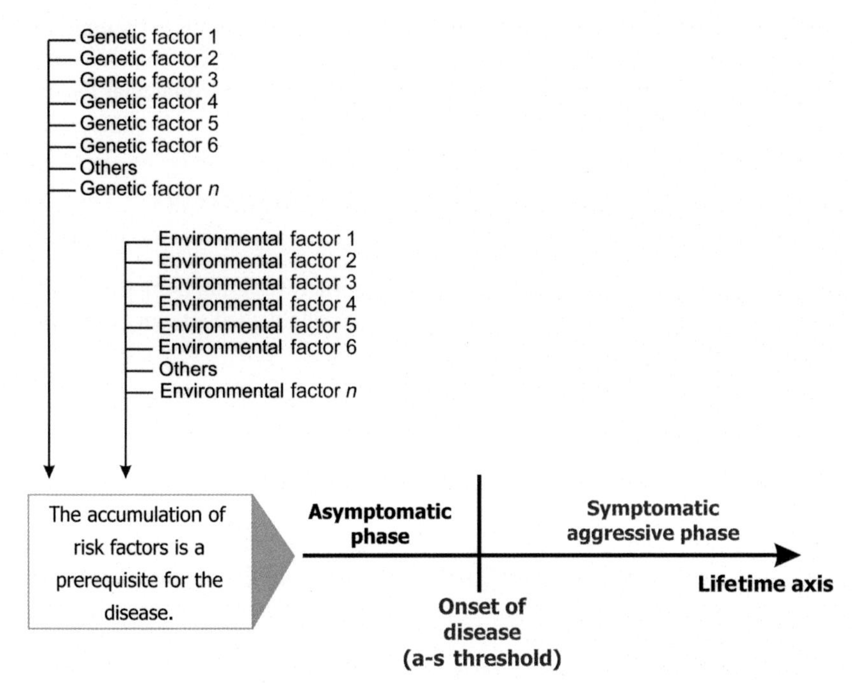

FIGURE 37 The accumulation theory under consideration of genetic and environmental factors. Genetic and environmental factors can influence the individual at any time point on the lifetime axis.

a factor of 12 in the smoking monozygotic twin compared with the nonsmoking monozygotic twin, which is a situation that controls the genetic influence.[1323] The risk remains elevated for several years after smoking cessation.[1324] This demonstrates that smoking is a very strong risk factor.

Newer models described by Lars Klareskog show that smoking may trigger HLA-DR-restricted immune reactions to autoantigens modified by citrullination (recall, this is a mechanism of posttranslational protein modification leading to autoantigenic neo-self epitopes).[1325] Similarly, alcohol—an environmental factor—protects from developing rheumatoid arthritis (immunosuppressive role of alcohol).[1326] Intake of oily fish was associated with a modestly decreased risk of developing rheumatoid arthritis.[1327] Another important risk factor for rheumatoid arthritis and other autoimmune diseases is silica exposure as can happen in construction work (cement and demolition).[1328, 1329] It was discussed that silica is a stimulator of the immune system, which may trigger the autoimmune process.[1328, 1329]

Importantly, smoking is also linked to an increased risk to develop other chronic inflammatory systemic diseases such as ankylosing spondylitis,[1330] multiple sclerosis,[1331] Crohn's disease,[1332] and systemic lupus erythematosus.[1333, 1334]

However, smoking can also reduce the risk of developing chronic inflammatory systemic diseases such as type 1 diabetes mellitus,[1335] ulcerative colitis,[1332] and pemphigus.[1336] These examples clearly demonstrate that environmental factors are essential in the accumulation theory.

PLEIOTROPY THEORY OF CHRONIC INFLAMMATORY SYSTEMIC DISEASES

Williams, an evolutionary biologist, studied an interesting phenomenon in the context of aging research.[1337] Already in 1957, he established a theory that says the following:

> *It is necessary to postulate genes that have opposite effects on fitness at different ages, or, more accurately, in different somatic environments.*

This theory is now widely accepted in aging research, and it is called "antagonistic pleiotropy." By the way, the word pleiotropy comes from the Greek pleio, meaning "many," and trepein, meaning "influencing." Pleiotropic means that genes can have quite different functions at different time points in the lifetime axis. I state that the same theory applies to the situation in chronic inflammatory systemic diseases.

There may be some genes that were positively selected because they confer better reproduction, stronger skeletal muscles, or better fight-or-flight responses (stronger activation of the sympathetic nervous system or HPA axis; Walter Bradford Cannon, 1871-1945, introduced the term fight-or-flight response). In this case, selection was not hindered but positively supported because the gene carrier was more likely to mate or to win in competition fights.

However, such a gene can play a deleterious role in an elderly person suffering from a chronic inflammatory systemic disease. For example, a gene may confer an excellent immune response against infectious agents because it leads to a stronger activity of immune cells or better bacteria/virus recognition. This gene is advantageous in younger ages because the gene carrier has less childhood infections and better chances to mate and fight. However, when the gene carrier would suffer from a chronic inflammatory systemic disease later in life, the same gene might have an unfavorable role because it can overactivate the immune system and kill the carrier. The link between the HLA system, infection, and set of secreted cytokines was nicely summarized by Chella David's group from the Mayo Clinic.[108] Some more examples can be found in Table 27.

Let us make another example from the hormonal world. It is known that women with an early onset of menorrhea, with shorter cycle lengths and a higher number of ovulatory cycles, have an increased risk of early menopause.[1338, 1339] In order to understand this phenomenon, one has to recall that ovaries are equipped with a limited number of oocytes, and, with every ovulation, oocytes are released and the remaining number decreases gradually. Under

TABLE 27 Examples of Antagonistic Pleiotropy for Genes that Increase Risk or Severity of Chronic Inflammatory Diseases[1340]

Genes	Chronic Inflammatory Disease	Pleiotropic Meaning Outside of Chronic Inflammatory Diseases (with Selection Advantage)	References
HLA-DR4 (DRB1*04)	Rheumatoid arthritis and other autoimmune diseases	Decreased risk of dengue hemorrhagic fever (defense against infectious agents)	1341
HLA-B27	Ankylosing spondylitis and other axial forms of spondyloarthritis	Decrease in viral infection (defense against infectious agents)	1342, 343
PTPN22 1858 C>T[a]	Many autoimmune diseases	Higher body mass index, higher waist-to-hip ratio in women (storage of energy-rich fuels)	1344
CTLA4 49 A>G	Many autoimmune diseases	Better defense against hepatitis B virus and *Helicobacter pylori* (defense against infectious agents)	1345, 1346
NOD2/CARD15	Crohn's disease	Hypertension (activation of the sympathetic nervous system and, thus, the fight-or-flight response)	1347

[a]*PTPN22 1858 C>T is associated with many autoimmune diseases but is also linked to a higher risk of infection. This seems to contradict the theory of antagonistic pleiotropy. Until today, nobody has focused on a possible selection advantage in the context of reproduction. It might well be that this mutation is related to a decreased T cell-dependent rejection of the semiallogenic fetus. This would support reproduction. The PTPN22 1858 C>T mutation is linked to a risk of endometriosis, which already demonstrates a role in the context of reproduction.*[1348]

consideration of an early onset of menorrhea and shorter ovulatory cycles, more oocytes will be released until age 40, which causes earlier menopause. This is advantageous for reproduction because the carrier of the trait can mate earlier and more often.

It seems to be clear that genetic reasons can exist for an early onset of menorrhea and premature adrenarche,[1349] which can tell us that this trait is genetically determined. But what has it to do with chronic inflammatory systemic diseases? It is widely accepted that the incidence rate of rheumatoid arthritis in women largely increases after menopause (Figure 38).[756]

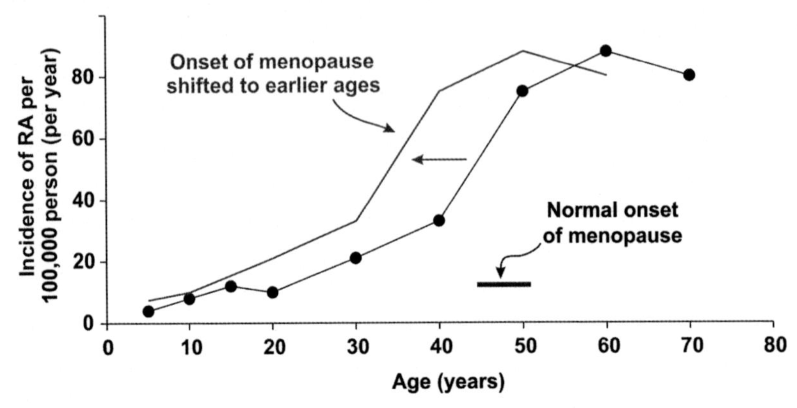

FIGURE 38 Incidence rate of rheumatoid arthritis (RA). The black line indicates the normal situation with a menopause starting at 45 years of age. The red line demonstrates a fictitious situation with accelerated menopause years before.

When the menopause is accelerated, rheumatoid arthritis can develop earlier in a woman at risk. Thus, a gene conferring an early onset of menorrhea, which might be favorable for sexual function and reproduction early in life, would lead to early menopause and an earlier onset of rheumatoid arthritis. Indeed, there exist links between early menarche and the risk of rheumatoid arthritis.[1320]

There is another handsome example that is directly related to the genetic risk factor HLA-DR4 (DRB1*04) known in rheumatoid arthritis and other chronic inflammatory systemic diseases. In a Mexican population, it was demonstrated that HLA-DR4 (DRB1*04) was highly negatively associated with the risk of dengue hemorrhagic fever.[1341] HLA-DR4 homozygous individuals were 11.6 times less likely to develop dengue hemorrhagic fever in comparison with DR4 negative persons. This clearly demonstrates that HLA-DR4 (DRB1*04) is a genetic factor that is protective against this infectious disease.[1341] Thus, it might well be that HLA-DR4 (DRB1*04) has been evolutionarily positively selected to fight off dengue hemorrhagic fever or other microbes, but later in life, this particular surface molecule supports the development of a chronic inflammatory systemic disease. Some may argue that dengue infection does not exist in many parts of the world. However, since we are all descendants of African *Homo sapiens*, dengue was relevant some 100,000 years ago when our ancestors lived in Africa.

The Fc-γ receptor IIIA is responsible for the recognition of immunoglobulin G leading to immune cell activation. As is known, this recognition of immunoglobulin is important to detect bacteria and viruses, which are bound to immunoglobulin G (called opsonization). After binding to the cell surface, Fc-γ receptor IIIA and immunoglobulin G together with bacteria or viruses are taken up, and the cell gets stimulated to destroy the microbe. Interestingly, the Fc-γ receptor IIIA with a polymorphism in position 158, homozygous for the amino acid valine, is associated with a higher risk to develop rheumatoid arthritis.[1350] Interestingly, this Fc-γ receptor IIIA (158 valine/valine) has a high binding affinity for immunoglobulin G when compared with other Fc-γ receptors.[1351]

In addition, patients with poliomyelitis were less likely to have this particular Fc-γ receptor IIIA (158 valine/valine),[1352] which indicates that this receptor confers protection against the poliovirus. Again, it seems that the genotype 158 valine/valine of the Fc-γ receptor IIIA has been evolutionarily positively selected to fight off infectious agents by increasing microbe clearance, for example, of the poliovirus. This trait was most probably advantageous in early ages, but the same surface molecule assists the development of a chronic inflammatory systemic disease in later years on the lifetime axis.

Pleiotropy occurs when a single gene influences multiple phenotypic traits. Depending on the "somatic environment," such a gene can be largely advantageous for the carrier in earlier years in reproductive time, but it can also be highly disadvantageous in later ages in another somatic environment such as aging or autoimmune activation. These considerations might open our eyes when we study genetic risk factors in patients with chronic inflammatory systemic diseases. It also explains why a risk gene is most probably not specific for a given chronic inflammatory systemic disease, because the gene was evolutionarily positively selected for a different function. Such a gene had an important role for the survival of the species in early ages, but the question appears whether it was always positive for the individual.

ARE EVOLUTIONARILY POSITIVELY SELECTED MECHANISMS ALWAYS ADVANTAGEOUS FOR THE INDIVIDUAL?

Years ago, I had a debate with a colleague in the department. The story was going like this: Are evolutionarily positively selected genes always advantageous for the individual gene carrier? My colleague was convinced that it must be always advantageous: "Our body knows what to do. Evolutionarily positively selected genes always serve the carrier in a beneficial way." This debate was started because my colleague thought that specific genetic programs are available, which attempt to get rid of a specific chronic inflammatory systemic disease. This necessarily implies that specific genes were evolutionarily positively selected during the symptomatic phase of a chronic inflammatory systemic disease, meaning after the a-s threshold of Figure 34. As it was discussed before, it is highly unlikely that natural selection has permitted such programs to be retained. The debate was fired by experiments of nature, which showed that evolutionarily positively selected genes can even be disadvantageous—or even fatal—for the individual but highly advantageous for the species. Here are some examples.

Antechinus is a marsupial the size of a large mouse, with a pointed snout and a short-haired, medium-long tail living in Australia. The reproduction of this species is relatively unique, and is an important part of the life cycle of this animal, as the males die after mating in their first year and the females survive to mate for more years. Mating takes place once a year and the short mating season is apparently stimulated by a certain increase in daylight during the second half of winter (Australian summer). During this time, males travel extensively between communal nests in a hectic mating frenzy. Mating can take up to 12h, with the death of the males shortly after copulation. The reason for this is that with all the male's attention and energy taken up with sex, rather than feeding, stress and sex hormones strip the body of protein and fat (glucocorticoids and testosterone).[1353] The result is a breakdown in the animal's immune system and death within 2 weeks. This male die-off is largely brought on by severe infection, particularly, owing to parasites. Removing the male *Antechinus* from the natural habitat leads to a life expectancy similar to female animals. This is not a good beneficial program for an individual male *Antechinus*, isn't it?

A second example demonstrates a similar behavior of another species. Salmon spend part of their life as freshwater fish (youth), part as saltwater fish (adulthood), and part as freshwater animals again during reproduction (before death). When they return to rivers from the ocean to give birth, salmon do not eat. Spawning salmon live solely on accumulated skeletal muscle and fat reserves for several months. In rivers brimming with food, 20 pound salmon will drive themselves to the edge of starvation. Over the 3-5 months between arriving from the ocean, migrating upstream, and spawning in the late fall, salmon expose themselves to predators. Salmon in rivers will jump and roll and swirl on the surface of pools, for no apparent reason, even though this reveals their location to predators and makes them vulnerable to death or injury during that critical period before they mate. After mating, salmon die due to exhaustion and parasitic infections.

These stunning examples show that genetically inherited programs are fatal for the individual but highly advantageous for the species. It makes very clear that not all genetic programs are good, and, sometimes, the program might even be fatal.

BORROWED GENES FOR CHRONIC INFLAMMATORY SYSTEMIC DISEASES: ENERGY CONSUMPTION VERSUS ENERGY PROTECTION

At this point, we recall that "disease-related genes, whether advantageous (by attenuating the disease) or disadvantageous (by boosting the disease), can only be transferred to the progeny if the transfer is not blocked." We have learned that genetic programs are not specifically positively selected for the symptomatic phase of a chronic inflammatory systemic disease because the transfer of respective genes to the offspring most probably was blocked.

This nonspecificity is well known in clinical medicine because symptoms in various diseases are surprisingly similar. Think of the C-reactive protein, which most often climbs independent of the type of acute inflammation or chronic inflammatory systemic disease. Or think of fatigue that often accompanies acute inflammation or chronic inflammatory systemic diseases. Many more common symptoms are summarized in Table 28.

And when we come to the heart of inflammation looking into tissues, we also see very similar phenomena on cellular and molecular levels. In our department, we had the fortune to study different chronic inflammatory systemic diseases in parallel. There were gastroenterologists with a foible for inflammatory bowel diseases and liver cirrhosis, rheumatologists with a weakness for rheumatoid arthritis and other rheumatic diseases, endocrinologists with a focus on

TABLE 28 Common Symptoms in Chronic Inflammatory Systemic Diseases

Overt Symptoms	Change
Body temperature and sweating	Increases
Erythrocyte sedimentation rate	Increases
Interleukin-6 serum levels (one example of a cytokine in the blood)	Increases
Hemoglobin per erythrocyte	Decreases
Serum albumin	Decreases
Heart rate	Increases
Protein in the urine	Increases
Physical activity	Decreases
Food intake, appetite (finally body weight)	Decreases
Disposition to pain (skeletal muscle, joints)	Increases
Headache	Increases
Libido, erectile dysfunction	Decreases
Amenorrhea	Increases
Fatigue	Increases
Sleeping problems	Increases
Weakness	Increases
Vertigo	Increases
Numbness	Increases
Avolition	Increases
Symptoms of depression	Increases

Think of a situation without adequate therapy showing the full picture. This list is certainly not complete.

fat tissue and proinflammatory adipokines, and finally clinical infectiologists. Sometimes, it was startling that diseases such as Crohn's disease and rheumatoid arthritis had many phenomena in common. Initially, we expected different pathophysiological pathways but, more and more, the things amalgamated leading to a common mechanistic view. For me, this was a strong stimulus to consider a more unifying concept underlying these similarities. Inevitably, I had to think of common genetic programs used in symptomatic chronic inflammatory systemic diseases. Where are they coming from?

Earlier, it was demonstrated that a highly activated immune/repair system cannot be switched on for a long time because this would be very energy-consuming (section "Total Consumption Time" of Chapter III).[1255] A highly activated immune system is accompanied by sickness behavior and anorexia that prevents adequate food intake and necessitates life on stored reserves (inflammation-induced anorexia). Under systemic inflammatory conditions, breaking down all reserves takes 19-43 days (section "Total Consumption Time" of Chapter III, Table 20) [1255]. A highly activated immune/repair system needs huge amounts of energy, which is exemplified in the case of extensive burn wounds (up to 20,000 kJ/d [4776 kcal/d]).[1255] Although this aspect demonstrates the extreme of the spectrum, it indicates that energy consumption is a critical factor during evolution.

I hypothesize that energy consumption and energy storage are the most critical determinants in evolution to undergo either negative selection or positive selection, respectively. If alterations of homeostasis lead to marked energy consumption, the situation cannot be chronic—it must be acute. Since the total consumption time ranges between 19 and 43 days,[1255] an acute energy-consuming change of homeostasis must be started and terminated in this time frame.

A very good example for this time window is the germinal center reaction of B cell expansion and contraction that happens within approximately 21-28 days.[1354] Most acute disease states are terminated within this time frame such as infectious diseases, wound healing, and repair, but also strong mental activation in stressful situations must be terminated because they are energy-consuming exemplified in short-term stress (see also Table 20).[1355] During evolution, respective homeostatic networks were positively selected for short-lived acute energy-consuming responses, but not for long-standing chronic inflammatory systemic diseases or chronic mental illness. These chronic situations generated a huge negative selection pressure.

In contrast, if mutations were helpful to protect energy reserves, they were positively selected during evolution. This is true for memory responses because immediate reaction of an educated system can spare energy reserves. This is exemplified by the immune memory that leads to shorter, more effective, and, finally, less energy-consuming reactions toward microbes. Importantly, acquisition of immune memory during the primary contact must fit into the above specified time frame of 19-43 days (and this happens as exemplified by the germinal center reaction in secondary lymphoid organs). In this context, tolerance versus harmless foreign antigens of microbes on body surfaces (see gut, skin, respiratory tract, urogenital tract) or harmless autoantigens is a memory function that spares energy reserves. Sometimes, microbes such as *Mycobacterium tuberculosis*, *Mycobacterium leprae*, and viruses enable or mimic tolerant immune responses leading to long-standing infection but finally leading to death due to emaciation.

Similarly, neuronal memory can largely decrease time to accomplish successful foraging in the wild.[1356] Neuronal memory systems are tuned to ancestral priorities in the context of foraging and other Paleolithic tasks.[1357] Additionally, tool making, invention of language and writing, and storage of data on computer hard disks protect time and, thus, energy for otherwise necessary reinvention.

Another example of positively selected gene variants are observed for food intake and fat storage, both of which are important in determining the abovementioned total consumption time. Indeed, a female *Australopithecus afarensis* had a consumption time of approximately 19 days, while a modern female *H. sapiens* can rely on 43 days (Table 19, Chapter III) [1255]. Particularly, fat storage has been markedly increased over the last 3-4 million years of human evolution. Not surprisingly, the latest meta-analysis of genome-wide association studies of obesity and the metabolic syndrome found polymorphisms in genes relevant for food intake such as *FTO* (fat mass and obesity associated), *MC4R* (melanocortin-4 receptor), *POMC* (proopiomelanocortin, the precursor of melanocortin), and genes relevant for fat storage such as the insulin-stimulating *GIPR* (gastric inhibitory polypeptide receptor).[1358]

Another important indication for positive selection of fat storage networks is given by the fact that the number of adipocytes in humans is determined before puberty.[1359] After puberty, the total number of adipocytes stay constant with an annual exchange rate of 10%.[1359] If evolution led to a phenomenon relevant before reproduction time, it will be easily transferred to offspring when it is an advantageous trait. Since the phenomenon still exists in modern children,[1359] we expect that fat storage was an important factor during evolution.

Similarly, humans can deposit large amounts of fat already *in utero* and are consequently one of the fattest species at birth.[1360] In addition, newborn humans devote roughly 70% of growth expenditure to fat deposition during early postnatal months, which reduces the risk of energy stress during possible postnatal infections.[1360] If they are not able to store large amounts of fat tissue *in utero* or if malnutrition is a problem in fetal life, it seems that a postnatal program is switched on that supports obesity already during childhood and adolescence. This was called the thrifty phenotype theory.[1361, 1362] Again, this is an indication that important positively selected gene variants exist that serve storage of energy. Going back to the immune system, genes and networks are positively selected under defined conditions (Table 29).

The evolutionary principle of replication with variation and selection is undeniably fundamental and it has history. With the words of the physicist Paul Davies, in his book Cosmic Jackpot, "life as we observe it today is 1% physics and 99%

TABLE 29 Positively Selected Immune Mechanisms under Defined Conditions of (A) Acute, Highly Energy-Consuming Responses Terminated Within 3-8 Weeks and (B) Long-Standing, Energy-Protective Responses

A. Positively Selected for Acute, Highly Energy-Consuming Situations	B. Positively Selected to Protect Energy Stores
Immune response due to infection	Tolerogenic immune reactions
Immune response to foreign bodies	Control of inner and outer body surfaces
Clonal expansion and apoptosis	Memory of the immune system
Wound healing, burn wounds	Replacement of cells and tissue (physiological regeneration and degeneration)
Implantation of stem cells into injured tissue	Implantation of a blastocyst into the uterine epithelium
Specific immunoglobulin production and affinity maturation	Immune phenomena facilitating semiallogenic pregnancy
High production rate of cytokines and chemokines	Allergic reactions (preformed response to clear or block threats on body surfaces)
Increased rate of phagocytosis	
Immune-stimulated neoangiogenesis and wound healing	

The list is not complete.

history." It was a successful history of positive selection, which can only happen under circumstances of unrestricted gene transfer to the progeny. Wouldn't it be highly unprofitable if for every type of inflammation new mechanisms are to be invented? Or can you imagine an IL-6 for herpes virus infections such as shingles and a slightly different IL-6 for *Staphylococcus* infections and another one for mumps or rheumatoid arthritis? No! This is a waste of time and energy.

In conclusion, genes and networks are positively selected if they serve acute, highly energy-consuming situations, which are terminated within 3-8 weeks (inflammatory episodes), or gene variants are positively selected if they protect energy stores, which are relevant during their entire life (beyond 3-8 weeks).

SUMMARY

- Disease-related genes, whether advantageous (by attenuating the disease) or disadvantageous (by boosting the disease), can only be transferred to the progeny if the transfer is not blocked.

- In chronic inflammatory systemic diseases, gene transfer can happen before the a-s threshold, during the asymptomatic phase, but not in the symptomatic phase.
- Chronic inflammatory systemic diseases are not "programmed" but result largely from the accumulation of genes that confer an increased risk. Don't forget the environmental influence, which is higher than the genetic influence!
- In addition, there may be adverse gene actions at older ages arising from purely deleterious genes that escaped the force of natural selection or from pleiotropic genes that trade benefit at an early age against harm during the course of chronic inflammatory systemic diseases.
- Evolutionarily positively selected genetic programs can be disadvantageous for the individual.
- Genes and their programs used in symptomatic chronic inflammatory systemic diseases were derived either from acute, highly energy-consuming situations, which are terminated within 3-8 weeks (inflammatory episodes), or from programs that protect energy stores. Sometimes, microbes imitate immunotolerant functions so that chronic smoldering inflammation is a consequence.

Chapter V

Origin of Typical Disease Sequelae

Beauty is the proper conformity of the parts to one another and to the whole.

Werner Heisenberg

Chapter Contents

The Origin of Chronic Inflammatory Systemic Diseases and their Sequelae.
http://dx.doi.org/10.1016/B978-0-12-803321-0.00005-7

The reader reached the critical point of the book because, now, the different elements are combined to get a more complete explanation of systemic disease sequelae related to chronic inflammatory systemic diseases. These sequelae are not induced by autoimmunity itself. This means that recognition of autoimmune targets by T cells or B cells or their products is not responsible for the observed phenomena (e.g., circulating autoantibodies or immune complexes do not play a direct role).

This chapter advances the hypothesis that systemic sequelae in the above-mentioned sense result from the prolonged use of adaptive programs positively selected for short-lived inflammatory episodes. To integrate emerging data from studies in metabolism, neuroendocrinology, and immunology, I am proposing a new model to explain sometimes baffling symptomatology and to find a common denominator of those systemic disease sequelae given in Table 30.

SICKNESS BEHAVIOR, FATIGUE, AND DEPRESSIVE SYMPTOMS

All of us have experienced episodes of viral or bacterial infection. Particularly, in childhood, these episodes are prolonged because our immune system is not equipped with memory T and B lymphocyte that would protect the body from usual infectious childhood illnesses. During episodes of infection, we confine ourselves to a safe place in order to protect our body from further harm (predators) and in order to preserve energy stores. Typically, muscular work and brain work are switched off at the expense of an activated immune system. This principle has been recognized as an adaptive program to protect the body.[1251–1253, 1363] In a recent excellent review, Robert Dantzer and colleagues linked sickness behavior to major depression.[1279] These authors hypothesized that major depression is a form of exaggerated sickness behavior that can lead to chronic

TABLE 30 Sequelae of Chronic Inflammatory Systemic Diseases
Sickness behavior/fatigue/depressive symptoms
Sleep disturbances
Anorexia
Malnutrition
Muscle wasting—cachexia
Cachectic obesity
Insulin resistance with hyperinsulinemia (insulin-like growth factor 1 resistance)
Dyslipidemia
Increase of adipose tissue in the proximity of inflammatory lesions
Alterations of steroid hormone axes
Disturbances of the hypothalamic-pituitary-gonadal (HPG) axis
Elevated sympathetic tone and local sympathetic nerve fiber loss
Hypertension, volume expansion
Decreased parasympathetic tone
Inflammation-related anemia
Bone loss
Hypercoagulability
Circadian rhythms of symptoms
Pregnancy immunosuppression and postpartum inflammation
Stress exacerbates disease

decompensation in vulnerable patients.[1279] While these authors mainly focused on inflammatory events, additional problems like chronic pain and chronic sleep alterations can add to the inflammatory pathway.

It is well known that patients with chronic inflammatory systemic diseases show signs of chronic fatigue and depression.[1364–1371] In recent years, it has been demonstrated that circulating cytokines and activation of sensory nerve fibers in the periphery are most probably involved in these central nervous alterations (see also Figure 7).[1363, 1372] Cytokines induce so-called sickness behavior in the form of malaise, fatigue, numbness, coldness, skeletal muscle and joint aches, reduced appetite, anxiety, and depressive mood (Figure 39). Injection of lipopolysaccharide into healthy controls leads to a significant increase of depression scores.[1373] Importantly, lipopolysaccharide injection can also increase

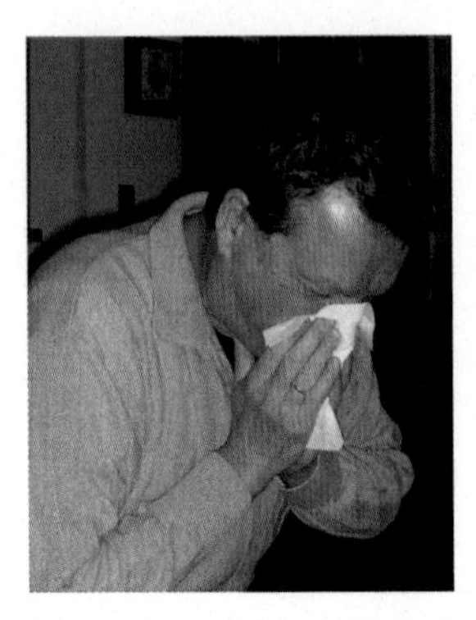

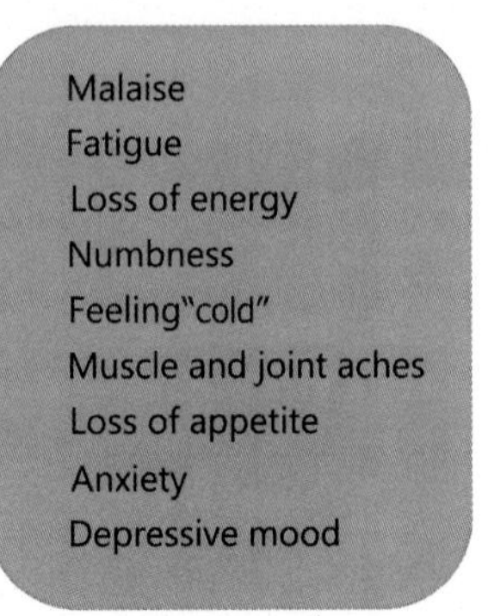

FIGURE 39 Components of sickness behavior. The behavioral components of sickness represent, together with the fever response and the associated neuroendocrine changes, a highly organized strategy evolutionarily positively selected and beneficial during acute infections.

hyperalgesia.[1374] Therapeutic administration of IFN-α markedly increased depression scores in hepatitis C patients indicating the enormous role of circulating cytokines for depression.[1374, 1375]

In addition, patients with rheumatoid arthritis under anti-TNF therapy, similar to other immunosuppressive drugs, demonstrate a marked reduction in fatigue scores.[1365, 1376–1379] These findings show that elevated circulating cytokines have an important impact on brain function.

When we take all information together, it becomes clear that sickness behavior is a positively selected program that is applied to chronic inflammatory systemic diseases. Due to the fact that this program for short episodes of inflammation is used for a long time in chronic inflammatory systemic disease, we may call it a program utterly misguided. The adequate treatment is consequent reduction of systemic inflammation and pain. At the moment, we do not know why patients can develop overt major depression but it might depend on the above-discussed vulnerability in some individuals.[1279]

SLEEP DISTURBANCES

Patients with chronic inflammatory systemic diseases often demonstrate sleep disorders (Table 31). Sleep alterations have been linked to clinical symptoms like fatigue and increased pain in these patients.[1380–1387] Thus, sleep quality is a clinically important subject in patients with chronic inflammatory systemic diseases.

In addition, treatment of sleep disorders improves quality of life and reduces daytime fatigue in chronic inflammatory systemic diseases.[1385, 1386, 1395, 1396]

Sleep and HPA Axis Hormones

While sleep was usually regarded as a brain-driven phenomenon without interferences from the periphery,[1404] research between 1970 and 1990 has

TABLE 31 Sleep Disturbance in Chronic Inflammatory Systemic Diseases

Disease	References
Rheumatoid arthritis	1388–1390
Systemic lupus erythematosus	1383, 1383, 1391, 1392
Sjögren syndrome	1383, 1393
Systemic sclerosis	1394
Ankylosing spondylitis	1387, 1395–1397
Behçet's disease	1398
Inflammatory bowel disease	1399, 1400
Multiple sclerosis	1385, 1401, 1402
Psoriasis vulgaris	1386, 1403

This list is not complete.

demonstrated a bidirectional interaction between sleep and HPA axis activity. Sleep influences the activity of the HPA axis, and HPA axis hormones influence sleep patterns in humans and animals.[1405–1408] The connections materialize in diseases with hypercortisolism such as Cushing's disease or major depression where patients suffer from sleep problems.[1405]

Nowadays, it is accepted that CRH decreases slow wave sleep and increases light sleep and wake time after sleep onset, which might depend on inhibition of the growth hormone surge after sleep onset.[1409] Unfortunately, CRH cannot be measured in serum without the marked influence of other peripheral CRH sources (gut and activated immune system; Ref. 796). Thus, one cannot relate hypothalamic-hypophyseal portal CRH levels to the sleep patterns observed in patients with chronic inflammatory systemic diseases.

The influence of cortisol on sleep is more complex as compared to CRH because hormonal feedback mechanisms between the periphery and the hypothalamus determine sleep-relevant CRH release. In addition, differential effects of endogenous (or, similarly, exogenous) glucocorticoids on sleep exist via low-affinity glucocorticoid and high-affinity mineralocorticoid receptors, respectively.[1408, 1410, 1411] More recent reviews on the topic suggested that glucocorticoids inhibit or enhance deep sleep depending on the dose of exogenous glucocorticoids and the location of influence.[1408] Low doses of glucocorticoids increase deep sleep and high doses of glucocorticoids decrease deep sleep.[1408] When glucocorticoids are low (high), CRH in the hypothalamus is high (low), which should decrease (increase) deep sleep.

Another HPA axis hormone is ACTH that may directly influence sleep.[1405–1408] Nocturnal infusions of ACTH suppressed REM sleep in normal controls.[1412]

These examples demonstrate that HPA axis hormones have an important role in sleep architecture and quality.

Sleep and Immune System

When we talk about a bidirectional interaction of sleep and HPA axis, we can also talk about a bidirectional interaction between sleep and activity of the immune system (late examples, Ref. 1413 and reviews in Refs. 1414, 1415). Complex interactions between sleep and immune system are important for normal sleep.[628, 1414–1418] Further findings demonstrated that elevation of cytokine concentrations deteriorates sleep and declarative memory.[1419] Thus, sleep problems in chronic inflammatory systemic diseases may be generated on different levels of these supersystems (neuronal, endocrine, and immune). Since the immune system seems to play an important role for normal sleep, changes in inflammatory load may influence sleep quality.[1414, 1415]

This is best demonstrated during infections with excess sleep reported by patients or studied using polysomnography (reviewed in Ref. 1414). In this context, cytokines like IL-1β, IL-6, and TNF are critical in changing sleep quality not only in animals but also in humans.[1414, 1415] In addition, injection of lipopolysaccharide into humans reduces wakefulness, reduces REM sleep, and increases slow wave sleep.[1416] Chronic IFN-α administration to patients with hepatitis C disrupts sleep continuity and depth, which is associated with increased fatigue.[1417]

In patients with rheumatoid arthritis and ankylosing spondylitis, TNF was linked to sleep problems because anti-TNF therapy improved sleep efficiency and decreased awakening after sleep onset.[1395, 1396, 1420] Similarly, patients with psoriasis vulgaris had marked sleep disorders, and anti-TNF treatment improved sleep outcomes and other patient-reported outcomes including health-related quality of life, work productivity, daily activity, and disease-related pain.[1386] Anti-IL-6 receptor therapy with tocilizumab improved subjectively reported sleep quality in rheumatoid arthritis.[1421] Inhibition of the immune system with abatacept, an inhibitor of T-cell co-stimulation, improved self-reported sleep problems.[1422, 1423] It has been demonstrated that TNF neutralization decreased sleepiness in patients with sleep apnea.[1424]

Particularly, therapy with biologics clearly demonstrated the influence of a single cytokine or a distinct immune pathway on objective and subjective sleep parameters.

Sleep, HPA Axis Hormones, and Immune Activation

The tripartite interrelation between serum cortisol, sleep, and peripheral inflammation was subject of a recent unpublished study in patients with rheumatoid arthritis of our group and the group of Drs. Buttgereit and Burmester in Berlin. In the presence of inflammation at baseline of the follow-up study, serum cortisol was negatively related to total sleep time and sleep efficiency. This negative interrelation was eliminated after correction for C-reactive protein. In addition, this negative interrelation of cortisol and deep sleep changed towards the opposite direction after anti-inflammatory treatment for 16 weeks with anti-TNF therapy. This indicates that the presence of cortisol and the absence of proinflammatory load are related to good sleep quality. In other words, the arousal effect of cortisol at baseline depends on concomitant inflammation. After control of inflammation, higher levels of serum cortisol are linked to deep sleep and less awakening.

These findings may show a disrupted feedback of cortisol to the hypothalamus. At similar serum levels of cortisol (baseline vs. week 16), at baseline when inflammation was high, serum cortisol was negatively related to total sleep time and sleep efficiency. This situation resembles a situation with CRH injection, because CRH decreases deep sleep and stimulates awakening. Thus, at baseline, cortisol might not lead to proper negative feedback on hypothalamic CRH. Central CRH stays high and disrupts sleep in our patients with high inflammation. However, when inflammation was controlled by 16 weeks of anti-inflammatory anti-TNF therapy, cortisol at the same serum levels inhibited hypothalamic CRH. Now, low central CRH led to improved sleep. These findings might be another indication of how anti-inflammatory therapy, particularly with anti-TNF strategies, corrects HPA axis function (compare section "The link between TNF and hormone axes" of Chapter II).

At this point, we should add another important factor, namely, noradrenaline, that is produced in the locus coeruleus in the brain stem. The locus coeruleus is an important center of the sympathetic nervous system, which has reciprocal excitatory interactions with the CRH-secreting paraventricular nucleus of the hypothalamus. Thus, it is not surprising that high central noradrenaline should decrease slow wave sleep and should increase light sleep.[1425] Since patients with chronic inflammatory systemic diseases have an elevated sympathetic nervous tone, this can add to the sleeping problems observed in these patients.

Conclusions

Chronic inflammatory systemic diseases are accompanied by sleep problems, daytime fatigue, and decreased daytime activity. Similar phenomena are present during acute inflammation best demonstrated during infection, cytokine administration, or lipopolysaccharide injection. This adaptive program has been positively selected to increase the time spent at rest in order to deviate energy-rich fuels from the brain/skeletal muscles to the immune system. Anticytokine therapies unambiguously demonstrate that peripheral inflammation is the key element that induces sleep problems. Long-term application in chronic inflammatory systemic diseases is a misguided adaptive program.

ANOREXIA AND MALNUTRITION

Since, in chronic inflammatory systemic disease, sickness behavior increases the time spent at rest,[1279, 1426] and energy-rich fuels are allocated to the immune system (and not to active abdominal organs or skeletal muscles), intake of energy-rich fuels should be lower than expected. Indeed, poor nutrient status in patients with rheumatic diseases has been reported several times (reviewed in Ref. 1427). Intake of energy-rich fuels is reduced in rheumatoid arthritis,[1428, 1429] systemic sclerosis,[1430] multiple sclerosis,[1431] juvenile idiopathic arthritis,[1432] and other chronic inflammatory systemic diseases. It should be stressed that decreased fuel intake is often linked to lowered levels of vitamin D and other lipid-soluble vitamins, iron, zinc, copper, magnesium, and others. Figure 40 makes an example for total body magnesium in patients with rheumatoid arthritis.

It was demonstrated that intake of energy-rich fuels in rheumatoid arthritis patients was inversely related to stimulated IL-1β production from peripheral blood cells,[1433] which demonstrates that the "energy appeal reaction" by spillover inflammation is the driving force of anorexia. IL-1β is an important cytokine for sickness behavior,[1279, 1426] and chronic sickness behavior has a deleterious effect on fuel provision leading to anorexia and depressive-like symptoms.

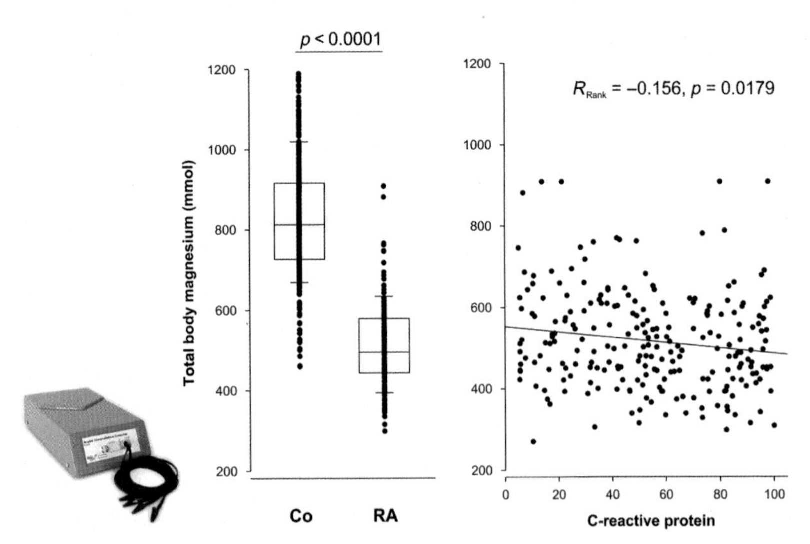

FIGURE 40 Decreased total body magnesium in patients with rheumatoid arthritis (RA, $n=156$) compared to healthy controls (Co, $n=430$). The patients were matched according to age, gender, height, weight, body mass index, and serum creatinine (was normal in all subjects). Patients with rheumatoid arthritis were treatment-naive. Total body magnesium was measured by bioimpedance analysis (BIA-ACC, Biotekna, Marcon, Venezia, Italy, according to the technique published in Ref. 1435). This clearly demonstrates the decreased magnesium load in patients with chronic inflammatory systemic disease, and this was linked to an increased C-reactive protein (right panel of RA patients).

This program evolved to support allocation of energy-rich fuels to the immune system in short-lasting inflammatory diseases (not longer than 3–5 weeks) since exercise in the course of food acquisition would have deviated energy-rich fuels to the skeletal muscles and brain. In order to get a better picture of the energy-saving value of anorexia, it can be important to study energy expenditure and intake in otherwise healthy people under natural conditions in the wild.

In traditional Pygmy hunter-gatherers in Cameroon (they need less energy than tall Caucasians!), the energy balance on three typical days was negative between total energy intake per day (men: 7200 kJ/d (1719 kcal/d); women: 6700 kJ/d (1600 kcal/d)) and total energy expenditure per day (men: 8100 kJ/d (1934 kcal/d); women: 7300 kJ/d (1743 kcal/d)).[1434] In this observational study, under natural conditions in noninflamed subjects, individuals spent more energy than they consumed. It tells that normal life in the wild with typical foraging behavior is pretty energy-demanding (and one expects that there are also better days for these Pygmies). The question arises as to what would happen when a hunter-gatherer needs extra energy during acute inflammation.

With an activated immune system during systemic mild to moderate inflammation, a human needs approximately 400-500 kJ (96-119 kcal) more per day as compared to a situation without inflammation.[1152] Indeed, this amount of extra energy consumption is a strong stimulus to induce an energy redistribution program. Particularly, increased sleep and anorexia would save energy. Alone during sleep (resting brain and resting skeletal muscles), humans need 25% less energy compared to the wake state.[1244, 1245] During the wake state, it was measured that humans need 6430 kJ (1535 kcal) in 16 h.[1245] If one assumes that a similar amount would be needed during the remaining 8 h of sleep, we obtain the number of 0.5×6430 kJ (1535 kcal) (=3215 kJ [768 kcal]). However, in reality, the number is much lower because only 2000 kJ (478 kcal) is needed during 8 h of sleep.[1245] This indicates that we can save a lot of energy during sleep because the brain and skeletal muscles are in a dormant state.

From the perspective of acute inflammation, sickness behavior with an increased time spent at rest and sleeping and with anorexia to avoid energy-consuming foraging conserves enough energy to nourish the activated immune system for a while. However, the consequence of this adaptive program during acute infection/inflammation would be deficiency of energy-rich fuels (carbohydrates, free fatty acids, and proteins); loss of essential ions such as calcium, phosphorus, magnesium, and iron; and lower levels of essential vitamins like vitamin D. This can lead to malnutrition, which is a typical finding in many inflammatory diseases.[1436] Under anorexia conditions, the body is absolutely dependent on stored reserves.

All these programs evolved to cope with transient inflammatory episodes, but prolonged use of these programs in chronic inflammatory systemic diseases leads to pathology. Now that we know that anorexia can happen in chronic inflammatory systemic diseases, we should offer to the patient adequate amounts of essential nutrients. In addition, the key to remove anorexia is consequent treatment of inflammation.

MUSCLE WASTING, CACHEXIA, AND CACHECTIC OBESITY

As early as 1950, generalized skeletal muscle wasting was described in patients with rheumatoid arthritis.[1437] In the 1990s, Roubenoff et al. described a highly increased catabolic situation in chronic inflammatory systemic diseases.[1438] At this time, they linked cachexia to glucocorticoid therapy.[1438] Later, the same authors demonstrated that rheumatic cachexia was also driven by proinflammatory cytokines independent of glucocorticoids.[1433] Rheumatic cachexia was present at elevated resting energy expenditure (12% higher) and lower physical activity.[1433] Higher energy expenditure in the presence of low physical activity is indicative of energy utilization by the immune system (not the brain and skeletal muscles).

In addition, accelerated whole-body protein catabolism was observed (cachexia). In a recent review, Roubenoff summarized, "...low physical activity predisposes to fat gain and is believed to precipitate a negative reinforcing cycle of skeletal muscle loss, reduced physical function, and fat gain in rheumatoid arthritis, which leads to cachectic obesity.[1439]" Cachectic obesity denotes the relative increase of fat mass in relation to lean mass (skeletal muscle) without large changes in body mass index. Here, we recognize that the body mass index is probably not a good marker of body composition and health. The question appears why cachectic obesity is a necessary consequence of the sickness behavior program.

In a period of starvation, during the first day, glycogen (first half day) and protein (from the first to third day, mainly from the skeletal muscle) are used to maintain glucose homeostasis.[1440] In a full starvation program (from day 3 onward), usually liver-derived ketone bodies are used, which are glucose substitutes in the brain, skeletal muscle, and immune cells leading to skeletal muscle sparing.

I hypothesize that in the chronic phase of chronic inflammatory systemic diseases, the above-discussed anorexia program leads to a negative effect on skeletal muscle proteins because the full starvation—ketone—program is not switched on. Inflammation-induced increase of circulating catabolic cytokines such as TNF[1441, 1442] and IL-1β,[1442] the loss of anabolic androgenic hormones, insulin-like growth factor 1 resistance, loss of insulin-like growth factor 1,[840] and increased levels of myostatin serve an important glucose homeostasis program in anorexia. It is important to emphasize that the long-known loss of adrenal androgens in rheumatic diseases[763, 773, 774] is part of this fuel allocation program because skeletal muscle growth depends on androgens (and insulin-like growth factor 1).

Under consideration of energy aspects, leptin-induced inhibition of adrenal androgen generation[877] can be another important part of the glucose homeostasis program in inflammatory anorexia. Fat gain during chronic inflammation relative to lean mass[1433] would lead to relatively elevated leptin levels that support the negative reinforcing cycle of androgen deficiency, skeletal muscle loss, and fat gain ultimately leading to cachectic obesity.

Finally, Figure 41 makes an example of cachectic obesity in patients with rheumatoid arthritis as measured by bioimpedance analysis.

There is a negative correlation between body mass index at study entry and radiographic disease progression or mortality rate in patients with rheumatoid

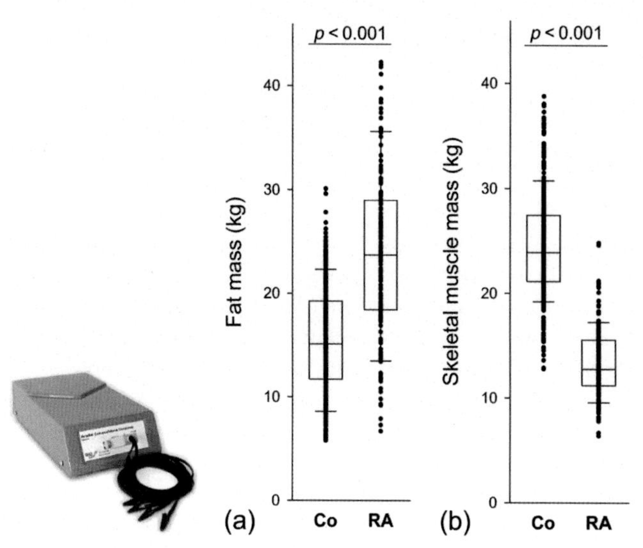

FIGURE 41 Cachectic obesity in patients with rheumatoid arthritis (RA, $n=156$) compared to healthy controls (Co, $n=430$). The patients were matched according to age, gender, height, weight, body mass index, and serum creatinine (was normal in all subjects). Patients with rheumatoid arthritis were treatment-naive. Fat mass (a) and skeletal muscle mass (b) were measured by bioimpedance analysis (BIA-ACC, Biotekna, Marcon, Venezia, Italy, according to the technique published in Ref. 1435). This clearly demonstrates the decreased skeletal muscle in the presence of higher fat mass in patients with a chronic inflammatory systemic disease.

arthritis.[1443–1446] This was called a paradoxical effect because a higher mortality rate might be predicted in patients with a higher body mass index. However, under consideration of fuel allocation to the activated immune system, patients with initially high body mass index had lower inflammatory activity, which was demonstrated in two studies.[1443, 1444] It tells us that patients with a higher body mass index at the beginning have a less severe course of the inflammatory disease (with less erosions) leading to less allocation of energy-rich fuels from skeletal muscle/fat to the immune system. I hypothesize that a milder disease is related to a higher body mass.

INSULIN RESISTANCE[a]

Introduction

In 1916, diabetologist Elliott P. Joslin recognized that "hyperglycemic situations appear after infectious diseases, painful conditions such as gall stones, and trauma."[1447] Already in 1920, Pemberton and Foster described impaired

---

a. This subject on insulin resistance was presented as an online contribution to Arthritis Research & Therapy under the Creative Commons Attribution License 4.0. Straub RH. Insulin resistance, selfish brain, and selfish immune system: an evolutionarily positively selected program used in chronic inflammatory diseases. Arthritis Research & Therapy 2014, 16(Suppl 2):S4.

glucose regulation in soldiers with arthritis.[1448] In 1924, Rabinowitch observed that diabetic patients need much more insulin during infection.[1449] In 1929, Root summarized the presence of an inadequately high need of insulin in different diseases, and he called the phenomenon "insulin resistance".[1450]

Over the last century, insulin resistance was found in physiological states, disease states, and diseases such as diabetes mellitus type 2, obesity, infection, sepsis, arthritis of different types (including rheumatoid arthritis), systemic lupus erythematosus, ankylosing spondylitis, trauma, painful states such as postoperative pain and migraine, schizophrenia, major depression, and mental stress to name the most important ones (chronology of events is summarized in Table 32). Thus, insulin resistance seems to be present in many medical conditions outside the field of diabetology or—more specifically—exterior of inherited insulin resistance syndromes (called the type A syndrome of insulin resistance) and also beyond autoantibodies to insulin or insulin receptor (type B syndrome of insulin resistance).[1451]

When considering these diseases and disease states, one observes two major clusters of clinical entities that are linked to insulin resistance: (1) inflammation with an activated immune/repair system and (2) increased mental activation. In this clearly defining distinction, obesity and type 2 diabetes mellitus can be integrated into item #1 due the inflammatory aspect of insulin resistance in these entities.[1452–1455] However, obesity and consequently also type 2 diabetes mellitus might also be integrated into item #2 because chronic mental stress is a well-known forerunner of obesity in approximately 40% of investigated stressed subjects.[1456–1461] At this point, the question appears why these two disease clusters are linked to insulin resistance, which will be addressed in this section of this chapter.

Since chronic inflammatory systemic diseases such as arthritis were among the first to be linked to insulin resistance,[1437, 1448] newer work in rheumatology recognized insulin resistance in many chronic inflammatory systemic diseases,[837, 838, 1489, 1490] cytokine-neutralizing strategies decrease insulin resistance in chronic inflammatory systemic diseases,[1490, 1495, 1496] and patients with chronic inflammatory systemic diseases are at an increased risk to develop type 2 diabetes mellitus,[1497] the special view from rheumatology to insulin resistance is understandable and necessary. The reader will see that insulin resistance is not an endocrine disorder *per se*, but more a disorder of several systems, better tackled from an interdisciplinary standpoint of neuroendocrine immunology.

Features of Insulin Resistance and Pathophysiology

Originally, insulin resistance was defined as a subnormal biological response to a certain insulin concentration, whereby the word "subnormal" already suggests "illness" (though it must not be illness). In the late 1950s, Yalow and Berson developed the radioimmunoassay to measure circulating insulin in the blood.[1467]

TABLE 32 History of Insulin Resistance (IR) from Different Perspectives of Research in the Fields of Diabetology, Infection/Inflammation, Pain, Mental Activation, Trauma, and Rheumatology

Year	Authors	Phenomena	References
1916	Joslin	Hyperglycemia with infection[a], painful gall stones,[b] trauma[c]	1447
1920	Pemberton and Foster	Impaired glucose regulation in soldiers with arthritis[a]	1448
1924	Rabinowitch	Enormous doses of insulin needed in infected diabetic patients[a]	1449
1929	Root	IR in the context of different diseases[a,b,c]	1450
1936	Himsworth	Insulin-sensitive and insulin-insensitive diabetes	1462
1938	Thomsen	Traumatic diabetes[c]	1463
1938	Graham	β-Cell defects in elderly long-standing diabetic patients	1464
1950	Liefmann	IR in rheumatoid arthritis (combined glucose and insulin test)[a]	1437
1956	Arendt and Pattee	IR in obese subjects	1465
1957	Collins	IR in schizophrenia	1466
1960	Yalow and Berson	IR in diabetic subjects (high glucose despite high insulin)	1467
1963	Randle et al.	Fatty acids support IR	1468
1965	van Praag and Leijnse	Major depression induces IR[d]	1469
1965	Butterfield and Wichelow	Forearm insulin sensitivity test	1470
1970	Shen et al.	The quadruple insulin sensitivity test	1471
1979	DeFronzo et al.	Euglycemic insulin clamp technique in combination with radioisotope turnover, limb catheterization, indirect calorimetry, and skeletal muscle biopsy	1472
1979	Wolfe et al.	Review: sepsis and trauma-induced IR[a,b,c]	1473
1982	Kasuga et al.	Insulin induces tyrosine phosphorylation of the insulin receptor	1474

(Continued)

TABLE 32 History of Insulin Resistance (IR) from Different Perspectives of Research in the Fields of Diabetology, Infection/Inflammation, Pain, Mental Activation, Trauma, and Rheumatology—Cont'd

Year	Authors	Phenomena	References
1982	Ciraldi et al.	Reduced insulin-stimulated glucose uptake in type 2 diabetes	1475
1984	Grunberger et al.	Dissociation between normal insulin binding and defective tyrosine kinase activity of the insulin receptor	1476
1986	Garvey et al.	Hyperinsulinemia induces insulin receptor desensitization	1477
1987	Svenson et al.	IR in rheumatoid arthritis (Re-discovery)	837
1988	Krieger and Landsberg	Hypertension, hyperinsulinemia, insulin resistance, and SNS	1478
1988	DeFronzo	Hyperglycemia decreases glucose transport and inhibits β-cell function (glucotoxicity)	1479
1988	DeFronzo and Reaven	Increased free fatty acids play key role in IR, β-cell dysfunction, and hepatic gluconeogenesis (lipotoxicity)	1479, 1480
1988	Uchida et al. and Greisen et al.	Pain influences IR via the HPA axis and SNS[b]	1481, 1482
1992	Feingold and Grunfeld	Cytokines like TNF play a role in hyperlipidemia and diabetes[a]	1483
1993	Hotamisligil	TNF critically influences IR[a]	461
1994	Moberg et al.	Mental stress induces acute IR in type 1 diabetic patients[d]	1484
1996	Keltikangas-Jarvinen et al.	Mental stress is accompanied by IR in nondiabetic people[d]	1485
1999	Björntorp	IR as a consequence of exaggerated HPA axis and SNS activation (CNS stress is the trigger)	1486
2000	Chrousos	Mental stress-induced hypercortisolism induces IR (the pseudo-Cushing's state)[d]	1487
2000	Seematter et al.	Mental stress acutely increases insulin-stimulated glucose utilization in healthy lean humans but not in obese nondiabetic humans[d]	1488

TABLE 32 History of Insulin Resistance (IR) from Different Perspectives of Research in the Fields of Diabetology, Infection/Inflammation, Pain, Mental Activation, Trauma, and Rheumatology—Cont'd

Year	Authors	Phenomena	References
2004	Tso et al.	Patients with systemic lupus erythematosus demonstrate IR independent of autoantibodies to insulin receptor[a]	1489
2005	Kiortsis et al. and Stagakis et al.	Patients with ankylosing spondylitis and rheumatoid arthritis have IR, which is reduced after anti-TNF therapy	1490, 1491
2007	Larsen et al.	IL-1ra improved β-cell secretory function in type 2 diabetic patients (no influence on IR)[e]	754
2008	Fleischman et al. and Goldfine et al.	Salsalate improved insulin sensitivity in young obese adults and in type 2 diabetic patients	1492, 1493
2010	Schultz et al.	Patients with rheumatoid arthritis show IR, which can be reduced by blocking IL-6	1494
2012	DIAGRAM et al.	Human gene polymorphisms link both inflammation and metabolic disease	1875
2014	Fall and Ingelsson	Human gene polymorphisms link both inflammation and metabolic disease	1358

[a]*Insulin resistance as a consequence of infection or inflammation.*
[b]*Insulin resistance as a consequence of pain.*
[c]*Insulin resistance as a consequence of trauma.*
[d]*Insulin resistance as a consequence of mental activation.*
[e]*Approved by the US Food and Drug Administration for patients with type 2 diabetes mellitus.*
Abbreviations: CNS, central nervous system; HPA axis, hypothalamic-pituitary-adrenal axis; OGTT, oral glucose tolerance test; SNS, sympathetic nervous system.

Already in this early paper, they described a state of insulin resistance in type 2 diabetes mellitus patients "…[there is a] lack of responsiveness of blood sugar, in the face of apparently adequate amounts of insulin secreted … ." The classical characteristics of insulin resistance are given in Table 33. Elements given in this table work together to induce clinically observed hyperglycemia and VLDL hyperlipidemia (triglycerides) despite elevated insulin levels.

Insulin resistance is measured by different techniques, whereby the gold standard is hyperinsulinemic euglycemic clamp and the silver standard the frequently sampled intravenous glucose tolerance test (Table 34). To study insulin resistance or insulin sensitivity in chronic inflammatory systemic diseases, often, simple fasting indexes are used such as the homeostasis model assessment insulin resistance (called HOMA-IR) and the Quicki (Table 34), which are adequate when applied in larger clinical studies.

TABLE 33 The Classical Signs of Insulin Resistance Until 1995[1451, 1479, 1480, 1486]

Structure, Organ	Observed Change
Insulin receptor	Inhibited
Insulin receptor signaling cascade	Inhibited
Skeletal muscle	
Glycogen synthase	Inhibited
Hexokinase II	Inhibited
Pyruvate dehydrogenase	Inhibited
Liver	
Hepatic glucose production (gluconeogenesis, glycogenolysis)	Stimulated
Insulin clearance	Stimulated
Adipose tissue	
Free fatty acid mobilization	Stimulated
Signs in circulating blood	
Hyperglycemia	Yes
Hyperlipidemia (TG-FFA-VLDL-TG cycle)	Yes
Glucagon	Increased

Abbreviations: FFA, free fatty acid; TG, triglycerides; VLDL, very low-density lipoprotein.

Pathophysiology of Insulin Resistance: A Chronology of Models

The first viable theory on insulin resistance was presented by Randle, who suggested that insulin resistance in skeletal muscles and adipose tissue is based on the glucose-fatty acid cycle.[1468] He suggested that insulin resistance is a consequence of an increased presence of circulating fatty acids and ketone bodies that leads to defects in glucose utilization and an ever-increasing insensitivity to insulin. The biochemical principles of this model are still valid and useful today.

Further clarification throughout the 1960s and 1970s came from endocrine diseases that were accompanied by insulin resistance. The explanatory power of hormones is particularly obvious in diseases with a high overproduction of distinct glucogenic hormones such as in Cushing's syndrome (cortisol), acromegaly (growth hormone), pheochromocytoma (catecholamines), glucagonoma, thyrotoxicosis (thyroxine and triiodothyronine), and insulinoma (insulin resistance as a consequence of insulin receptor desensitization).[1451] Since these diseases were

TABLE 34 Methods to Measure Insulin Resistance

Technique	Notes	References
Reference methods		
Hyperinsulinemic euglycemic clamp	Gold standard Highly invasive	1472
Frequently sampled intravenous glucose tolerance test	Silver standard Invasive	1504
Oral glucose tolerance test		
Insulin sensitivity glycemic index = 1 + 2/(INSp × GLYp)	Most commonly used Little invasive	1505
Whole body insulin sensitivity	Little invasive	1505
Skeletal muscle IS = (delta glucose/delta time)/mean plasma insulin[a]	Little invasive	1506
Hepatic IS = glucose 0-30 min (AUC) × insulin 0-30 min (AUC)[b]	Little invasive	1506
Fasting simple indexes		
Homeostasis model assessment insulin resistance (HOMA-IR)	Noninvasive	1505
Newer version of the HOMA-IR (HOMA2-S)	Noninvasive	1505
FGIR = fasting glucose (mg/dL)/fasting insulin (mU/L)	Noninvasive	1505
Quicki = 1/(log fasting insulin (mU/L) + log fasting glucose (mg/dL))	Noninvasive	1505
Biochemical markers of insulin resistance		
Sex hormone-binding globulin	Noninvasive	1505
Insulin-like growth factor binding protein 1	Noninvasive	1505
Other markers: YKL-40, α-hydroxybutyrate, soluble CD36, leptin, resistin, interleukin-18, retinol binding protein-4, chemerin, and FGF21	Noninvasive	1507

[a]*The rate of decay of plasma glucose concentration from its peak value to its nadir (delta glucose/delta time) during the oral glucose tolerance test.*
[b]*The product of total area under curve (AUC) for glucose and insulin during the first 30 min of the oral glucose tolerance test. Abbreviations: AUC, area under the curve; GLYp, area under glucose curve; INSp, area under the insulin curve; IS, insulin sensitivity.*

accompanied by insulin resistance, the respective hormones became the focus of insulin resistance research ("called the insulin antagonists"; not to speak of antibodies to insulin or insulin receptor).

Physiological and medical conditions with upregulated stress hormones were found to be accompanied by insulin resistance such as in psychological stress, psychiatric disease, starvation, fasting, and others (Table 32). The activation of stress axes is very closely related to the above-mentioned cluster of mental activation. For example, an overactive stress system has been described in different forms of insulin resistance.[1498, 1499] Stress system activation is an explanatory model for insulin resistance, still *in vogue*,[1486, 1487, 1500–1503] but in 1993—slowly but surely—the mainstream of research turned to inflammation-related insulin resistance (see paragraphs after the next paragraph).[461]

In addition, several authors indicated the central role of the brain because it dictates nutrient intake and foraging behavior. Excess energy intake *per se* would be an important factor for obesity and, thus, a possible cause of subsequently developing insulin resistance. This has been demonstrated in humans to play a role in congenital severe obesity with congenital leptin deficiency[1508] or a mutation in the melanocortin receptor type 4.[1509] There is a highly delicate system of hypothalamic regulation of satiety versus food intake, which is influenced by distinct pathways within the brain and from the periphery.[1501, 1510] Close relationships exist with psychological components comprising mood disturbances, altered reward perception and motivation, or addictive behavior.[1511] The interested reader is referred to comprehensive reviews of the subject.[1501, 1511, 1512]

Nowadays, inflammation-mediated insulin resistance is another important explanatory platform of insulin resistance in adipocytes, myocytes, and hepatocytes.[461, 1453, 1513, 1514] Disruption of insulin signaling at the level of insulin receptor substrate (IRS)-1 and IRS-2 and further downstream by TNF signaling, toll-like receptor signaling, NF-κB (NF-κB) and inhibitor of NF-κB kinase, and FoxO1 activation are key elements of inflammation-related insulin resistance.[1452, 1513, 1515] Crucial cytokines in insulin resistance are TNF, IL-1β, IL-6, IL-18, and adipokines. Although the concept behind inflammation-related insulin resistance is convincing, neutralization of TNF or IL-1β had no influence on insulin resistance in obesity or type 2 diabetes mellitus.[1513] This might depend on the redundancy of cytokine pathways because, typically, only one cytokine is neutralized, while many cytokines act in parallel. This might be overcome by a broader inhibition of proinflammatory signaling pathways, which has been shown for salsalate[b] therapy that reduced insulin resistance in patients with type 2 diabetes mellitus and in obese young adults.[1492, 1493]

In patients with chronic inflammatory systemic diseases, TNF and IL-6 neutralizing strategies reduced insulin resistance.[1490, 1491, 1494] Until now, it is not clear why the neutralizing strategies perfectly improve insulin sensitivity in chronic inflammatory systemic diseases but not in patients without these chronic diseases. This discrepancy will be discussed in a model of insulin resistance that

b. Salsalate, 2-(2-Hydroxybenzoyl)oxybenzoic acid, belongs to the class of salicylic acids or nonsteroidal anti-inflammatory drugs.

integrates the findings of patients with chronic inflammatory systemic diseases (see section, "The new model of insulin resistance").

In addition to the cytokine-centered theory of insulin resistance, a relatively new aspect is nutrient-induced inflammation that leads to endoplasmic reticulum stress, activation of c-Jun-N-terminal kinase (JNK), and inhibition of insulin receptor substrate 1 (IRS-1) and AKT (v-akt murine thymoma viral oncogene homologue 1) and, thus, insulin resistance in liver and adipose tissue.[1452] In this model of "metaflammation" (metabolic inflammation), free fatty acids can activate toll-like receptors, and free fatty acids and glucose undergoing oxidation in mitochondria stimulate free radical production, both of which inhibit insulin signaling.[1452, 1516] The theory describes that nutrient overload in our modern society of affluence gradually increases the involvement of immune system pathways. This leads to ongoing inflammation, mainly in fat tissue as substantiated by leukocyte infiltration (the macrophage is the big player). In consequence, involvement of these inflammatory pathways intensifies the inhibition of metabolic pathways.[1452] In addition, in patients with obesity, changes of the gut microbiota were observed, which in itself can be an inflammatory factor that contributes to insulin resistance.[1517–1519]

In this short pathophysiology collection of insulin resistance, we recognize again the two clusters linked to insulin resistance: (1) inflammation with an activated immune/repair system and (2) increased mental activation (mood, food intake, stress, and stress axes). However, the appearance of the two clusters is not yet explained by the interplay of the above-mentioned pathophysiological elements. Possibly, published theories on insulin resistance with an evolutionary perspective might help to shed light on the two clusters.

Evolutionary Medicine: Theories of Insulin Resistance 1962-2015

The **thrifty genotype hypothesis** of 1962 states that a gene has been positively selected for an exceptionally efficient intake and utilization of food, which was good for hunter-gatherers in a feast/famine environment but not for modern people in a world of plenty. In the original theory, a single gene was made responsible for rapid postprandial insulin release that supported quick storage of energy-rich substrates (called the "quick insulin trigger").[1520, 1521] While the original theory focused on the "quick insulin trigger," an alternative model focused on possible genes involved in insulin resistance.[1522] Today, we know that obesity and insulin resistance are based on a polygenic background with many single-nucleotide polymorphisms with small effect sizes. Selection on such mutations would probably be very weak because the individual advantages they would confer would be very small. The theory has been criticized due to modest support by genetic analyses, it has been even rejected, but it is still in use, and it has been adapted by researchers in the field of eating disorders.[1523]

Another theory of **starvation-induced insulin resistance** proposes that insulin resistance of the skeletal muscle during fasting is a positively selected program to maintain high circulating glucose levels in order to protect skeletal muscle from proteolysis during starvation.[1522, 1524] In addition, during starvation, lipolysis is switched on leading to provision of free fatty acids and then ketone bodies that can be used by the brain. Both mechanisms spare glucose and glucogenic amino acids in the skeletal muscle. Insulin resistance in the context of starvation is a special form of insulin resistance because insulin levels are very low, no inflammation accompanies starvation, and counterregulatory hormones such as glucagon and cortisol are continuously upregulated. This situation does not apply to insulin resistance observed in chronic inflammatory systemic diseases and obesity because hyperinsulinemia and inflammation are a hallmark.

Another important theory of insulin resistance is the **thrifty phenotype hypothesis.**[1361, 1362] This model is based on the important observations that underweight babies develop more often insulin resistance and obesity as compared to normal weight children. In this theory, intrauterine malnutrition and other fetal constraints induce insulin deficiency (lack of the growth-promoting activities of insulin) and a postnatal state of regulatory insulin resistance, which leads to rapid postnatal increase of adipose tissue that remains stable throughout life (accompanied to cardiovascular disease in the elderly).[1359] In many studies all over the world, epidemiological findings were very supportive of the model.[1361] The theory proposes that environmental factors are the dominant cause of obesity and that epigenetic intrauterine programming plays the critical role.[1525, 1526] This theory has been refined in the **predictive adaptive response model**. In this supplement to the original theory, the relative difference in nutrition between pre- and postnatal environments, rather than an absolute level of nutrition, determines the risk of insulin resistance.[1527] Both thrifty phenotype theories are accepted in insulin resistance research because they have been confirmed in many studies in humans and animals. In these days, it is amazing that a nongenetic theory has received so much support and attention (it is epigenetic!).

Based on the thrifty genotype hypothesis, insulin resistance and immune activation were recognized as an adaptive positively selected program to combat infections (I call it the **fight infections theory of insulin resistance**). The activation of the immune system during infectious disease and inflammation induces insulin resistance, which leads to redirection of glucose to the activated immune system.[1528] In a modern form, this was integrated into the concept of immune cell activation by pathogen- and nutrient-sensing pathways (with cytokines, toll-like receptors, c-Jun-N-terminal kinase, etc.).[1529] Here, even nutrients can induce an inflammatory state that can support insulin resistance, which is probably a dilemma after exaggerated food intake where nutrients cannot be adequately stored in fat tissue and elsewhere (nutrient overload).

Similarly based on the thrifty genotype theory is the **breakdown of robustness theory**, which says that a robust glucose control system developed during evolution. The breakdown of this robust glucose control system induces positive disease-stabilizing feedback loops leading to insulin resistance. The critical determinant of the breakdown is TNF.[1530] This theory incorporates many accepted aspects but TNF is not the sole pathophysiological factor.

With the discovery of leptin, a negative feedback loop between adipose tissue and food intake was discovered. While in earlier times, many have argued that energy homeostasis operates primarily to defend against weight loss, the discovery of the leptin negative feedback loop speaks for homeostatic mechanisms that inhibit uncontrolled weight gain. The **central resistance model** states that central hypothalamic pathways are defective (resistant to leptin and others such as insulin). This leads to increased food intake and the resulting obesity induces insulin resistance.[1531] This theory has much value because it added the central regulation of food intake to the peripheral pathophysiological pathways.

Finally, the **good calories-bad calories theory** explains that our present food is markedly different from Paleolithic food. Particularly, high-energy-density carbohydrates are consumed too often, which induces inadequate hyperinsulinemia.[1532] Long-term hyperinsulinemia is the platform for obesity and disease sequelae. Others hypothesized that disparities between Paleolithic food and contemporary food might be important factors underlying the etiology of common Western diseases.[1533] Typically, the type of ingested lipids and the relative amount of carbohydrates/lipids versus proteins are a problem.

In conclusion, the theories already indicate that insulin resistance can be an important aspect to support the brain and the activated immune system. As such, insulin resistance can be seen as a positively selected program to support the brain or immune system. The theories are summarized in Table 35. In the following sections, this concept is further developed by including aspects of energy regulation.

Energetic Benefits of Insulin Resistance for Non-insulin-Dependent Tissue

At this point, I recapitulate that insulin resistance increases circulating glucose and free fatty acids that are not taken up in the adipose tissue, liver, and skeletal muscles and are now freely available to all non-insulin-dependent tissues. The two main profiteers of hyperglycemia are the central nervous system and the immune system because either glucose, free fatty acids (not the brain), or ketone bodies are energetic substrates. Both organs do not become insulin-resistant. In contrast, the immune system profits from insulin because it is an important growth factor for leukocytes and, with the help of insulin, major glucose transporters like glucose transporter 3 (GLUT3) and GLUT4 are upregulated on all leukocyte subpopulations.[834] In answering the question whether, for example, hepatic glucose production really provides higher levels of circulating energy, the following simple calculations are presented for glucose (similar calculations can be done for free fatty acids).

One important factor of insulin resistance is overproduction of hepatic glucose.[1534] In normal subjects, hepatic glucose production after an overnight fast

TABLE 35 Characteristics of Theories on Insulin Resistance as Observed from an Evolutionary Medicine Standpoint

Theories of Insulin Resistance	Year	References
Thrifty genotype hypothesis: quick hyperinsulinemia after food intake to store energy in fat tissue and elsewhere (quick insulin trigger)	1962 1999	1520 1521
(Not so) thrifty genotype hypothesis: starvation induces a special form of IR in order to conserve nitrogen (amino acids from skeletal muscle and elsewhere)[a]	1979	1522, 1524
Thrifty phenotype hypothesis: intrauterine constraints induce IR and insulin deficiency, which allows the organism to survive long enough to reproduce in a nutritionally deprived environment but which leads to obesity in a world of plenty; maternal constraints support IR (small mother, first baby, many babies in parallel, maternal undernutrition, and similar)	1992 2001	1361 1362
Based on the thrifty genotype hypothesis: an insulin resistance genotype and a cytokine genotype exist (much IR, high cytokine response); IR is helpful for infections	1999	1528
Refined thrifty phenotype theory: predictive adaptive response model; the relative difference in nutrition between pre- and postnatal environments, rather than an absolute level of nutrition, determines the risk of IR	2004	1527
Central resistance model: there exists a homeostatic regulation of weight gain versus weight loss but defects in the weight loss system lead to obesity (e.g., insulin and leptin signaling, SOCS3, PTB-1B)	2004	1531
Thrifty genotype plus breakdown of robustness: the basis is the thrifty genotype model; a robust glucose control system evolved during evolution, the breakdown of which induces positive disease-stabilizing feedback loops (TNF)	2004	1530
Thrifty genotype: integration of cellular pathogen- and nutrient-sensing pathways (cytokines, TLRs, JNK, Ikkβ, PKC, ER stress)	2006	1529
Good calories-bad calories hypothesis: wrong nutrients, particularly carbohydrates, lead to obesity and IR; a Paleolithic diet has quite different qualities that prevent obesity and Western diseases	2008 2010	1532 1533

Abbreviations: ER, endoplasmic reticulum; Ikkβ, inhibitor of NF-kB kinase β; IR, insulin resistance; JNK, c-Jun-N-terminal kinase; n.i., not included; PKC, protein kinase C; PTB-1B, protein tyrosine phosphatase 1B; SOCS3, suppressor of cytokine signaling 3; TLR, toll-like receptor; TNF, tumor necrosis factor.

[a] *This is a special form of insulin resistance without hyperinsulinemia on the basis of a strong response of counterregulatory hormones. It is questionable to call it insulin resistance because of missing hyperinsulinemia and missing inflammation. In addition, activity of the sympathetic nervous system is low, while activity of the HPA axis is high in the typical nadir.*

is approximately 2.0 mg/kg/min. Under a situation with insulin resistance, for example, in type 2 diabetes mellitus patients, insulin is 2.5-fold increased and the rate of fasting glucose production can increase to 2.5 mg/kg/min.[1534] After an overnight fast during an observation period of 12 h, the liver of a normal person of 80 kg produces 115 g glucose. Using above given numbers, a person with insulin resistance produces 144 g glucose, leading to a plus of 29 g in 12 h. A plus of 2 × 29 g = 58 g glucose in 24 h corresponds to 974 kJ (233 kcal) in 24 h, which is a pretty high number in light of the normal metabolic rate of 10,000 kJ/d (2388 kcal/d) of an 80 kg person (sedentary way of life; see Table 16). Indeed, the 974 kJ (233 kcal) represents approximately 39% of the total energy need of the normally active central nervous system, or it represents 61% of the energy requirements of all resting immune cells (Table 16). Thus, insulin resistance is a perfect way to support the activity of the central nervous system, the immune system, and/or other insulin-independent tissues (e.g., heart; Table 16).

In conclusion, while insulin resistance is most often regarded as a pathological state to be treated, these numbers and the fact that insulin resistance is linked to so many diseases and disease states are indicative of a beneficial role of insulin resistance. While the value of insulin resistance can be estimated from above given numbers, the generation of the two disease clusters is not yet clear.

The Selfish Brain and the Selfish Immune System Independently Demand Energy

This section demonstrates aspects of hypothetical character, and the reader is advised to critically judge the theoretical model. The basal metabolic rate of the entire body is determined when the following conditions are met: awake, lying, after overnight fast, thermoneutral (no heat production due to low/high temperature), and no emotional stress.[1225] Under these conditions, a person with 80 kg and 1.80 m needs approximately 10,000 kJ (2388 kcal)/d (Table 16).

The so-called minimal metabolic rate is lower than the basal metabolic rate because 15% of energy is spared during sleep, so that a 24-h sleeping person of above size needs 8500 kJ/d (2030 kcal). The minimal metabolic rate is not up for negotiation between the different organs. The delta between this last number and the maximum of voluntary daily energy uptake in the gut (20,000 kJ/d (4776 kcal/d); see Table 16) is 11,500 kJ/d (2746 kcal/d). In this example, the 11,500 kJ/d (2746 kcal/d) is the controllable amount of energy (CAEN), because allocation of CAEN to different organs is controlled by the interplay of these organs. CAEN is up for negotiations. The question appears which organs are dominant in regulating CAEN? Dominance can be judged when looking on Table 16 that shows the main users of energy, but it can also be derived from simple theoretical considerations.

For example, if a Paleolithic hunter experiences tissue trauma with infection, the immune/repair system becomes strongly activated. In this life-threatening situation, regulation of CAEN allocation to the immune/repair system must be independent of other organs and immediate (hierarchically, the highest level of control to survive). In this situation, circulating cytokines and activated sensory nerve fibers are responsible for the immediate reallocation of CAEN to the activated immune system that increases energy consumption (Table 16).[1152] I called this reaction "energy appeal reaction."[1152]

Similarly, if the brain is active during, for example, hard forest work over 6 h, then, also the skeletal muscles, heart, lungs/diaphragm, and liver are active, but most other organs are on minimal metabolic levels. This is particularly true for the gastrointestinal tract and the immune system. In this example of 6-h forest work, a person with above given size would need 18,500 kJ (4418 kcal) for the entire day (calculated using data from Ref. 1535). The brain controls the CAEN of additional 10,000 kJ (2388 kcal) when there is a need for forest work. Likewise, if a Paleolithic hunter needs to escape from a severe dangerous threat, the brain must control CAEN. In such a life-threatening situation, the control of CAEN by the brain must be independent of other organs and immediate (again, the highest level of hierarchical control to survive). We might easily add the example of the iron man participant, who voluntarily spends 42,000 kJ (10,029 kcal) in 8 h.[1223] This amount of energy is way above the maximum of daily energy uptake in the gut (20,000 kJ/d (4776 kcal)). For an iron man participant, hierarchy towards the brain, skeletal muscles, heart, and diaphragm is apparent, but most other organs such as the immune system are on minimal metabolic levels. Such a behavior can make a hard iron man much more vulnerable to microbes.

With trauma/infection or fight/flight response, the activity of most organs depends on either the immune/repair system or the central nervous system, respectively. We recently delineated that allocation of CAEN to the brain and skeletal muscles happens mainly during daytime, while allocation of CAEN to immune/repair systems happens at night.[1152] From these theoretical considerations, it becomes clear that either immune/repair system or central nervous system is a dominant regulator of CAEN. One can easily ask the question who else in the body can become so dominant?

Coming back to the introduction of this section, with this model, the two clusters of clinical entities linked to insulin resistance become understandable in terms of energy regulation. One recognizes two independent organs: (1) the "selfish immune system" and (2) the "selfish brain,"[1510, 1536] related to the above-mentioned clusters of (1) inflammation with an activated immune/repair system and (2) increased mental activation.

With the look on chronic inflammatory and chronic mental diseases that induce insulin resistance (listed in Table 32), the question arises whether, or not, brain-supporting and immune system-supporting insulin resistance has been positively selected for acute disease or chronic disease. Such a distinction is not included in the available theories of insulin resistance, but it might be helpful to understand the role of insulin resistance in general.

A Difference Between Acute and Chronic Diseases

While an acute response is often adaptive and physiological to correct alterations of homeostasis, a chronic disease process is often accompanied by the wrong program.[1152, 1255] Looking on simple readout parameters, this can be demonstrated for immune/repair system activation and mental activation.

The acute activation of the immune/repair system is crucial to fight acute infections and trauma. However, long-standing inflammation in chronic inflammatory systemic diseases leads to severe disease sequelae as summarized in this particular chapter of the book.[1152, 1255] Thus, immune activation is perfect for short-lived inflammatory episodes but not for long-standing chronic inflammatory diseases.

Considering mental activation, we can also separate acute versus chronic. In the acute situation of emergency of a loved one, family members show strong mental activation that can lead to a higher state of activity, a better readiness to take action, and poor sleep and symptoms of anxiety.[1537, 1538] Similarly, a student's examination stress can lead not only to higher state of activity but also to poor sleep and acute increase in anxiety scores.[1539, 1540] Acute examination stress increased intake of highly palatable food in an unproportional manner.[1355] In these acute situations, mental activation, poor sleep, and increase in food intake are important to overcome the challenging situation.

However, long-term caregivers of, for example, patients with Alzheimer disease are more often obese than noncaregivers, demonstrate alterations typical of the metabolic syndrome, show a higher risk to develop major depression, and have a long-term increase in proinflammatory markers.[1541–1546] Similarly, chronically stressed students in a highly competitive university environment showed an increased risk of obesity.[1460] A dose-response relationship was found between chronic work stress and risk of general and central obesity that was largely independent of covariates such as age, sex, and social position,[1457] supported in other large studies.[1458, 1459] Moreover, chronic job stress was related to an increased risk of the metabolic syndrome and even type 2 diabetes mellitus.[1547–1549] Chronically poor sleep is related to metabolic risk factors, obesity, and inflammation.[1550]

This little collection demonstrates that activation of the immune/repair and central nervous systems is successful in acute emergency but that it is dangerous when applied chronically leading to typical signs of obesity, metabolic derangement with insulin resistance, chronic inflammation, and increased risk for cardiovascular events.[1551] The question appears why there is such a clear distinction between acute and chronic, which determines the full picture of the metabolic syndrome and insulin resistance.

Evolutionary Medicine: Acute Physiological Response Versus Chronic Disease

It is repeated from Chapter IV: Networks are positively selected if they serve acute, highly energy-consuming situations, which are terminated within

3-8 weeks (recall the total consumption time). We perceive a chronic disease when it lasts for more than 6 weeks as used in classification criteria in rheumatoid arthritis and juvenile idiopathic arthritis.[1257] In addition, gene variants are positively selected if they protect energy stores, which is relevant during the entire life (beyond weeks 3-8). Networks that lead to insulin resistance serve the acute activation of the selfish immune system or the selfish brain, but do not belong to networks that protect energy stores. In contrast, insulin resistance leads to loss of energy-rich substrates because it is a catabolic process (energy-rich fuels are consumed by non-insulin-dependent organs or simply excreted).

If the hypothesis of the "acute insulin resistance program" is correct, then chronic insulin resistance in chronic inflammation, in chronic inflammatory systemic diseases, and in chronic mental activation or chronic mental disease is a misguided acute program. In contrast to insulin resistance, food intake and storage of energy-rich substrates in adipose tissue *per se* is not a misguided program. In other words, "obesity is not dangerous" and obesity is not a disease.[1552] Yet obesity becomes a problem, if additional factors are switched on that usually serve acute energy-consuming situations (mental activation or inflammation). Per Björntorp once wisely noticed that "some disease-generating factors, in addition to the basic condition of central obesity, is required for associated diseases to become manifest.[1552]"

The New Model of Insulin Resistance

With all this information, one can generate a new model of insulin resistance that builds upon the existing theories. The new model includes four new aspects: (1) It much more respects the immune/repair system, whose energy requirements are enormous (Table 16)[1152]; (2) it juxtaposes selfish brain and selfish immune system on a similar hierarchical level in terms of energy demand and requirements (Table 16); (3) it respects that energy requirements convey an evolutionary pressure (highly energy-consuming states are acute (negative selection pressure), and energy storage is beneficial (positive selection pressure)); and (4) it accepts that either immune system activation or mental activation is equally important in inducing insulin resistance. On the basis of these elements, a new model of insulin resistance is presented in Figure 42. This model states that insulin resistance is an acute catabolic program to serve the selfish immune system or the selfish brain, positively selected (1) for inflammation with an activated immune/repair system and (2) for increased mental activation.

The Drivers of Insulin Resistance in Chronic Inflammatory and Mental Diseases

Under consideration of the new model in Figure 42, we immediately recognize the problem of continuous inflammation in chronic inflammatory systemic diseases. I repeat here: Chronic inflammatory systemic diseases are accompanied

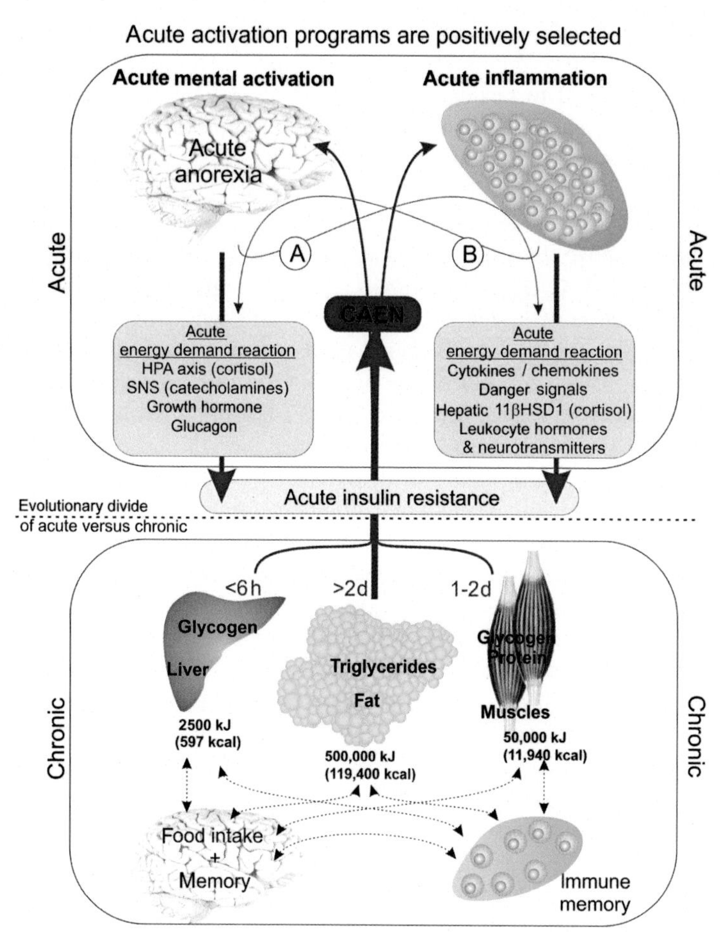

FIGURE 42 Pathophysiology of insulin resistance (IR) according to the new theory. Upper box: Acute activation programs were positively selected for short-lived activation of either the brain or the immune system. Hierarchically, the brain and immune system are on the same level. Activation of the brain mainly stimulates stress axes hormones and triggers the sympathetic nervous system (SNS). This is supported by a mild inflammatory process that is paralleled by mental activation (A). Activation of the immune system induces cytokines, chemokines, and danger signals. In addition, the inflammatory process uncouples the locally inflamed area from the control of the brain by cytokine-induced hormone/neurotransmitter production in the periphery independent of superordinate stress pathways. This leads to hepatic cortisol secretion,[742] ACTH-independent cortisol secretion,[635] and production of leukocyte hormones[1553] and leukocyte neurotransmitters.[607] The activation of the immune system is accompanied by a mild stimulation of the HPA axis (albeit inadequately low in relation to inflammation) and a somewhat stronger stimulation of the SNS (B). Despite activation of the SNS, anti-inflammatory neurotransmitters of sympathetic nerve fibers do not reach the uncoupled inflamed tissue.[1554] Inflammatory and mental activation are often accompanied by anorexia and sickness behavior, which aggravates energy shortage. The dashed black line between upper and lower boxes separates the programs positively selected for acute (catabolic) versus chronic states (storage and memory). Lower box: Chronic energy storage and memory programs

Continued

by markedly elevated serum levels of IL-6 ranging from 40.0 pg/ml before anti-TNF therapy to 8.0 pg/ml after anti-TNF therapy.[737] Thus, the levels are much higher as compared to healthy subjects (1-2 pg/ml; Ref. 741). According to studies in healthy subjects,[736] untreated patients with rheumatoid arthritis should increase daily energy expenditure by 750 kJ/d (179 kcal) (basal metabolic rate: 10,000 kJ/d (2388 kcal/d)). This number of 750 kJ/d (179 kcal) is remarkably similar to the glucose-derived energy obtained by hepatic insulin resistance as calculated above (974 kJ/d (233 kcal/d)). Since we expect that several cytokines like TNF, IL-6, IFN-γ, and IFN-α can drive a similar energy reallocation program, elevation of systemic cytokines explains why patients with chronic inflammatory systemic diseases do not need any other factor to provoke insulin resistance. They do not need the activation of the brain and, thus, activation of stress axes to induce insulin resistance. The brain is silenced in chronic inflammatory systemic diseases (sickness behavior). Insulin resistance can be stimulated by a direct influence of cytokines on hepatocytes, adipocytes, and myocytes. Now, we understand why cytokine-neutralizing therapies work perfectly well in, for example, rheumatoid arthritis because the key factors of insulin resistance are removed. When cytokine-neutralizing strategies do not work in obesity or type 2 diabetes mellitus, other parallel factors must exist.

The inflammatory load is remarkably different in the situation of chronic mental illness or psychological stress, where mild peripheral inflammation probably plays a small supportive role. When one compares serum levels of IL-6 as measured with the identical quantitative high-sensitivity ELISA technique, healthy subjects show mean values between 1 and 2 pg/ml,[741] caregivers show a mean value of 5.5 pg/ml,[1555] and subjects who report a high level of perceived hopelessness show a mean value of 3.0 pg/ml.[1556] These levels correspond to mild activation of the immune system, but they would not lead to an energy reallocation program.[736] Thus, in mental activation, stress axes must play the major role for the observed insulin resistance (cortisol, adrenaline, growth hormone, and glucagon). It is expected that neutralization of one cytokine would not change insulin resistance in these mentally activated people.

FIGURE 42—CONT'D were positively selected. The major storage organs are fat tissue (glycerol or free fatty acids) and skeletal muscles (proteins). The liver is more a switchboard to interchange and renew energetic substrates. The main storage factor is insulin so that insulin resistance can be seen as a catabolic program induced by catabolic pathways (upper box). The numbers in red give the typical time of energy provision by the respective organ (amino acids from skeletal muscle are spared from day 3 onward). Storage is mainly supported by a positively selected program of food intake/foraging behavior and memory. Memory is outstandingly important to spare energy-rich fuels (the brain and immune system). The dashed black arrows in the lower box demonstrate real and hypothetical connections between respective organs. The black numbers give a typical figure of stored energy in the respective organs.

Abbreviations: CAEN, controllable amount of energy (this is the energy that is regulated and negotiated between organs); HPA, hypothalamic-pituitary-adrenal axis; 11βHSD1, 11β-hydroxy steroid dehydrogenase type 1.[742]

Furthermore, when cytokine-neutralizing strategies do not work in type 2 diabetes mellitus, several factors in parallel are expected to drive insulin resistance. It is interesting that salsalate had a positive impact on insulin resistance in type 2 diabetes mellitus,[1493] but this type of drug and other nonsteroidal anti-inflammatory drugs can also inhibit mental activation in various chronic psychiatric diseases,[1557–1559] which is most likely related to reduced stress axes activation.

Conclusions

In chronic inflammatory systemic diseases, insulin resistance is an unfavorable factor because it supports the already activated immune system. It is a direct consequence of the proinflammatory load. Thus, insulin resistance should be treated by neutralizing inflammatory cytokines or by inhibiting the immune system with disease-modifying anti-inflammatory drugs in a more general way (like salsalate in type 2 diabetes mellitus). Since insulin resistance is a very direct consequence of immune system activation, the primary goal is anti-inflammatory treatment. In chronic inflammatory systemic diseases, it is expected that further treatment of insulin resistance beyond good inflammatory control is expected not to be needed. Since insulin resistance is a perfect diagnostic marker of an activated energy reallocation program (inflammation and mental activation), measuring insulin resistance might be a suitable biomarker to study the control of systemic inflammation in chronic inflammatory systemic diseases. Since several cytokines induce insulin resistance in a redundant manner, insulin resistance might be a more integral systemic diagnostic marker than C-reactive protein, erythrocyte sedimentation rate, or single cytokines.

In addition to aspects of insulin resistance in chronic inflammatory systemic diseases, this chapter demonstrates an extended theory of insulin resistance that classifies insulin resistance as a beneficial positively selected program to support activation of the immune/repair system and the brain. Insulin resistance makes sense in acute alterations of homeostasis in the context of short-lived diseases but is a misguided program in chronic inflammatory and mental activation.

Summary

- Insulin resistance is a consequence of mental activation (neuroendocrine axes) or inflammation that is a consequence of "selfishness" of the brain or "selfishness" of the immune system.
- Insulin resistance has been positively selected during evolution for short-lived energy-consuming activation of the brain or immune system.
- Long-term insulin resistance supports mental and chronic inflammatory diseases because energy-rich fuels are provided to these non-insulin-dependent tissues (continuous activation).
- Insulin resistance in chronic inflammatory diseases is treated by consequent reduction of the proinflammatory load.

- Treatment of insulin resistance in morbid obesity and type 2 diabetes mellitus is more complex because both inflammatory and neuroendocrine pathways need to be targeted. The pleiotropic anti-inflammatory and central nervous effects of salsalate constitute the first positive anti-inflammatory drug therapy of insulin resistance in type 2 diabetes mellitus.

DYSLIPIDEMIA

Insulin resistance is causally linked to dyslipidemia with the lipid triad of high levels of plasma triglycerides, low levels of high-density lipoprotein (HDL) cholesterol, appearance of small dense LDL cholesterol particles (sdLDL), and excessive postprandial lipemia.[1560] Although the pattern of dyslipoproteinemia may vary between different chronic inflammatory systemic diseases, which depends on severity of the disease, concomitant therapy, and diagnostic techniques, a common phenomenon in all chronic inflammatory systemic diseases is a low level of HDL cholesterol and/or apoA-I.[1561–1566] HDL is instrumental in removing cholesterol from the tissue (called reverse cholesterol transport). Importantly, the loss of HDL cholesterol and appearance of a proinflammatory subfraction of HDL with decreased apoA-I and apoA-II and increased serum amyloid A and ceruloplasmin were augmented in chronic inflammatory systemic diseases.[1567] Chronic inflammatory systemic disease-related transition of normal HDL to proinflammatory HDL becomes understandable in the context of acute inflammatory episodes.

Already in the late 1980s and early 1990s, it was described that elevated circulating cytokines can contribute to the decrease of HDL fraction, which was part of the acute-phase reaction of lipid metabolism (reviewed in Ref. 1568). Human subjects injected with TNF also demonstrated a rapid loss of plasma HDL.[1569] In a recent excellent review, the group of Grunfeld and Feingold demonstrated that acute-phase HDL is characterized by low levels of apoA-I, apoA-II, and lecithin-cholesterol acyltransferase (LCAT), but higher levels of serum amyloid A.[1570] Importantly, higher levels of serum amyloid A in HDL and increased inflammation-related secretory phospholipase A2 both support uptake of cholesterol into macrophages.[1570] Upregulation of secretory phospholipase A2 also supports the uptake of cholesteryl esters into the adrenal gland, presumably for increased steroid hormone synthesis.[1570] Moreover, the HDL transport protein ABCA1 responsible for reverse cholesterol transport is also downregulated by proinflammatory cytokines.[1570] Thus, the acute-phase HDL facilitates the delivery of cholesterol and other lipids to the macrophage due to a general downregulation of reverse cholesterol transport.

In context of acute inflammatory systemic diseases (e.g., infection), this response is appropriate because it increases allocation of energy-rich fuels to activated immune cells, particularly the macrophage. In addition, acute-phase HDL has many direct antimicrobial effects as recently summarized.[1570] In chronic inflammatory systemic diseases, this continuously present aspect of the acute-phase lipid response is deleterious leading to an increased risk of atherosclerosis.

INCREASE OF ADIPOSE TISSUE IN THE PROXIMITY OF INFLAMMATORY LESIONS

The detailed studies of Pond and Mattacks in the early 1990s have demonstrated that nearly all large lymph nodes, and many smaller ones, always occur embedded in the adipose tissue.[1571] The majority of smaller adipose depots enclose one or more lymph nodes.[1572] After activation of lymph nodes by lipopolysaccharide or other proinflammatory stimuli, lipolysis and glycerol release were increased from perinodal fat depots.[1572] It was suggested that lymph node-associated adipose tissue is important as a local source of energy-rich fuels since free fatty acids can be used by immune cells.

Several chronic inflammatory systemic diseases including rheumatoid arthritis, osteoarthritis, Crohn's disease, mesenteric panniculitis, and Graves' ophthalmopathy are characterized by selective hypertrophy of adipose depots in the proximity of inflammatory lesions.[459] Besides energy-rich fuels (fatty acids), adipocytes produce not only proinflammatory adipokines such as leptin, high-molecular-weight isoforms of adiponectin, TNF, IL-1β, MCP-1, and IL-6 but also anti-inflammatory factors like trimeric adiponectin.[459] Thus, the fine-tuned balance of pro- and anti-inflammatory factors supports processes in neighboring lymph nodes such as clonal expansion of antigen-specific lymphocytes.

In addition, fat tissue in the proximity of chronic inflammatory lesions most probably supports the local inflammatory milieu. It is important to mention that adipocytes differentiate from pluripotent mesenchymal stem cells, which readily enter chronically inflamed tissue.[459] Thus, adipose tissue adjacent to inflammatory lesions can develop locally, if stimuli for adipocyte differentiation are available. Regional adipose tissue represents a local store of energy-rich fuels necessary to nourish the neighboring inflammatory process.

In one of our recent publications,[1573] we were lucky to demonstrate that sympathetic nerve fibers appear at higher innervation density in the surrounding fat of inflamed synovial tissue in patients with rheumatoid arthritis compared to less inflamed osteoarthritis. In addition, the density of sympathetic nerve fibers was also higher in the fat tissue surrounding lymph nodes in collagen type II arthritic mice when compared to control animals.[1573] The higher density in fat tissue was accompanied by a stronger activation of fat tissue, which indicates that sympathetic nerve fibers stimulate lipolysis. This can be an important way to support inflammation locally by energy-rich free fatty acids.

A hypothetical extrapolation of this concept might be the presence of the greater omentum (omentum majus) in the abdomen. In this sense, visceral fat might similarly serve the active immune system in the gut, liver, spleen, and mesenteric lymph nodes. Since these organs are at the portal of entry of intestinal microbes, rapid provision of energy-rich fuels can be an important function of visceral fat. This can be the reason why visceral fat accumulation and degradation are much faster when compared to a situation in subcutaneous fat tissue.

ALTERATIONS OF STEROID HORMONE AXES

Alterations of the homeostatic steroid regulating axes have been extensively described in Chapter II: at present, the consensus of alterations is demonstrated in Table 36.

From the point of energy regulation and allocation of energy-rich fuels to the immune system, mildly elevated cortisol and severely downregulated anabolic androgens support muscle breakdown and gluconeogenesis. The cortisol-to-androgen preponderance is catabolic. Since cortisol and adrenaline support each other's signaling pathways, the concomitant increase of both hormones potentiates gluconeogenesis. In addition, the stimulation of lipolysis by cortisol at physiological levels is enhanced by growth hormone, leading to increased allocation of energy-rich fuels to the immune system.[1577] All this happens in the presence of cortisol resistance of immune cells.[1574, 1575] Thus, activated immune cells are not inhibited by somewhat increased cortisol levels because the anti-inflammatory influence of cortisol is not available in the presence of cortisol resistance.

Another alteration of homeostatic steroid axes are normal serum levels of estrogens in the presence of low androgens.[591] This phenomenon is attributed to (A) the increased activity of the aromatase complex in inflamed tissue and spill-over of estrogens into circulation[1553] and (B) a decreased androgen production in adrenal and gonadal glands. Since 17β-estradiol can stimulate the female type of subcutaneous lipid accumulation,[1578, 1579] increased levels of estrogens (also

TABLE 36 Alterations of the Steroid Hormone Axes

Cortisol serum levels are normal to somewhat elevated in untreated patients with chronic inflammatory systemic diseases, but cortisol levels are inadequately low in relation to inflammation (disproportion principle)
Low-dose glucocorticoid therapy can be a substitution therapy in the presence of relative adrenal insufficiency (e.g., polymyalgia rheumatica)
TNF can be an important cytokine that negatively influences the HPA axis in a chronic situation
Responsivity of the HPA axis is visibly disturbed when subtle stress tests are applied
It exists a state of cortisol resistance[1574, 1575]
Adrenal and gonadal androgen production is nearly switched off at the expense of cortisol production[591, 1576] (since androgens have anti-inflammatory capacities, this loss is a proinflammatory unfavorable development)
Androgens are converted to proinflammatory 16α-hydroxyestrogens by cytokine-stimulated aromatase and other enzymes in inflamed tissue
The hypothalamic-pituitary-gonadal axis is downregulated (See section "Disturbances of the hypothalamic-pituitary-gonadal (HPG) axis")

in men) relative to androgens can support female fat distribution, the aspect of cachectic obesity, and even generation of regional fat depots in the proximity of inflammation (due to high local estrogens). In addition, the relative increase of estrogens to androgens aggravates skeletal muscle wasting as substantiated in transsexuals.[1579]

All these alterations of the homeostatic steroid axes evolved to cope optimally with short-lasting inflammatory episodes. The major effect is skeletal muscle (for gluconeogenesis) and bone breakdown (for provision of expedients such as calcium). Their long-term use in chronic inflammatory systemic diseases is deleterious because these alterations support allocation of energy-rich fuels to the activated immune system.

DISTURBANCES OF THE HYPOTHALAMIC-PITUITARY-GONADAL (HPG) AXIS

The first experimental evidence that proinflammatory cytokines can interfere with the global function of the HPG axis came from IL-1β injection studies in rodents.[1580] Similarly, HPG axis abnormalities were observed in human studies of male rheumatoid arthritis patients, reflected as elevated follicle-stimulating hormone and luteinizing hormone levels, compared to controls.[769] This finding was observed in male patients with ankylosing spondylitis and systemic lupus erythematosus[1581, 1582] and in patients with Sjögren syndrome, when administered a luteinizing hormone-releasing hormone (LHRH) function test.[1583]

While elevated follicle-stimulating hormone and luteinizing hormone levels indicate peripheral gonadal dysfunction, lower than normal levels of these pituitary hormones are a sign of central abnormalities. Lower levels of luteinizing hormone were observed in patients with rheumatoid arthritis and systemic lupus erythematosus.[1584, 1585] If both peripheral and central alterations can exist in conjunction, levels of follicle-stimulating hormone and luteinizing hormone might not deviate at all, although the peripheral sex hormones are actually low.[1586]

As demonstrated in the last decade, abnormalities of the HPG axis can translate into a decrease of fertility in male and female patients, as tested in systemic lupus erythematosus,[782, 1582, 1585] ankylosing spondylitis,[1587] and other human chronic inflammatory systemic diseases.[1588–1590] In conclusion, inflammation-related hypogonadism is an often observed phenomenon in patients with chronic inflammatory systemic diseases, and it is a clinically relevant issue because of fertility problems. Moreover, it can be an important problem due to the diminished production of anti-inflammatory androgens (see above).

With the new information given in the section before, the loss of the gonadal function can be seen as a mechanism to (A) stop courtship behavior and respective energy expenditure (recall *Antechinus* of Chapter IV) and (B) stop skeletal muscle and bone growth (cachexia and bone loss), which provides either proteins for gluconeogenesis or calcium/phosphorus (see section below "Bone loss"). Both mechanisms have been evolutionarily positively selected to overcome short-lived inflammatory episodes. The long-standing use in chronic inflammatory systemic diseases is deleterious because these alterations support allocation of energy-rich fuels to the activated immune system, play a proinflammatory role, and destroy the body.

ELEVATED SYMPATHETIC TONE AND LOCAL NERVE FIBER LOSS

The sympathetic nervous system with its two effector arms, the adrenal medulla and peripheral sympathetic nerve fibers, is the major regulatory component of glycogenolysis, gluconeogenesis, and lipolysis. While the adrenal medulla is more the general stimulator of splanchnic organs, sympathetic nerve fibers can stimulate distinct adipose tissue regions in the body. Provision of energy-rich fuels by the sympathetic nervous system depends on β2-adrenergic receptor signaling. For example, catecholamines via β2-adrenergic receptors activate the hormone-sensitive lipase in order to breakdown triglycerides into glycerol (used in gluconeogenesis) and free fatty acids. The α2-adrenergic subtype of the receptor exerts antilipolytic effects.[1578] Thus, an elevated systemic activity of the sympathetic nervous system supports lipolysis, gluconeogenesis, and glycogenolysis. In addition, a high firing rate of sympathetic nerve fibers to adipose tissue would support local lipolysis. This would happen in the presence of high noradrenaline concentrations that exert β-adrenergic effects ($\geq 10^{-7}$ mol/l).

Importantly, many chronic inflammatory systemic diseases are accompanied by increased levels of circulating proinflammatory cytokines and an elevated activity of the sympathetic nervous system, which can be a substantial risk factor for cardiovascular disease.[1166–1169, 1591–1593] The circulating cytokines directly stimulate the sympathetic nervous system in the hypothalamus and locus coeruleus. In addition, hyperinsulinemia is related to increased sympathetic nervous system activity because insulin stimulates the sympathetic nervous system.[1594] Chronic inflammatory systemic diseases are accompanied by hyperinsulinemia and the reasons for this phenomenon have been discussed above.

Under consideration of systemic energy regulation in chronic inflammatory systemic diseases, the somewhat higher activity of the sympathetic nervous system is important to sustain allocation of energy-rich fuels to the immune system and to maintain systemic circulation. Indeed, denervation of the sympathetic nervous system largely decreased inflammation in animal models.[1042, 1116, 1595] The aggravating influence of late denervation of the sympathetic nervous system, recently demonstrated,[1116] is most probably not related to sympathetic nerve fibers but, possibly, to the manipulation of anti-inflammatory tyrosine hydroxylase-positive sympathetic cells in inflamed tissue (see also Chapter II).[607, 1121, 1122]

It is important to recognize that many components of the immune system such as macrophages, dendritic cells, neutrophils, NK cells, and T helper type 1 cells are inhibited via β2-adrenergic stimulation. The parallel activation of

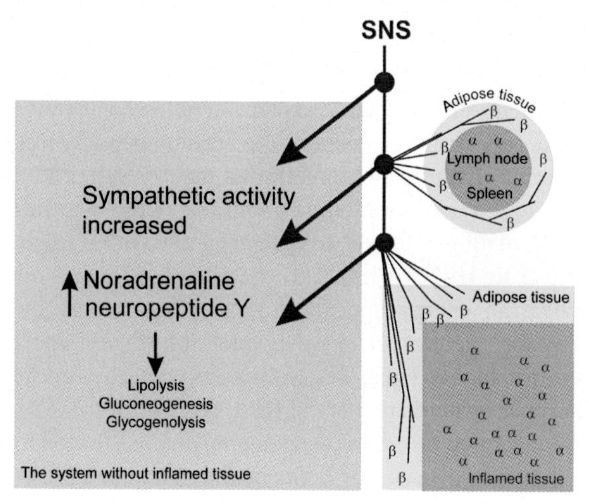

FIGURE 43 Chronic inflammation increases systemic activity of the sympathetic nervous system (SNS) and leads to parallel loss of sympathetic nerve fibers from inflamed tissue and in secondary lymphoid organs (lymph nodes and spleen). While density of sympathetic nerve fibers is very low in inflamed tissue and draining lymph nodes, the surrounding adipose tissue is perfectly innervated.[1573] When the SNS activity is increased, this allows for systemic release of energy-rich fuels from stores and energy-rich fuels from adipose tissue to nourish the inflammatory process (orange area), but it does not impede the inflammatory process via β2-adrenergic signaling.

β2-adrenergic receptors to stimulate lipolysis on one hand and to inhibit immune function on the other seems to be contradictory because the sympathetic nervous system should serve the activated immune system. The two contrasting functions of adrenergic neurotransmitters can be explained by compartmentalization (Figure 43). If sympathetic nerve fibers are lost in inflamed tissue but remain present in the adipose tissue, the two sympathetic functions can happen in parallel. Indeed, sympathetic nerve fibers get rapidly lost in inflamed tissue of patients with rheumatoid arthritis and animals with experimental arthritis and in the activated spleen and lymph nodes in arthritic animals.[606, 1042, 1044, 1119] However, sympathetic nerve fibers remain present in adjacent adipose tissue near the inflamed tissue and surrounding the draining lymph node.[1573] This compartmentalization allows parallel lipolysis (β2) and immune activation (α2 or α1) (Figure 43).

In conclusion, elevated systemic sympathetic activity as the consequence of an "energy appeal reaction" and local retraction of sympathetic nerve fibers are of importance to serve an acutely activated immune system either in inflamed tissue or in draining lymph nodes/spleen. It also serves water retention via activation of the renin-angiotensin-aldosterone system (next section), which is a pretty negative signal due to volume expansion and the proinflammatory influence of angiotensin II (see Chapter II). All these mechanisms were positively selected for short-lived inflammatory episodes, but they are deleterious in chronic inflammatory systemic disease.

HYPERTENSION

In Chapter III, hormonal and neuronal systems were described that are instrumental in water retention and volume expansion in the context of inflammation. Chronic inflammatory systemic diseases are accompanied by mild/moderate to severe inflammation. Even if a chronic inflammatory systemic disease is well treated, a state of low-grade smoldering chronic systemic inflammation persists. In this setting, circulating cytokine levels can be 5-10 times higher than normal (e.g., IL-6 from 2 to 10-20 pg/ml), and the overall stimulus might be strong enough to induce an energy appeal reaction and—as described in Chapter III—systemic water retention. The water retention system and the energy allocation system for highly energy-consuming processes are identical: (1) the sympathetic nervous system and (2) the HPA axis. Since these two systems for highly energy-consuming programs were not positively selected to serve chronic inflammatory systemic diseases as delineated in Chapter IV,[738, 1596] genes, signaling pathways, and networks that are operative in this setting are similar to those that determine acute energy-consuming episodes of inflammation.

Under conditions of local inflammation with exposed outer and inner surfaces, circulating cytokines from inflamed tissue activate the energy appeal and water retention reaction. Since the local inflammatory process leads to water loss over local surfaces, locally formed water—by degradation of energy-rich fuels in activated inflammatory cells (Figure 44, red circular area)—can be rap-

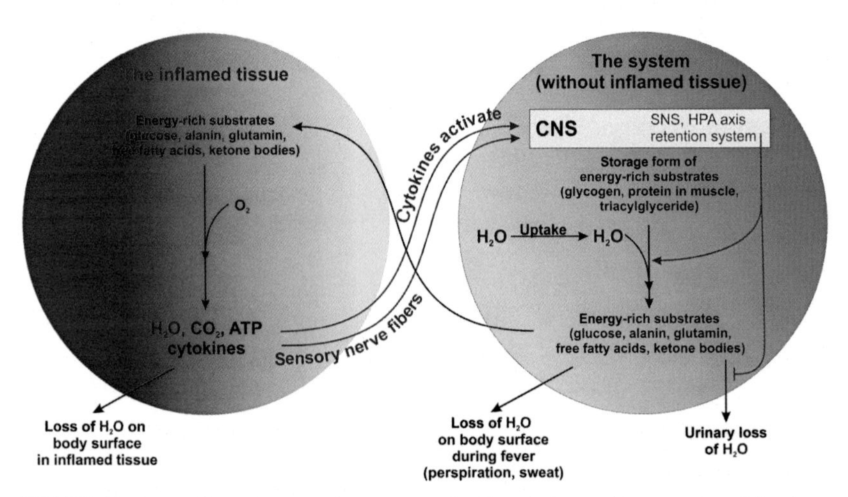

FIGURE 44 Water fluxes between the system and inflamed tissue with systemic and local water loss in a transient acute inflammatory episode. The inflamed tissue (red orange circular area) releases cytokines or stimulates sensory nerve endings in order to induce an energy appeal and water retention reaction in the system (pink blue circular area). In the system, urinary water loss is inhibited by the sympathetic nervous system (SNS), the hypothalamic-pituitary-adrenal axis (HPA axis), and the renin-angiotensin-aldosterone system, because water is needed for many important reactions for provision of energy-rich substrates and immune cell proliferation (see Figure 33 in Chapter III). In the acute situation, water is lost from the system and from inflamed tissue via outer and inner surfaces. The water fluxes are in a balance.

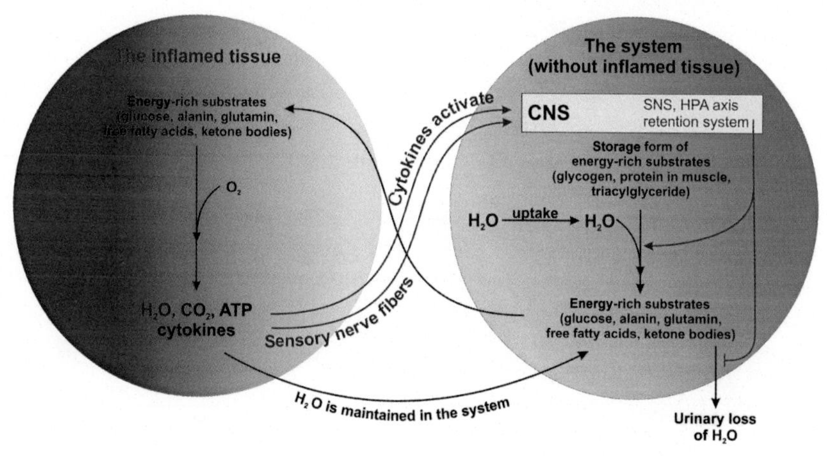

FIGURE 45 Water fluxes between the system and inflamed tissue without water loss in chronic inflammation. Similar as in Figure 44, the inflamed tissue activates water retention and energy appeal reaction. However, due to low-grade inflammation without fever and due to a lack of inflamed exposed surface areas, water is not lost from the system and from inflamed tissue. Water can recirculate between activated immune cells and the system (H_2O is maintained in the system). In this scenario, it does not matter how the inflamed tissue looks like. It might be a separated tissue such as an inflamed joint or an inflamed segment of the aorta. However, it might also be the sum of disseminated activated inflammatory cells in any tissue (red circular area). It might also be the sum of disseminated an activated cells in different parts of the body, which might happen during aging when different organs present higher levels of inflammatory activity. The logic behind the water flux between the system and inflamed cells remains the same. It is only important that secreted cytokines of these inflammatory cells activate the energy appeal and water retention reaction and that water loss via inner and outer surfaces is not increased. Chronic inflammation accompanied by essential hypertension is a necessary consequence.

idly lost via the exposed surface. It can be hypothesized that, in the presence of local and systemic water loss, the water retention system will not induce hypertension or volume overload under these acute inflammatory conditions.

The situation of water retention might be very different depending on the setting and operation of different mechanisms of water loss. These mechanisms would include the following events, which may be present during disease: local water loss through outer/inner surfaces, water loss due to systemic effects such as fever, recirculation of formed water from activated inflammatory cells, and cytokine-activated water retention and energy appeal reaction. Depending on the operation and extent of these mechanisms, systemic hypertension could result from the continuous activation of a water retention and energy appeal reaction but without the usual water loss known from acute inflammation (Figure 45).

It was already discussed that in chronic inflammatory diseases, both the sympathetic nervous system and the HPA axis are chronically activated.[729, 737, 1168, 1597] In addition, there are obvious signs of volume overload in chronic inflammatory diseases because serum levels of atrial natriuretic factor and NT-brain natriuretic factor are markedly increased in different diseases.[1598–1601] NT-brain

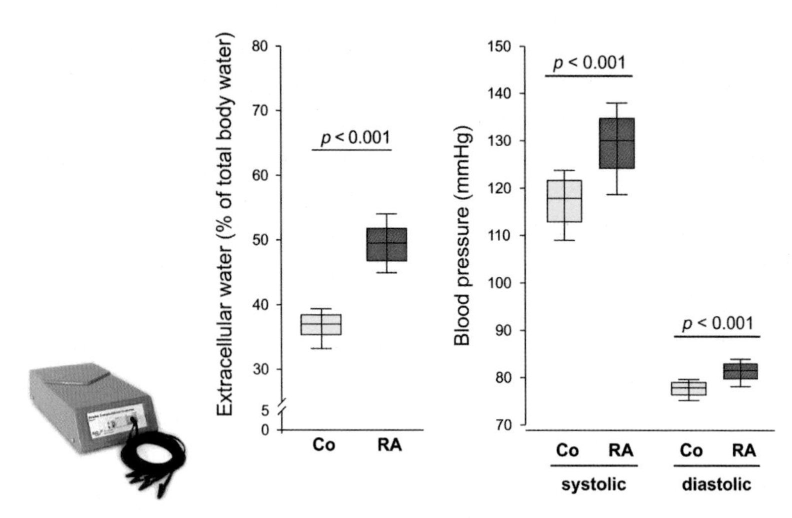

FIGURE 46 Increased extracellular water in patients with rheumatoid arthritis (RA, $n = 156$) compared to healthy controls (Co, $n = 430$). The patients were matched according to age, gender, height, weight, body mass index, and serum creatinine (was normal in all subjects). Patients with rheumatoid arthritis were treatment-naive. Extracellular water was measured by bioimpedance analysis (BIA-ACC, Biotekna, Marcon, Venezia, Italy, according to the technique published in Ref. 1435). This clearly demonstrates the increased volume load in patients with chronic inflammatory systemic disease, and this was linked to an increased systolic and diastolic blood pressure.

natriuretic factor decreased with TNF neutralization therapy but increased under therapy with volume retention-inducing glucocorticoids.[1598, 1600]

The theory predicts that the operation of these mechanisms will not depend on the site of inflammation and whether it occurs in the arteries as atherosclerosis or vasculitis, in the kidneys, in the gastrointestinal tract, in the joints, or at various tissues with inflammatory cells. The key feature relates to the production of cytokines of inflammatory cells, their activation of the energy appeal and water retention reaction, and the extent of water loss via inner and outer surfaces. The magnitude of these changes would determine any detrimental effect on the induction of hypertension. In a recent example of patients with rheumatoid arthritis, we recognized an increase of extracellular water volume and elevated systolic and diastolic blood pressure when compared to healthy controls (Figure 46).

As noted above, the energy appeal reaction and water retention reaction appear to be the result of selection in evolution for their role in transient inflammatory episodes. Induction of energy appeal reaction and water retention reaction provides a survival value if used for a short period of time. However, prolonged operation of these adaptive programs such as in chronic inflammatory systemic diseases of today can in itself become pathogenic because there is no program to counteract continuous water retention and energy appeal reaction during long-term age-related inflammation and in chronic inflammatory diseases.

DECREASED PARASYMPATHETIC TONE

Before I discuss the decreased parasympathetic tone in chronic inflammatory diseases, the role of the vagus nerve in energy regulation is recapitulated.

In the fasting situation, vagal afferents are important in transferring hepatoportal information on low blood glucose and low levels of other nutrients to the dorsal vagal complex leading to hunger signals, inhibition of sympathetic activation, inhibition of efferent vagal activation, hypometabolism, and hypothermia.[1602–1604] This is supported by gastrointestinal hormones such as ghrelin.

In the feeding situation, vagal afferents together with gastrointestinal hormones such as cholecystokinin transmit signals to the dorsal vagal complex leading to satiety signals, activation of the sympathetic nervous system (pretty short-lived postprandial thermogenesis and hypermetabolism), activation of efferent vagal nerve fibers (propulsive motility and secretion of exocrine and endocrine pancreatic and gastrointestinal factors, mainly, insulin), hyperinsulinemia, and, thus, storage of energy-rich fuels.[1602–1605] The parasympathetic nervous system also improves insulin sensitivity to allow storage of fuels in the liver, fat tissue, and skeletal muscles.[1606, 1607] The parasympathetic nervous system stimulates hepatic glycogen storage and hepatic vagal denervation decreases hepatic glucose uptake.[1608]

Although the sympathetic nervous system postprandially induces short-lived hypermetabolism, the net effect of the vagus nerve leads to energy storage mainly due to hyperinsulinemia. In the words of Thorens, one can summarize that "the net functional role of vagal afferents is to dampen food intake [note of author: or to induce a hypometabolic hunger constellation], whereas intact vagal afferents are necessary to defend a normal high body weight."[1603, 1605] This is supported by an important experiment in the early 1990s. Lesioning the headquarter of the sympathetic nervous system in the hypothalamus, the ventromedial hypothalamic nucleus, leads to hypoactivity of the sympathetic nervous system and hyperactivity of the efferent vagus nerve, the consequence of which is hyperinsulinemia and obesity.[1609] Thus, in the fasting and in the feeding situation, the vagus nerve is responsible for energy storage.

Importantly, fat tissue and skeletal muscles are not innervated by parasympathetic nerve fibers,[1153] which implies that the vagal influence on fuel storage is mainly directed towards hepatic glucose metabolism and insulin secretion. The vagus nerve constitutes the branch of the autonomic nervous system relevant for uptake and storage of energy-rich fuels. Thus, the parasympathetic nervous system would have the function to withhold energy-rich fuels from an active immune system, which is completely opposite to the sympathetic nervous system (see section above "Elevated sympathetic tone and local nerve fiber loss").

In consequence, one might expect an anti-inflammatory influence of the parasympathetic nervous system because this part of the autonomic nervous system would not support allocation of energy-rich fuels to activated immune cells. Indeed, publications of Kevin Tracey's group delineated the anti-inflammatory effects of the parasympathetic neurotransmitter acetylcholine via the vagal cholinergic anti-inflammatory pathway.[1174] While these investigators demonstrated a direct anti-inflammatory effect via α7 nicotinergic receptors, the overall energy-relevant role of the parasympathetic nervous system can add to the direct effects on this particular receptor. This has been described in Chapter II.

Consistent with an increased activity of the sympathetic nervous system,[1166–1169, 1591–1593] a decreased activity of the parasympathetic nervous system can occur during the course of chronic inflammatory systemic diseases.[1610–1614] The sympathetic nervous system and the parasympathetic nervous system are often working in opposite directions, which was called the yin and yang of the sympathetic and parasympathetic drive.[1615] In chronic inflammatory systemic diseases, the sympathetic nervous system is switched on at the expense of the parasympathetic nervous system, which is linked to reduced gastrointestinal activity, decreased nutrient uptake, diminished glucose uptake into the liver, and a more proinflammatory situation due to the loss of the cholinergic anti-inflammatory influence.[694]

As in the case of other systemic events described in this chapter, the program has been evolutionarily positively selected for transient inflammatory episodes but becomes unfavorable in chronic inflammatory systemic diseases.

INFLAMMATION-RELATED ANEMIA

Inflammation-related anemia is frequently observed in chronic inflammatory systemic diseases.[1616] Anemia in chronic inflammatory systemic diseases is a mild to moderate normocytic-normochromic anemia characterized by decreased serum iron and transferrin and increased iron stores observed as increased circulating ferritin—the storage protein of iron.[1616, 1617]

The following pathophysiological causes were described: (1) allocation of iron to monocytes/macrophages and other cells of the reticuloendothelial system (RES) stimulated by hepcidin (Figure 47), (2) reduced intestinal iron resorption initiated by hepcidin (Figure 47), (3) disturbed erythropoiesis due to proinflammatory cytokines and reduced half-lives of erythrocytes (phagocytosis by macrophages!) (Figure 47), (4) reduced influence of erythropoietin on erythropoiesis (little production and/or hormone resistance),[1616] (5) chronic blood loss, (6) hemolysis, and (7) vitamin deficiencies. Many processes are stimulated by circulating and local proinflammatory cytokines such as TNF, IL-1β, and IL-6, which can be studied using anticytokine therapy in patients with chronic inflammatory systemic diseases that reduce inflammation-related anemia.[1618, 1619]

Increased iron uptake and accumulation in macrophages is necessary, for example, for peroxide-generating enzymes and nitric oxide-generating enzymes, which are extremely important in bacterial killing.[1620, 1621] During the process of coevolution of bacteria and vertebrates, microbes and host always competed for iron (Figure 47).[1621, 1622] As part of innate immunity, the body modifies iron metabolism in order to allocate iron to the reticuloendothelial system, which makes iron less available to microorganisms (Figure 47).[1620–1622]

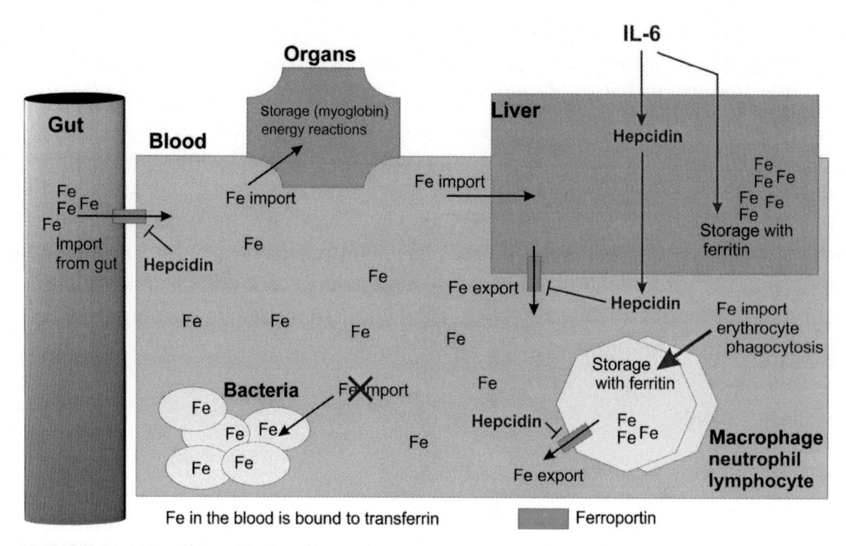

FIGURE 47 Hepcidin-stimulated iron allocation to host cells instead of bacteria. Cytokines such as IL-6 stimulate hepcidin, and hepcidin blocks the iron transporter ferroportin, which is active in the gut (iron import from gut lumen), the liver (iron export into the blood), and iron export from macrophages. When hepcidin is stimulated, the amount of iron in the circulation is reduced and iron is stored in form of ferritin in liver and macrophages. This method and inhibition of iron import into bacteria remove iron from bacteria. The visible consequence is inflammation-related anemia, which is a positively selected program for the competition with bacteria. Abbreviations: Fe, iron; IL-6, interleukin-6.

In addition, iron is an essential metal for the respiratory chain complex—the cytochromes—to handle the necessary protons in the context of ATP generation. We recall that 2300 ATP molecules are needed for the synthesis of one typical protein, for example, a cytokine (Figure 30 in Chapter III). In the case of high turnover rates of immune cells and clonal expansion of lymphocytes, iron availability is important for proper function of the immune system.[1623]

The question appears as to how decreased levels of serum iron and increased ferritin might be linked to overall energy regulation: (1) Anemia is accompanied by reduced energy expenditure for erythropoiesis, which can be a significant aspect as substantiated in sickle cell anemia.[1624] Children with asymptomatic sickle cell anemia had 52% higher protein turnover, which cost 460 kJ/d (110 kcal/d) more as compared to healthy children.[1624] (2) The development of mild anemia limits the oxygen transport capacity in general, which will increase the time spent at rest (immune cells switch to anaerobic glycolytic metabolism). (3) Increased ferritin is linked to insulin resistance, peripheral hyperinsulinemia, liver insulin resistance, and increased gluconeogenesis, which support glucose allocation to the immune system.[1625] (4) Lower oxygen tension leads to decreased physical activity and cardiovascular performance.[1616] And (5) reduced iron levels are followed by decreased myoglobin levels in skeletal muscle, which influence muscular performance.[1620] All these effects of anemia corroborate allocation of energy-rich fuels to the activated immune system.

It is important that mild anemia, as it can happen in chronic inflammatory systemic diseases, is often accompanied by an increase of the sympathetic nervous system activity and glucose production.[1626] Thus, we hypothesize that mild inflammation-related anemia is another important factor that can drive the sympathetic nervous system and the HPA axis in order to allocate energy-rich fuels to the activated immune system. This has been positively selected in the context of acute inflammatory episodes or hemorrhage, but not for long-standing chronic inflammatory systemic diseases. Here, we recognize it as another misguided adaptive program.

BONE LOSS[c]

Introduction

From the perspective of acute inflammation, sickness behavior with an increased time spent at rest and sleeping and with anorexia to avoid energy-consuming foraging, enough energy is conserved to nourish the activated immune system for a while (consumption time; Table 20). However, the consequence of this adaptive program during acute infection/inflammation would be deficiency of energy-rich fuels (carbohydrates, free fatty acids, and proteins). Intake of energy-rich fuels is reduced in rheumatoid arthritis,[1428, 1429] systemic sclerosis,[1430] multiple sclerosis,[1431] juvenile idiopathic arthritis,[1432] and other chronic inflammatory systemic diseases. It should be stressed that decreased fuel intake is often linked to loss of essential ions such as calcium, phosphorus, magnesium, iron, zinc, and copper and lowered levels of vitamin D and other lipid-soluble vitamins (discussed above). This can lead to malnutrition, which is a typical finding in many inflammatory diseases.[1436] Under anorexia conditions, the body is absolutely dependent on stored reserves, which is also true for calcium and phosphorus.

Calcium Balance Under Normal and Inflamed Conditions

The normal calcium balance is depicted in Figure 48. The average adult human body contains approximately 1 kg of calcium, which corresponds to 25 mol calcium.[1627] A total of 99% is stored in the bone and teeth. Most of the calcium is inaccessible to most physiological processes.[1627] Approximately 1% of the stored calcium (10 g or 0.25 mol) is immediately accessible.[1627] This fraction serves many different roles, which range from intracellular signaling and maintenance of membrane integrity to skeletal muscle contraction, neuronal transmission, and immune cell function. Immune cells depend on calcium as in-

c. This subject on BONE LOSS was presented as a contribution to Seminars in Arthritis and Rheumatism. Straub RH, Cutolo M, Pacifici R: Evolutionary medicine and bone loss in chronic inflammatory diseases – a theory of inflammation-related osteopenia. Sem Arthritis Rheum, 2015, in press.

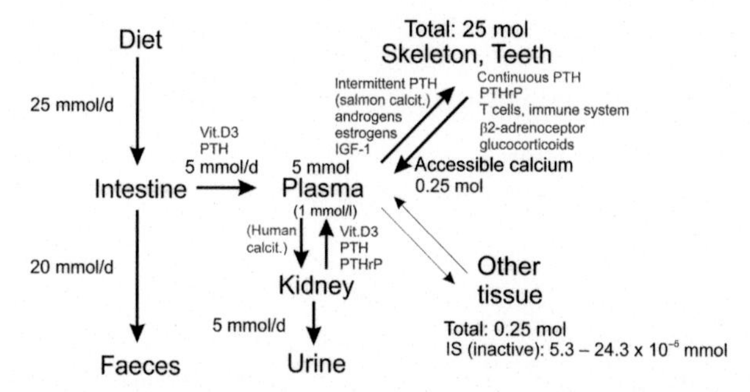

FIGURE 48 Calcium homeostasis in physiology. The major hormonal regulators are given in red. The numbers for the immune system (IS) demonstrate the total amount of calcium within all immune cells. The numbers in millimole per day are turnover rates, while numbers in mole or millimole are the total amount in the mentioned compartment. Numbers are derived from information of Refs. 1152, 1627–1630. Abbreviations: calcit., calcitonin; IGF-1, insulin-like growth factor 1; IS, immune system; PTH, parathyroid hormone; PTHrP, PTH-related peptide; Vit. D3, vitamin D3.

dicated by *calcium release-activated calcium* (CRAC) channel channelopathies that lead to severe immunodeficiency in affected individuals.[1628] At this point, the question appears whether immune cells really need most of the accessible calcium for proper function in acute inflammation.

Assume an immune cell has a diameter of 8-20 μm, and the classical formula for a perfect sphere is used, the volume of one cell is 2.7×10^{-13} to $4.2 \times 10^{-12}\,m^3$ (268-4188 fl; 10^{-15} l). Multiplied with the total number of leukocytes ($\approx 5.8 \times 10^{11}$), we obtain a volume of 0.16-2.43 l. With this number, the total amount of calcium in all immune cells together is 5.3-24.3 $\times 10^{-5}$ mmol, which is miniscule in relation to the accessible calcium store of 250 mmol (Figure 48). With activation of immune cells as in acute inflammation, intracellular calcium concentration increases from 10^{-7} to 10^{-6} mol/l.[1628] This factor of 10 increases the amount of calcium in all immune cells in a minimal way (from 5.3-24.3 $\times 10^{-5}$ to 5.3-24.3 $\times 10^{-4}$ mmol, if all immune cells would be activated, which is not the case).

From these considerations, it is clear that quantitative calcium redistribution from the bone to activated immune cells is unlikely. In the presence of inflammation-induced anorexia, bone loss is directly stimulated by inflammatory factors from activated immune cells (Table 37). This uniform response of the immune system is surprisingly strong given the fact that immune cells do not need so much calcium. Thus, it seems that immune-mediated bone loss is a program to supply calcium to the rest of the nonskeleton tissue. This is particularly true when renal calcium loss or glomerular filtration rate remains normal or even increase during acute and chronic inflammations when there is no structural kidney injury.[1629–1632] The nonskeleton tissue comprises all organs and skeletal muscles.

Indeed, a simple calculation from numbers in Figure 48 shows that in the absence of dietary calcium intake but with a remaining renal loss of 5 mmol/d, the nonskeleton tissue store of 0.25 mol would be emptied within 50 days (250 mmol

TABLE 37 Typical Inflammation-Related Factors of Bone Loss (Recently Reviewed in Refs. 1633–1638)

Inflammation-Related Factor
CD4+ T cells[a] via TNF, RANKL, IFN-γ, LIGHT, IL-15, IL-17
CD8+ T cells[a] via TNF, RANKL
Macrophages via TNF, IL-1β, IL-6
Neutrophils via reactive oxygen species
Immunologically stimulated bone stromal cells via RANKL and M-CSF

[a]In contrast, T regulatory cells and B cells suppress bone resorption in ovariectomy-induced bone loss and in models of rheumatoid arthritis by secreting antiosteoclastogenic factors like osteoprotegerin.

divided by 5 mmol/d). This number is surprisingly similar to the total consumption time of a modern human (see Chapter III). It seems as if accessible amount of stored calcium was evolutionarily positively selected to serve calcium provision during this critical time of acute inflammation with sickness behavior and anorexia. Although the activated immune system only needs miniscule amounts of accessible calcium, it is the driving system that starts in the first place observable bone resorption. During sickness behavior and anorexia, immune system-activated bone resorption and calcium provision can be seen as an offered service to the rest of the nonskeleton body. Figure 49 summarizes the situation in inflammation.

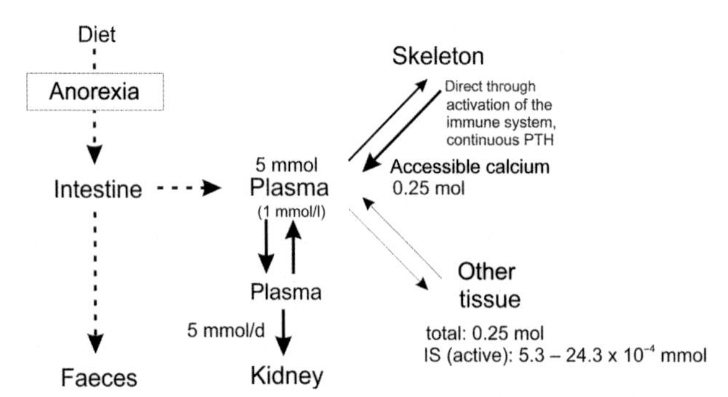

FIGURE 49 Calcium homeostasis in inflammation. The major hormonal regulators are given in red. Dotted arrows indicate strong reduction of calcium shifts due to anorexia. The numbers for the immune system (IS) demonstrate the total amount of calcium within all immune cells. Note that the activated immune system contains 10 times more calcium compared to the inactive immune system when all cells would be involved (which is not the case in reality). The numbers in millimole per day are turnover rates, while numbers in mole or millimole are the total amount in the mentioned compartment. Numbers are derived from information of Refs. 1152, 1627–1630. Abbreviations: calcit., calcitonin; IS, immune system; PTH, parathyroid hormone.

Phosphate Balance Under Normal and Inflamed Conditions

Similar as with calcium, plasma concentration of inorganic phosphorus (Pi) depends on dietary intake, passive intestinal absorption, release from the bone and nonskeleton tissue, and renal excretion (the picture would look like Figure 48). Usually, 30 mmol is excreted with the urine, and dietary intake equals renal and fecal Pi excretion.

However, the situation is pretty different with Pi compared to calcium because Pi reabsorption in the kidneys can be almost 100%.[1639] This adaptation happens very fast in conditions of low phosphate intake, and this adaptation is mainly independent of hormonal regulation.[1640] In the case of acute inflammation, sickness behavior, and anorexia, the kidneys prevent phosphate loss. Given the above-mentioned 0.25 mol accessible calcium from skeleton tissue, approximately 0.14 mol of phosphate is also accessible, because the molar ratio of calcium to Pi in the bone is ≈1.7:1 (Ref. 1535). In summary, during acute inflammation with anorexia, Pi deficiency is a minor problem due to the nearly complete reabsorption of Pi in the kidneys and the huge amount of accessible Pi from the bone.

However, states of low phosphate serum levels can exist, which typically appear as refeeding hypophosphatemia after states of anorexia. This form of phosphate deficiency is not an absolute loss of Pi but rather a rapid transmembrane shift of Pi from extracellular to intracellular space, which is well known in the context of alkalosis or hyperglycemia/hyperinsulinemia.[1641] This may lead to deficits in intracellular ATP in certain organs, which can lead to cardiac arrhythmias and heart failure, muscular fatigue (respiratory problems), even rhabdomyolysis, neurological manifestations (confusion, seizure, weakness, and fatigue), and hemolytic anemia.[1641] Such a situation might appear after refeeding after prolonged anorexia, and it can even be a critical factor in patients with anorexia nervosa.[1642] Thus, there might exist a relative Pi deficiency after inflammation-induced anorexia due to refeeding after the anorectic phase. Inflammation-induced bone resorption would counterbalance this form of Pi deficiency (think of the 0.14 mol of accessible Pi).

Similarly as discussed for calcium, after sickness behavior, anorexia, and subsequent refeeding, immune system-activated bone resorption and Pi provision can be seen as an offered service to the rest of the nonskeleton body.

Other Pathways of Inflammation-Stimulated Bone Loss

Due to energy requirements of the activated immune system in the presence of anorexia, fatigue, skeletal muscle wasting (leading to a relative increase of fat over muscular tissue called cachectic obesity), and inflammation-related anemia are ways to support immobility. While immobility reduces energy requirements of the activated brain and skeletal muscles (redistribution to the activated immune system), immobility also stimulates bone loss.[1643] Bone loss happens already during the first 4-6 weeks of immobilization.[1644, 1645] In addition, immobilization leads to hypercalcemia and subsequent hypercalciuria that strongly increased from week 1 to 4[1646]. The fact that immobilization-related bone loss happens within 4-6 weeks is another good indication of an adaptive program, because it well fits into the time window of 6 weeks given by the total consumption time (see Chapter III).

Another aspect of acute inflammation is hypogonadism. This phenomenon has been described in rats after cytokine or endotoxin injection[1647] and in humans with infectious diseases.[1648–1651] I described it in the section "Disturbances of the hypothalamic-pituitary-gonadal axis" (this chapter). Subcutaneous injection of IL-6 into human volunteers rapidly decreased serum-free testosterone within 24 h (Figure 35).[1307] Since gonadal hormones play an instrumental role in bone generation, inflammation-related acute loss of gonadal hormones supports bone resorption.

Two other important systems change bone homeostasis during acute inflammation. Both the HPA axis and the sympathetic nervous system are activated during acute inflammation.[611] The major hormones of the two systems are cortisol and adrenaline/noradrenaline, whose role on bone resorption is well known. While negative effects of cortisol or glucocorticoids on the bone are known since long, recent studies clearly demonstrated that the sympathetic nervous system via β2-adrenergic receptors has strong bone-resorbing effects, too.[1652, 1653]

In addition, anorexia-induced hypovitaminosis D (and K) and hypomagnesemia are well-known factors in acute inflammation and critical illness[1654, 1655] but similarly in chronic inflammation (summarized in Ref. 1656). Such a situation would aggravate bone resorption during inflammatory episodes with anorexia.

Importantly, parathyroid hormone (PTH) is often elevated in acute inflammation such as septic shock,[1657] sepsis,[1658] or after endotoxin injection,[1659] but levels are normal, high, or low in chronic inflammation.[1660–1664] Catabolic effects of PTH on the bone might be largely determined by parallel inflammation[1633] or by an increase in parallel glucocorticoids.[1665] Under inflammatory conditions, PTH might lose its bone-anabolic activities to become bone-resorbing.[1633] In addition, inflamed tissue and activated immune cells can produce parathyroid hormone-related peptide (PTHrP), which can be elevated in acute and chronic inflammation.[1666–1668] Lipopolysaccharide can directly induce PTHrP secretion,[1669] and blockade of PTHrP reduces bone loss in experimental arthritis.[1670] Thus, both PTH and PTHrP do not protect the bone in acute and chronic inflammations, which most probably depends on the circumstances of high inflammation and appearance of other bone-resorbing factors.

Figure 50 summarizes the different factors leading to inflammation-related bone loss. It is hypothesized that all these elements serve an adaptive program for acute inflammatory illness but not for chronic inflammatory systemic diseases.

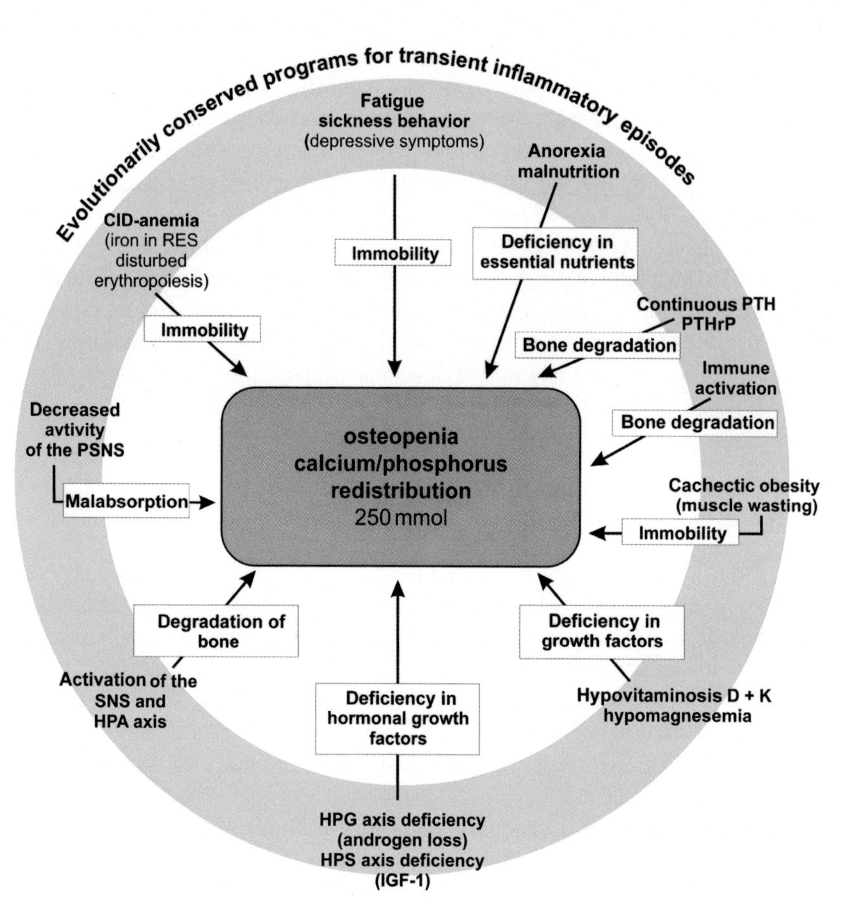

FIGURE 50 Disease sequelae of inflammation leading to osteopenia. The number in the pink middle area (250 mmol) represents the accessible amount of calcium provided from the bone.[1627] Abbreviations: CID-anemia, inflammation-related anemia of chronic inflammatory diseases; HPG axis, hypothalamic-pituitary-gonadal axis; HPA axis, hypothalamic-pituitary-adrenal axis; HPS axis, hypothalamic-pituitary-somatic axis (growth hormone, IGF-1); IGF-1, insulin-like growth factor 1; PSNS, parasympathetic nervous system; PTH, parathyroid hormone; PTHrP, PTH-related peptide; RES, reticuloendothelial system; SNS, sympathetic nervous system.

When we observe homeostatic regulation in chronic inflammation, we expect that they have not been evolutionarily positively selected in the context of chronic energy-consuming inflammation but for acute inflammation (see Chapter IV). In acute inflammation, observed programs of Figure 50 are adaptive programs. They have been positively selected and maintained for eons to serve a short-lived energy-consuming inflammatory episode.

In contrast, if programs were helpful to protect energy reserves, they were positively selected during evolution and maintained in modern species (see also Chapter IV). It is hypothesized that this is true for energy storage of nutrients in storage organs (e.g., adipose tissue), for immunoregulatory mechanisms (T regulatory cells), for immune memory (T and B cells), and for brain memory (hippocampal neurogenesis), because memory saves energy. Another organ that stores important expedients for cellular energy reactions is the bone. The bone stores calcium and phosphorus, and a small part of it can be used as accessible pool (250 mol of the total 25 mol; Figure 48).

In conclusion, bone resorption stimulated by immune-mediated pathways (Table 37) is an adaptive program to provide calcium and phosphorus (and magnesium) to the nonskeleton tissue essential for survival over 3-6 weeks. At this point, the question appears as to why the immune system serves calcium to the rest of the nonskeleton body although it only needs miniscule amounts of calcium as demonstrated above.

Why Does the Immune System Serve Calcium to the Rest of the Nonskeleton Body?

In an environment without any calcium and phosphate stores, in the absence of an exoskeleton (in crustaceans or mussels) or endoskeleton (in bony fish or tetrapods), bone degradation does not exist. For example, this is true for the tunicate sea squirt (*Ciona intestinalis*), a sessile animal permanently attached to a rock under seawater (Protochordates). In need, *Ciona intestinalis* receives calcium from the surrounding seawater (seawater has a calcium concentration of 10 mmol/l; serum calcium in humans is approximately 2.4 mmol/l).

However, sea squirts already have a primitive immune system with eukaryotic orthologs of such important bone-degrading factors such as TNF, IL-17, and receptor activator of nuclear factor-kappa B (RANK).[1671] In addition, they possess the critical bone-degrading enzyme of osteoclasts, the tartrate-resistant acid phosphatase.[1671] It was observed that sea squirts produce a TNF-like cytokine upon stimulation with lipopolysaccharide.[1672] Thus, it is highly likely that *Ciona intestinalis* used TNF to overcome infectious disease. *Ciona intestinalis* also possess neutrophils and primitive ancestors of modern antigen receptors.[1673]

The role of an orthologous protein to RANK in *Ciona intestinalis* is presently not known. Such a *Ciona* RANK molecule might have something to do with insulin signaling because insulin has been found in *Ciona intestinalis*[1674] and the RANK/RANKL system has been critically linked to insulin resistance in mammals.[1675] It was discussed that insulin resistance is an adaptive program to deviate energy-rich fuels from stores to the activated immune system (see above).[1152] A similar mechanism might exist in *Ciona intestinalis* but this needs to be elucidated.

Nevertheless, it tells us that *Ciona intestinalis* orthologs of bone-degrading cytokines already existed many million years before the development of an endoskeleton in an animal without the bone or calcium stores. When the endoskeleton appeared during the process of evolution, the already existing bone-degrading cytokines were not counterproductive, but most probably necessary for an inflammation-dependent short-lived burst release of calcium from the endoskeleton in terrestrial animals. In other words, without this already existing calcium provision system, terrestrial life with endoskeletons probably would not have happened.

Low Bone Mass is a Typical Disease Sequelae of Many Chronic Inflammatory Systemic Diseases

While we learned that episodic bone resorption is an adaptive program for acute inflammatory illness, long-term application of this program might induce low bone mass. Indeed, bone loss is absolutely typical in rheumatoid arthritis, psoriasis vulgaris, ankylosing spondylitis, systemic lupus erythematosus, multiple sclerosis, inflammatory bowel diseases, pemphigus vulgaris, and others (recently reviewed in Refs. 1676–1682). It is also typical in transplantation-related inflammation.[1633] Figure 51 demonstrates the example for rheumatoid arthritis.

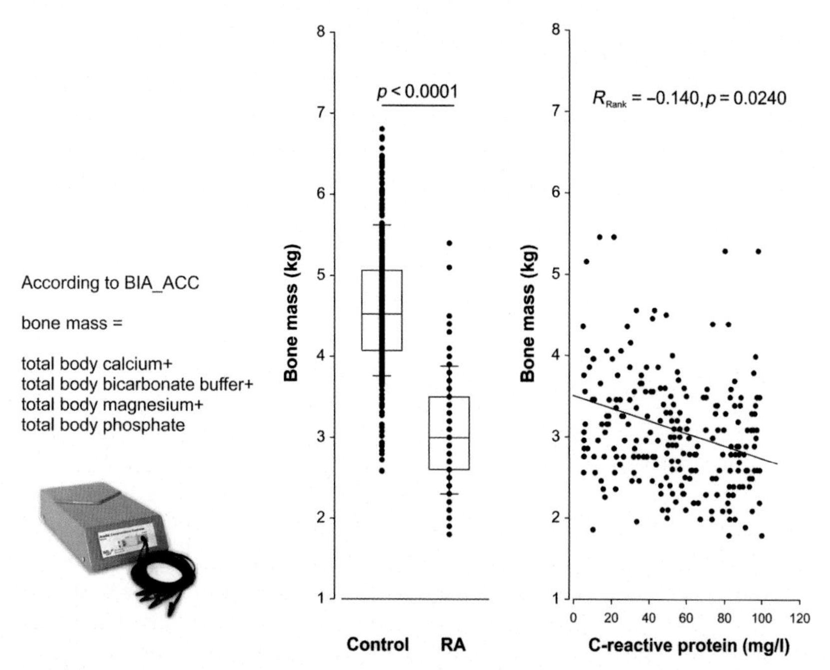

FIGURE 51 Bone loss in patients with rheumatoid arthritis (RA, $n=156$) compared to healthy controls (Co, $n=430$). The patients were matched according to age, gender, height, weight, body mass index, and serum creatinine (was normal in all subjects). Patients with rheumatoid arthritis were treatment-naive. Bone mass was measured by bioimpedance analysis (BIA-ACC, Biotekna, Marcon, Venezia, Italy, according to the technique published in Ref. 1435). This clearly demonstrates the decreased bone mass in patients with a chronic inflammatory systemic disease, which was dependent on higher inflammation measured by C-reactive protein (right panel).

Conclusions

The three pillars of evolutionary medicine, of energy regulation, and of neuroendocrine regulation of homeostasis and immune function explain that adaptive programs for acute inflammation are wrongly used in chronic inflammation. One of these adaptive programs in acute inflammation is provision of calcium and phosphorus from the bone to the nonskeleton tissues. Thus, an acute burst of bone loss is not an accident but an important program.

In order to steer calcium/phosphorus provision, many specific immune-mediated pathways are switched on (Table 37). Several specific immune-mediated pathways can induce bone resorption independent of classical hormonal pathways. This is indicative of adaptation to acute inflammation (a nonhormonal, immunologic bypass). While these considerations might explain the deeper meaning of bone resorption in inflammation, we are critically dependent on finding the major molecular pathways to better treat these disease sequelae. This section should also stimulate the idea that sometimes, the critical pathways leading to osteopenia can be outside usual mainstream thinking. It also demonstrates that osteopenia must be a multifactorial affair (Figure 50).

HYPERCOAGULABILITY

Patients with chronic inflammatory systemic diseases demonstrate hypercoagulability.[1683–1687] This is associated with an increased risk of atherosclerosis.[1567, 1688–1709] While inflammation plays a key role of hypercoagulability, other important factors of stress axes are also involved. This section summarizes the causal effects of inflammation, stress, the sympathetic nervous system, and the renin-angiotensin-aldosterone system on hypercoagulability. Finally, these findings are discussed with an evolutionary perspective.

Inflammation Induces Blood Coagulation

Tissue factor is the primary initiator of blood coagulation *in vivo*.[1710] Tissue factor contributes to the thrombotic complications associated with atherosclerosis.[1711] The proinflammatory cytokines TNF, IL-1β, and more directly lipopolysaccharide induce tissue factor in monocytic and endothelial cells, which is mediated by various intracellular signaling pathways and the transcription factors NF-κB, AP-1, and Egr-1 (Ref. 1712). Anti-TNF therapy can normalize hypercoagulability in patients with rheumatoid arthritis.[1713, 1714] Thus, an elevated proinflammatory load can directly lead to tissue factor expression and atherothrombosis. There are apparent links between the acute inflammatory response in infection and increased coagulability.[1715] It is not surprising that patients with chronic inflammatory systemic diseases, who demonstrate an increased proinflammatory load, can activate tissue factor and the coagulation system in vulnerable vessels.

Stress and the Sympathetic Nervous System Induce Hypercoagulability

An increased activity of the sympathetic nervous system is an important feature of chronic inflammatory systemic diseases (see above). Thus, investigation of stressful states with an increase of sympathetic nervous system activity can be helpful in understanding increased hypercoagulability in chronic inflammatory systemic diseases.

Physical exercise leads to the activation of blood coagulation that results in the formation of thrombin and fibrin.[1716] Immediately after, and 2, 8, and 21 h after a triathlon lasting 128-163 min, one can observe a moderate activation of blood coagulation resulting in thrombin and fibrin formation, which is accompanied by a greatly enhanced plasmin generation.[1717] Thrombin is formed, in particular, when associated with anaerobic metabolism, and platelets are activated during high-intensity exercise.[1718] This increase in exercise-induced blood coagulation is accompanied by higher catecholamine plasma levels.[1719]

Increased noradrenaline and adrenaline levels induce changes in hemostasis and fibrinolysis associated with exercise.[1720] Adrenaline can promote both hemostasis and fibrinolysis via β-adrenergic pathways.[1721] Adrenaline, at physiological concentrations, enhances von Willebrand factor-dependent platelet interaction with a collagen-coated surface under blood flow.[1722] This might depend on enhanced high shear-induced platelet aggregation.[1723] Adrenaline increases factor VIII activity, von Willebrand factor, tissue-type plasminogen, and platelet aggregation. Catecholamines can have a detrimental function, particularly, when endothelial function is impaired (reviewed in Ref. 1724). Pheochromocytoma can induce venous thrombosis (reviewed in Ref. 1724). Thus, stimulation of the sympathetic nervous system with increased release of catecholamines accelerates thrombus formation in atherosclerotic arteries after exposure of subendothelial vascular wall.[1719]

There are some other forms of stressful events that also increase coagulability. Already in the 1950s, the effect of work stress was demonstrated to accelerate clotting time.[1725] Hypoglycemia stress increases factors relevant for hypercoagulability.[1726] Whole-body hyperthermia using the warm water blanket technique to reach 41.8 °C is related to consumption coagulopathy and activated coagulation.[1727] Heatstroke is a well-known factor that increases not only fibrinolysis but also hypercoagulability.[1728] This might also depend on changes in bodily water content, which is regulated by the sympathetic nervous system and the renin-angiotensin-aldosterone system.

We recall that angiotensin II and other hormones of the renin-angiotensin-aldosterone system are stimulated by the sympathetic nervous system and they are proinflammatory. Thus, it is not surprising that angiotensin II induces the expression of tissue factor in human monocytes and a procoagulant activity of monocytes.[1729] Angiotensin II stimulates tissue factor expression both *in vitro* and *in vivo*, which is inhibited by angiotensin-converting enzyme inhibitors or angiotensin II receptor inhibition (reviewed in Ref. 1730). Similarly, renin-angiotensin system blockers like aliskiren, a direct renin inhibitor, are able to modulate tissue factor expression in monocytes and vascular endothelial cells activated by inflammatory cytokines (reviewed in Ref. 1730).

In addition, vasopressin—another hormone of the stress response system—can induce hypercoagulability by direct effects on clotting factors and platelet aggregation.[1731–1733] Since chronic inflammation is related to an upregulation of the vasopressin system at the expense of the CRH-stimulated HPA axis,[1734, 1735] the vasopressin-induced hypercoagulation might play a more important role in chronic inflammation.

In conclusion, stressful states with an activation of the sympathetic nervous system and the water retention system through vasopressin, renin, angiotensin II, and aldosterone activate the coagulation system.

Other Hormones Induce Blood Coagulation

It was already mentioned that parathyroid hormone-related peptide is often increased in systemic inflammation and cancer. Inflamed tissue and activated immune cells can produce parathyroid hormone-related peptide, which can be elevated in acute and chronic inflammation.[1666–1668] Endotoxin can directly induce PTHrP secretion.[1669] I discussed that this can be related to bone loss because blockade of parathyroid hormone-related peptide reduces bone loss in experimental arthritis.[1670] Importantly, human platelets express the PTH1 receptor and parathyroid hormone-related peptide enhances platelet activation induced by typical stimulators through MAP kinase-dependent pathways.[1736] This would be an additional factor in how a state of hypercoagulation might develop in chronic inflammatory systemic diseases.

Evolutionary Considerations

Throughout our evolutionary history, wounding and hemorrhage was a continuous threat, which led to a positively selected perfect system of hemostasis. Our earliest ancestors already suffered from injuries that can lead to exsanguination. Thus, it is not surprising that mice, rats, dogs, pigs, chicken, zebra fish, turtles, frogs, and even *Ciona intestinalis* possess functional orthologs of the tissue factor, to make one important example of clotting factors.[1671] Russell Doolittle nicely elaborated on the evolutionary value and development of hemostasis from lampreys to humans.[1737]

Importantly, the fight-or-flight response is often accompanied by the risk of injury, wounding, and exsanguination, which is best demonstrated on the battlefield in combat, because hemorrhage is the most common cause of death.[1738] Thus, it is not surprising that inflammatory factors (TNF, IL-1β, etc.) and the sympathetic nervous system together with angiotensin II and vasopressin are strong stimulators of blood coagulation. However, long-term increase of blood coagulation was not a good program. Hypercoagulability plays a role in an acute situation of injury and hemorrhage. The long-term use in chronic inflammatory systemic diseases is a misguided adaptive program.

CIRCADIAN RHYTHMS OF SYMPTOMS

The term circadian rhythm refers to the 24-h cycles in the physiological processes of living organism. While circadian rhythms are a fundamental feature of the biology of animals, they are usually not considered as important determinants of disease, although their influence is profound. Since symptomatology in chronic inflammatory systemic diseases can have a pronounced diurnal cycle with a maximum in the morning, the underlying causal mechanisms are relevant for pathophysiology of chronic inflammatory systemic diseases, for clinical patient care, and for optimizing treatment strategies.

In the early 1970s, studies using brain-lesion techniques as well as metabolic and electrophysiological experiments demonstrated that the central circadian oscillator is located in the hypothalamic suprachiasmatic nucleus.[1739] Since the circadian rhythm is generated in the suprachiasmatic nucleus of the hypothalamus, learning from these observations can provide new clues to understand neuroendocrine immune pathways relevant to chronic inflammatory systemic diseases. Since these aspects have been best elaborated in rheumatoid arthritis, the findings in this disease are discussed here.

Because of the favorable effects of anti-TNF and anti-IL-6 therapy in large clinical trials, rheumatologist recognized the central role of these proinflammatory cytokines. These cytokines are also important for symptoms such as stiffness, joint swelling, pain, and mood (Table 38). These symptoms are worst in the early morning hours.

An important question therefore is whether cytokines demonstrate a circadian oscillation. Indeed, both IL-6 and TNF display a circadian rhythm not only in healthy subjects but also in patients with chronic inflammatory systemic diseases (Figure 52). Thus, it is not surprising that symptoms mentioned in Table 38 are worst in the early morning. In view of these observations, the key question is what is driving the circadian oscillation of serum levels of cytokines?

The adrenal hormone cortisol is an endogenous steroid with anti-inflammatory actions. In the 1950s and 1960s, shortly after the discovery of cortisol, the circadian rhythm of this hormone was described. The cycle of the HPA axis has a maximum in the early morning hours at 8 AM and a nadir at midnight (Figure 52). Detailed analyses of the circadian curves of cytokines and cortisol have revealed a lag time of cortisol rise in relation to the increase of cytokines of approximately 60-120 min.[739] From this study and a recent analysis of several independent investigations, it appears that morning cytokines drive the increase of cortisol.[565] What then is the factor that drives the nightly cytokine surge?

At present, it is thought that the cortisol nadir at midnight and the parallel increase of proinflammatory hormones such as growth hormone, prolactin, and melatonin (see above) drive the increase of nightly TNF and IL-6. Then, cytokines drive cortisol, and the cortisol increase finally dampens the cytokine surge, and so forth. The influence of one system on the other system can be

TABLE 38 How Do Cytokines Modulate Disease Manifestations?

Symptom	Mechanism
Pain	Proinflammatory cytokines, bradykinin, prostaglandins, and other mediators bind to nerve terminals of afferent pain fibers in the inflamed tissue (see Figure 21). Upon binding of these factors to the nerve terminal, pain centers in the central nervous system are informed of local activation of pain fibers
Swelling of joints	Morning swelling of joints is the consequence of fluid transfer from intravascular to intra-articular space. It is a consequence of vasodilatation and vessel leaking, stimulated by proinflammatory cytokines, prostaglandins, bradykinin, and other mediators
Stiffness	Morning stiffness results from plasma extravasation into the interstitial space of skeletal muscles and connective tissue. Plasma extravasation is a consequence of vasodilatation and vessel leaking, which are stimulated by proinflammatory factors
Bad mood	Cytokines can induce sickness behavior and depression-like symptoms (Cytokines Sing the Blues).[1363] Anti-TNF therapy in patients with rheumatoid arthritis can decrease depression-like symptoms

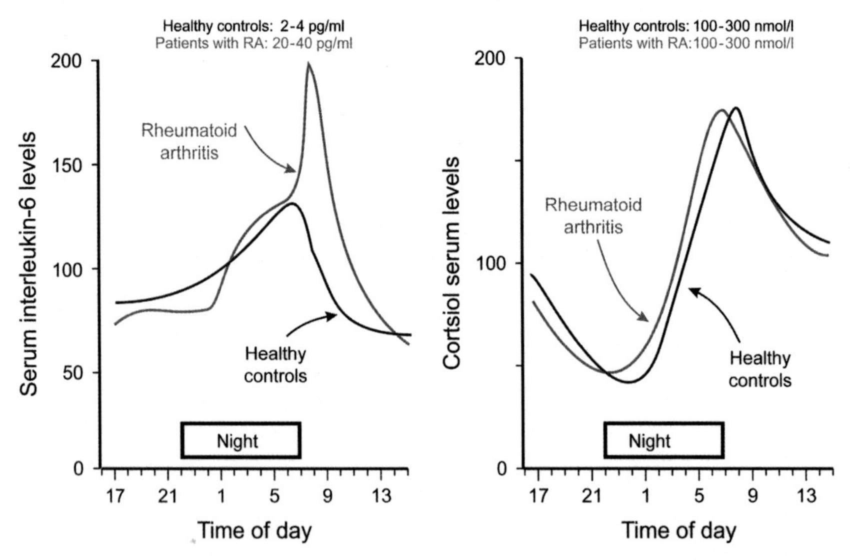

FIGURE 52 Circadian rhythm of serum interleukin-6 (IL-6) and cortisol in healthy subjects and patients with rheumatoid arthritis. The data are given in percent of the 24 h mean (similar as in Ref. 565). The concentration range of serum cortisol and serum IL-6 is given above the main graphs. Despite the markedly increased serum levels of IL-6 in rheumatoid arthritis compared to healthy subjects, the circadian rhythm of cortisol is pretty similar in both groups (in amplitude and period). This phenomenon is called "inadequate secretion of cortisol in relation to inflammation."

viewed as an infinite loop similar to that depicted in the painting of Maurits C. Escher (Dutch painter), where the right hand paints the left and *vice versa*.

Since the circadian rhythm is generated in higher brain centers of the endocrine and nervous systems (in the hypothalamus), the circadian rhythms of cytokines (and symptoms) mirror the activity of neuroendocrine centers in the brain. These findings provide a strong indication that neuroendocrine pathways influence disease-related pathophysiology since symptoms of chronic inflammatory systemic diseases have very similar rhythms. Nighttime release of therapeutic glucocorticoids has beneficial effects in rheumatoid arthritis patients that can be explained by understanding circadian rhythms of cytokine increase in the night.[565, 1740] The more fundamental question appears why immune system activation and cytokine increase do happen in the night.

We already learned that energy regulation is steered in a circadian fashion (Chapter III). The immune and repair system and growth-related phenomena happen during the night, while the brain and skeletal muscles are mainly active during the day. Circadian allocation of energy-rich fuels to daytime and nighttime consumers is a mechanism to distribute energy in the presence of an upper limit of available energy. The hormonal and neuronal systems have developed in a way to provide immune activation during the night. The main features were demonstrated in Figure 31, Chapter III.

Form these considerations in healthy individuals, it becomes understandable that the nighttime increase of immune activation is a positively selected program. The same program is also used in different chronic inflammatory systemic diseases leading to the famous problem of early morning symptoms. Table 39 gives some examples of chronic inflammatory systemic diseases with early morning symptoms.

TABLE 39 Chronic Inflammatory Systemic Diseases with Nocturnal or Early Morning Symptoms

Disease with Nocturnal or Early Morning Symptoms	References
Rheumatoid arthritis	565
Ankylosing spondylitis	1741
Polymyalgia rheumatica	1741
Pruritus	1742
Migraine headache	1743
Cardiovascular and cerebrovascular disease (atherosclerosis as an inflammatory process)	1744
Atopic dermatitis	1745
Bronchial asthma and allergic rhinitis	1746

PREGNANCY IMMUNOSUPPRESSION AND POSTPARTUM INFLAMMATION

Pregnancy is often accompanied by anti-inflammatory effects in different chronic inflammatory systemic diseases. Pregnancy levels of 17β-estradiol suppress many cytotoxic and innate immune responses, whereas antibody production, neoangiogenesis, and growth-associated phenomena can be stimulated.[329] One of the most important immunologic modifications during pregnancy is the increase of B-cell responses due to the progressive increase of progesterone and estrogens during pregnancy, which reach their peak level in the third trimester of gestation.[1747] It is important to mention that during pregnancy, several steroid hormones increase such as cortisol, estrogens, and progesterone. The role of cortisol as an anti-inflammatory hormone is well known; the importance of progesterone is shortly summarized in Table 40; and effects of estrogens were reviewed extensively.[329]

This section shortly summarizes present knowledge in exemplary chronic inflammatory systemic diseases, namely, systemic lupus erythematosus, rheumatoid arthritis, and multiple sclerosis. Systemic lupus erythematosus tends to flare during pregnancy and the puerperium and, not rarely, systemic lupus erythematosus starts during pregnancy. Most flares are mild, and cutaneous and joint disease are the most common manifestations.[755] Systemic lupus erythematosus flares are associated with increased prematurity,[1748] and active nephritis is an independent factor for fetal mortality.[1749] Systemic lupus erythematosus disease activity scores varied during pregnancy, being increased in the second trimester and decreased in the third trimester.[1750] This indicates that the chronic inflammatory systemic disease of lupus is most probably estrogen-/progesterone-driven due to the supportive effects of these two hormones on B-cell immunity.[329]

Observations in the nineteenth century (Trousseau, 1871; Charcot, 1881; Bannatyne, 1896) indicated that pregnancy is favorable in rheumatoid arthritis as summarized in the Nobel Prize Lecture of Philip S. Hench.[1759] The first indication that T-cell pathways are suppressed in pregnant patients with rheumatoid arthritis was provided by Russell et al. in 1997.[1760] Lipopolysaccharide-stimulated whole blood cultures were demonstrated to release less IL-2 but more soluble TNF receptor p55 and p75 during pregnancy. In this study, serum TNF and IL-1β concentrations were unchanged.[1761] Others demonstrated decreased IFN-γ and IL-12 production by lymphocytes after phytohemagglutinin stimulation in third trimester pregnant women with rheumatoid arthritis as compared to healthy pregnant woman.[1762] This study was complemented by data in rheumatoid arthritis patients (third trimester pregnancy), which demonstrated that *ex vivo* monocytic IL-12 production was about threefold and TNF production was approximately 40% lower than postpartum values.[1763] In a prospective longitudinal study, elevated IL-10 levels were found in pregnant women with rheumatoid arthritis and systemic lupus erythematosus as

TABLE 40 Immunomodulating Effects of Progesterone

Effect	References
Anti-inflammatory effects	
Inhibits phagocytosis of rat macrophages	1751
Inhibits IL-1β secretion from human peripheral blood lymphocytes	1752
Inhibits IL-6 from decidual stromal cells	1753
Inhibits IL-12 secretion from mouse mitogen-activated lymphocytes	1754
Inhibits IFN-γ secretion of concanavalin A-stimulated lymphocytes	1755
Inhibits NK cell activity	1756
Inhibits stimulated proliferation of human lymphocytes	1757
T helper type 2 stimulating effects	
Stimulates IL-4 secretion from lymphocytes	1754, 1758
Stimulates IL-5 secretion from lymphocytes	1752
Stimulates IL-10 secretion from lymphocytes	1754

compared to healthy pregnant controls,[1764] which also fits to the stimulating effects of 17β-estradiol and progesterone on this particular anti-inflammatory cytokine.

Significantly higher concentrations of soluble TNF receptor 2 and anti-inflammatory IL-1ra were measured in pregnant compared to nonpregnant women.[1765] An increase of IL-1ra from the second to the third trimester correlated with improvement of disease activity in both rheumatoid arthritis and ankylosing spondylitis patients. Compared to nonpregnant patients and to other pregnant women, rheumatoid arthritis patients showed markedly elevated levels of soluble CD30 during pregnancy.[1765] CD30 (Ki-1) antigen has been considered to be expressed not only in hematopoietic cells (e.g., Reed-Sternberg cells of Hodgkin's disease) but also in nonhematopoietic cells such as human decidual stromal cells. It is thought that CD30 is a relatively specific marker for T helper type 2 cells.[1752] All these data in pregnant women with inflammatory arthritis demonstrate a T helper type 2 dominance, which might be even more pronounced as compared to healthy pregnant women. These studies also demonstrate that important anti-inflammatory factors such as soluble TNF receptors and IL-1ra are additionally upregulated.[1761, 1765] Furthermore, these studies indicate the favorable changes during pregnancy. These effects are most probably mediated by estrogens and progesterone together.[329]

Other authors studied 254 women with multiple sclerosis during 269 pregnancies in 12 European countries.[1766] The women were followed during their pregnancies and for up to 12 months after delivery. In women with multiple sclerosis, the rate of relapses declined during pregnancy, especially in the third trimester, and increased during the first three months postpartum before returning to the prepregnancy rate.[1766] The postpartum flare-up of chronic inflammatory systemic diseases is also seen in thyroiditis, which is called postpartum thyroiditis,[1767] or in Crohn's disease.[1768]

In conclusion, on the basis of effects of 17β-estradiol and progesterone on innate and adaptive immune responses, it is expected that B-cell- or antibody-driven diseases exacerbate, whereas NKT cell-driven and T-cell-driven diseases with cytotoxic and innate immune responses improve. From an evolutionary standpoint, estrogen-/progesterone-induced inhibition of NK, NKT, and T cells is highly valuable because these cells would attack the semiallogeneic fetus. Now, it becomes understandable that many chronic inflammatory systemic diseases profit from pregnancy, while B-cell-driven diseases can exacerbate.

In the postpartum phase, prolactin is increased, which was positively selected for breast-feeding. Since prolactin is an immunostimulatory hormone (Chapter II), it supports breast-feeding and, fortunately, also antimicrobial immunity (think of the problem of mastitis of the mother and general immune incompetence of the child that receives immunoglobulins via breast milk). Since these immune-stimulating mechanisms are used in chronic inflammatory systemic disease, we expect postpartum flares in many of these diseases (Table 41).

TABLE 41 Chronic Inflammatory Systemic Diseases with Postpartum Flare-up of Disease Activity

Disease	References
Rheumatoid arthritis	1769
Systemic lupus erythematosus	1770
Ankylosing spondylitis	1771
Psoriatic arthritis	1771
Juvenile idiopathic arthritis	1771
Multiple sclerosis	1772
Pemphigus	1773
Autoimmune thyroiditis	1774
Autoimmune hepatitis	1775

MENTAL STRESS AND CHRONIC INFLAMMATORY SYSTEMIC DISEASES

One can separate two vulnerable phases in chronic inflammatory systemic diseases. One phase occurs before outbreak of the disease (asymptomatic phase where stress might be a provoking factor?), while the other phase starts after disease onset (symptomatic phase where stress might be a modulating factor?). With respect to the asymptomatic phase, several studies demonstrated that childhood factors and perinatal characteristics influence development of rheumatoid arthritis in adulthood.[1776–1778] Patients with rheumatoid arthritis reported lower childhood household education, food insecurity, and young maternal age, which is perceived as long-term stressful.[1776] In another study in adult rheumatoid arthritis women, childhood trauma like child abuse and emotional neglect was more prevalent compared to controls as measured with the self-report Childhood Trauma Questionnaire.[1777]

In a retrospective cohort study, adverse childhood experiences were studied like childhood physical, emotional, or sexual abuse; witnessing domestic violence; and growing up with household substance abuse, mental illness, parental divorce, and/or an incarcerated household member. The event rate (per 10,000 person-years) for a first hospitalization with any autoimmune disease was 31.4 in women and 34.4 in men. First hospitalizations for any autoimmune disease increased with an increasing number of adverse childhood experiences. The authors concluded that childhood traumatic stress increased the likelihood of hospitalization with a diagnosed autoimmune disease decades into adulthood.[1778] Further studies on childhood or adolescent trauma are reported for juvenile idiopathic arthritis, multiple sclerosis, systemic sclerosis, and other autoimmune diseases.[1779–1782]

With respect to the second vulnerable phase after the onset of chronic inflammatory disease, some studies described that long-lasting major stress can decrease severity, but most clinical studies showed that psychological stress can aggravate the disease (Figure 53) (e.g., Ref. 1779). These results have recently been confirmed in different investigations between 2000 and 2014. However, these findings seem to be in disagreement with the immunosuppressive role of stress claimed during the decades between 1980 and 2000. In recent years, the stress-induced immunosuppression paradigm changed in the following way: Under healthy conditions, acute low time-integral stress is accompanied by enhancement of immune function, whereas sustained high time-integral stress is linked to immunosuppression.[724, 1012, 1783] The high time-integral stress in adults does not often appear in real life, but some studies demonstrated that diseases like rheumatoid arthritis are ameliorated (reviewed in Ref. 1779).

The immunostimulating role is delineated, for example, in healthy humans and control experimental animals, where stress is linked to an increase of inflammation as objectified by increased serum cytokines and other immune factors.[1788] It is best delineated when looking on serum levels of IL-1β and IL-6.[1788]

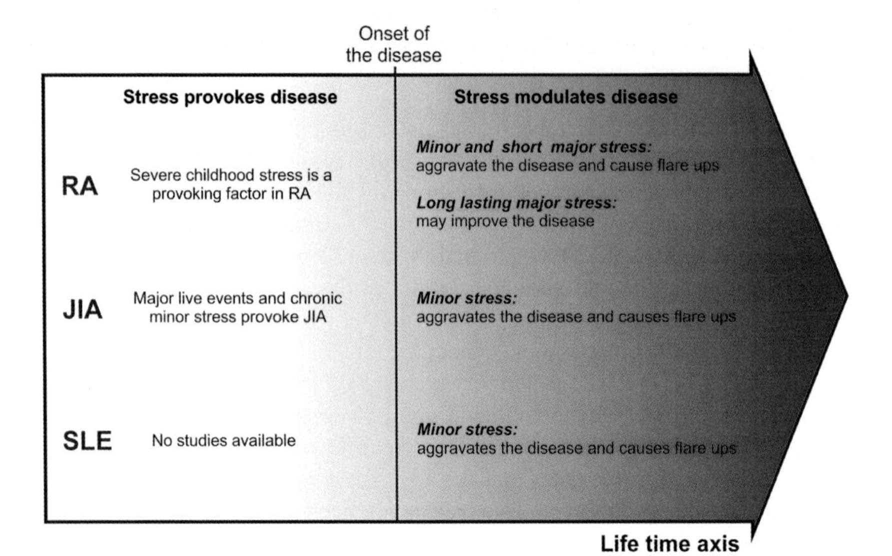

FIGURE 53 Influence of minor or major stress on the onset and course of rheumatoid arthritis (RA), juvenile idiopathic arthritis (JIA), systemic lupus erythematosus (SLE),[1779] and other chronic inflammatory systemic diseases.[1784–1787]

Stress activates inflammation via stimulation of NF-κB,[1789] induction of endothelial dysfunction, accelerated aging of cells, switching on of cell redistribution in the body and β-adrenergic stimulation of myelopoiesis, activation of homeostatic stress axes like the sympathetic nervous system, brain activation and sleep disturbance (increase of cytokines), and rise in hyperalgesia with possible increased proinflammatory substance P signaling (Figure 54). While many of these elements have been studied under acute stressful events, research in long-standing stress demonstrated the proinflammatory negative sequelae of chronic exposure.[1790]

When a proinflammatory effect of stress happens in healthy subjects and animals, one would expect that several of the mentioned pathways are also used in stressed patients with chronic inflammatory systemic disease. One might argue that in a situation with an already existing chronic inflammatory systemic disease, the situation can be different compared to healthy individuals. Indeed, one has to expect the presence of a preactivated immune system and an already present proinflammatory state as described by elevated serum cytokines. Thus, a stress-induced increase of a proinflammatory signal (e.g., increase of IL-1β, Ref. 1791) might more easily concur with a proinflammatory signal of the chronic inflammatory systemic disease. Together, both might constitute a "two-hit model" of stress-induced inflammation in chronic inflammatory systemic diseases.

For example, patients with a homozygous missense mutation (Phe127Cys) in the tumor necrosis factor α-induced protein 3 (TNFAIP3 is one of the strongest susceptibility genes in rheumatoid arthritis[496] and other chronic

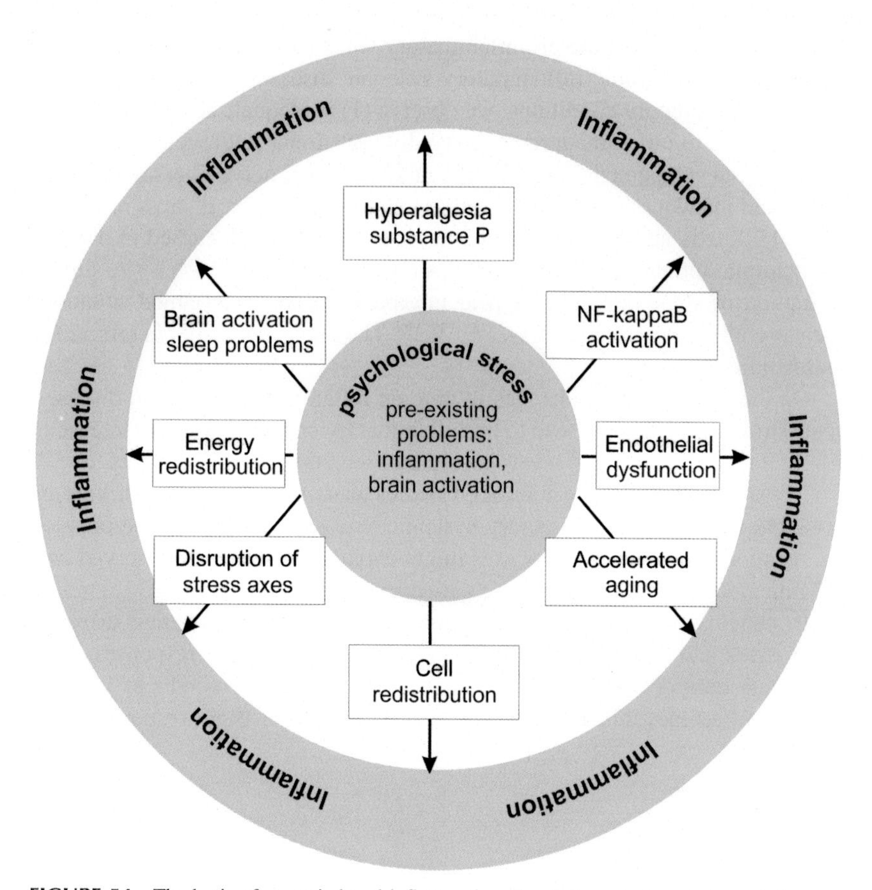

FIGURE 54 The basis of stress-induced inflammation. Psychological stress induces inflammation via several pathways. Preexisting inflammation and mental stress add to the problem. Thus, the picture might also be represented by ripples after a stone touched water. The center starts with the preexisting genetically determined platform, which induces psychological stress that induces inflammation, which induces psychological stress that induces inflammation and so on. It could also be an opposite way: The center starts with the preexisting genetically determined platform, which induces inflammation that induces psychological stress, which induces inflammation that induces psychological stress and so on. Abbreviations: NF, nuclear factor.

inflammatory systemic diseases; see also Table 3 in Chapter I) more easily activate the proinflammatory signaling cascade through IL-1 receptor, TNF receptor type 1, toll-like receptors (TLR), and other proinflammatory receptors, because TNFAIP3 is a strong inhibitory element of signaling through NF-κB.[1792] When stress through other mechanisms activates the same proinflammatory pathway,[1789] the presence of the missense mutation in TNFAIP3 would add to the problem leading to a stronger proinflammatory response (two hits, Ref. 1793). Thus, in chronic inflammatory systemic diseases, the described dichotomous effects of stress can change into a proinflammatory direction.

Stress can also become proinflammatory when stress axes are defective as demonstrated in chronic inflammatory systemic diseases (Chapter II): Under chronic inflammatory conditions, we observe (1) inadequate secretion of cortisol in relation to inflammation, (2) severe loss of adrenal androgens, and (3) increased basal sympathetic tone and loss of sympathetic nerve fibers in inflamed tissue, and many more described above. In addition, the typical stress responses of the HPA axis and the sympathetic nervous system are disturbed in patients with chronic inflammatory systemic diseases (Figure 55).[722, 1794, 1795] A similar disruption of stress axes may appear in other chronically inflamed situations such as chronic caregiving stress.[1546, 1790, 1796] The interested reader is referred to multivolume books.[1797]

On the basis of these changes in chronic inflammatory systemic diseases, acute stress responses can lead to proinflammatory episodes with an increase of IL-1β, IL-6, IL-8, and TNF levels in serum or secretion from leukocytes.[1791, 1793, 1798–1802] Although no exactly controlled studies on sustained stress conditions are available in chronic inflammatory systemic diseases (ethically questionable), the mentioned defects of stress axes might explain the clinically observed proinflammatory role of stress in these patients.

In conclusion, in patients with chronic inflammatory systemic diseases, acute stress and also sustained stress evoke proinflammatory responses. With an evolutionary perspective, the phenomenon of acute stress-induced immune activation becomes understandable, because the fight-or-flight response might

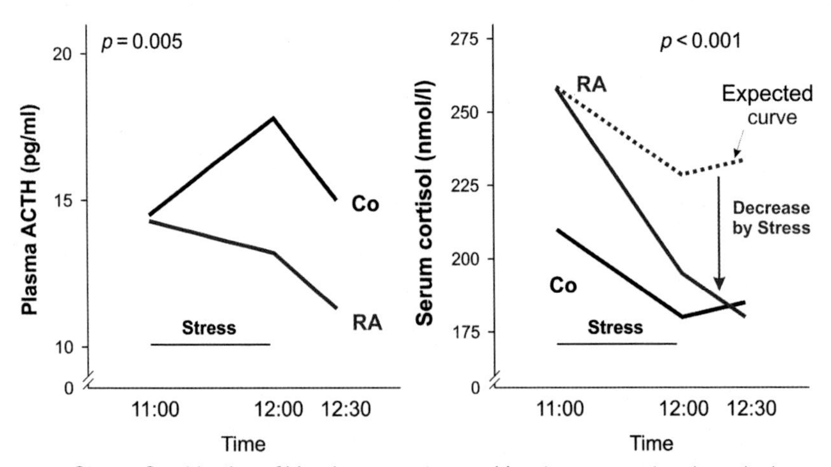

FIGURE 55 Influence of a combined stress test on responsivity of the HPA axis in patients with rheumatoid arthritis (RA) compared to healthy controls (Co). Figure schematically redrawn using information from Ref. 722. Left panel: The response of adrenocorticotropic hormone (ACTH) is lower in RA compared to controls. Right panel: The response of cortisol is decreased in relation to controls (red continuous line compared to the red dotted line). These data demonstrate an inadequate secretion of HPA axis hormones under mental stress. The *p*-value indicates the difference between the two curves.

be harmful leading to wounding. Injury needs a better immunosurveillance and immune activation. As Firdaus Dhabhar of Stanford University puts it, "short-term stress induces a redistribution of leukocytes from the barracks, through the boulevards, and to potential battlefields." Leukocyte redistribution enhances immune function in compartments to which immune cells traffic during stress. This leads to a significant enhancement of tissue-directed immunity.[1783] This evolutionary positively selected program is also used in chronic inflammatory systemic disease, but here, it is an unfavorable misguided program because it stimulates inflammation.

THE COMMON DENOMINATOR

Looking on all disease sequelae of this chapter, we identify three important aspects: (1) The sequelae belong to an adaptive program positively selected for acute inflammatory episodes; (2) the adaptive programs mentioned serve a reallocation program that deviates energy-rich fuels from the brain/skeletal muscles to the activated immune system, which is accompanied by a water retention and calcium delivery program; and (3) in chronic inflammatory systemic diseases, the same adaptive programs are used as long as systemic inflammation is high (cytokines in circulation or cytokines activate sensory nerve endings, i.e., pain), because no specific programs were evolutionarily positively selected for chronic inflammatory systemic diseases.

In the past, several notable features of chronic inflammatory systemic diseases were considered as independent phenomena in the absence of a unifying model to explain both their prominence and coexistence. As a common denominator, the redirection of energy-rich fuels from energy stores to the activated immune system and the accompanying water retention program can explain many systemic disease sequelae in different chronic inflammatory systemic diseases.

Chapter VI

Aging-Related Sequelae: Inflamm-Aging

Chapter Contents

When we look on disease sequelae given in Table 30 at the beginning of Chapter V, we recognize that many consequences of chronic inflammatory systemic diseases are also observed in elderly people. While the common denominator in chronic inflammatory systemic diseases is an energy reallocation program to nourish the activated immune system, it is questionable whether, or not, such an energy reallocation program exists in elderly people. To better understand this phenomenon, one needs to discuss the aspect of inflamm-aging.[1803]

INFLAMM-AGING OF THE ELDERLY

When the immune system is encountered with a foreign antigen, at the beginning of the disease, a typical polyclonal immune response is initiated, which is needed for optimization of the immune attack.[1804] In contrast, a long-term encounter with a known antigen such as during chronic infection with EBV or CMV[1805] leads to adaptive changes of the immune system resulting in an oligoclonal response.[1806–1808] The most prominent alteration during aging is shrinkage of the repertoire and oligoclonal expansion of T cells most probably due to continuous encounter with person-specific foreign antigens.[1807, 1809] The oligoantigenic load leads to centering of the repertoire, which leads to a reduction of the immunologic space in peripheral lymphoid organs.

Concomitantly, immune cells of elderly people have a senescent phenotype with a dramatic decrease in the telomere length paralleled by a loss of naive T cells and thymic output.[1810] Furthermore, the adaptive changes comprise a loss of the costimulatory molecule CD28 (on CD8+ T cells during aging), an increase in the natural killer cell type of T cells (with granzyme B, perforin, and killer cell inhibitory receptors), and an unspecific activation

The Origin of Chronic Inflammatory Systemic Diseases and their Sequelae.
http://dx.doi.org/10.1016/B978-0-12-803321-0.00006-9

TABLE 42 Immunosenescence in Healthy Subjects

Immune System Aberrations	Healthy Aging (Duration: Four Decades)	References
Oligoantigenic load	↑	1807, 1809
Repertoire shrinkage with oligoclonal expansion (T and B cells)	↑	1806–1808
Memory T cells	↑	1807, 1811
Naive T cells as identified by TRECs	↓	1812
Loss of the costimulatory molecule CD28	↑	1807, 1813
Natural killer cell type of T cell (granzyme B, perforin)	↑	1814, 1815
Increase in monocyte/macrophage activity (serum and tissue levels of TNF, IL-6, MCP-1, β2-microglobulin, and neopterin)	↑	741, 1816–1820
Telomere shortening in T cells	↑	547, 1810

↑, Indicates an increase; ↓, indicates a decrease; TRECs, αβT cell receptor rearrangement excision circles.

of macrophages (Table 42). The continuous monocyte/macrophage activation and smoldering systemic inflammation are evident as an increase in serum TNF, serum IL-6, serum MCP-1, serum β2-microglobulin, serum neopterin, and many others (Table 42). This continuous proinflammatory load may be the most important factor for the subsequent deterioration of other global systems such as the endocrine system, the metabolic system, and the nervous/muscular system. However, the question arises whether the increase in the proinflammatory load is high enough to induce an energy reallocation program in the elderly.

IS THE PROINFLAMMATORY LOAD HIGH ENOUGH TO INDUCE AN ENERGY REALLOCATION PROGRAM?

I focus on the serum levels of IL-6, because this cytokine has been studied very often and because it is a strong stimulator of C-reactive protein, which was in the focus of aging research. When one compares serum levels of IL-6 as measured with the identical quantitative high-sensitivity ELISA technique, healthy elderly subjects range from 2.0 to 3.5 pg/ml (60-80 years).[741] These levels correspond to mild activation of the immune system, but they might not lead to an energy reallocation program as studied in the following seminal investigation.[736]

The interrelation between dose of subcutaneously injected recombinant human IL-6 (rhIL-6), serum levels of IL-6, and increase in energy expenditure was studied in healthy volunteers.[736] Already in Chapter III, this study was demonstrated, and it is shortly recapitulated here. It was shown that injection of 0.1 μg rhIL-6/kg b.w. increased serum levels of IL-6 to approximately 10-15 pg/ml, a dose of 1.0 μg rhIL-6/kg b.w. led to 45 pg/ml, 3.0 μg stimulated a serum level of 250 pg/ml, and 10 μg/kg b.w. was accompanied by an IL-6 serum concentration of more than 1000 pg/ml. In parallel, the maximal increase in metabolic rate in percent of basal metabolic rate was 4%, 7.5%, 18%, and 25%, respectively.[736] This means that a visible influence on energy regulation was observed at an IL-6 serum level of 10-15 pg/ml, but the effect was small in these healthy volunteers. Thus, an increase in serum IL-6 from 1.0 to 3.5 pg/ml as in healthy subjects[741] should not induce a marked energy expenditure program.

From this point of view, it is questionable whether an increase in inflammation during aging can lead to substantial changes in energy regulation. However, as Table 42 demonstrates, other proinflammatory factors are also increased. These factors might have an independent influence on energy distribution so that additive or even synergistic effects on energy regulation can exist. This might be demonstrated by lipopolysaccharide injection where several cytokines are upregulated in parallel.

Indeed, endotoxin injection, which stimulates many proinflammatory cytokines (but has also direct effects via toll-like receptor 4), increased energy expenditure by 25-40%.[1821] In this study, using a bioassay for IL-6 levels, endotoxin-induced IL-6 increase was 10 times the baseline value, which can be transferred to an increase from 1.5 pg/ml (serum level in healthy subjects; Ref. 741) to 15 pg/ml when using the modern high-sensitivity ELISA technique discussed above. Since energy expenditure rises by 4% (IL-6 at 10-15 pg/ml) with subcutaneous rhIL-6 alone[736] and by 25-40% with endotoxin,[1821] a parallel increase in different cytokine pathways—not only IL-6—seems to lead to a higher energy expenditure compared with a single cytokine injection. In addition, there are other important ways to increase inflammation in aged people.

For example, chronic mental stress is more prevalent in the elderly than in young people. Using the identical quantitative high-sensitivity ELISA technique for serum IL-6, it was reported that aged caregivers show a mean value of 5.5 pg/ml,[1555] and subjects who report a high level of perceived hopelessness show 3.0 pg/ml.[1556] Again, the values are not very high, but—together with other age-related stimuli—stressful life events can be an additional factor to start the energy reallocation program.

Another factor for an elderly person can be pain, which can increase cytokine levels *per se*, but inflammation can also stimulate pain, which is best delineated after injection of endotoxin into humans.[1374, 1822–1824] Since elderly people often have chronic pain such as in osteoarthritis of multiple joints, this can constitute another source of proinflammatory load because pain signals induce cytokine upregulation in the periphery and in the central nervous system (recently summarized in Ref. 1825). Thus, the afferent sensory nervous pathway may be more important to install a proinflammatory situation in the central nervous system than circulating inflammatory factors. Central cytokines can activate stress axes in the brain,[611, 1826] which lead to an energy reallocation program regardless of circulating cytokine levels in the blood. This might be the reason why fatigue (recall sickness behavior) is perfectly linked to pain but not to circulating cytokines or peripheral inflammation.[1380, 1827]

Another important factor that upregulates cytokines is sleep disturbance typical in elderly people. While sleep disturbance induces changes in circulating cytokines, a more direct effect of sleep alterations on central nervous inflammation exists. This can directly impact on stress axes and, thus, the energy reallocation program.

CONCLUSIONS

Several mechanisms can induce a situation that stimulates stress axes in the aged brain (peripheral inflammation, psychological stress, pain, sleep problems, continuous smoldering infections, smoldering peripheral inflammation, etc.). Chronically, this can increase stress axis activity and the related energy reallocation program demonstrated in Chapter III. This can happen as a consequence of peripheral inflammation and elevated circulating cytokines or more directly within the central nervous system as a direct stimulation of stress axes (mental stress). This bimodal stimulating effect was already discussed in section "Insulin Resistance" of Chapter V. Thus, it is not surprising that the sympathetic nervous system is the most upregulated stress axis in the aged person, even ahead of the HPA axis,[1597] but that the somatotropic axis (insulin-like growth factor 1 and growth hormone) and the hypothalamic-pituitary-gonadal axis are downregulated at the same time. Table 43 gives the changes of the endocrine system and Table 44 the changes of the nervous system in normal aging people. On the basis of these considerations, age-related central nervous system activation and concomitant peripheral inflammation support similar disease sequelae as given for chronic inflammatory systemic diseases in Table 30.

At this point, one needs to recall that aging on a population level, in men and women, beyond postmenopausal age is a new phenomenon in the last 200 years. Thus, no positive selection pressure existed supporting healthy aging or longevity.[1828] The inflammatory process that appears during aging is new to the biological evolution of humans and not relevant in our close or distant ancestors. Since we know that acute inflammation or acute stressful events can stimulate many adaptive programs, it is not surprising that we see similar sequelae during the process of aging similar as in chronic inflammatory systemic diseases. However, the marked difference between aging and chronic

TABLE 43 Aging of the Endocrine System

Endocrine and Metabolic System Aberrations	Healthy Aging (Duration: Four Decades)	References
Cortisol relative to other adrenal and gonadal hormones	↑	1597
Adrenal androgens (DHEAS, DHEA, androstenedione)	↓	741, 1829–1831
Testosterone	↓	1829
Progesterone	↓	1597, 1829
IGF-1	↓	1832
Vitamin D	↓	1833–1835

↑, Indicates an increase; ↓, indicates a decrease; DHEA, dehydroepiandrosterone; DHEAS, DHEA sulfate; IGF-1, insulin-like growth factor 1.

TABLE 44 Aging of the Nervous System

Nervous System Aberrations	Healthy Aging (Duration: Four Decades)	References
Autonomic nervous reflexes	↓	1836, 1837
Sympathetic nervous tone (blood pressure, heart rate)	↑	1838–1841
Sympathetic response to stimuli	↑	1597, 1842
Plasma noradrenaline or neuropeptide Y (relative to cortisol and other steroid hormones)	↑	1597, 1842–1846
Peripheral autonomic nervous innervation	↓	Reviewed in Refs. 1847–1849

↑, Indicates an increase; ↓, indicates a decrease.

inflammatory systemic diseases is the velocity of the appearance of problems. While one observes the changes in aging over decades, one can see the changes in chronic inflammatory systemic diseases within months. Some of the changes in chronic inflammatory systemic diseases are reversible, but some of them are long-standing (even imprinted; similarly, imprinted than in aging). Thus, aging is a smoldering chronic inflammatory systemic disease.

Chapter VII

Continuation and Desynchronization

In a congregation of flashing fireflies, everyone is continually sending and receiving signals, shifting the rhythms of others and being shifted by them in turn.

Strogatz, a SYNC researcher.[1850]

Chapter Contents

When the reader reached this level of the book, it might be evident that the origin of disease sequelae in chronic inflammatory diseases—but also of similar age-related phenomena—can be explained by (1) changes in systemic energy and volume regulation, (2) concepts from evolutionary medicine, and (3) the neuroendocrine immune cross talk of supersystems (nervous, endocrine, and immune systems). However, it is not so clear whether, or not, disease-related changes of endocrine and neuronal systems can have an important influence on perpetuation of chronic inflammatory systemic diseases. In addition, it is not clear how changes of the endocrine or nervous system during aging can lead to changes of the immune system and, possibly, chronic inflammatory systemic disease. This chapter has hypothetical character, but it can be an important platform to trigger new experiments or to stimulate epidemiological studies. The ideas behind this inductive part of the book must stand the test of time.

The Origin of Chronic Inflammatory Systemic Diseases and their Sequelae.
http://dx.doi.org/10.1016/B978-0-12-803321-0.00007-0

CLASSICAL FACTORS FOR THE CONTINUATION OF CHRONIC INFLAMMATORY SYSTEMIC DISEASES

Before I provide novel mechanisms on the continuation of chronic inflammatory systemic diseases using the perspective of neuroendocrine immunology, classical factors of disease prolongation are shortly recapitulated. One of the most important arguments of perpetuation is immunologic memory of T cells, B cells (also plasma cells), NK cells, or NKT cells. We learned that memory cells can induce a much faster secondary response to a foreign antigen, easily observed after vaccination or as lifelong protection from typical childhood infectious diseases.

Memory was discussed in Chapter IV and again in Chapter V ("Insulin resistance") as a positively selected adaptive program in order to protect energy stores, which is true for immunologic memory, neuronal memory, and energy memory (stored energy in fat tissue). Thus, it does not come as a surprise when the same immunologic memory is continuously used in the context of autoimmunologic memory (recent clinical example for pemphigus in Ref. 1851). As it was already discussed in Figure 2 (Chapter I), immunologic memory is bad for 5% of the population when it comes to autoimmunity, while 95% are happy with immunologic memory in the context of secondary responses to infectious diseases or continuous anti-autoimmune regulation (T and B regulatory cell memory, e.g., Ref. 1852). One can summarize that this positive adaptive program is used in autoimmunity in a misguided harmful way when autoantigens or harmless foreign antigens are the target. Sometimes, the autoantigen such as collagen type II or myelin is produced in larger numbers as a consequence of the ongoing inflammatory process. This would stir up the autoimmune attack. Similarly, continuous attack of harmless foreign microbes on body surfaces would stimulate a long-standing attack because microbes cannot be eliminated (think of bacteria in the gut).

There is another important way how a chronic inflammatory systemic disease can persist for a long time. With the sensational discovery of the TNF receptor type 1-associated periodic syndromes (called TRAPS) by McDermott and colleagues in 1999, we recognized that germ line mutations of important proinflammatory pathways can lead to continuous immune activation.[1853] Continuous immune activation goes along with a chronic inflammatory systemic disease. Today, a good portion of chronic inflammation can be explained by these monogenic autoinflammatory diseases (recent reviews of the subject in Refs. 506, 1854).

A third pathway leading to prolongation of chronic inflammatory systemic diseases is a defect in apoptotic pathways. If the uptake of apoptotic cell debris is impaired, persistent immune activation can lead to autoimmunity.[401] Defects in apoptosis lead to perpetuation (see also section "Apoptosis" of Chapter I). The autoimmunity-like Canale-Smith syndrome is recalled (defective Fas-induced apoptosis).

Fourth, when the degeneration of tissue reached a certain level of destruction, the appearance of neoself-epitopes can perpetuate an inflammatory disease. This seems to be true for neoself-epitopes of cyclic citrullinated peptides that play a role in rheumatoid arthritis and multiple sclerosis.[310, 527]

A fifth mechanism is linked to continuous infection with microbes.[1855] For example, continuous infection with the Epstein-Barr virus was documented in patients with rheumatoid arthritis.[1856] The Epstein-Barr virus infection is particularly evident in ectopic lymphoid structures, the so-called tertiary follicles within the synovial tissue. Within these follicles, the Epstein-Barr virus infection of anticyclic citrullinated peptide antibody-producing plasma cells was observed.[1856] Others described active Epstein-Barr virus infection in the thymus of patients with myasthenia gravis.[1857] Similarly, the presence of bacteria on body surfaces stimulates a continuous inflammatory process as discussed in inflammatory bowel disease.[1858] All infectious stimuli can perpetuate an inflammatory disease.

Since immunodeficiency diseases are often accompanied by autoimmunity, a slower and less strong inflammatory activity and a mild and slower paced anti-inflammatory feedback response may start and perpetuate inflammation. Usually, the immune response toward acute insults is strong, but due to immunodeficient pathways, less robust attack and feedback responses are to be expected. In the words of Germain, "a more prolonged nature of such responses … could represent the first step in the establishment of a chronic state of immune activity to the pathogen or the self-component."[1859]

Furthermore, epigenetic programming of immune cells and supportive cells like fibroblasts may be involved in the perpetuation of chronic inflammatory systemic diseases. In rheumatoid arthritis, the aggressive nature of synovial fibroblasts is maintained *in vitro* for a long time,[431, 1860] and it was discussed that epigenetic changes in these cells lead to an inflammatory memory independent of the adaptive immunologic memory.[1861] These epigenetic changes might lead to the expression of proto-oncogenes and antiapoptotic molecules and a lack of tumor suppressor genes.[1862]

Another pathway was discussed in the context of multiple sclerosis research where one form of multiple sclerosis might be a degenerative form. Degeneration affects oligodendrocytes or astrocytes that undergo cell death without much inflammation.[278, 1863] Such a degenerative process may lead to a long-standing smoldering inflammatory course and even disease.

A further mechanism is microchimerism, which describes the presence of a small cell population in one individual obtained from another genetically distinct individual. Naturally acquired microchimerism happens between mother and fetus, and iatrogenic microchimerism happens after transplantation or blood transfusion. Natural microchimerism was linked to autoimmunity and perpetuation of the disease,[1864] but exact pathways are still to be investigated. The individual elements of this paragraph are summarized in Table 45.

TABLE 45 Mechanisms of the Continuation of Chronic Inflammatory Systemic Diseases

Mechanisms of Continuation
Immunologic memory (T cells, B cells, plasma cells, NK cells, and NKT cells) toward autoantigens or harmless foreign antigens, partly in tertiary lymphoid follicles
Monogenetically inherited autoinflammatory immune activation
Defects in apoptotic pathways (e.g., complement deficiency and the Canale-Smith syndrome)
Appearance of neoself-epitopes and epitope spreading
Continuous viral infection of tertiary follicles perpetuates an inflammatory process
Chronic presence of the wrong bacteria on body surfaces stirs a continuous inflammatory process (NOD2/CARD15 example of Crohn's disease)
Slower-paced anti-inflammatory feedback (e.g., in immunodeficiency diseases)
Epigenetic imprinting of inflammatory memory and tumorlike pathways
Degeneration of cellular elements stirs a continuous process
Microchimerism stirs autoimmunity

Apart from these classical explanations of disease perpetuation, I consider changes of the interplay of the supersystems (nervous, endocrine, and immune systems) as a possible cause of disease continuation. The three supersystems talk to each other using mutually exchangeable factors (neurotransmitters/neuropeptides, hormones, and cytokines). The crosstalk is mediated by connectors in feedback loops, and as mentioned in the "Preface," research can be called the *science of connections* or *interactions*. The crosstalk happens via nerve fibers or factors transported in circulation in time-dependent synchronized ways. Supposing that these factors or their receptors are deficient or timing of feedback is inadequate or defective, the cross talk might be severely hampered and chronic disease might develop. These considerations are the platform of the following sections.

THE CONCEPT OF SELF-ORGANIZATION, SYNCHRONIZATION, AND COOPERATION IN PHYSIOLOGY

Before I move to the questions of synchronization, some theoretical considerations on the nature of self-organization in biology are necessary. The term self-organization was introduced in modern science in 1947 by the psychiatrist and engineer Ross Ashby. Later, it was used by researchers working with general systems theory in the 1960s (Wiener, von Bertalanffy, Kauffman, and others), and it was adopted by physicists and researchers in the field of complex systems theory in the 1970s and 1980s.

One of the most beautiful examples of self-organization is oscillating chemical systems, called the Belousov-Zhabotinsky reactions, which can be found in the Internet under "www.youtube.com" (enter "belousov-zhabotinsky" in the search field). These reactions are irreversible and they exchange energy and matter in a thermodynamically open system, which is operating far from thermodynamic equilibrium. According to Prigogine (Nobel prize 1977, Ref. 1865), who was a pioneer of nonequilibrium thermodynamics theory, these systems are called dissipative systems. Here, dissipation means that one form of energy (e.g., mechanical modes such as waves and oscillations) is converted into another form of energy usually leading to heat (and an increase of entropy).

In biology and medicine, we most often deal with dissipative or self-organized systems that are operating far from thermodynamic equilibrium. The concept of self-organization is an important tool to describe the behavior of biological systems, from the subcellular to the ecosystem level. For example, a bacterium in culture broth incorporates glucose as an energy-rich fuel to build ATP, which is used to support the self-organized life of this bacterium. For the time being, we do not want to elaborate on how self-organization evolved in bacteria but a system such as a living bacterium is self-organizing because, left alone under adequate conditions, it remains organized (it keeps its own delicate organization; it lives and reproduces).

From a physicist or chemist point of view, this is a counterintuitive property because one expects that things get disorganized leading to an increase of entropy. When we observe a very high degree of order or a rise in order, our intuition is that something or someone is responsible. We look for a causal agent (the argument of "intelligent design"). However, this expectation is wrong, because we know that things can start in a highly random state and, without being modified from the outside, become more and more organized. Self-organization is one of the most interesting concepts in modern science, and it relies on four basic ingredients:

- Positive feedback and negative feedback (self-reference)
- Autonomy (although energy and matter are exchanged with the outside)
- Multiple interactions (complexity)
- Redundancy (many parts of the system can be modulators of the system)

Such a self-organized system can lose its stability by fluctuation. In the case of dividing bacteria, decreasing glucose in the culture broth would be such a fluctuation. The novel situation gives rise to a new self-organized system under altered conditions, which is an energy-consuming process because reorganization of the system produces entropy. At any time point, fluctuations can influence self-organized systems from the outside. This is usually described as given in Figure 56.

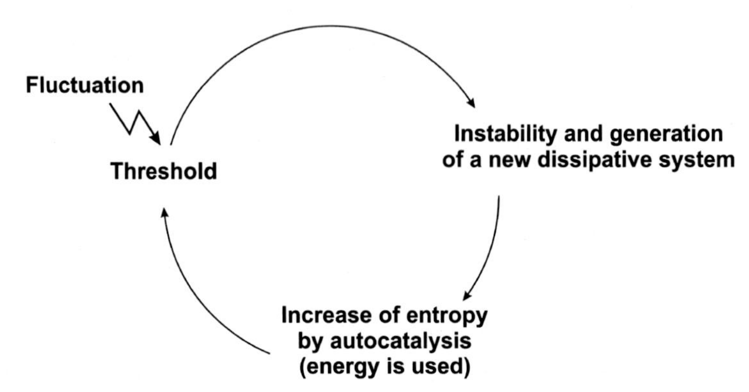

FIGURE 56 Transition from one self-organized dissipative system to another system by fluctuation. When fluctuation is not absorbed (or buffered) by the dissipative system, a threshold is reached where the system becomes instable. Instability is followed by the generation of a new dissipative self-organized system, and this always leads to an increase of energy expenditure by an autocatalytic process leading to an increase of entropy.[1866]

Thus, one self-organized dissipative system can switch to a new system depending on fluctuation. Looking on a bacterium, this happens all the time in the life cycle of a bacterium when fluctuation appears. Similarly, subcellular molecular systems and an entire cell in our body are self-organized dissipative systems, which undergo continuous transitions from one state to another. Fluctuations can happen by chance, but the reaction of a dissipative system is a well-organized and evolutionarily positively selected response. A fluctuation can be relatively small but, nevertheless, a small fluctuation can have surprisingly strong effects on the entire system.

On the level of physics and chemistry, self-organized dissipative systems can be mathematically described in an exact way (examples are the Belousov-Zhabotinsky reactions such as the "Brusselator" developed by Ilya Prigogine, who worked in Brussels, Figure 57). Exact differential equations describe the dissipative system and, thus, self-organization on a molecular level. The complexity can be increased by describing interwoven molecular systems with given hierarchical structures.[1867] But when it comes to more complex biochemical or biological systems of self-organization, it becomes increasingly difficult to determine whether the same label applies. Self-organization, despite its intuitive simplicity as a concept, has proven notoriously difficult to define and pin down formally or mathematically. This is particularly true when it comes to complex biological systems such as the human body or the ecosystem because nonlinearity dominates.[1868]

The concept of self-organization in biomedicine and ecology was used to explain diverse phenomena such as

- the spontaneous folding of proteins and other macromolecules,
- the formation of lipid bilayer membranes,

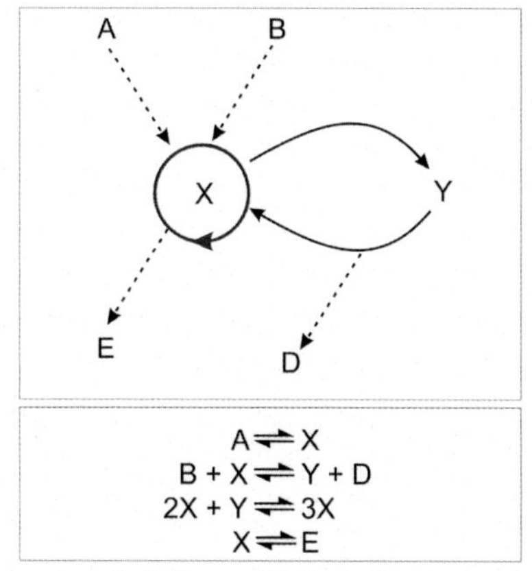

FIGURE 57 Schematic drawing of a Belousov-Zhabotinsky reaction (upper panel) and the respective equations (lower panel). This example is the "Brusselator" according to Prigogine.[1865] A and B are initial products and D and E are final products, which are maintained constant. X and Y are intermediate components that change in time. The red arrow loop indicates the autocatalytic production of X as given in the third reaction equation. Go to "www.youtube.com" and enter "brusselator" in the search field.

- the origin of life itself from self-organizing chemical systems,
- pattern formation and morphogenesis/embryogenesis,
- the coordination of human movement,
- homeostasis (the self-maintaining nature of systems from the cell to the whole organism),
- the creation of structures by social animals (bees, ants, termites, and many mammals),
- flocking behavior (birds, fish, etc.),
- the organization of Earth's biosphere (Gaia hypothesis).

Although knowing that, at present, an exact formal description of self-organization in a human body would be difficult due to the lack of respective data and complexity, the concept might be used as an extrapolation similarly as in other biological systems given above.

In a healthy person, homeostatic mechanisms serve the entire body so that self-organization is maintained (this is really an ultralevel of hierarchical self-organization when compared to the reaction in Figure 57). One might label this system in a healthy human body "self-organization H" (H stands for health). Fortunately, "self-organization H" is a relatively stable dissipative system without producing too much entropy (we feel well). The mechanisms of "self-organization H" have been evolutionarily positively selected in order to maintain normal homeostasis (recall Walter Cannon). Little fluctuations do not jeopardize "self-organization H" in the healthy situation, although there are many fluctuations on the microlevel.

In our first example, an infectious disease is a fluctuation that changes "self-organization H" so that an autocatalytic process is switched on to generate "self-organization I" (I stands for infection). The autocatalytic mechanism leading to the new self-organization is the activation of the immune, endocrine, nervous, and other systems, which is an energy-consuming and entropy-producing process (Figure 58). The "self-organization I" is a new dissipative system installed to overcome infection. During long-standing coevolution with infectious agents, "self-organization I" has been positively selected to overcome infections. Under consideration of the expression *hypercycle* formulated by Eigen et al.,[1867] one might call this an evolutionary positively selected ultracycle. "Self-organization I" leads to the complete elimination of infectious agents, which is itself a fluctuation to "self-organization I." Upon this fluctuation, the ultracycle is closed leading to a return back to "self-organization H" (Figure 58).

In our second example, implantation of the embryonic blastocyst demonstrates an important fluctuation in a healthy female body in "self-organization H" (similar to Figure 58). Implantation and placenta generation lead to a transition from "self-organization H" to "self-organization P" in the mother (P stands for pregnancy). This process includes changes of different systems leading to a favorable environment for the semiallogenic embryo. At the end of pregnancy, presently not fully understood factors terminate "self-organization P" (birth giving as a fluctuation), leading to a transition back to the initial "self-organization H" in the mother. Similarly, as for the infection-related ultracycle, one might talk about an evolutionarily positively selected pregnancy-related ultracycle.

There are other fluctuations leading to a transition from "self-organization H" to a new "self-organization ILL" (ILL stands for a disease). It simply needs a respective trigger that functions as a fluctuation. The threshold is physically,

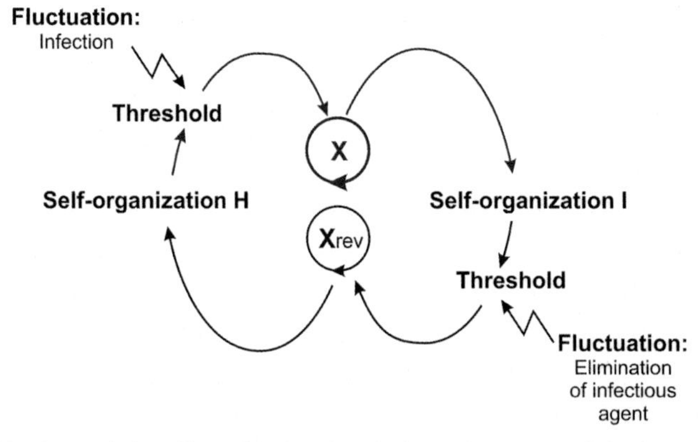

FIGURE 58 An evolutionarily positively selected ultracycle. Upon an infection, a fluctuation leads to a new self-organized system I (I stands for infection, and H for health). This transition needs an autocatalytic process X, which is the activation of different bodily systems (red colors: innate and adaptive immune system, endocrine system, nervous system, etc.). After the elimination of the infectious agent, the loss of the microbe and its antigens is a fluctuation leading to a transition from self-organization I to H (reversed process X, Xrev). This reversal is similarly an energy-consuming process (apoptosis, removal of cellular debris, and normalization of systems).

chemically, or biologically predefined and spanning a certain range, but it can be fine-tuned by, for example, the plasticity phase *in utero* (see section "Timescale" of Chapter I). The transition from "self-organization H" to "self-organization ILL" is never a problem when the cycle is not blocked: it amounts to an acute short-lived disease and confinement to bed. However, it is a dilemma when the way there and back is blocked.

As discussed above, the trigger of a chronic inflammatory systemic disease might be an antigen (self or foreign), and the clonal expansion of B and T cells represents the energy-consuming autocatalytic process of the immune system (but also triggers and autocatalytic reactions in other systems are imaginable as we have learned). Similar immune-activating mechanisms are in place to induce the mentioned transition as demonstrated in Figure 58. In addition, the complex network with the endocrine, nervous, reproductive, and local cellular support systems is also involved to induce a transition from "self-organization H" to "self-organization ILL" because it is never a game of only one system.

So far so good, but what happens when the trigger is indelible like an autoantigen? The important fluctuation necessary for the transition of "self-organization ILL" back to "self-organization H" is missing. The way back is blocked, and the usual mechanisms of the evolutionarily positively selected ultracycle are not effective anymore or even wrong. One might also say that "self-organization ILL" is frozen. This can happen when the immune system is attacking an autoantigen that can never be removed (but also triggers in other systems are imaginable; see Chapter I).

A further important aspect of an ultracycle is the evolutionarily positively selected interplay of all systems. As already indicated above, one bodily system never plays a lonely game. Thus, the transition from "self-organization H" to "self-organization ILL" is accompanied by a delicate interplay of all systems selected for short inflammatory episodes (see Chapter IV). One might say that the entire ultracycle is evolutionarily positively selected so that a person after recovery from illness finds himself in "self-organization H." We want to call this positively selected ultracycle a cooperative synchronization. Cooperative means that factors in the body (between organs, within organs, or within cells) work together, and synchronization means that these factors work together at the same time leading to cooperative effects (an example was given in Chapter II, section Neuroimmunology, "Cooperation of the sympathetic nervous system and HPA axis").

We more and more recognize that synchronization is a central aspect in biology. When looking on signaling cascades in a cell, the importance of synchronization and cooperation is immediately clear because without synchronization and cooperation, nothing would work. For example, the estrogen receptor type α, which is activated by cooperative factors, can lead to the generation of very heterogeneous complexes that activate distinct estrogen-dependent genes (Figure 59).

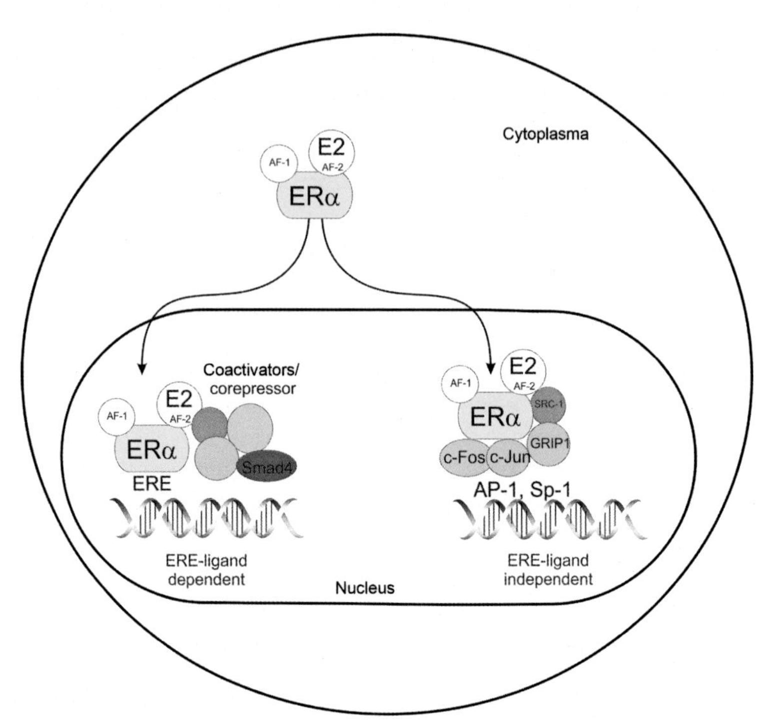

FIGURE 59 Cooperation and synchronization in the context of estrogen receptor α (ERα) activation. ERα is activated by 17β-estradiol (E2), the major estrogen. Upon activation, the complex of ERα and E2 translocates to the nucleus where new heterogeneous complexes can be generated. On the left side, the complex binds to the classical estrogen receptor response element (ERE) on the promoter of an estrogen-dependent gene. On the right side, the given complex binds to areas on the promoter independent of ERE (these are AP-1 and Sp-1 areas of the promoter). The presence of all factors is necessary to elicit estrogen effects, which is indicative of cooperation and synchronization (modified with information from Ref. 1869). AF, activating function of the ERα; E2, 17β-estradiol. Smad4, c-FOS, c-Jun, GRIP1, and SRC-1 are transcription factors. AP-1 and Sp-1 are binding sites on the promoter of respective responsive genes.

Without the simultaneous presence of these factors, no such complex would appear and no estrogen effects would be elicited. This clearly shows cooperation and synchronization. People might argue that these factors are present all the time and, thus, synchronization is not mandatory. However, this is not correct because the generation of proteins is always energy-consuming[1234] and, thus, continuous production of all possible proteins is impossible. It is clear that some important proteins are available all the time because they are needed for maintenance functions (housekeeping proteins), but the number of these proteins is small. Most proteins need to be produced when they are needed, and this is suggestive of synchronization (coupling).

Another important aspect of synchronization and cooperation is visible in the circadian rhythm of many different endogenous compounds. The circadian rhythm is controlled by the central nervous system, in the so-called nucleus suprachiasmaticus (see also section "Circadian Rhythms of Symptoms" of Chapter V). This brain area guides the body in order to maintain a synchronized rhythm, which is fine-tuned by external light.[565] For example, the secretion of cortisol and noradrenaline is highly synchronized with a maximum in the early morning hours (acrophase) and a nadir at midnight (see Figure 27). These rhythms are important for cooperation because both hormones support each other's function.[565] Most receptors in our body are G protein-coupled receptors, which are highly sensitive to desensitization (meaning downregulation of signaling in the presence of a specific ligand). Thus, G protein-coupled receptor pathways would be insensitive in the presence of constant concentrations of the ligand, for example, *in vitro* in a culture dish. In the presence of circadian changes of ligand concentrations, G protein-coupled receptors become sensitized and desensitized, respectively (Figure 60). Sensitization is important to keep the system in a responsive state, which would not happen in the presence of constant ligand concentrations. Although many researchers are working with cell cultures *in vitro*, this information should lead to better experimental systems such as superfusion or perfusion systems.

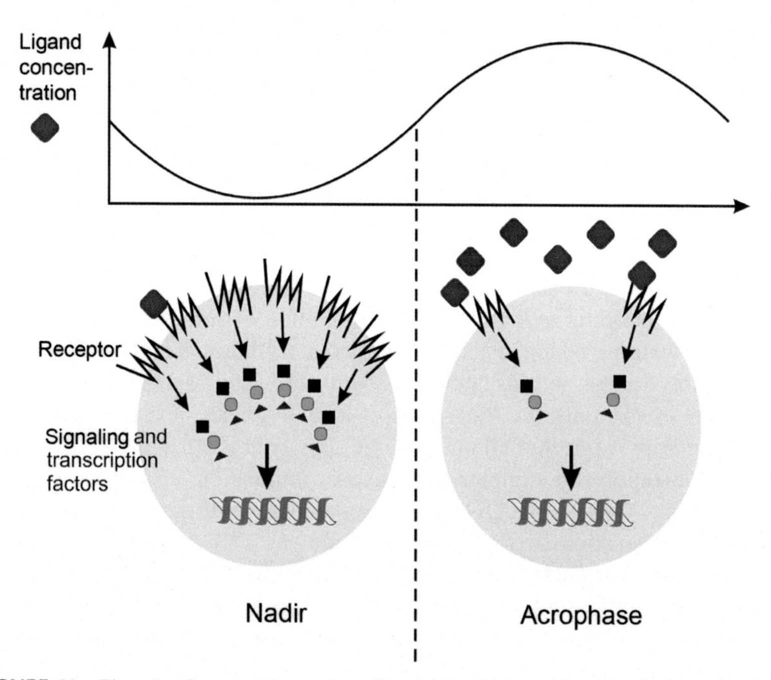

FIGURE 60 The role of cooperation and synchronization during a circadian rhythm. The circadian rhythm with its minimum (nadir) and maximum (acrophase), respectively, leads to sensitization and desensitization of G protein-coupled receptor-dependent signaling pathways. During the nadir, ligand concentrations are low and a number of receptors and signaling molecules are high, whereas it is opposite during the acrophase. This circadian up and down of signaling pathways is important for normal physiological function, for responsiveness of receptors and signaling pathways.

In conclusion, this section introduced the ultracycle (the transition of "self-organization H" to "self-organization ILL" and back again) that was positively selected for short-lived episodes (see Chapter IV). In the case of a chronic inflammatory systemic disease, the ultracycle is disturbed and "self-organization H" cannot be installed again. The body is frozen in "self-organization ILL"; it is a chronic disease. The normal ultracycle includes aspects of cooperation and synchronization (homeostasis), which are necessary for normal functioning of the body. The circadian rhythm and also other rhythms are instrumental for responsiveness and physiological function. The disruption of the ultracycle also destroys cooperation and synchronization, the consequence of which is desynchronization of otherwise synchronized factors. Desynchronization is an uncontrolled process because the desynchronized game of the different systems has not been positively selected during evolution.

DESYNCHRONIZATION AS A CONSEQUENCE OF A CHRONIC INFLAMMATORY SYSTEMIC DISEASES

In section "The Link Between TNF and Hormone Axes" of Chapter II, I already made an example how desynchronization can appear. It was described that anti-TNF therapy in patients with rheumatoid arthritis over several weeks can improve the HPA axis.[748, 750] It was demonstrated that patients with improvement demonstrated an increase of serum cortisol, which was opposite in patients with no or little improvement. This study in a human chronic inflammatory systemic disease demonstrates that inflammation-induced TNF interferes with HPA axis integrity, which is linked to the disease outcome (see Figure 12).[750] TNF-dependent worsening of the HPA axis is a perfect example of an *a posteriori* cause mentioned in Table 7, a fluctuation for the purpose of this chapter.

In this particular example, TNF is an immunologic disruptor that changes the HPA axis leading to the continuous inhibition of cortisol secretion. Since cortisol is a strong endogenous anti-inflammatory hormone,[714] the TNF break disrupts the normal anti-inflammatory feedback. Since anti-TNF therapy removes the break from steroidogenesis of adrenal glands, and because this results in an improvement of clinical disease activity, it is clear that TNF-induced desynchronization is a stimulus for disease continuation or, at least, aggravation. In other words, with high levels of serum TNF, "self-organization ILL" remains frozen, and this can be changed by removing TNF and, thereby, inducing normal HPA axis function. In the above-cited study,[750] this mechanism applies to about half of investigated patients, but pathways in the rest of the study group remain unclear (other immune pathways and cytokines might be relevant, we don't know).

There are other examples of desynchronization: the two hormonal pathways of the HPA axis and the sympathetic nervous system are synchronized (Figure 27). It has been demonstrated that in patients with Crohn's disease and ulcerative colitis, synchronization of the sympathetic nervous system and HPA axis is disturbed.[1173] Under continuous inflammatory stress in chronic inflammatory systemic diseases, the sympathetic nervous tone increases, whereas the HPA axis tone remains relatively normal or slightly elevated (see also Chapter V). This triggers a new state in "self-organization ILL" that may contribute to continuation.

A very similar phenomenon with an increased tone of the sympathetic nervous system and a low normal activity of the HPA axis was observed in patients with rheumatoid arthritis and systemic lupus erythematosus.[1592] Similarly, desynchronization between the two axes was also detected in patients with chronic liver cirrhosis independent of severity of liver disease.[1870] In inflammatory diseases, coupling of the HPA axis and the sympathetic nervous system is important to dampen peripheral inflammation due to cooperativity between cortisol and noradrenaline, which has recently been demonstrated in patients with rheumatoid arthritis (Figure 27-29).[581] In the state of "self-organization ILL," desynchronization is maintained and can contribute to chronic illness.

Another example of desynchronization is the change of neuronal innervation in inflamed tissue (Chapters II and V). It was described that sympathetic nerve fibers get lost while proinflammatory sensory nerve fibers undergo sprouting and hyperinnervation. In synovial tissue, the normal ratio of nerve fiber density between sympathetic and sensory nerve fibers is 1:1.[1871] This number largely changes to approximately 1:8 in inflamed tissue of patients with rheumatoid arthritis.[1871] Under these conditions, the proinflammatory substance P prevails relative to possibly anti-inflammatory neurotransmitters of the sympathetic nerve ending (noradrenaline, adenosine, and endogenous opioids at high concentrations). Furthermore, the loss of efferent nerve fibers is a typical constellation that contributes to the isolation of inflamed tissue from the rest of the body, particularly isolation from the central nervous system. In other words, locally inflamed tissue, whether in secondary lymphoid organs or elsewhere, gets uncoupled from the brain.

Desynchronization happens often in chronic inflammatory systemic diseases. The following scenario describes the possible sequence of events: At the beginning of chronic inflammatory systemic diseases (in a presymptomatic phase), immune tolerance against self-antigens is broken or an inadequate immune response to harmless foreign antigens is triggered. T cells, B cells, and antigen-presenting cells are hidden players of the disease process in the presymptomatic phase of the disease (Figure 61).

Then, the disease, probably because of an additional trigger, strides into the symptomatic phase that involves many other cell types. Continuous aggression against an antigen leads to a dissociation between the immune system and the other supersystems (Figure 61). Now, chronic inflammatory systemic diseases become chronic due to the aberrant expression of programs that have been preserved to cope only with acute events during transient inflammatory reactions (desynchronization) (Figure 61). The program "self-organization ILL" is frozen due to the wrong programs used.

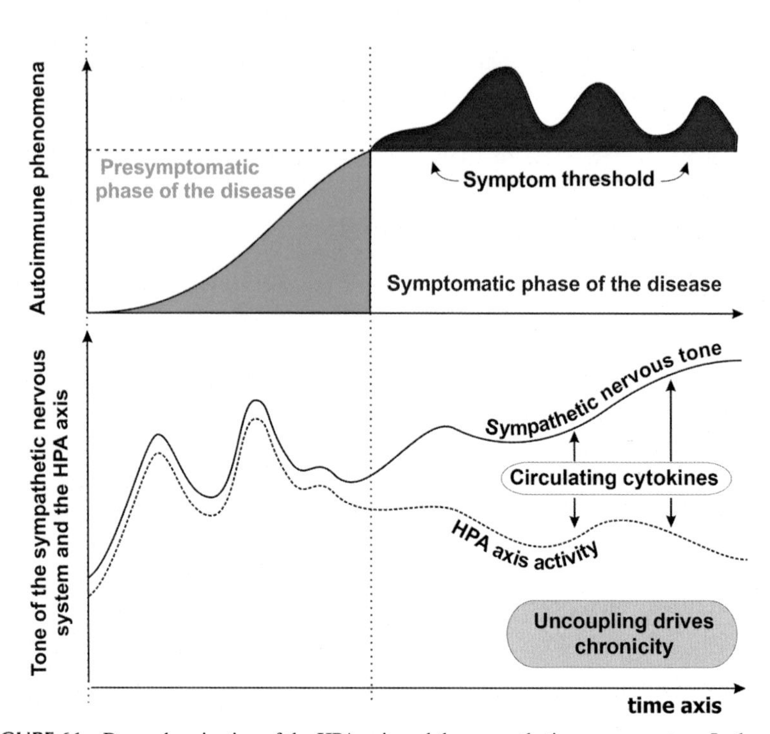

FIGURE 61 Desynchronization of the HPA axis and the sympathetic nervous system. In the presymptomatic phase without overt inflammation, the sympathetic nervous tone is synchronized with the tone of the HPA axis. In the symptomatic phase of the disease, the activity of the two axes is uncoupled. I hypothesize that desynchronization is an important factor that drives chronicity.

Even the removal of the initial disruptor might not lead to the normalization of the neuroendocrine immune crosstalk as has been demonstrated with anti-TNF therapy. Anti-TNF therapy over 12-16 weeks cannot normalize the hypothalamic-pituitary-gonadal axis, the increased activity of the sympathetic nervous system, and elevated serum levels of leptin.[748] It seems that these alterations are imprinted for a much longer period of time. Similarly, anti-IL-6 therapy can only improve some aspects of steroidogenesis in patients with chronic inflammatory systemic disease.[749] Imprinting of disruption can lead to a continuous problem so that the state of "self-organization ILL" can never be reverted; it is frozen. This can also happen in surviving patients with severe sepsis who need lifelong glucocorticoid substitution therapy.[747]

It is interesting that important hormonal and neuronal factors change during the course of chronic inflammatory systemic diseases so that in the chronic phase, the systemic situation and the local micromilieu in inflamed tissue typically demonstrate predominantly proinflammatory desynchronized patterns. Desynchronization in this sense means a change relative to the initial healthy situation (Figure 62).

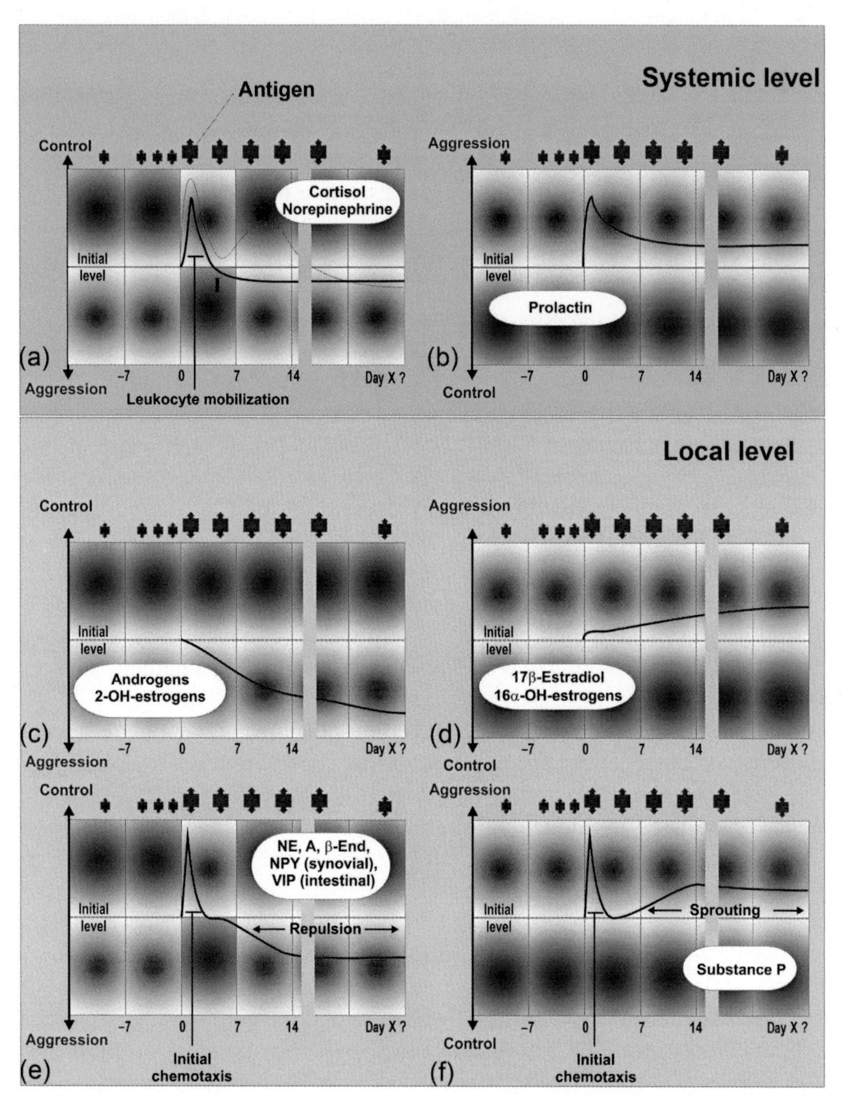

FIGURE 62 Changes of the supersystems related to acute immune challenges, whose maintenance in chronic inflammatory systemic diseases induces an overall proinflammatory situation. Time point zero indicates the starting point of the overt inflammatory process (the symptomatic phase). Day X indicates a later time point in the symptomatic chronic inflammatory systemic disease. Changes include inadequate low levels of systemic cortisol, two different time courses of cortisol increase and decrease are demonstrated (a), decreased local androgens and 2-hydroxyestrogens (c), decreased local sympathetic neurotransmitters due to sympathetic nerve fiber loss (e), elevated systemic levels of prolactin (b), locally increased 17β-estradiol and 16-hydroxylated estrogens (d), and high local levels of substance P due to sensory hyperinnervation (nerve fiber sprouting) (f). In relation to the initial levels, the systemic or local level of all mentioned factors changes into an aggressive direction (red zone) and never into an anti-inflammatory direction (blue zone). The larger symbol for the antigen indicates that sometimes access to a self-antigen or foreign antigen increases during the symptomatic phase of a disease (think of collagen type II or myelin). β-END, β-endorphin (from sympathetic nerve terminals); NE, norepinephrine; NPY, neuropeptide Y; OH, hydroxyl group; VIP, vasoactive intestinal peptide.

Another way to induce synchronization in a cell is coordinated regulation along circadian rhythms that are installed in many peripheral cells such as myocardial or hepatic cells. Circadian rhythm molecules generate synchronization within peripheral cells. Also, immune cells seem to have intracellular rhythms, but the desynchronization of these rhythms seems to happen in chronic inflammatory systemic diseases.[1872]

DESYNCHRONIZATION AS A CONSEQUENCE OF AGING

Finally, desynchronization is also a phenomenon observed during aging. When we studied serum levels of major hormones in healthy women and men ranging in age between 18 and 75 years, important relative changes of hormone levels to each other were observed.[1597, 1873] During aging, adrenal cortisol production remains relatively stable, whereas the production of adrenal and gonadal androgens as well as of gonadal estrogens continuously decreases (more obvious in women than in men, particularly after the menopause).[1597, 1873] In addition, plasma levels of noradrenaline increase in relation to other steroid hormones including cortisol.[1597] Aging is accompanied by a decrease of sympathetic innervation and a general increase of the sympathetic nervous tone.[1597, 1847]

Several other forms of desynchronization phenomena, which are beyond the scope of this chapter, are apparent during the aging process. During aging, desynchronization is a slowly progressive process (Figure 63), which may be paralleled by hidden autoimmune phenomena as mentioned above. The autoaggressive "sleepers" are sitting somewhere and wait until they get the endocrine or neuronal call. Recall that autoimmunity seems to be a rather normal phenomenon in a healthy person, which is supported by a specific genetic background and largely controlled in the presymptomatic phase of chronic inflammatory systemic diseases. Regulatory elements outweigh autoaggressive elements. In addition, we know that the autoimmune process is long apparent before the outbreak of the disease (see Figure 3 in Chapter I). It is hypothesized that the desynchronization of homeostatic programs during the process of aging is strong enough to drive a process to a symptomatic disease and chronicity, with or without an additional environmental trigger (Figure 63).

This hypothesis is based on observations in animal models of chronic inflammatory systemic diseases, which develop spontaneously or after adrenalectomy such as in obese strain chickens (spontaneous thyroiditis due to a defect of the HPA axis)[715] and in rats of the Lewis strain (experimental autoimmune encephalomyelitis and experimental colitis due to defect of the HPA axis).[524, 716–718] We recently found a very similar outbreak of experimental arthritis in nonsusceptible middle-aged PVG rats after adrenalectomy.[1874] These animal models demonstrate that the loss of an important endocrine system leads to the development of spontaneous or antigen-induced chronic inflammatory systemic

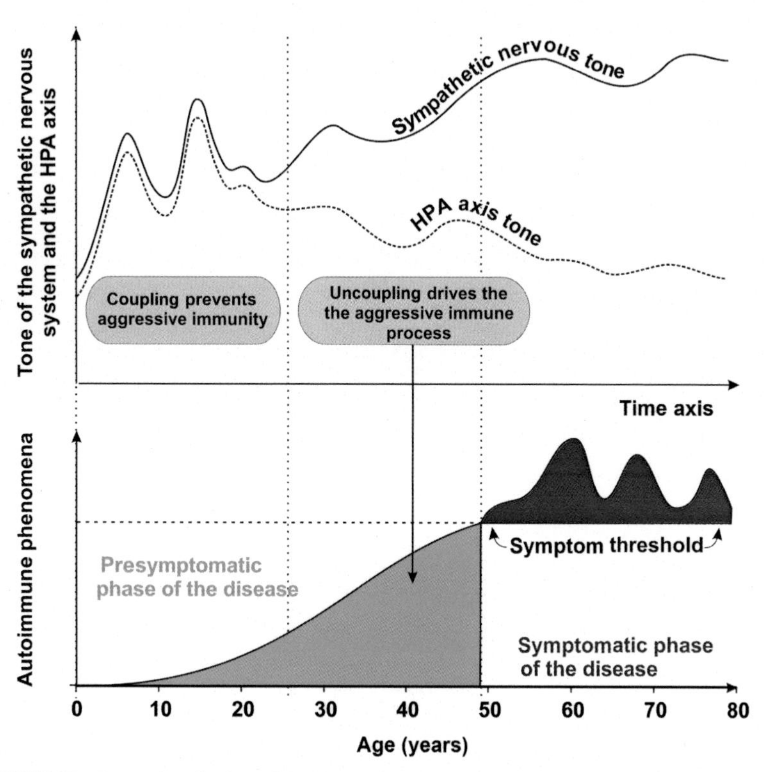

FIGURE 63 Desynchronization of the HPA axis and the sympathetic nervous system during the process of aging. This example shows that during early life (until 25 years of age in this example), the activities of the HPA axis and the sympathetic nervous system are synchronized. Evolutionarily positively selected synchronization of the response of these two systems is necessary in order to provide cooperative effects, which prevent overt immune reactions against self-antigens or foreign antigens. The desynchronization of the activity of the two systems during aging steadily drives the aggressive immune process. At a certain time point, with or without an exogenous trigger, the symptomatic phase of the disease starts, and the desynchronization of the two supersystems gets boosted leading to further increased desynchronization phenomena.

disease. From this point of view, adrenalectomy or a deficit of the HPA axis is a trigger that induces "self-organization ILL," but it can also be a block that does not allow the return to "self-organization H (healthy)."

CONCLUSIONS

While we usually look for classical immunologic factors responsible for the continuation of chronic inflammatory diseases, the desynchronization of neuronal and endocrine systems may similarly lead to chronicity. In addition, since aging induces desynchronization on many levels, evolutionarily not positively selected, the process of aging itself can trigger chronic inflammatory systemic diseases by simple desynchronization.

References

1. Eichmann K. *The network collective—rise and fall of a scientific paradigm*. Basel: Birkhäuser; 2008.
2. Kuhn TS. *The structure of scientific revolutions*. Chicago, IL: The University of Chicago Press; 1962.
3. Fleck L. *Genesis and development of a scientific fact*. Chicago: The University of Chicago Press; 1935.
4. Schaefer H. Die Stellung der Regeltheorie im System der Wissenschaft. In: Mittelstaedt H, editor. Regelungsvorgänge in der Biologie (Beihefte zur Regelungstechnik). München: R. Oldenbourg; 1956. p. 44–54.
5. Silverstein AM. *A history of immunology*. Amsterdam: Elsevier; 2009.
6. Bywaters EG. Historical aspects of the aetiology of rheumatoid arthritis. *Br J Rheumatol* 1988;**27**(Suppl. 2):110–5.
7. Hughes RA. Focal infection revisited. *Br J Rheumatol* 1994;**33**:370–7.
8. Arrhenius S. *Immunochemistry*. New York: Macmillan; 1907.
9. Arthus NM. Injections répétées de serum du cheval chez le lapin. *C R Seances Soc Biol Fil* 1903;**55**:817–20.
10. von Pirquet C, Schick B. Die Serumkrankheit. Wien: F. Deuticke; 1905.
11. Bergmann K-C, Ring J. *History of allergy*. Basel: Karger; 2014.
12. Hunter W. The role of sepsis and antisepsis in medicine. *Lancet* 1911;**1**:79–86.
13. Fox RF, Van Breemen J. *Chronic rheumatism: causation and treatment*. London: J. & A. Churchill Ltd.; 1934.
14. Levinsky WJ, Lansbury J. An attempted to transmit rheumatoid arthritis to humans. *Proc Soc Exp Biol Med* 1951;**78**:325–6.
15. Gorer PA, Lyman S, Snell GD. Studies on the genetic and antigenic basis of tumour transplantation: linkage between a histocompatibility gene and "fused" in mice. *Proc R Soc Lond Ser B* 1948;**135**:499–505.
16. Snell GD. Methods for the study of histocompatibility genes. *J Genet* 1948;**49**:87–108.
17. Dausset J. Iso-leuko-antibodies. *Acta Haematol* 1958;**20**:156–66.
18. Levine BB, Ojeda A, Benaceraff B. Studies on artificial antigens. III. The genetic control of the immune response to hapten-poly-L-lysine conjugates in guinea pigs. *J Exp Med* 1963;**118**:953–7.
19. McDevitt HO, Sela M. Genetic control of the antibody response. I. Demonstration of determinant-specific differences in response to synthetic polypeptide antigens in two strains of inbred mice. *J Exp Med* 1965;**122**:517–31.
20. Medawar PB. The behaviour and fate of skin autografts and skin homografts in rabbits: a report to the War Wounds Committee of the Medical Research Council. *J Anat* 1944;**78**:176–99.

21. Billingham RE, Brent L, Medawar PB. Actively acquired tolerance of foreign cells. *Nature* 1953;**172**:603–6.
22. Bruton OC. Agammaglobulinemia. *Pediatrics* 1952;**9**:722–8.
23. Burnet FM. *The conal selection theory of acquired immunity*. Nashville: Cambridge University Press; 1959.
24. Lederberg J. Genes and antibodies. *Science* 1959;**129**:1649–53.
25. Hozumi N, Tonegawa S. Evidence for somatic rearrangement of immunoglobulin genes coding for variable and constant regions. 1976 [classical article]. *J Immunol* 2004;**173**:4260–4.
26. Woodruff MF, Forman B. Evidence for the production of circulating antibodies by homografts of lymphoid tissue and skin. *Br J Exp Pathol* 1950;**31**:306–15.
27. Jerne NK, Nordin AA. Plaque formation in agar by single antibody-producing cells. *Science* 1963;**140**:405.
28. Glick B, Chang TS, Jaap RG. The bursa of Fabricius and antibody production. *Poult Sci* 1956;**35**:224–5.
29. Miller JF, Mitchell GF. Cell to cell interaction in the immune response. I. Hemolysin-forming cells in neonatally thymectomized mice reconstituted with thymus or thoracic duct lymphocytes. *J Exp Med* 1968;**128**:801–20.
30. Mitchell GF, Miller JF. Cell to cell interaction in the immune response. II. The source of hemolysin-forming cells in irradiated mice given bone marrow and thymus or thoracic duct lymphocytes. *J Exp Med* 1968;**128**:821–37.
31. Nossal GJ, Cunningham A, Mitchell GF, Miller JF. Cell to cell interaction in the immune response. 3. Chromosomal marker analysis of single antibody-forming cells in reconstituted, irradiated, or thymectomized mice. *J Exp Med* 1968;**128**:839–53.
32. Davies AJ. The thymus and the cellular basis of immunity. *Transplant Rev* 1969;**1**:43–91.
33. Claman HN, Chaperon EA. Immunologic complementation between thymus and marrow cells—a model for the two-cell theory of immunocompetence. *Transplant Rev* 1969;**1**:92–113.
34. Claman HN, Chaperon EA, Triplett RF. Thymus-marrow cell combinations. Synergism in antibody production. *Proc Soc Exp Biol Med* 1966;**122**:1167–71.
35. Warner NL, Szenberg A. Immunological recativity of bursaless chickens in graft versus host reactions. *Nature* 1963;**199**:43–4.
36. Warner NL, Szenberg A. Effect of neonatal thymectomy on the immune response in the chicken. *Nature* 1962;**196**:784–5.
37. Cooper MD, Peterson RD, Good RA. Delineation of the thymic and bursal lymphoid systems in the chicken. *Nature* 1965;**205**:143–6.
38. Raff MC. T and B lymphocytes and immune responses. *Nature* 1973;**242**:19–23.
39. Rajewsky K, Schirrmacher V, Nase S, Jerne NK. The requirement of more than one antigenic determinant for immunogenicity. *J Exp Med* 1969;**129**:1131–43.
40. Mitchison NA. The carrier effect in the secondary response to hapten-protein conjugates. II. Cellular cooperation. *Eur J Immunol* 1971;**1**:18–27.
41. Gershon RK, Cohen P, Hencin R, Liebhaber SA. Suppressor T cells. *J Immunol* 1972;**108**:586–90.
42. Janeway Jr CA, Sharrow SO, Simpson E. T-cell populations with different functions. *Nature* 1975;**253**:544–6.
43. Cerottini JC, Nordin AA, Brunner KT. Specific in vitro cytotoxicity of thymus-derived lymphocytes sensitized to alloantigens. *Nature* 1970;**228**:1308–9.

44. Cerottini JC, Nordin AA, Brunner KT. In vitro cytotoxic activity of thymus cells sensitized to alloantigens. *Nature* 1970;**227**:72–3.
45. Raff M. Theta isoantigen as a marker of thymus-derived lymphocytes in mice. *Nature* 1969;**224**:378–9.
46. Coombs RR, Feinstein A, Wilson AB. Immunoglobulin determinants on the surface of human lymphocytes. *Lancet* 1969;**2**:1157–60.
47. Köhler G, Milstein C. Continuous cultures of fused cells secreting antibody of predefined specificity. *Nature* 1975;**256**:495–7.
48. Bonner WA, Hulett HR, Sweet RG, Herzenberg LA. Fluorescence activated cell sorting. *Rev Sci Instrum* 1972;**43**:404–9.
49. Kincade PW, Lawton AR, Bockman DE, Cooper MD. Suppression of immunoglobulin G synthesis as a result of antibody-mediated suppression of immunoglobulin M synthesis in chickens. *Proc Natl Acad Sci U S A* 1970;**67**:1918–25.
50. Wu TT, Kabat EA. An analysis of the sequences of the variable regions of Bence Jones proteins and myeloma light chains and their implications for antibody complementarity. *J Exp Med* 1970;**132**:211–50.
51. Kabat EA, Wu TT. Construction of a three-dimensional model of the polypeptide backbone of the variable region of kappa immunoglobulin light chains. *Proc Natl Acad Sci U S A* 1972;**69**:960–4.
52. Shevach EM, Rosenthal AS. Function of macrophages in antigen recognition by guinea pig T lymphocytes. II. Role of the macrophage in the regulation of genetic control of the immune response. *J Exp Med* 1973;**138**:1213–29.
53. Katz DH, Hamaoka T, Dorf ME, Benacerraf B. Cell interactions between histoincompatible T and B lymphocytes. The H-2 gene complex determines successful physiologic lymphocyte interactions. *Proc Natl Acad Sci U S A* 1973;**70**:2624–8.
54. Kindred B, Shreffler DC. H-2 dependence of co-operation between T and B cells in vivo. *J Immunol* 1972;**109**:940–3.
55. Zinkernagel RM, Doherty PC. Restriction of in vitro T cell-mediated cytotoxicity in lymphocytic choriomeningitis within a syngeneic or semiallogeneic system. *Nature* 1974;**248**:701–2.
56. Blanden RV, Doherty PC, Dunlop MB, Gardner ID, Zinkernagel RM, David CS. Genes required for cytotoxicity against virus-infected target cells in K and D regions of H-2 complex. *Nature* 1975;**254**:269–70.
57. Koszinowski U, Ertl H. Lysis mediated by T cells and restricted by H-2 antigen of target cells infected with vaccinia virus. *Nature* 1975;**255**:552–4.
58. Gordon RD, Simpson E, Samelson LE. In vitro cell-mediated immune responses to the male specific(H-Y) antigen in mice. *J Exp Med* 1975;**142**:1108–20.
59. von Boehmer H, Haas W, Jerne NK. Major histocompatibility complex-linked immune-responsiveness is acquired by lymphocytes of low-responder mice differentiating in thymus of high-responder mice. *Proc Natl Acad Sci U S A* 1978;**75**:2439–42.
60. Bevan MJ. Interaction antigens detected by cytotoxic T cells with the major histocompatibility complex as modifier. *Nature* 1975;**256**:419–21.
61. Babbitt BP, Allen PM, Matsueda G, Haber E, Unanue ER. Binding of immunogenic peptides to Ia histocompatibility molecules. *Nature* 1985;**317**:359–61.
62. Buus S, Colon S, Smith C, Freed JH, Miles C, Grey HM. Interaction between a "processed" ovalbumin peptide and Ia molecules. *Proc Natl Acad Sci U S A* 1986;**83**:3968–71.
63. Buus S, Sette A, Colon SM, Jenis DM, Grey HM. Isolation and characterization of antigen-Ia complexes involved in T cell recognition. *Cell* 1986;**47**:1071–7.

64. Townsend AR, Rothbard J, Gotch FM, Bahadur G, Wraith D, McMichael AJ. The epitopes of influenza nucleoprotein recognized by cytotoxic T lymphocytes can be defined with short synthetic peptides. *Cell* 1986;**44**:959–68.
65. Maryanski JL, Pala P, Corradin G, Jordan BR, Cerottini JC. H-2-restricted cytolytic T cells specific for HLA can recognize a synthetic HLA peptide. *Nature* 1986;**324**:578–9.
66. Rötzschke O, Falk K, Deres K, Schild H, Norda M, Metzger J, et al. Isolation and analysis of naturally processed viral peptides as recognized by cytotoxic T cells. *Nature* 1990;**348**:252–4.
67. Lurquin C, Van PA, Mariame B, De PE, Szikora JP, Janssens C, et al. Structure of the gene of tum- transplantation antigen P91A: the mutated exon encodes a peptide recognized with Ld by cytolytic T cells. *Cell* 1989;**58**:293–303.
68. Bjorkman PJ, Saper MA, Samraoui B, Bennett WS, Strominger JL, Wiley DC. Structure of the human class I histocompatibility antigen, HLA-A2. *Nature* 1987;**329**:506–12.
69. Bjorkman PJ, Saper MA, Samraoui B, Bennett WS, Strominger JL, Wiley DC. The foreign antigen binding site and T cell recognition regions of class I histocompatibility antigens. *Nature* 1987;**329**:512–8.
70. Garcia KC, Degano M, Stanfield RL, Brunmark A, Jackson MR, Peterson PA, et al. An alphabeta T cell receptor structure at 2.5 A and its orientation in the TCR-MHC complex. *Science* 1996;**274**:209–19.
71. Garboczi DN, Ghosh P, Utz U, Fan QR, Biddison WE, Wiley DC. Structure of the complex between human T-cell receptor, viral peptide and HLA-A2. *Nature* 1996;**384**:134–41.
72. Jorgensen JL, Esser U, Fazekas de St GB, Reay PA, Davis MM. Mapping T-cell receptor-peptide contacts by variant peptide immunization of single-chain transgenics. *Nature* 1992;**355**:224–30.
73. Arrigo AP, Tanaka K, Goldberg AL, Welch WJ. Identity of the 19S 'prosome' particle with the large multifunctional protease complex of mammalian cells (the proteasome). *Nature* 1988;**331**:192–4.
74. Michalek MT, Grant EP, Gramm C, Goldberg AL, Rock KL. A role for the ubiquitin-dependent proteolytic pathway in MHC class I-restricted antigen presentation. *Nature* 1993;**363**:552–4.
75. Rock KL, Gramm C, Rothstein L, Clark K, Stein R, Dick L, et al. Inhibitors of the proteasome block the degradation of most cell proteins and the generation of peptides presented on MHC class I molecules. *Cell* 1994;**78**:761–71.
76. Harding CV, Collins DS, Slot JW, Geuze HJ, Unanue ER. Liposome-encapsulated antigens are processed in lysosomes, recycled, and presented to T cells. *Cell* 1991;**64**:393–401.
77. Bevan MJ. Cross-priming for a secondary cytotoxic response to minor H antigens with H-2 congenic cells which do not cross-react in the cytotoxic assay. *J Exp Med* 1976;**143**:1283–8.
78. Kurts C, Robinson BW, Knolle PA. Cross-priming in health and disease. *Nat Rev Immunol* 2010;**10**:403–14.
79. Kurts C, Kosaka H, Carbone FR, Miller JF, Heath WR. Class I-restricted cross-presentation of exogenous self-antigens leads to deletion of autoreactive CD8(+) T cells. *J Exp Med* 1997;**186**:239–45.
80. Steinman RM, Cohn ZA. Identification of a novel cell type in peripheral lymphoid organs of mice. I. Morphology, quantitation, tissue distribution. *J Exp Med* 1973;**137**:1142–62.
81. Steinman RM, Moberg CL. Zanvil Alexander Cohn 1926–1993. *J Exp Med* 1994;**179**:1–30.
82. Allison JP, Lanier LL. Structure, function, and serology of the T-cell antigen receptor complex. *Annu Rev Immunol* 1987;**5**:503–40.
83. Kronenberg M, Davis MM, Early PW, Hood LE, Watson JD. Helper and killer T cells do not express B cell immunoglobulin joining and constant region gene segments. *J Exp Med* 1980;**152**:1745–61.

84. Hedrick SM, Nielsen EA, Kavaler J, Cohen DI, Davis MM. Sequence relationships between putative T-cell receptor polypeptides and immunoglobulins. *Nature* 1984;**308**:153–8.
85. Hedrick SM, Cohen DI, Nielsen EA, Davis MM. Isolation of cDNA clones encoding T cell-specific membrane-associated proteins. *Nature* 1984;**308**:149–53.
86. Yanagi Y, Yoshikai Y, Leggett K, Clark SP, Aleksander I, Mak TW. A human T cell-specific cDNA clone encodes a protein having extensive homology to immunoglobulin chains. *Nature* 1984;**308**:145–9.
87. Saito H, Kranz DM, Takagaki Y, Hayday AC, Eisen HN, Tonegawa S. Complete primary structure of a heterodimeric T-cell receptor deduced from cDNA sequences. *Nature* 1984;**309**:757–62.
88. Hannum CH, Kappler JW, Trowbridge IS, Marrack P, Freed JH. Immunoglobulin-like nature of the alpha-chain of a human T-cell antigen/MHC receptor. *Nature* 1984;**312**:65–7.
89. Chien Y, Becker DM, Lindsten T, Okamura M, Cohen DI, Davis MM. A third type of murine T-cell receptor gene. *Nature* 1984;**312**:31–5.
90. Saito H, Kranz DM, Takagaki Y, Hayday AC, Eisen HN, Tonegawa S. A third rearranged and expressed gene in a clone of cytotoxic T lymphocytes. *Nature* 1984;**312**:36–40.
91. Acuto O, Fabbi M, Smart J, Poole CB, Protentis J, Royer HD, et al. Purification and NH2-terminal amino acid sequencing of the beta subunit of a human T-cell antigen receptor. *Proc Natl Acad Sci U S A* 1984;**81**:3851–5.
92. Lanier LL, Phillips JH, Hackett Jr J, Tutt M, Kumar V. Natural killer cells: definition of a cell type rather than a function. *J Immunol* 1986;**137**:2735–9.
93. Kärre K, Ljunggren HG, Piontek G, Kiessling R. Selective rejection of H-2-deficient lymphoma variants suggests alternative immune defence strategy. *Nature* 1986;**319**:675–8.
94. Karlhofer FM, Ribaudo RK, Yokoyama WM. MHC class I alloantigen specificity of Ly-49+ IL-2-activated natural killer cells. *Nature* 1992;**358**:66–70.
95. Lanier LL. NK cell recognition. *Annu Rev Immunol* 2005;**23**:225–74.
96. Isaacs A, Lindenmann J. Virus interference. I. The interferon. *Proc R Soc Lond B Biol Sci* 1957;**147**:258–67.
97. Isaacs A, Lindenman J, Valentine RC. Virus interference. II. Some properties of interferon. *Proc R Soc Lond B Biol Sci* 1957;**147**:268–73.
98. Bloom BR, Bennett B. Mechanism of a reaction in vitro associated with delayed-type hypersensitivity. *Science* 1966;**153**:80–2.
99. David JR. Delayed hypersensitivity in vitro: its mediation by cell-free substances formed by lymphoid cell-antigen interaction. *Proc Natl Acad Sci U S A* 1966;**56**:72–7.
100. Robb RJ, Smith KA. Heterogeneity of human T-cell growth factor(s) due to variable glycosylation. *Mol Immunol* 1981;**18**:1087–94.
101. Auron PE, Webb AC, Rosenwasser LJ, Mucci SF, Rich A, Wolff SM, et al. Nucleotide sequence of human monocyte interleukin 1 precursor cDNA. *Proc Natl Acad Sci U S A* 1984;**81**:7907–11.
102. Lomedico PT, Gubler U, Hellmann CP, Dukovich M, Giri JG, Pan YC, et al. Cloning and expression of murine interleukin-1 cDNA in Escherichia coli. *Nature* 1984;**312**:458–62.
103. Beutler B, Greenwald D, Hulmes JD, Chang M, Pan YC, Mathison J, et al. Identity of tumour necrosis factor and the macrophage-secreted factor cachectin. *Nature* 1985;**316**:552–4.
104. Yokota T, Otsuka T, Mosmann T, Banchereau J, DeFrance T, Blanchard D, et al. Isolation and characterization of a human interleukin cDNA clone, homologous to mouse B-cell stimulatory factor 1, that expresses B-cell- and T-cell-stimulating activities. *Proc Natl Acad Sci U S A* 1986;**83**:5894–8.
105. Lee F, Yokota T, Otsuka T, Meyerson P, Villaret D, Coffman R, et al. Isolation and characterization of a mouse interleukin cDNA clone that expresses B-cell stimulatory factor 1 activities and T-cell- and mast-cell-stimulating activities. *Proc Natl Acad Sci U S A* 1986;**83**:2061–5.

106. Hirano T, Yasukawa K, Harada H, Taga T, Watanabe Y, Matsuda T, et al. Complementary DNA for a novel human interleukin (BSF-2) that induces B lymphocytes to produce immunoglobulin. *Nature* 1986;**324**:73–6.
107. Mosmann TR, Cherwinski H, Bond MW, Giedlin MA, Coffman RL. Two types of murine helper T cell clone. I. Definition according to profiles of lymphokine activities and secreted proteins. *J Immunol* 1986;**136**:2348–57.
108. Mangalam AK, Taneja V, David CS. HLA class II molecules influence susceptibility versus protection in inflammatory diseases by determining the cytokine profile. *J Immunol* 2013;**190**:513–8.
109. Murray JS. How the MHC selects Th1/Th2 immunity. *Immunol Today* 1998;**19**:157–63.
110. Kerr JF, Wyllie AH, Currie AR. Apoptosis: a basic biological phenomenon with wide-ranging implications in tissue kinetics. *Br J Cancer* 1972;**26**:239–57.
111. Laster SM, Wood JG, Gooding LR. Tumor necrosis factor can induce both apoptic and necrotic forms of cell lysis. *J Immunol* 1988;**141**:2629–34.
112. Trauth BC, Klas C, Peters AM, Matzku S, Moller P, Falk W, et al. Monoclonal antibody-mediated tumor regression by induction of apoptosis. *Science* 1989;**245**:301–5.
113. Yonehara S, Ishii A, Yonehara M. A cell-killing monoclonal antibody (anti-Fas) to a cell surface antigen co-downregulated with the receptor of tumor necrosis factor. *J Exp Med* 1989;**169**:1747–56.
114. Itoh N, Yonehara S, Ishii A, Yonehara M, Mizushima S, Sameshima M, et al. The polypeptide encoded by the cDNA for human cell surface antigen Fas can mediate apoptosis. *Cell* 1991;**66**:233–43.
115. Nagata S, Hanayama R, Kawane K. Autoimmunity and the clearance of dead cells. *Cell* 2010;**140**:619–30.
116. Lavrik IN, Eils R, Fricker N, Pforr C, Krammer PH. Understanding apoptosis by systems biology approaches. *Mol Biosyst* 2009;**5**:1105–11.
117. Raff MC, Megson M, Owen JJ, Cooper MD. Early production of intracellular IgM by B-lymphocyte precursors in mouse. *Nature* 1976;**259**:224–6.
118. Schatz DG, Oettinger MA, Baltimore D. The V(D)J recombination activating gene, RAG-1. *Cell* 1989;**59**:1035–48.
119. Oettinger MA, Schatz DG, Gorka C, Baltimore D. RAG-1 and RAG-2, adjacent genes that synergistically activate V(D)J recombination. *Science* 1990;**248**:1517–23.
120. Melchers F. The pre-B-cell receptor: selector of fitting immunoglobulin heavy chains for the B-cell repertoire. *Nat Rev Immunol* 2005;**5**:578–84.
121. von Boehmer H. Unique features of the pre-T-cell receptor alpha-chain: not just a surrogate. *Nat Rev Immunol* 2005;**5**:571–7.
122. LeBien TW, Tedder TF. B lymphocytes: how they develop and function. *Blood* 2008;**112**:1570–80.
123. von Boehmer H, Melchers F. Checkpoints in lymphocyte development and autoimmune disease. *Nat Immunol* 2010;**11**:14–20.
124. Geissmann F, Manz MG, Jung S, Sieweke MH, Merad M, Ley K. Development of monocytes, macrophages, and dendritic cells. *Science* 2010;**327**:656–61.
125. Springer TA, Dustin ML, Kishimoto TK, Marlin SD. The lymphocyte function-associated LFA-1, CD2, and LFA-3 molecules: cell adhesion receptors of the immune system. *Annu Rev Immunol* 1987;**5**:223–52.
126. Shaw S, Luce GE, Quinones R, Gress RE, Springer TA, Sanders ME. Two antigen-independent adhesion pathways used by human cytotoxic T-cell clones. *Nature* 1986;**323**:262–4.

127. Simmons D, Makgoba MW, Seed B. ICAM, an adhesion ligand of LFA-1, is homologous to the neural cell adhesion molecule NCAM. *Nature* 1988;**331**:624–7.
128. Lasky LA, Singer MS, Yednock TA, Dowbenko D, Fennie C, Rodriguez H, et al. Cloning of a lymphocyte homing receptor reveals a lectin domain. *Cell* 1989;**56**:1045–55.
129. Cybulsky MI, Fries JW, Williams AJ, Sultan P, Eddy R, Byers M, et al. Gene structure, chromosomal location, and basis for alternative mRNA splicing of the human VCAM1 gene. *Proc Natl Acad Sci U S A* 1991;**88**:7859–63.
130. Elices MJ, Osborn L, Takada Y, Crouse C, Luhowskyj S, Hemler ME, et al. VCAM-1 on activated endothelium interacts with the leukocyte integrin VLA-4 at a site distinct from the VLA-4/fibronectin binding site. *Cell* 1990;**60**:577–84.
131. Granger DN, Kubes P. The microcirculation and inflammation: modulation of leukocyte-endothelial cell adhesion. *J Leukoc Biol* 1994;**55**:662–75.
132. McEver RP. Selectins: novel receptors that mediate leukocyte adhesion during inflammation. *Thromb Haemost* 1991;**65**:223–8.
133. Tamkun JW, DeSimone DW, Fonda D, Patel RS, Buck C, Horwitz AF, et al. Structure of integrin, a glycoprotein involved in the transmembrane linkage between fibronectin and actin. *Cell* 1986;**46**:271–82.
134. Kishimoto TK, O'Connor K, Lee A, Roberts TM, Springer TA. Cloning of the beta subunit of the leukocyte adhesion proteins: homology to an extracellular matrix receptor defines a novel supergene family. *Cell* 1987;**48**:681–90.
135. Petri B, Phillipson M, Kubes P. The physiology of leukocyte recruitment: an in vivo perspective. *J Immunol* 2008;**180**:6439–46.
136. McCutcheon M. Studies on the locomotion of leucocytes: III. The normal rate of locomotion of human lymphocytes in vitro. *Am J Physiol* 1924;**69**:279–89.
137. Shin HS, Snyderman R, Friedman E, Mellors A, Mayer MM. Chemotactic and anaphylatoxic fragment cleaved from the fifth component of guinea pig complement. *Science* 1968;**162**:361–3.
138. Anisowicz A, Bardwell L, Sager R. Constitutive overexpression of a growth-regulated gene in transformed Chinese hamster and human cells. *Proc Natl Acad Sci U S A* 1987;**84**:7188–92.
139. Schall TJ, Jongstra J, Dyer BJ, Jorgensen J, Clayberger C, Davis MM, et al. A human T cell-specific molecule is a member of a new gene family. *J Immunol* 1988;**141**:1018–25.
140. Matsushima K, Morishita K, Yoshimura T, Lavu S, Kobayashi Y, Lew W, et al. Molecular cloning of a human monocyte-derived neutrophil chemotactic factor (MDNCF) and the induction of MDNCF mRNA by interleukin 1 and tumor necrosis factor. *J Exp Med* 1988;**167**:1883–93.
141. Poncz M, Surrey S, LaRocco P, Weiss MJ, Rappaport EF, Conway TM, et al. Cloning and characterization of platelet factor 4 cDNA derived from a human erythroleukemic cell line. *Blood* 1987;**69**:219–23.
142. Bachelerie F, Ben-Baruch A, Burkhardt AM, Combadiere C, Farber JM, Graham GJ, et al. International Union of Basic and Clinical Pharmacology. [corrected]. LXXXIX. Update on the extended family of chemokine receptors and introducing a new nomenclature for atypical chemokine receptors. *Pharmacol Rev* 2013;**66**:1–79.
143. Gowans JL. The life-history of lymphocytes. *Br Med Bull* 1959;**15**:50–3.
144. Gowans JL, Knight EJ. The route of re-circulation of lymphocytes in the rat. *Proc R Soc Lond B Biol Sci* 1964;**159**:257–82.
145. Lowes MA, Bowcock AM, Krueger JG. Pathogenesis and therapy of psoriasis. *Nature* 2007;**445**:866–73.

146. Schlesinger M. Cell surface receptors and lymphocyte migration. *Immunol Commun* 1976;**5**:775–93.
147. Springer TA, Anderson DC. The importance of the Mac-1, LFA-1 glycoprotein family in monocyte and granulocyte adherence, chemotaxis, and migration into inflammatory sites: insights from an experiment of nature. *Ciba Found Symp* 1986;**118**:102–26.
148. Jalkanen S, Reichert RA, Gallatin WM, Bargatze RF, Weissman IL, Butcher EC. Homing receptors and the control of lymphocyte migration. *Immunol Rev* 1986;**91**:39–60.
149. Gallatin M, St John TP, Siegelman M, Reichert R, Butcher EC, Weissman IL. Lymphocyte homing receptors. *Cell* 1986;**44**:673–80.
150. MacKay CR. T-cell memory: the connection between function, phenotype and migration pathways. *Immunol Today* 1991;**12**:189–92.
151. Westermann J, Engelhardt B, Hoffmann JC. Migration of T cells in vivo: molecular mechanisms and clinical implications. *Ann Intern Med* 2001;**135**:279–95.
152. Woodland DL, Kohlmeier JE. Migration, maintenance and recall of memory T cells in peripheral tissues. *Nat Rev Immunol* 2009;**9**:153–61.
153. Tokoyoda K, Hauser AE, Nakayama T, Radbruch A. Organization of immunological memory by bone marrow stroma. *Nat Rev Immunol* 2010;**10**:193–200.
154. Hammarlund E, Lewis MW, Hansen SG, Strelow LI, Nelson JA, Sexton GJ, et al. Duration of antiviral immunity after smallpox vaccination. *Nat Med* 2003;**9**:1131–7.
155. Amanna IJ, Carlson NE, Slifka MK. Duration of humoral immunity to common viral and vaccine antigens. *N Engl J Med* 2007;**357**:1903–15.
156. Sallusto F, Lenig D, Forster R, Lipp M, Lanzavecchia A. Two subsets of memory T lymphocytes with distinct homing potentials and effector functions. *Nature* 1999;**401**:708–12.
157. Jameson SC, Masopust D. Diversity in T cell memory: an embarrassment of riches. *Immunity* 2009;**31**:859–71.
158. Ahmed R, Gray D. Immunological memory and protective immunity: understanding their relation. *Science* 1996;**272**:54–60.
159. Dorner T, Radbruch A. Antibodies and B cell memory in viral immunity. *Immunity* 2007;**27**:384–92.
160. Vieira P, Rajewsky K. Persistence of memory B cells in mice deprived of T cell help. *Int Immunol* 1990;**2**:487–94.
161. Radbruch A, Muehlinghaus G, Luger EO, Inamine A, Smith KG, Dorner T, et al. Competence and competition: the challenge of becoming a long-lived plasma cell. *Nat Rev Immunol* 2006;**6**:741–50.
162. Arstila TP, Casrouge A, Baron V, Even J, Kanellopoulos J, Kourilsky P. A direct estimate of the human alphabeta T cell receptor diversity. *Science* 1999;**286**:958–61.
163. Bachmann MF, Kundig TM, Kalberer CP, Hengartner H, Zinkernagel RM. How many specific B cells are needed to protect against a virus? *J Immunol* 1994;**152**:4235–41.
164. Janeway Jr CA. Approaching the asymptote? Evolution and revolution in immunology. *Cold Spring Harb Symp Quant Biol* 1989;**54**:1–13.
165. Lemaitre B, Nicolas E, Michaut L, Reichhart JM, Hoffmann JA. The dorsoventral regulatory gene cassette spatzle/Toll/cactus controls the potent antifungal response in Drosophila adults. *Cell* 1996;**86**:973–83.
166. Medzhitov R, Preston-Hurlburt P, Janeway Jr CA. A human homologue of the Drosophila Toll protein signals activation of adaptive immunity. *Nature* 1997;**388**:394–7.
167. Medzhitov R. Approaching the asymptote: 20 years later. *Immunity* 2009;**30**:766–75.
168. Poltorak A, He X, Smirnova I, Liu MY, Van HC, Du X, et al. Defective LPS signaling in C3H/HeJ and C57BL/10ScCr mice: mutations in Tlr4 gene. *Science* 1998;**282**:2085–8.

169. Miller JF. Immunological function of the thymus. *Lancet* 1961;**2**:748–9.
170. Miller JF. Role of the thymus in immunity. *Br Med J* 1963;**2**:459–64.
171. MacDonald HR, Hengartner H, Pedrazzini T. Intrathymic deletion of self-reactive cells prevented by neonatal anti-CD4 antibody treatment. *Nature* 1988;**335**:174–6.
172. Sha WC, Nelson CA, Newberry RD, Kranz DM, Russell JH, Loh DY. Positive and negative selection of an antigen receptor on T cells in transgenic mice. *Nature* 1988;**336**:73–6.
173. Hengartner H, Odermatt B, Schneider R, Schreyer M, Walle G, MacDonald HR, et al. Deletion of self-reactive T cells before entry into the thymus medulla. *Nature* 1988;**336**:388–90.
174. Zepp F, Staerz UD. Thymic selection process induced by hybrid antibodies. *Nature* 1988;**336**:473–5.
175. MacDonald HR, Lees RK, Schneider R, Zinkernagel RM, Hengartner H. Positive selection of CD4+ thymocytes controlled by MHC class II gene products. *Nature* 1988;**336**:471–3.
176. Lorenz RG, Allen PM. Thymic cortical epithelial cells can present self-antigens in vivo. *Nature* 1989;**337**:560–2.
177. Zuniga-Pflucker JC, Longo DL, Kruisbeek AM. Positive selection of CD4-CD8+ T cells in the thymus of normal mice. *Nature* 1989;**338**:76–8.
178. Bill J, Palmer E. Positive selection of CD4+ T cells mediated by MHC class II-bearing stromal cell in the thymic cortex. *Nature* 1989;**341**:649–51.
179. Finkel TH, Cambier JC, Kubo RT, Born WK, Marrack P, Kappler JW. The thymus has two functionally distinct populations of immature alpha beta + T cells: one population is deleted by ligation of alpha beta TCR. *Cell* 1989;**58**:1047–54.
180. Nagamine K, Peterson P, Scott HS, Kudoh J, Minoshima S, Heino M, et al. Positional cloning of the APECED gene. *Nat Genet* 1997;**17**:393–8.
181. Finnish-German APECED Consortium. An autoimmune disease, APECED, caused by mutations in a novel gene featuring two PHD-type zinc-finger domains. *Nat Genet* 1997;**17**:399–403.
182. Anderson MS, Venanzi ES, Klein L, Chen Z, Berzins SP, Turley SJ, et al. Projection of an immunological self shadow within the thymus by the aire protein. *Science* 2002;**298**:1395–401.
183. Liston A, Lesage S, Wilson J, Peltonen L, Goodnow CC. Aire regulates negative selection of organ-specific T cells. *Nat Immunol* 2003;**4**:350–4.
184. Kyewski B, Klein L. A central role for central tolerance. *Annu Rev Immunol* 2006;**24**:571–606.
185. Yurasov S, Nussenzweig MC. Regulation of autoreactive antibodies. *Curr Opin Rheumatol* 2007;**19**:421–6.
186. Katz DH, Benacerraf B. The regulatory influence of activated T cells on B cell responses to antigen. *Adv Immunol* 1972;**15**:1–94.
187. Gershon RK, Lance EM, Kondo K. Immuno-regulatory role of spleen localizing thymocytes. *J Immunol* 1974;**112**:546–54.
188. Baker PJ, Reed ND, Stashak PW, Amsbaugh DF, Prescott B. Regulation of the antibody response to type 3 pneumococcal polysaccharide. I. Nature of regulatory cells. *J Exp Med* 1973;**137**:1431–41.
189. Sakaguchi S, Sakaguchi N, Asano M, Itoh M, Toda M. Immunologic self-tolerance maintained by activated T cells expressing IL-2 receptor alpha-chains (CD25). Breakdown of a single mechanism of self-tolerance causes various autoimmune diseases. *J Immunol* 1995;**155**:1151–64.
190. Sakaguchi S. Regulatory T, cells: key controllers of immunologic self-tolerance. *Cell* 2000;**101**:455–8.
191. Takahashi T, Tagami T, Yamazaki S, Uede T, Shimizu J, Sakaguchi N, et al. Immunologic self-tolerance maintained by CD25(+)CD4(+) regulatory T cells constitutively expressing cytotoxic T lymphocyte-associated antigen 4. *J Exp Med* 2000;**192**:303–10.

192. Hori S, Nomura T, Sakaguchi S. Control of regulatory T cell development by the transcription factor Foxp3. *Science* 2003;**299**:1057–61.
193. Ohkura N, Kitagawa Y, Sakaguchi S. Development and maintenance of regulatory T cells. *Immunity* 2013;**38**:414–23.
194. Edinger M, Hoffmann P, Ermann J, Drago K, Fathman CG, Strober S, et al. $CD4^+CD25^+$ regulatory T cells preserve graft-versus-tumor activity while inhibiting graft-versus-host disease after bone marrow transplantation. *Nat Med* 2003;**9**:1144–50.
195. Koenecke C, Czeloth N, Bubke A, Schmitz S, Kissenpfennig A, Malissen B, et al. Alloantigen-specific de novo-induced Foxp3+ Treg revert in vivo and do not protect from experimental GVHD. *Eur J Immunol* 2009;**39**:3091–6.
196. Hoffmann P, Boeld TJ, Eder R, Huehn J, Floess S, Wieczorek G, et al. Loss of FOXP3 expression in natural human $CD4^+CD25^+$ regulatory T cells upon repetitive in vitro stimulation. *Eur J Immunol* 2009;**39**:1088–97.
197. Riley JL, June CH, Blazar BR. Human T regulatory cell therapy: take a billion or so and call me in the morning. *Immunity* 2009;**30**:656–65.
198. Sutherland EW, Oye I, Butcher RW. The action of epinephrine and the role of the adenyl cyclase system in hormone action. *Recent Prog Horm Res* 1965;**21**:623–46.
199. Rosenfeld MG, Abrass JB, Mendelsohn J, Ross BA, Boone RF, Garren LD. Control of transcription of RNA rich in polyadenylic acid in human lymphocytes. *Proc Natl Acad Sci U S A* 1972;**69**:2306–11.
200. Webb DR, Stites DP, Perlman JD, Luong D, Fudenberg HH. Lymphocyte activation: the dualistic effect of cAMP. *Biochem Biophys Res Commun* 1973;**53**:1002–8.
201. Zurier RB, Weissmann G, Hoffstein S, Kammerman S, Tai HH. Mechanisms of lysosomal enzyme release from human leukocytes. II. Effects of cAMP and cGMP, autonomic agonists, and agents which affect microtubule function. *J Clin Invest* 1974;**53**:297–309.
202. Bokoch GM, Gilman AG. Inhibition of receptor-mediated release of arachidonic acid by pertussis toxin. *Cell* 1984;**39**:301–8.
203. Monroe JG, Cambier JC. B cell activation. III. B cell plasma membrane depolarization and hyper-Ia antigen expression induced by receptor immunoglobulin cross-linking are coupled. *J Exp Med* 1983;**158**:1589–99.
204. Monroe JG, Niedel JE, Cambier JC. B cell activation. IV. Induction of cell membrane depolarization and hyper-I-A expression by phorbol diesters suggests a role for protein kinase C in murine B lymphocyte activation. *J Immunol* 1984;**132**:1472–8.
205. Bird PI, Trapani JA, Villadangos JA. Endolysosomal proteases and their inhibitors in immunity. *Nat Rev Immunol* 2009;**9**:871–82.
206. Alcover A, Thoulouze MI. Vesicle traffic to the immunological synapse: a multifunctional process targeted by lymphotropic viruses. *Curr Top Microbiol Immunol* 2010;**340**:191–207.
207. Lanzavecchia A, Sallusto F. From synapses to immunological memory: the role of sustained T cell stimulation. *Curr Opin Immunol* 2000;**12**:92–8.
208. Dustin ML, Cooper JA. The immunological synapse and the actin cytoskeleton: molecular hardware for T cell signaling. *Nat Immunol* 2000;**1**:23–9.
209. Nanney DL. Epigenetic control systems. *Proc Natl Acad Sci U S A* 1958;**44**:712–7.
210. Allfrey VG, Faulkner R, Mirsky AE. Acetylation and methylation of histones and their possible role in regulation of RNA synthesis. *Proc Natl Acad Sci U S A* 1964;**51**:786–94.
211. Friedman OM, Mahapatra GN, Stevenson R. The methylation of deoxyribonucleosides by diazomethane. *Biochim Biophys Acta* 1963;**68**:144–6.
212. Seyfert VL, McMahon SB, Glenn WD, Yellen AJ, Sukhatme VP, Cao XM, et al. Methylation of an immediate-early inducible gene as a mechanism for B cell tolerance induction. *Science* 1990;**250**:797–800.

213. Kohge T, Gohda E, Okamura T, Yamamoto I. Promotion of antigen-specific antibody production in murine B cells by a moderate increase in histone acetylation. *Biochem Pharmacol* 1998;**56**:1359–64.
214. Gasson JC, Ryden T, Bourgeois S. Role of de novo DNA methylation in the glucocorticoid resistance of a T-lymphoid cell line. *Nature* 1983;**302**:621–3.
215. Storb U, Arp B. Methylation patterns of immunoglobulin genes in lymphoid cells: correlation of expression and differentiation with undermethylation. *Proc Natl Acad Sci U S A* 1983;**80**:6642–6.
216. Peterlin BM, Gonwa TA, Stobo JD. Expression of HLA-DR by a human monocyte cell line is under transcriptional control. *J Mol Cell Immunol* 1984;**1**:191–200.
217. Richardson B, Kahn L, Lovett EJ, Hudson J. Effect of an inhibitor of DNA methylation on T cells. I. 5-Azacytidine induces T4 expression on T8+ T cells. *J Immunol* 1986;**137**:35–9.
218. Zheng Y, Josefowicz S, Chaudhry A, Peng XP, Forbush K, Rudensky AY. Role of conserved non-coding DNA elements in the Foxp3 gene in regulatory T-cell fate. *Nature* 2010;**463**:808–12.
219. Pan F, Yu H, Dang EV, Barbi J, Pan X, Grosso JF, et al. Eos mediates Foxp3-dependent gene silencing in CD4+ regulatory T cells. *Science* 2009;**325**:1142–6.
220. van Loosdregt J, Vercoulen Y, Guichelaar T, Gent YY, Beekman JM, Van BO, et al. Regulation of Treg functionality by acetylation-mediated Foxp3 protein stabilization. *Blood* 2010;**115**:965–74.
221. Lagos-Quintana M, Rauhut R, Lendeckel W, Tuschl T. Identification of novel genes coding for small expressed RNAs. *Science* 2001;**294**:853–8.
222. Lau NC, Lim LP, Weinstein EG, Bartel DP. An abundant class of tiny RNAs with probable regulatory roles in Caenorhabditis elegans. *Science* 2001;**294**:858–62.
223. Lee RC, Ambros V. An extensive class of small RNAs in Caenorhabditis elegans. *Science* 2001;**294**:862–4.
224. Chen CZ, Li L, Lodish HF, Bartel DP. MicroRNAs modulate hematopoietic lineage differentiation. *Science* 2004;**303**:83–6.
225. Fazi F, Rosa A, Fatica A, Gelmetti V, De Marchis ML, Nervi C, et al. A minicircuitry comprised of microRNA-223 and transcription factors NFI-A and C/EBPalpha regulates human granulopoiesis. *Cell* 2005;**123**:819–31.
226. Hochberg MC, Silman AJ, Smolen JS, Weinblatt ME, Weisman MH. *Rheumatology*. Philadelphia, PA: Elsevier Mosby; 2015.
227. Firestein GS, Kelley WN, Budd RC, Gabriel SE, McInnes IB, O'Dell JR. *Kelley's textbook of rheumatology*. Philadelphia, PA: Elsevier/Saunders; 2013.
228. Watts RA, Conaghan P, Denton C, Foster H, Isaacs J, Müller-Ladner U. *Oxford textbook of rheumatology*. Oxford: Oxford University Press; 2013.
229. Bijlsma JW, da Silva JA, Hachulla E, Doherty M, Cope A, Lioté F. *EULAR textbook on rheumatic diseases*. London: BMJ Group; 2012.
230. Coombs RRA, Mourant AE, Race RR. In-vivo isosensitisation of red cells in babies with haemolytic disease. *Lancet* 1946;**1**:264–6.
231. Coons AH, Creech HJ, Jones RN, Berliner E. The demonstration of pneumococcal antigens in tissues by the use of fluorescent antibody. *J Immunol* 1942;**45**:157–70.
232. Beutner EH. Immunofluorescent staining: the fluorescent antibody method. *Bacteriol Rev* 1961;**25**:49–76.
233. Freund J, Sommer HE, Walter AW. Immunization against malaria: vaccination of ducks with killed parasites incorporated with adjuvants. *Science* 1945;**102**:200–2.
234. Witebsky E, Rose NR, Terplan K, Paine JR, Egan RW. Chronic thyroiditis and autoimmunization. *J Am Med Assoc* 1957;**164**:1439–47.
235. Rivers TM, Sprunt DH, Berry GP. Observations on attempts to produce acute disseminated encephalomyelitis in monkeys. *J Exp Med* 1933;**58**:39–53.

236. Kabat EA, Wolf A, Bezer AE. Rapid production of acute disseminated encephalomyelitis in rhesus monkeys by injection of brain tissue with adjuvants. *Science* 1946;**104**:362–3.
237. Smadel JE. Experimental nephritis in rats induced by injection of anti-kidney serum. I. Preparation and immunological studies of nephrotoxin. *J Exp Med* 1936;**64**:921–42.
238. Smadel JE, Farr LE. Experimental nephritis in rats induced by injection of anti-kidney serum. II. Clinical and functional studies. *J Exp Med* 1937;**65**:527–40.
239. Solomon DH, Gardella JW. Nephrotoxic nephritis in rats; evidence for the glomerular origin of the kidney antigen. *J Exp Med* 1949;**90**:267–72.
240. Trentham DE, Townes AS, Kang AH. Autoimmunity to type II collagen an experimental model of arthritis. *J Exp Med* 1977;**146**:857–68.
241. Pearson CM. Development of arthritis, periarthritis and periostitis in rats given adjuvants. *Proc Soc Exp Biol Med* 1956;**91**:95–101.
242. Rose NR, Witebsky E. Studies on organ specificity. V. Changes in the thyroid glands of rabbits following active immunization with rabbit thyroid extracts. *J Immunol* 1956;**76**:417–27.
243. Grodsky GM, Feldman R, Toreson WE, Lee JC. Diabetes mellitus in rabbits immunized with insulin. *Diabetes* 1966;**15**:579–85.
244. Mackay IR, Gajdusek DC. An autoimmune reaction against human tissue antigens in certain acute and chronic diseases. II. Clinical correlations. *AMA Arch Intern Med* 1958;**101**:30–46.
245. Lange K, Gold MM, Weiner D, Simon V. Autoantibodies in human glomerulonephritis. *J Clin Invest* 1949;**28**:50–5.
246. Witebsky E, Rose NR. Studies on organ specificity. IV. Production of rabbit thyroid antibodies in the rabbit. *J Immunol* 1956;**76**:408–16.
247. Campbell PN, Doinach D, Hudson RV, Roitt IM. Auto-antibodies in Hashimoto's disease (lymphadenoid goitre). *Lancet* 1956;**271**:820–1.
248. Miescher P, Cooper NS, Benaceraff B. Experimental production of antinuclear antibodies. *J Immunol* 1960;**85**:27–36.
249. Deicher HR, Holman HR, Kunkel HG. Anti-cytoplasmic factors in the sera of patients with systemic lupus erythematosus and certain other diseases. *Arthritis Rheum* 1960;**3**:1–15.
250. Cepellini R, Polli E, Celada F. A DNA-reacting factor in serum of a patient with lupus erythematosus diffusus. *Proc Soc Exp Biol Med* 1957;**96**:572–4.
251. Waaler E. On the occurrence of a factor in human serum activating the specific agglutintion of sheep blood corpuscles. *Acta Pathol Microbiol Scand* 1940;**17**:172–88.
252. Rose HM, Ragan C. Differential agglutination of normal and sensitized sheep erythrocytes by sera of patients with rheumatoid arthritis. *Proc Soc Exp Biol Med* 1948;**68**:1–6.
253. Ziff M. The agglutination reaction in rheumatoid arthritis. *J Chronic Dis* 1957;**5**:644–67.
254. Epstein W, Johnson A, Ragan C. Observations on a precipitin reaction between serum of patients with rheumatoid arthritis and a preparation (Cohn fraction II) of human gamma globulin. *Proc Soc Exp Biol Med* 1956;**91**:235–7.
255. Bläß S, Engel JM, Burmester GR. The immunologic homunculus in rheumatoid arthritis. *Arthritis Rheum* 1999;**42**:2499–506.
256. Archelos JJ, Storch MK, Hartung HP. The role of B cells and autoantibodies in multiple sclerosis. *Ann Neurol* 2000;**47**:694–706.
257. Sherer Y, Gorstein A, Fritzler MJ, Shoenfeld Y. Autoantibody explosion in systemic lupus erythematosus: more than 100 different antibodies found in SLE patients. *Semin Arthritis Rheum* 2004;**34**:501–37.
258. Koulu L, Kusumi A, Steinberg MS, Klaus-Kovtun V, Stanley JR. Human autoantibodies against a desmosomal core protein in pemphigus foliaceus. *J Exp Med* 1984;**160**:1509–18.

259. Edgington TS, Glassock RJ, Dixon FJ. Autologous immune-complex pathogenesis of experimental allergic glomerulonephritis. *Science* 1967;**155**:1432–4.
260. Germuth Jr FG, Senterfit LB, Pollack AD. Immune complex disease. I. Experimental acute and chronic glomerulonephritis. *Johns Hopkins Med J* 1967;**120**:225–51.
261. Lerner RA, Glassock RJ, Dixon FJ. The role of anti-glomerular basement membrane antibody in the pathogenesis of human glomerulonephritis. *J Exp Med* 1967;**126**:989–1004.
262. Rankin JA, Matthay RA. Pulmonary renal syndromes. II. Etiology and pathogenesis. *Yale J Biol Med* 1982;**55**:11–26.
263. Lennon VA, Lindstrom JM, Seybold ME. Experimental autoimmune myasthenia: a model of myasthenia gravis in rats and guinea pigs. *J Exp Med* 1975;**141**:1365–75.
264. Aharonov A, Abramsky O, Tarrab-Hazdai R, Fuchs S. Humoral antibodies to acetylcholine receptor in patients with myasthenia gravis. *Lancet* 1975;**2**:340–2.
265. Meek JC, Jones AE, Lewis UJ, Vanderlaan WP. Charactertization of the long-acting thyroid stimulator of Graves' disease. *Proc Natl Acad Sci U S A* 1964;**52**:342–9.
266. Scott JS, Maddison PJ, Taylor PV, Esscher E, Scott O, Skinner RP. Connective-tissue disease, antibodies to ribonucleoprotein, and congenital heart block. *N Engl J Med* 1983;**309**:209–12.
267. Ho SY, Esscher E, Anderson RH, Michaelsson M. Anatomy of congenital complete heart block and relation to maternal anti-Ro antibodies. *Am J Cardiol* 1986;**58**:291–4.
268. Deng JS, Bair Jr LW, Shen-Schwarz S, Ramsey-Goldman R, Medsger Jr T. Localization of Ro (SS-A) antigen in the cardiac conduction system. *Arthritis Rheum* 1987;**30**:1232–8.
269. Krone KA, Allen KL, McCrae KR. Impaired fibrinolysis in the antiphospholipid syndrome. *Curr Rheumatol Rep* 2010;**12**:53–7.
270. Millet A, Pederzoli-Ribeil M, Guillevin L, Witko-Sarsat V, Mouthon L. Antineutrophil cytoplasmic antibody-associated vasculitides: is it time to split up the group? *Ann Rheum Dis* 2013;**72**:1273–9.
271. Cossio PM, Arana RM, Morteo OG, Hubscher O, Roux EB. Complement-fixing ability of antinuclear factors. Studies in adult and juvenile rheumatoid arthritis and systemic lupus erythematosus. *Ann Rheum Dis* 1971;**30**:640–4.
272. Wilson CL, Wojnarowska F, Dean D, Pasricha JS. IgG subclasses in pemphigus in Indian and UK populations. *Clin Exp Dermatol* 1993;**18**:226–30.
273. Herkel J, Heidrich B, Nieraad N, Wies I, Rother M, Lohse AW. Fine specificity of autoantibodies to soluble liver antigen and liver/pancreas. *Hepatology* 2002;**35**:403–8.
274. Boe AS, Bredholt G, Knappskog PM, Hjelmervik TO, Mellgren G, Winqvist O, et al. Autoantibodies against 21-hydroxylase and side-chain cleavage enzyme in autoimmune Addison's disease are mainly immunoglobulin G1. *Eur J Endocrinol* 2004;**150**:49–56.
275. Williams Jr RC, Emmons JD, Yunis EJ. Studies of human sera with cytotoxic activity. *J Clin Invest* 1971;**50**:1514–24.
276. Smolen JS, Steiner G. Are autoantibodies active players or epiphenomena? *Curr Opin Rheumatol* 1998;**10**:201–6.
277. Naparstek Y, Plotz PH. The role of autoantibodies in autoimmune disease. *Annu Rev Immunol* 1993;**11**:79–104.
278. Lucchinetti C, Bruck W, Parisi J, Scheithauer B, Rodriguez M, Lassmann H. Heterogeneity of multiple sclerosis lesions: implications for the pathogenesis of demyelination. *Ann Neurol* 2000;**47**:707–17.
279. Magalhaes R, Stiehl P, Morawietz L, Berek C, Krenn V. Morphological and molecular pathology of the B cell response in synovitis of rheumatoid arthritis. *Virchows Arch* 2002;**441**:415–27.

280. Yurasov S, Wardemann H, Hammersen J, Tsuiji M, Meffre E, Pascual V, et al. Defective B cell tolerance checkpoints in systemic lupus erythematosus. *J Exp Med* 2005;**201**:703–11.
281. Samuels J, Ng YS, Coupillaud C, Paget D, Meffre E. Impaired early B cell tolerance in patients with rheumatoid arthritis. *J Exp Med* 2005;**201**:1659–67.
282. Yurasov S, Tiller T, Tsuiji M, Velinzon K, Pascual V, Wardemann H, et al. Persistent expression of autoantibodies in SLE patients in remission. *J Exp Med* 2006;**203**:2255–61.
283. Prineas JW. Multiple sclerosis: presence of lymphatic capillaries and lymphoid tissue in the brain and spinal cord. *Science* 1979;**203**:1123–5.
284. Roxanis I, Micklem K, McConville J, Newsom-Davis J, Willcox N. Thymic myoid cells and germinal center formation in myasthenia gravis; possible roles in pathogenesis. *J Neuroimmunol* 2002;**125**:185–97.
285. Stott DI, Hiepe F, Hummel M, Steinhauser G, Berek C. Antigen-driven clonal proliferation of B cells within the target tissue of an autoimmune disease. The salivary glands of patients with Sjogren's syndrome. *J Clin Invest* 1998;**102**:938–46.
286. Lopez De Padilla CM, Vallejo AN, Lacomis D, McNallan K, Reed AM. Extranodal lymphoid microstructures in inflamed muscle and disease severity of new-onset juvenile dermatomyositis. *Arthritis Rheum* 2009;**60**:1160–72.
287. Grabner R, Lotzer K, Dopping S, Hildner M, Radke D, Beer M, et al. Lymphotoxin beta receptor signaling promotes tertiary lymphoid organogenesis in the aorta adventitia of aged ApoE−/− mice. *J Exp Med* 2009;**206**:233–48.
288. Qin Y, Duquette P, Zhang Y, Talbot P, Poole R, Antel J. Clonal expansion and somatic hypermutation of V(H) genes of B cells from cerebrospinal fluid in multiple sclerosis. *J Clin Invest* 1998;**102**:1045–50.
289. Owens GP, Ritchie AM, Burgoon MP, Williamson RA, Corboy JR, Gilden DH. Single-cell repertoire analysis demonstrates that clonal expansion is a prominent feature of the B cell response in multiple sclerosis cerebrospinal fluid. *J Immunol* 2003;**171**:2725–33.
290. Monson NL, Brezinschek HP, Brezinschek RI, Mobley A, Vaughan GK, Frohman EM, et al. Receptor revision and atypical mutational characteristics in clonally expanded B cells from the cerebrospinal fluid of recently diagnosed multiple sclerosis patients. *J Neuroimmunol* 2005;**158**:170–81.
291. Marrack P, Kappler JW. The role of H-2-linked genes in helper T cell function. VII. Expression of I region and immune response genes by B cells in bystander help assays. *J Exp Med* 1980;**152**:1274–88.
292. McKean DJ, Infante AJ, Nilson A, Kimoto M, Fathman CG, Walker E, et al. Major histocompatibility complex-restricted antigen presentation to antigen-reactive T cells by B lymphocyte tumor cells. *J Exp Med* 1981;**154**:1419–31.
293. Glimcher LH, Kim KJ, Green I, Paul WE. Ia antigen-bearing B cell tumor lines can present protein antigen and alloantigen in a major histocompatibility complex-restricted fashion to antigen-reactive T cells. *J Exp Med* 1982;**155**:445–59.
294. Lanzavecchia A. Antigen-specific interaction between T and B cells. *Nature* 1985;**314**:537–9.
295. Lund FE, Randall TD. Effector and regulatory B cells: modulators of CD4(+) T cell immunity. *Nat Rev Immunol* 2010;**10**:236–47.
296. Shlomchik MJ. Sites and stages of autoreactive B cell activation and regulation. *Immunity* 2008;**28**:18–28.
297. Edwards JC, Cambridge G. B-cell targeting in rheumatoid arthritis and other autoimmune diseases. *Nat Rev Immunol* 2006;**6**:394–403.
298. McFarlin DE, Hsu SC, Slemenda SB, Chou FC, Kibler RF. The immune response against myelin basic protein in two strains of rat with different genetic capacity to develop experimental allergic encephalomyelitis. *J Exp Med* 1975;**141**:72–81.

299. Esquivel PS, Rose NR, Kong YC. Induction of autoimmunity in good and poor responder mice with mouse thyroglobulin and lipopolysaccharide. *J Exp Med* 1977;**145**:1250–63.
300. Bernard CC, Mitchell GF, Leydon J, Bargerbos A. Experimental autoimmune orchitis in T-cell-deficient mice. *Int Arch Allergy Appl Immunol* 1978;**56**:256–63.
301. Roder JC, Bell DA, Singhal SK. T cell activation and cellular cooperation in autoimmune NZB/NZW F hybrid mice. *J Immunol* 1975;**115**:466–72.
302. Wekerle H. Immunological T-cell memory in the in vitro-induced experimental autoimmune orchitis: specificity of the reaction and tissue distribution of the autoantigens. *J Exp Med* 1978;**147**:233–50.
303. Gershon RK, Eardley DD, Naidorf K, Cantor H. Association of defective feedback suppressor T cell activity with autoimmunity in NZB mice. *Arthritis Rheum* 1978;**21**:S180–4.
304. Bolton WK, Benton FR, Lobo PI. Requirement of functional T-cells in the production of autoimmune glomerulotubular nephropathy in mice. *Clin Exp Immunol* 1978;**33**:474–7.
305. Horowitz SD, Borcherding W, Bargman GJ. Suppressor T cell function in diabetes mellitus. *Lancet* 1977;**310**:1291.
306. Rees AD, Lombardi G, Scoging A, Barber L, Mitchell D, Lamb J, et al. Functional evidence for the recognition of endogenous peptides by autoreactive T cell clones. *Int Immunol* 1989;**1**:624–30.
307. Lacour M, Rudolphi U, Schlesier M, Peter HH. Type II collagen-specific human T cell lines established from healthy donors. *Eur J Immunol* 1990;**20**:931–4.
308. Tarkowski A, Klareskog L, Carlsten H, Herberts P, Koopman WJ. Secretion of antibodies to types I and II collagen by synovial tissue cells in patients with rheumatoid arthritis. *Arthritis Rheum* 1989;**32**:1087–92.
309. Burkhardt H, Koller T, Engstrom A, Nandakumar KS, Turnay J, Kraetsch HG, et al. Epitope-specific recognition of type II collagen by rheumatoid arthritis antibodies is shared with recognition by antibodies that are arthritogenic in collagen-induced arthritis in the mouse. *Arthritis Rheum* 2002;**46**:2339–48.
310. Haag S, Schneider N, Mason DE, Tuncel J, Andersson IE, Peters EC, et al. Identification of new citrulline-specific autoantibodies, which bind to human arthritic cartilage, by mass spectrometric analysis of citrullinated type II collagen. *Arthritis Rheumatol* 2014;**66**:1440–9.
311. Szabo SJ, Kim ST, Costa GL, Zhang X, Fathman CG, Glimcher LH. A novel transcription factor, T-bet, directs Th1 lineage commitment. *Cell* 2000;**100**:655–69.
312. Zheng W, Flavell RA. The transcription factor GATA-3 is necessary and sufficient for Th2 cytokine gene expression in CD4 T cells. *Cell* 1997;**89**:587–96.
313. Carlsten H, Nilsson N, Jonsson R, Backman K, Holmdahl R, Tarkowski A. Estrogen accelerates immune complex glomerulonephritis but ameliorates T cell-mediated vasculitis and sialadenitis in autoimmune MRL lpr/lpr mice. *Cell Immunol* 1992;**144**:190–202.
314. Steinman L. A brief history of T(H)17, the first major revision in the T(H)1/T(H)2 hypothesis of T cell-mediated tissue damage. *Nat Med* 2007;**13**:139–45.
315. Wildin RS, Ramsdell F, Peake J, Faravelli F, Casanova JL, Buist N, et al. X-linked neonatal diabetes mellitus, enteropathy and endocrinopathy syndrome is the human equivalent of mouse scurfy. *Nat Genet* 2001;**27**:18–20.
316. Bennett CL, Christie J, Ramsdell F, Brunkow ME, Ferguson PJ, Whitesell L, et al. The immune dysregulation, polyendocrinopathy, enteropathy, X-linked syndrome (IPEX) is caused by mutations of FOXP3. *Nat Genet* 2001;**27**:20–1.
317. Cua DJ, Sherlock J, Chen Y, Murphy CA, Joyce B, Seymour B, et al. Interleukin-23 rather than interleukin-12 is the critical cytokine for autoimmune inflammation of the brain. *Nature* 2003;**421**:744–8.

318. Murphy CA, Langrish CL, Chen Y, Blumenschein W, McClanahan T, Kastelein RA, et al. Divergent pro- and antiinflammatory roles for IL-23 and IL-12 in joint autoimmune inflammation. *J Exp Med* 2003;**198**:1951–7.
319. Aggarwal S, Ghilardi N, Xie MH, de Sauvage FJ, Gurney AL. Interleukin-23 promotes a distinct CD4 T cell activation state characterized by the production of interleukin-17. *J Biol Chem* 2003;**278**:1910–4.
320. Nakae S, Nambu A, Sudo K, Iwakura Y. Suppression of immune induction of collagen-induced arthritis in IL-17-deficient mice. *J Immunol* 2003;**171**:6173–7.
321. Komiyama Y, Nakae S, Matsuki T, Nambu A, Ishigame H, Kakuta S, et al. IL-17 plays an important role in the development of experimental autoimmune encephalomyelitis. *J Immunol* 2006;**177**:566–73.
322. Koenders MI, Lubberts E, Oppers-Walgreen B, van den BL, Helsen MM, Di Padova FE, et al. Blocking of interleukin-17 during reactivation of experimental arthritis prevents joint inflammation and bone erosion by decreasing RANKL and interleukin-1. *Am J Pathol* 2005;**167**:141–9.
323. Koenders MI, Lubberts E, Oppers-Walgreen B, van den BL, Helsen MM, Kolls JK, et al. Induction of cartilage damage by overexpression of T cell interleukin-17A in experimental arthritis in mice deficient in interleukin-1. *Arthritis Rheum* 2005;**52**:975–83.
324. Ivanov II, McKenzie BS, Zhou L, Tadokoro CE, Lepelley A, Lafaille JJ, et al. The orphan nuclear receptor RORgammat directs the differentiation program of proinflammatory IL-17+ T helper cells. *Cell* 2006;**126**:1121–33.
325. Harrington LE, Mangan PR, Weaver CT. Expanding the effector CD4 T-cell repertoire: the Th17 lineage. *Curr Opin Immunol* 2006;**18**:349–56.
326. Littman DR, Rudensky AY. Th17 and regulatory T cells in mediating and restraining inflammation. *Cell* 2010;**140**:845–58.
327. Carter PH, Zhao Q. Clinically validated approaches to the treatment of autoimmune diseases. *Expert Opin Investig Drugs* 2010;**19**:195–213.
328. Burmester GR, Emmrich F. Anti-CD4 therapy in rheumatoid arthritis. *Clin Exp Rheumatol* 1993;**11**(Suppl. 8):S139–45.
329. Straub RH. The complex role of estrogens in inflammation. *Endocr Rev* 2007;**28**:521–74.
330. Thomas L. *Clinical laboratory diagnostics: use and assessment of clinical laboratory results*. Frankfurt, Germany: TH-Books; 1998.
331. Dubois PC, van Heel DA. Translational mini-review series on the immunogenetics of gut disease: immunogenetics of coeliac disease. *Clin Exp Immunol* 2008;**153**:162–73.
332. Tomfohrde J, Silverman A, Barnes R, Fernandez-Vina MA, Young M, Lory D, et al. Gene for familial psoriasis susceptibility mapped to the distal end of human chromosome 17q. *Science* 1994;**264**:1141–5.
333. Baranzini SE. The genetics of autoimmune diseases: a networked perspective. *Curr Opin Immunol* 2009;**21**:596–605.
334. Onengut-Gumuscu S, Concannon P. Recent advances in the immunogenetics of human type 1 diabetes. *Curr Opin Immunol* 2006;**18**:634–8.
335. Begovich AB, Carlton VE, Honigberg LA, Schrodi SJ, Chokkalingam AP, Alexander HC, et al. A missense single-nucleotide polymorphism in a gene encoding a protein tyrosine phosphatase (PTPN22) is associated with rheumatoid arthritis. *Am J Hum Genet* 2004;**75**:330–7.
336. Hutchinson D, Shepstone L, Moots R, Lear JT, Lynch MP. Heavy cigarette smoking is strongly associated with rheumatoid arthritis (RA), particularly in patients without a family history of RA. *Ann Rheum Dis* 2001;**60**:223–7.
337. Panayi GS. Unified concept of cell-mediated immune reactions. *Br Med J* 1970;**2**:656–8.

338. Rosenthal AS, Barcinski MA, Rosenwasser LJ. Function of macrophages in genetic control of immune responsiveness. *Fed Proc* 1978;**37**:79–85.
339. Sakashita T. Antitumor effect of lymphocyte and macrophage from mice immunized with Ehrlich ascites tumor cells. *Mie Med J* 1971;**20**:227–41.
340. Cruse JM, Whitten HD, Lewis GK, Watson ES. Facilitation of macrophage-mediated destruction of allogeneic fibrosarcoma cells by tumor-enhancing IgG 2 in vitro. *Transplant Proc* 1973;**5**:961–7.
341. Unkeless JC, Fleit H, Mellman IS. Structural aspects and heterogeneity of immunoglobulin Fc receptors. *Adv Immunol* 1981;**31**:247–70.
342. Adams DO, Hamilton TA. The cell biology of macrophage activation. *Annu Rev Immunol* 1984;**2**:283–318.
343. Degre M, Sonnenfeld G, Rollag H, Morland B. Effect of gamma interferon preparations on in vitro phagocytosis and degradation of Escherichia coli by mouse peritoneal macrophages. *J Interferon Res* 1981;**1**:505–12.
344. Kleinschmidt WJ, Schultz RM. Similarities of murine gamma interferon and the lymphokine that renders macrophages cytotoxic. *J Interferon Res* 1982;**2**:291–9.
345. Steeg PS, Moore RN, Johnson HM, Oppenheim JJ. Regulation of murine macrophage Ia antigen expression by a lymphokine with immune interferon activity. *J Exp Med* 1982;**156**:1780–93.
346. Van Rooijen N, van NR. Elimination of phagocytic cells in the spleen after intravenous injection of liposome-encapsulated dichloromethylene diphosphonate. An enzyme-histochemical study. *Cell Tissue Res* 1984;**238**:355–8.
347. Huitinga I, Van RN, de Groot CJ, Uitdehaag BM, Dijkstra CD. Suppression of experimental allergic encephalomyelitis in Lewis rats after elimination of macrophages. *J Exp Med* 1990;**172**:1025–33.
348. Jun HS, Yoon CS, Zbytnuik L, Van RN, Yoon JW. The role of macrophages in T cell-mediated autoimmune diabetes in nonobese diabetic mice. *J Exp Med* 1999;**189**:347–58.
349. Flora L. Comparative antiinflammatory and bone protective effects of two diphosphonates in adjuvant arthritis. *Arthritis Rheum* 1979;**22**:340–6.
350. Hartung HP, Toyka KV. T-cell and macrophage activation in experimental autoimmune neuritis and Guillain-Barre syndrome. *Ann Neurol* 1990;**27**(Suppl.):S57–63.
351. Schreiner GF. The role of the macrophage in glomerular injury. *Semin Nephrol* 1991;**11**:268–75.
352. Burmester GR, Stuhlmuller B, Rittig M. The monocyte/macrophage system in arthritis–leopard tank or Trojan horse? *Scand J Rheumatol Suppl* 1995;**101**:77–82.
353. Sherry B, Horii Y, Manogue KR, Widmer U, Cerami A. Macrophage inflammatory proteins 1 and 2: an overview. *Cytokines* 1992;**4**:117–30.
354. Mantovani A, Gray PA, Van DJ, Sozzani S. Macrophage-derived chemokine (MDC). *J Leukoc Biol* 2000;**68**:400–4.
355. Mosser DM, Edwards JP. Exploring the full spectrum of macrophage activation. *Nat Rev Immunol* 2008;**8**:958–69.
356. Lacey DL, Timms E, Tan HL, Kelley MJ, Dunstan CR, Burgess T, et al. Osteoprotegerin ligand is a cytokine that regulates osteoclast differentiation and activation. *Cell* 1998;**93**:165–76.
357. Yasuda H, Shima N, Nakagawa N, Yamaguchi K, Kinosaki M, Mochizuki S, et al. Osteoclast differentiation factor is a ligand for osteoprotegerin/osteoclastogenesis-inhibitory factor and is identical to TRANCE/RANKL. *Proc Natl Acad Sci U S A* 1998;**95**:3597–602.
358. Yun TJ, Chaudhary PM, Shu GL, Frazer JK, Ewings MK, Schwartz SM, et al. OPG/FDCR-1, a TNF receptor family member, is expressed in lymphoid cells and is up-regulated by ligating CD40. *J Immunol* 1998;**161**:6113–21.

359. Kotake S, Udagawa N, Takahashi N, Matsuzaki K, Itoh K, Ishiyama S, et al. IL-17 in synovial fluids from patients with rheumatoid arthritis is a potent stimulator of osteoclastogenesis. *J Clin Invest* 1999;**103**:1345–52.
360. Kong YY, Feige U, Sarosi I, Bolon B, Tafuri A, Morony S, et al. Activated T cells regulate bone loss and joint destruction in adjuvant arthritis through osteoprotegerin ligand. *Nature* 1999;**402**:304–9.
361. MacNaul KL, Hutchinson NI, Parsons JN, Bayne EK, Tocci MJ. Analysis of IL-1 and TNF-alpha gene expression in human rheumatoid synoviocytes and normal monocytes by in situ hybridization. *J Immunol* 1990;**145**:4154–66.
362. Brennan FM, Chantry D, Jackson AM, Maini RN, Feldmann M. Cytokine production in culture by cells isolated from the synovial membrane. *J Autoimmun* 1989;**2**(Suppl.):177–86.
363. Brennan FM, Chantry D, Jackson A, Maini R, Feldmann M. Inhibitory effect of TNF alpha antibodies on synovial cell interleukin-1 production in rheumatoid arthritis. *Lancet* 1989;**2**:244–7.
364. Elliott MJ, Maini RN, Feldmann M, Long-Fox A, Charles P, Katsikis P, et al. Treatment of rheumatoid arthritis with chimeric monoclonal antibodies to tumor necrosis factor alpha. *Arthritis Rheum* 1993;**36**:1681–90.
365. Bresnihan B, varo-Gracia JM, Cobby M, Doherty M, Domljan Z, Emery P, et al. Treatment of rheumatoid arthritis with recombinant human interleukin-1 receptor antagonist. *Arthritis Rheum* 1998;**41**:2196–204.
366. Nishimoto N, Yoshizaki K, Miyasaka N, Yamamoto K, Kawai S, Takeuchi T, et al. Treatment of rheumatoid arthritis with humanized anti-interleukin-6 receptor antibody: a multicenter, double-blind, placebo-controlled trial. *Arthritis Rheum* 2004;**50**:1761–9.
367. van den Berg WB, Miossec P. IL-17 as a future therapeutic target for rheumatoid arthritis. *Nat Rev Rheumatol* 2009;**5**:549–53.
368. Genovese MC, Durez P, Richards HB, Supronik J, Dokoupilova E, Mazurov V, et al. Efficacy and safety of secukinumab in patients with rheumatoid arthritis: a phase II, dose-finding, double-blind, randomised, placebo controlled study. *Ann Rheum Dis* 2013;**72**:863–9.
369. Mellman I, Steinman RM. Dendritic cells: specialized and regulated antigen processing machines. *Cell* 2001;**106**:255–8.
370. Banchereau J, Steinman RM. Dendritic cells and the control of immunity. *Nature* 1998;**392**:245–52.
371. Liu K, Victora GD, Schwickert TA, Guermonprez P, Meredith MM, Yao K, et al. In vivo analysis of dendritic cell development and homeostasis. *Science* 2009;**324**:392–7.
372. Corcoran L, Ferrero I, Vremec D, Lucas K, Waithman J, O'Keeffe M, et al. The lymphoid past of mouse plasmacytoid cells and thymic dendritic cells. *J Immunol* 2003;**170**:4926–32.
373. de Jong EC, Smits HH, Kapsenberg ML. Dendritic cell-mediated T cell polarization. *Springer Semin Immunopathol* 2005;**26**:289–307.
374. Bayry J, Lacroix-Desmazes S, Kazatchkine MD, Hermine O, Tough DF, Kaveri SV. Modulation of dendritic cell maturation and function by B lymphocytes. *J Immunol* 2005;**175**:15–20.
375. Lebre MC, Jongbloed SL, Tas SW, Smeets TJ, McInnes IB, Tak PP. Rheumatoid arthritis synovium contains two subsets of CD83-DC-LAMP- dendritic cells with distinct cytokine profiles. *Am J Pathol* 2008;**172**:940–50.
376. Cauli A, Pitzalis C, Yanni G, Awad M, Panayi GS. CD1 expression in psoriatic and rheumatoid arthritis. *Rheumatology (Oxford)* 2000;**39**:666–73.
377. Page G, Miossec P. RANK and RANKL expression as markers of dendritic cell-T cell interactions in paired samples of rheumatoid synovium and lymph nodes. *Arthritis Rheum* 2005;**52**:2307–12.

378. Radstake TR, van Lent PL, Pesman GJ, Blom AB, Sweep FG, Ronnelid J, et al. High production of proinflammatory and Th1 cytokines by dendritic cells from patients with rheumatoid arthritis, and down regulation upon FcgammaR triggering. *Ann Rheum Dis* 2004;**63**:696–702.
379. Rivollier A, Mazzorana M, Tebib J, Piperno M, Aitsiselmi T, Rabourdin-Combe C, et al. Immature dendritic cell transdifferentiation into osteoclasts: a novel pathway sustained by the rheumatoid arthritis microenvironment. *Blood* 2004;**104**:4029–37.
380. Lande R, Giacomini E, Serafini B, Rosicarelli B, Sebastiani GD, Minisola G, et al. Characterization and recruitment of plasmacytoid dendritic cells in synovial fluid and tissue of patients with chronic inflammatory arthritis. *J Immunol* 2004;**173**:2815–24.
381. Ueno H, Klechevsky E, Morita R, Aspord C, Cao T, Matsui T, et al. Dendritic cell subsets in health and disease. *Immunol Rev* 2007;**219**:118–42.
382. Lebre MC, Tak PP. Dendritic cells in rheumatoid arthritis: which subset should be used as a tool to induce tolerance? *Hum Immunol* 2009;**70**:321–4.
383. Ezzelarab M, Thomson AW. Tolerogenic dendritic cells and their role in transplantation. *Semin Immunol* 2011;**23**:252–63.
384. Phillips BE, Giannoukakis N, Trucco M. Dendritic cell mediated therapy for immunoregulation of type 1 diabetes mellitus. *Pediatr Endocrinol Rev* 2008;**5**:873–9.
385. Hilkens CM, Isaacs JD. Tolerogenic dendritic cell therapy for rheumatoid arthritis: where are we now? *Clin Exp Immunol* 2013;**172**:148–57.
386. Lünemann A, Lünemann JD, Münz C. Regulatory NK-cell functions in inflammation and autoimmunity. *Mol Med* 2009;**15**:352–8.
387. von Bubnoff D, Andres E, Hentges F, Bieber T, Michel T, Zimmer J. Natural killer cells in atopic and autoimmune diseases of the skin. *J Allergy Clin Immunol* 2010;**125**:60–8.
388. Vivier E, Tomasello E, Baratin M, Walzer T, Ugolini S. Functions of natural killer cells. *Nat Immunol* 2008;**9**:503–10.
389. Godfrey DI, Stankovic S, Baxter AG. Raising the NKT cell family. *Nat Immunol* 2010;**11**:197–206.
390. O'Leary JG, Goodarzi M, Drayton DL, von Andrian UH. T cell- and B cell-independent adaptive immunity mediated by natural killer cells. *Nat Immunol* 2006;**7**:507–16.
391. Sun JC, Beilke JN, Lanier LL. Adaptive immune features of natural killer cells. *Nature* 2009;**457**:557–61.
392. Cooper MA, Elliott JM, Keyel PA, Yang L, Carrero JA, Yokoyama WM. Cytokine-induced memory-like natural killer cells. *Proc Natl Acad Sci U S A* 2009;**106**:1915–9.
393. Paust S, Gill HS, Wang BZ, Flynn MP, Moseman EA, Senman B, et al. Critical role for the chemokine receptor CXCR6 in NK cell-mediated antigen-specific memory of haptens and viruses. *Nat Immunol* 2010;**11**:1127–35.
394. Burg ND, Pillinger MH. The neutrophil: function and regulation in innate and humoral immunity. *Clin Immunol* 2001;**99**:7–17.
395. Peng SL. Neutrophil apoptosis in autoimmunity. *J Mol Med* 2006;**84**:122–5.
396. Kain R, Exner M, Brandes R, Ziebermayr R, Cunningham D, Alderson CA, et al. Molecular mimicry in pauci-immune focal necrotizing glomerulonephritis. *Nat Med* 2008;**14**:1088–96.
397. Watanabe-Fukunaga R, Brannan CI, Copeland NG, Jenkins NA, Nagata S. Lymphoproliferation disorder in mice explained by defects in Fas antigen that mediates apoptosis. *Nature* 1992;**356**:314–7.
398. Drappa J, Vaishnaw AK, Sullivan KE, Chu JL, Elkon KB. Fas gene mutations in the Canale-Smith syndrome, an inherited lymphoproliferative disorder associated with autoimmunity. *N Engl J Med* 1996;**335**:1643–9.

399. Krammer PH, Kaminski M, Kiessling M, Gulow K. No life without death. *Adv Cancer Res* 2007;**97**:111–38.
400. Sheriff A, Gaipl US, Voll RE, Kalden JR, Herrmann M. Apoptosis and systemic lupus erythematosus. *Rheum Dis Clin North Am* 2004;**30**:505–27.
401. Lleo A, Selmi C, Invernizzi P, Podda M, Gershwin ME. The consequences of apoptosis in autoimmunity. *J Autoimmun* 2008;**31**:257–62.
402. Chaurio RA, Janko C, Munoz LE, Frey B, Herrmann M, Gaipl US. Phospholipids: key players in apoptosis and immune regulation. *Molecules* 2009;**14**:4892–914.
403. Ravichandran KS, Lorenz U. Engulfment of apoptotic cells: signals for a good meal. *Nat Rev Immunol* 2007;**7**:964–74.
404. Artemiadis AK, Anagnostouli MC. Apoptosis of oligodendrocytes and post-translational modifications of myelin basic protein in multiple sclerosis: possible role for the early stages of multiple sclerosis. *Eur Neurol* 2010;**63**:65–72.
405. Schmidt E, Waschke J. Apoptosis in pemphigus. *Autoimmun Rev* 2009;**8**:533–7.
406. Peng SL. Fas (CD95)-related apoptosis and rheumatoid arthritis. *Rheumatology (Oxford)* 2006;**45**:26–30.
407. Manganelli P, Fietta P. Apoptosis and Sjogren syndrome. *Semin Arthritis Rheum* 2003;**33**:49–65.
408. Kaneider NC, Leger AJ, Kuliopulos A. Therapeutic targeting of molecules involved in leukocyte-endothelial cell interactions. *FEBS J* 2006;**273**:4416–24.
409. Takeuchi O, Akira S. Pattern recognition receptors and inflammation. *Cell* 2010;**140**:805–20.
410. Matzinger P. The danger model: a renewed sense of self. *Science* 2002;**296**:301–5.
411. McCormack WJ, Parker AE, O'Neill LA. Toll-like receptors and NOD-like receptors in rheumatic diseases. *Arthritis Res Ther* 2009;**11**:243.
412. Guy B. The perfect mix: recent progress in adjuvant research. *Nat Rev Microbiol* 2007;**5**:505–17.
413. Hugot JP, Chamaillard M, Zouali H, Lesage S, Cezard JP, Belaiche J, et al. Association of NOD2 leucine-rich repeat variants with susceptibility to Crohn's disease. *Nature* 2001;**411**:599–603.
414. Ogura Y, Bonen DK, Inohara N, Nicolae DL, Chen FF, Ramos R, et al. A frameshift mutation in NOD2 associated with susceptibility to Crohn's disease. *Nature* 2001;**411**:603–6.
415. Holler E, Rogler G, Herfarth H, Brenmoehl J, Wild PJ, Hahn J, et al. Both donor and recipient NOD2/CARD15 mutations associate with transplant-related mortality and GvHD following allogeneic stem cell transplantation. *Blood* 2004;**104**:889–94.
416. Dorhoi A, Desel C, Yeremeev V, Pradl L, Brinkmann V, Mollenkopf HJ, et al. The adaptor molecule CARD9 is essential for tuberculosis control. *J Exp Med* 2010;**207**:777–92.
417. Gringhuis SI, den Dunnen J, Litjens M, van der Vlist M, Geijtenbeek TB. Carbohydrate-specific signaling through the DC-SIGN signalosome tailors immunity to Mycobacterium tuberculosis, HIV-1 and Helicobacter pylori. *Nat Immunol* 2009;**10**:1081–8.
418. Lee HM, Yuk JM, Shin DM, Jo EK. Dectin-1 is inducible and plays an essential role for mycobacteria-induced innate immune responses in airway epithelial cells. *J Clin Immunol* 2009;**29**:795–805.
419. Maeda N, Nigou J, Herrmann JL, Jackson M, Amara A, Lagrange PH, et al. The cell surface receptor DC-SIGN discriminates between Mycobacterium species through selective recognition of the mannose caps on lipoarabinomannan. *J Biol Chem* 2003;**278**:5513–6.
420. Raschi E, Borghi MO, Grossi C, Broggini V, Pierangeli S, Meroni PL. Toll-like receptors: another player in the pathogenesis of the anti-phospholipid syndrome. *Lupus* 2008;**17**:937–42.
421. Testro AG, Visvanathan K. Toll-like receptors and their role in gastrointestinal disease. *J Gastroenterol Hepatol* 2009;**24**:943–54.
422. Guandalini S, Setty M. Celiac disease. *Curr Opin Gastroenterol* 2008;**24**:707–12.

423. Lien E, Zipris D. The role of Toll-like receptor pathways in the mechanism of type 1 diabetes. *Curr Mol Med* 2009;**9**:52–68.
424. Bhat R, Steinman L. Innate and adaptive autoimmunity directed to the central nervous system. *Neuron* 2009;**64**:123–32.
425. Brentano F, Kyburz D, Schorr O, Gay R, Gay S. The role of Toll-like receptor signalling in the pathogenesis of arthritis. *Cell Immunol* 2005;**233**:90–6.
426. Andreakos E, Sacre S, Foxwell BM, Feldmann M. The toll-like receptor-nuclear factor kappaB pathway in rheumatoid arthritis. *Front Biosci* 2005;**10**:2478–88.
427. Avalos AM, Busconi L, Marshak-Rothstein A. Regulation of autoreactive B cell responses to endogenous TLR ligands. *Autoimmunity* 2010;**43**:76–83.
428. Schilling JA. Wound healing. *Physiol Rev* 1968;**48**:374–423.
429. Ross R, Odland G. Human wound repair. II. Inflammatory cells, epithelial-mesenchymal interrelations, and fibrogenesis. *J Cell Biol* 1968;**39**:152–68.
430. Fassbender HG. Two different types of pathologic anatomical tissue processes in primary chronic polyarthritis. *Verh Dtsch Ges Rheumatol* 1969;**1**:222–9.
431. Müller-Ladner U, Kriegsmann J, Franklin BN, Matsumoto S, Geiler T, Gay RE, et al. Synovial fibroblasts of patients with rheumatoid arthritis attach to and invade normal human cartilage when engrafted into SCID mice. *Am J Pathol* 1996;**149**:1607–15.
432. Bartok B, Firestein GS. Fibroblast-like synoviocytes: key effector cells in rheumatoid arthritis. *Immunol Rev* 2010;**233**:233–55.
433. Lefevre S, Knedla A, Tennie C, Kampmann A, Wunrau C, Dinser R, et al. Synovial fibroblasts spread rheumatoid arthritis to unaffected joints. *Nat Med* 2009;**15**:1414–20.
434. Maxwell DS, Kruger L. The fine structure of astrocytes in the cerbral cortex and their response to focal injury produced by heavy ionizing particles. *J Cell Biol* 1965;**25**:141–57.
435. Raine CS. Membrane specialisations between demyelinated axons and astroglia in chronic EAE lesions and multiple sclerosis plaques. *Nature* 1978;**275**:326–7.
436. Snyder DH, Valsamis MP, Stone SH, Raine CS. Progressive demyelination and reparative phenomena in chronic experimental allergic encephalomyelitis. *J Neuropathol Exp Neurol* 1975;**34**:209–21.
437. Frohman EM, Racke MK, Raine CS. Multiple sclerosis—the plaque and its pathogenesis. *N Engl J Med* 2006;**354**:942–55.
438. Abraham DJ, Eckes B, Rajkumar V, Krieg T. New developments in fibroblast and myofibroblast biology: implications for fibrosis and scleroderma. *Curr Rheumatol Rep* 2007;**9**:136–43.
439. Burke JP, Mulsow JJ, O'Keane C, Docherty NG, Watson RW, O'Connell PR. Fibrogenesis in Crohn's disease. *Am J Gastroenterol* 2007;**102**:439–48.
440. Cohen JN, Guidi CJ, Tewalt EF, Qiao H, Rouhani SJ, Ruddell A, et al. Lymph node-resident lymphatic endothelial cells mediate peripheral tolerance via Aire-independent direct antigen presentation. *J Exp Med* 2010;**207**:681–8.
441. Sozzani S, Rusnati M, Riboldi E, Mitola S, Presta M. Dendritic cell-endothelial cell crosstalk in angiogenesis. *Trends Immunol* 2007;**28**:385–92.
442. Trachtman H. Nitric oxide and glomerulonephritis. *Semin Nephrol* 2004;**24**:324–32.
443. McCandless EE, Klein RS. Molecular targets for disrupting leukocyte trafficking during multiple sclerosis. *Expert Rev Mol Med* 2007;**9**:1–19.
444. Heidenreich R, Rocken M, Ghoreschi K. Angiogenesis: the new potential target for the therapy of psoriasis? *Drug News Perspect* 2008;**21**:97–105.
445. van Zonneveld AJ, de Boer HC, Van der Veer EP, Rabelink TJ. Inflammation, vascular injury and repair in rheumatoid arthritis. *Ann Rheum Dis* 2010;**69**(Suppl. 1):57–60.

446. Distler JH, Beyer C, Schett G, Luscher TF, Gay S, Distler O. Endothelial progenitor cells: novel players in the pathogenesis of rheumatic diseases. *Arthritis Rheum* 2009;**60**:3168–79.
447. Szekanecz Z, Besenyei T, Paragh G, Koch AE. Angiogenesis in rheumatoid arthritis. *Autoimmunity* 2009;**42**:563–73.
448. Meroni PL, Tincani A, Sepp N, Raschi E, Testoni C, Corsini E, et al. Endothelium and the brain in CNS lupus. *Lupus* 2003;**12**:919–28.
449. Manoussakis MN, Kapsogeorgou EK. The role of epithelial cells in the pathogenesis of Sjogren's syndrome. *Clin Rev Allergy Immunol* 2007;**32**:225–30.
450. Gershwin ME, Mackay IR. The causes of primary biliary cirrhosis: convenient and inconvenient truths. *Hepatology* 2008;**47**:737–45.
451. Sturm A, Dignass AU. Epithelial restitution and wound healing in inflammatory bowel disease. *World J Gastroenterol* 2008;**14**:348–53.
452. Culton DA, Qian Y, Li N, Rubenstein D, Aoki V, Filhio GH, et al. Advances in pemphigus and its endemic pemphigus foliaceus (Fogo Selvagem) phenotype: a paradigm of human autoimmunity. *J Autoimmun* 2008;**31**:311–24.
453. Terhorst D, Kalali BN, Ollert M, Ring J, Mempel M. The role of toll-like receptors in host defenses and their relevance to dermatologic diseases. *Am J Clin Dermatol* 2010;**11**:1–10.
454. Veldman C, Feliciani C. Pemphigus: a complex T cell-dependent autoimmune disorder leading to acantholysis. *Clin Rev Allergy Immunol* 2008;**34**:313–20.
455. Casciola-Rosen LA, Anhalt G, Rosen A. Autoantigens targeted in systemic lupus erythematosus are clustered in two populations of surface structures on apoptotic keratinocytes. *J Exp Med* 1994;**179**:1317–30.
456. Fitch E, Harper E, Skorcheva I, Kurtz SE, Blauvelt A. Pathophysiology of psoriasis: recent advances on IL-23 and Th17 cytokines. *Curr Rheumatol Rep* 2007;**9**:461–7.
457. Wang KC, Kim JA, Sivasankaran R, Segal R, He Z. P75 interacts with the Nogo receptor as a co-receptor for Nogo, MAG and OMgp. *Nature* 2002;**420**:74–8.
458. Pernet V, Joly S, Christ F, Dimou L, Schwab ME. Nogo-A and myelin-associated glycoprotein differently regulate oligodendrocyte maturation and myelin formation. *J Neurosci* 2008;**28**:7435–44.
459. Schäffler A, Müller-Ladner U, Schölmerich J, Büchler C. Role of adipose tissue as an inflammatory organ in human diseases. *Endocr Rev* 2006;**27**:449–67.
460. Ahima RS, Flier JS. Adipose tissue as an endocrine organ. *Trends Endocrinol Metab* 2000;**11**:327–32.
461. Hotamisligil GS, Shargill NS, Spiegelman BM. Adipose expression of tumor necrosis factor-alpha: direct role in obesity-linked insulin resistance. *Science* 1993;**259**:87–91.
462. Lin Y, Lee H, Berg AH, Lisanti MP, Shapiro L, Scherer PE. The lipopolysaccharide-activated toll-like receptor (TLR)-4 induces synthesis of the closely related receptor TLR-2 in adipocytes. *J Biol Chem* 2000;**275**:24255–63.
463. Schäffler A, Schölmerich J. Innate immunity and adipose tissue biology. *Trends Immunol* 2010;**31**:228–35.
464. Lord GM, Matarese G, Howard JK, Baker RJ, Bloom SR, Lechler RI. Leptin modulates the T-cell immune response and reverses starvation-induced immunosuppression. *Nature* 1998;**394**:897–901.
465. Caldefie-Chezet F, Poulin A, Tridon A, Sion B, Vasson MP. Leptin: a potential regulator of polymorphonuclear neutrophil bactericidal action? *J Leukoc Biol* 2001;**69**:414–8.
466. Mancuso P, Gottschalk A, Phare SM, Peters-Golden M, Lukacs NW, Huffnagle GB. Leptin-deficient mice exhibit impaired host defense in Gram-negative pneumonia. *J Immunol* 2002;**168**:4018–24.

467. Matarese G, Sanna V, Lechler RI, Sarvetnick N, Fontana S, Zappacosta S, et al. Leptin accelerates autoimmune diabetes in female NOD mice. *Diabetes* 2002;**51**:1356–61.
468. Sanna V, Di GA, La CA, Lechler RI, Fontana S, Zappacosta S, et al. Leptin surge precedes onset of autoimmune encephalomyelitis and correlates with development of pathogenic T cell responses. *J Clin Invest* 2003;**111**:241–50.
469. Sordet C, Goetz J, Sibilia J. Contribution of autoantibodies to the diagnosis and nosology of inflammatory muscle disease. *Joint Bone Spine* 2006;**73**:646–54.
470. Dalakas MC. Inflammatory muscle diseases: a critical review on pathogenesis and therapies. *Curr Opin Pharmacol* 2010;**10**:346–52.
471. Howard OM, Dong HF, Yang D, Raben N, Nagaraju K, Rosen A, et al. Histidyl-tRNA synthetase and asparaginyl-tRNA synthetase, autoantigens in myositis, activate chemokine receptors on T lymphocytes and immature dendritic cells. *J Exp Med* 2002;**196**:781–91.
472. Schmidt J, Barthel K, Wrede A, Salajegheh M, Bahr M, Dalakas MC. Interrelation of inflammation and APP in sIBM: IL-1 beta induces accumulation of beta-amyloid in skeletal muscle. *Brain* 2008;**131**:1228–40.
473. Wiendl H, Hohlfeld R, Kieseier BC. Muscle-derived positive and negative regulators of the immune response. *Curr Opin Rheumatol* 2005;**17**:714–9.
474. Sawamukai N, Yukawa S, Saito K, Nakayamada S, Kambayashi T, Tanaka Y. Mast cell-derived tryptase inhibits apoptosis of human rheumatoid synovial fibroblasts via rho-mediated signaling. *Arthritis Rheum* 2010;**62**:952–9.
475. Magnusson SE, Pejler G, Kleinau S, Abrink M. Mast cell chymase contributes to the antibody response and the severity of autoimmune arthritis. *FASEB J* 2009;**23**:875–82.
476. Nakano S, Mishiro T, Takahara S, Yokoi H, Hamada D, Yukata K, et al. Distinct expression of mast cell tryptase and protease activated receptor-2 in synovia of rheumatoid arthritis and osteoarthritis. *Clin Rheumatol* 2007;**26**:1284–92.
477. Di GN, Indoh I, Jackson N, Wakefield D, McNeil HP, Yan W, et al. Human mast cell-derived gelatinase B (matrix metalloproteinase-9) is regulated by inflammatory cytokines: role in cell migration. *J Immunol* 2006;**177**:2638–50.
478. Woolley DE. The mast cell in inflammatory arthritis. *N Engl J Med* 2003;**348**:1709–11.
479. Olsson N, Ulfgren AK, Nilsson G. Demonstration of mast cell chemotactic activity in synovial fluid from rheumatoid patients. *Ann Rheum Dis* 2001;**60**:187–93.
480. Tetlow LC, Harper N, Dunningham T, Morris MA, Bertfield H, Woolley DE. Effects of induced mast cell activation on prostaglandin E and metalloproteinase production by rheumatoid synovial tissue in vitro. *Ann Rheum Dis* 1998;**57**:25–32.
481. Bridges AJ, Malone DG, Jicinsky J, Chen M, Ory P, Engber W, et al. Human synovial mast cell involvement in rheumatoid arthritis and osteoarthritis. Relationship to disease type, clinical activity, and antirheumatic therapy. *Arthritis Rheum* 1991;**34**:1116–24.
482. Gruber BL, Schwartz LB, Ramamurthy NS, Irani AM, Marchese MJ. Activation of latent rheumatoid synovial collagenase by human mast cell tryptase. *J Immunol* 1988;**140**:3936–42.
483. Gruber B, Ballan D, Gorevic PD. IgE rheumatoid factors: quantification in synovial fluid and ability to induce synovial mast cell histamine release. *Clin Exp Immunol* 1988;**71**:289–94.
484. Yoffe JR, Taylor DJ, Wooley DE. Mast cell products stimulate collagenase and prostaglandin E production by cultures of adherent rheumatoid synovial cells. *Biochem Biophys Res Commun* 1984;**122**:270–6.
485. Kneilling M, Hultner L, Pichler BJ, Mailhammer R, Morawietz L, Solomon S, et al. Targeted mast cell silencing protects against joint destruction and angiogenesis in experimental arthritis in mice. *Arthritis Rheum* 2007;**56**:1806–16.

486. Kurashima Y, Amiya T, Nochi T, Fujisawa K, Haraguchi T, Iba H, et al. Extracellular ATP mediates mast cell-dependent intestinal inflammation through P2X7 purinoceptors. *Nat Commun* 2012;**3**:1034. http://dx.doi.org/10.1038/ncomms2023.:1034.
487. Hamilton MJ, Sinnamon MJ, Lyng GD, Glickman JN, Wang X, Xing W, et al. Essential role for mast cell tryptase in acute experimental colitis. *Proc Natl Acad Sci U S A* 2011;**108**:290–5.
488. Raithel M, Winterkamp S, Pacurar A, Ulrich P, Hochberger J, Hahn EG. Release of mast cell tryptase from human colorectal mucosa in inflammatory bowel disease. *Scand J Gastroenterol* 2001;**36**:174–9.
489. Gelbmann CM, Mestermann S, Gross V, Kollinger M, Schölmerich J, Falk W. Strictures in Crohn's disease are characterised by an accumulation of mast cells colocalised with laminin but not with fibronectin or vitronectin. *Gut* 1999;**45**:210–7.
490. Xu X, Rivkind A, Pikarsky A, Pappo O, Bischoff SC, Levi-Schaffer F. Mast cells and eosinophils have a potential profibrogenic role in Crohn disease. *Scand J Gastroenterol* 2004;**39**:440–7.
491. Lorentz A, Schwengberg S, Mierke C, Manns MP, Bischoff SC. Human intestinal mast cells produce IL-5 in vitro upon IgE receptor cross-linking and in vivo in the course of intestinal inflammatory disease. *Eur J Immunol* 1999;**29**:1496–503.
492. Christy AL, Walker ME, Hessner MJ, Brown MA. Mast cell activation and neutrophil recruitment promotes early and robust inflammation in the meninges in EAE. *J Autoimmun* 2013;**42**:50–61.
493. Brown MA, Tanzola MB, Robbie-Ryan M. Mechanisms underlying mast cell influence on EAE disease course. *Mol Immunol* 2002;**38**:1373–8.
494. Rozniecki JJ, Hauser SL, Stein M, Lincoln R, Theoharides TC. Elevated mast cell tryptase in cerebrospinal fluid of multiple sclerosis patients. *Ann Neurol* 1995;**37**:63–6.
495. Christy AL, Brown MA. The multitasking mast cell: positive and negative roles in the progression of autoimmunity. *J Immunol* 2007;**179**:2673–9.
496. Okada Y, Wu D, Trynka G, Raj T, Terao C, Ikari K, et al. Genetics of rheumatoid arthritis contributes to biology and drug discovery. *Nature* 2014;**506**:376–81.
497. Beecham AH, Patsopoulos NA, Xifara DK, Davis MF, Kemppinen A, Cotsapas C, et al. Analysis of immune-related loci identifies 48 new susceptibility variants for multiple sclerosis. *Nat Genet* 2013;**45**:1353–60.
498. Plagnol V, Howson JM, Smyth DJ, Walker N, Hafler JP, Wallace C, et al. Genome-wide association analysis of autoantibody positivity in type 1 diabetes cases. *PLoS Genet* 2011;**7**:e1002216.
499. Bradfield JP, Qu HQ, Wang K, Zhang H, Sleiman PM, Kim CE, et al. A genome-wide meta-analysis of six type 1 diabetes cohorts identifies multiple associated loci. *PLoS Genet* 2011;**7**:e1002293.
500. Barrett JC, Clayton DG, Concannon P, Akolkar B, Cooper JD, Erlich HA, et al. Genome-wide association study and meta-analysis find that over 40 loci affect risk of type 1 diabetes. *Nat Genet* 2009;**41**:703–7.
501. Rullo OJ, Tsao BP. Recent insights into the genetic basis of systemic lupus erythematosus. *Ann Rheum Dis* 2013;**72**(Suppl. 2):ii56–61.
502. Franke A, McGovern DP, Barrett JC, Wang K, Radford-Smith GL, Ahmad T, et al. Genome-wide meta-analysis increases to 71 the number of confirmed Crohn's disease susceptibility loci. *Nat Genet* 2010;**42**:1118–25.
503. Banting FG, Best CH, Collip JB, Campbell WR, Fletcher AA. Pancreatic extracts in the treatment of diabetes mellitus. *Can Med Assoc J* 1922;**12**:141–6.
504. Dicke WK, Weijers HA, Van der Kamer JH. Coeliac disease. II. The presence in wheat of a factor having a deleterious effect in cases of coeliac disease. *Acta Paediatr Scand* 1953;**42**:34–42.

505. Van de Kamer JH, Weijers HA. Coeliac disease. V. Some experiments on the cause of the harmful effect of wheat gliadin. *Acta Paediatr* 1955;**44**:465–9.
506. Russo RA, Brogan PA. Monogenic autoinflammatory diseases. *Rheumatology (Oxford)* 2014;**53**:1927–39.
507. Straub RH. Concepts of evolutionary medicine and energy regulation contribute to the etiology of systemic chronic inflammatory diseases. *Brain Behav Immun* 2011;**25**:1–5.
508. Straub RH, Schölmerich J, Cutolo M. The multiple facets of premature aging in rheumatoid arthritis. *Arthritis Rheum* 2003;**48**:2713–21.
509. Doll R, Hill AB. Smoking and carcinoma of the lung; preliminary report. *Br Med J* 1950;**2**:739–48.
510. Baka Z, Buzas E, Nagy G. Rheumatoid arthritis and smoking: putting the pieces together. *Arthritis Res Ther* 2009;**11**:238.
511. Tanda ML, Piantanida E, Lai A, Lombardi V, Dalle Mule I, Liparulo L, et al. Thyroid autoimmunity and environment. *Horm Metab Res* 2009;**41**:436–42.
512. Nancy AL, Yehuda S. Prediction and prevention of autoimmune skin disorders. *Arch Dermatol Res* 2009;**301**:57–64.
513. Neidhart M, Rethage J, Kuchen S, Kunzler P, Crowl RM, Billingham ME, et al. Retrotransposable L1 elements expressed in rheumatoid arthritis synovial tissue: association with genomic DNA hypomethylation and influence on gene expression. *Arthritis Rheum* 2000;**43**:2634–47.
514. Damian RT. Molecular mimicry: antigen sharing by parasite and host and its consequences. *Am Nat* 1964;**98**:129–49.
515. Getts MT, Miller SD. 99th Dahlem conference on infection, inflammation and chronic inflammatory disorders: triggering of autoimmune diseases by infections. *Clin Exp Immunol* 2010;**160**:15–21.
516. Kirvan CA, Swedo SE, Heuser JS, Cunningham MW. Mimicry and autoantibody-mediated neuronal cell signaling in Sydenham chorea. *Nat Med* 2003;**9**:914–20.
517. Yuki N, Taki T, Inagaki F, Kasama T, Takahashi M, Saito K, et al. A bacterium lipopolysaccharide that elicits Guillain-Barre syndrome has a GM1 ganglioside-like structure. *J Exp Med* 1993;**178**:1771–5.
518. Valdimarsson H, Thorleifsdottir RH, Sigurdardottir SL, Gudjonsson JE, Johnston A. Psoriasis—as an autoimmune disease caused by molecular mimicry. *Trends Immunol* 2009;**30**:494–501.
519. Amedei A, Bergman MP, Appelmelk BJ, Azzurri A, Benagiano M, Tamburini C, et al. Molecular mimicry between Helicobacter pylori antigens and H+, K+ −-adenosine triphosphatase in human gastric autoimmunity. *J Exp Med* 2003;**198**:1147–56.
520. Wucherpfennig KW, Strominger JL. Molecular mimicry in T cell-mediated autoimmunity: viral peptides activate human T cell clones specific for myelin basic protein. *Cell* 1995;**80**:695–705.
521. Agmon-Levin N, Blank M, Paz Z, Shoenfeld Y. Molecular mimicry in systemic lupus erythematosus. *Lupus* 2009;**18**:1181–5.
522. Anderton SM. Post-translational modifications of self antigens: implications for autoimmunity. *Curr Opin Immunol* 2004;**16**:753–8.
523. Schellekens GA, de Jong BA, van den Hoogen FH, van de Putte LB, van Venrooij WJ. Citrulline is an essential constituent of antigenic determinants recognized by rheumatoid arthritis-specific autoantibodies. *J Clin Invest* 1998;**101**:273–81.
524. Sternberg EM, Young WS, Bernardini R, Calogero AE, Chrousos GP, Gold PW, et al. A central nervous system defect in biosynthesis of corticotropin- releasing hormone is associated with susceptibility to streptococcal cell wall-induced arthritis in Lewis rats. *Proc Natl Acad Sci U S A* 1989;**86**:4771–5.

525. Nielen MM, van Schaardenburg D, Reesink HW, van de Stadt RJ, van der Horst-Bruinsma IE, de Koning MH, et al. Specific autoantibodies precede the symptoms of rheumatoid arthritis: a study of serial measurements in blood donors. *Arthritis Rheum* 2004;**50**:380–6.
526. Rantapää-Dahlqvist S, de Jong BA, Berglin E, Hallmans G, Wadell G, Stenlund H, et al. Antibodies against cyclic citrullinated peptide and IgA rheumatoid factor predict the development of rheumatoid arthritis. *Arthritis Rheum* 2003;**48**:2741–9.
527. van Venrooij WJ, Pruijn GJ. Citrullination: a small change for a protein with great consequences for rheumatoid arthritis. *Arthritis Res* 2000;**2**:249–51.
528. Mastronardi FG, Wood DD, Mei J, Raijmakers R, Tseveleki V, Dosch HM, et al. Increased citrullination of histone H3 in multiple sclerosis brain and animal models of demyelination: a role for tumor necrosis factor-induced peptidylarginine deiminase 4 translocation. *J Neurosci* 2006;**26**:11387–96.
529. Koivula MK, Heliovaara M, Ramberg J, Knekt P, Rissanen H, Palosuo T, et al. Autoantibodies binding to citrullinated telopeptide of type II collagen and to cyclic citrullinated peptides predict synergistically the development of seropositive rheumatoid arthritis. *Ann Rheum Dis* 2007;**66**:1450–5.
530. Russell AS, Devani A, Maksymowych WP. The role of anti-cyclic citrullinated peptide antibodies in predicting progression of palindromic rheumatism to rheumatoid arthritis. *J Rheumatol* 2006;**33**:1240–2.
531. Stuart JM, Cremer MA, Townes AS, Kang AH. Type II collagen-induced arthritis in rats. Passive transfer with serum and evidence that IgG anticollagen antibodies can cause arthritis. *J Exp Med* 1982;**155**:1–16.
532. Stuart JM, Dixon FJ. Serum transfer of collagen-induced arthritis in mice. *J Exp Med* 1983;**158**:378–92.
533. Rudick RA. Evolving concepts in the pathogenesis of multiple sclerosis and their therapeutic implications. *J Neuroophthalmol* 2001;**21**:279–83.
534. Barakat SO, Fernandez Perez MJ, Benavente FA, Garcia Moreno JM, Ruiz Pena JL, Fajardo GJ, et al. The use of magnetic resonance imaging in the study of asymptomatic familial multiple sclerosis patients. *Rev Neurol* 2003;**37**:811–4.
535. Miller D, Barkhof F, Montalban X, Thompson A, Filippi M. Clinically isolated syndromes suggestive of multiple sclerosis, part I: natural history, pathogenesis, diagnosis, and prognosis. *Lancet Neurol* 2005;**4**:281–8.
536. Greaves ML, Pochapin M. Asymptomatic ileitis: past, present, and future. *J Clin Gastroenterol* 2006;**40**:281–5.
537. Silva LM, Chavez J, Canalli MH, Zanetti CR. Determination of IgG subclasses and avidity of antithyroid peroxidase antibodies in patients with subclinical hypothyroidism—a comparison with patients with overt hypothyroidism. *Horm Res* 2003;**59**:118–24.
538. Seissler J, Gluck M, Speck U, Yassin N, Fetzer A, Bornstein S, et al. Epidemiologic studies of the recognition of the preclinical phase of type I diabetes in school children. *Dtsch Med Wochenschr* 1990;**115**:689–94.
539. Harrison LC, De AH, Loudovaris T, Campbell IL, Cebon JS, Tait BD, et al. Reactivity to human islets and fetal pig proislets by peripheral blood mononuclear cells from subjects with preclinical and clinical insulin-dependent diabetes. *Diabetes* 1991;**40**:1128–33.
540. De Aizpurua HJ, Wilson YM, Harrison LC. Glutamic acid decarboxylase autoantibodies in preclinical insulin-dependent diabetes. *Proc Natl Acad Sci U S A* 1992;**89**:9841–5.
541. Knip M, Vahasalo P, Karjalainen J, Lounamaa R, Akerblom HK. Natural history of preclinical IDDM in high risk siblings. Childhood Diabetes in Finland Study Group. *Diabetologia* 1994;**37**:388–93.

542. Baekkeskov S, Landin M, Kristensen JK, Srikanta S, Bruining GJ, Mandrup-Poulsen T. Antibodies to a 64,000 Mr human islet cell antigen precede the clinical onset of insulin-dependent diabetes. *J Clin Invest* 1987;**79**:926–34.
543. Joossens S, Reinisch W, Vermeire S, Sendid B, Poulain D, Peeters M, et al. The value of serologic markers in indeterminate colitis: a prospective follow-up study. *Gastroenterology* 2002;**122**:1242–7.
544. Sharief MK, Thompson EJ. The predictive value of intrathecal immunoglobulin synthesis and magnetic resonance imaging in acute isolated syndromes for subsequent development of multiple sclerosis. *Ann Neurol* 1991;**29**:147–51.
545. Masjuan J, varez-Cermeno JC, Garcia-Barragan N, az-Sanchez M, Espino M, Sadaba MC, et al. Clinically isolated syndromes: a new oligoclonal band test accurately predicts conversion to MS. *Neurology* 2006;**66**:576–8.
546. Ziegler AG, Herskowitz RD, Jackson RA, Soeldner JS, Eisenbarth GS. Predicting type I diabetes. *Diabetes Care* 1990;**13**:762–5.
547. Koetz K, Bryl E, Spickschen K, O'Fallon WM, Goronzy JJ, Weyand CM. T cell homeostasis in patients with rheumatoid arthritis. *Proc Natl Acad Sci U S A* 2000;**97**:9203–8.
548. Shanks N, Lightman SL. The maternal-neonatal neuro-immune interface: are there long-term implications for inflammatory or stress-related disease? *J Clin Invest* 2001;**108**:1567–73.
549. Matsumoto A, Arai Y, Urano A, Hyodo S. Molecular basis of neuronal plasticity to gonadal steroids. *Funct Neurol* 1995;**10**:59–76.
550. Barker DJ, Winter PD, Osmond C, Margetts B, Simmonds SJ. Weight in infancy and death from ischaemic heart disease. *Lancet* 1989;**2**:577–80.
551. Barker DJ, Martyn CN. The maternal and fetal origins of cardiovascular disease. *J Epidemiol Community Health* 1992;**46**:8–11.
552. Reyes TM, Coe CL. Prenatal manipulations reduce the proinflammatory response to a cytokine challenge in juvenile monkeys. *Brain Res* 1997;**769**:29–35.
553. Shanks N, Windle RJ, Perks PA, Harbuz MS, Jessop DS, Ingram CD, et al. Early-life exposure to endotoxin alters hypothalamic-pituitary-adrenal function and predisposition to inflammation. *Proc Natl Acad Sci U S A* 2000;**97**:5645–50.
554. Kandel ER. *In search of memory—the emergence of a new science of mind*. New York: W. W. Norton & Company; 2006.
555. Mitchell M. *Complexity*. New York: Oxford University Press; 2009.
556. Hench PS, Slocumb CH, Holley HF, Kendall EC. Effect of cortisone and pituitary adrenocorticotropic hormone (ACTH) on rheumatic diseases. *J Am Med Assoc* 1950;**1327–35**.
557. Stern K, Davidsohn I. Effect of estrogen and cortisone on immune hemoantibodies in mice of inbred strains. *J Immunol* 1955;**74**:479–84.
558. Woods AC. Clinical and experimental observation on the use of ACTH and cortisone in ocular inflammatory disease. *Trans Am Ophthalmol Soc* 1950;**48**:259–96.
559. Hayes SP. The effect of cortisone on local antibody formation. *J Immunol* 1953;**70**:450–3.
560. Masi AT, Feigenbaum SL, Chatterton RT. Hormonal and pregnancy relationships to rheumatoid arthritis: convergent effects with immunologic and microvascular systems. *Semin Arthritis Rheum* 1995;**25**:1–27.
561. Cutolo M, Wilder R. Different roles for androgens and estrogens in the susceptibility to autoimmune rheumatic diseases. *Rheum Dis Clin North Am* 2000;**26**:825–39.
562. Huong DL, Wechsler B, Vauthier-Brouzes D, Duhaut P, Costedoat N, Lefebvre G, et al. Importance of planning ovulation induction therapy in systemic lupus erythematosus and antiphospholipid syndrome: a single center retrospective study of 21 cases and 114 cycles. *Semin Arthritis Rheum* 2002;**32**:174–88.

563. Cutolo M, Straub RH, Chrousos GP. *Neuroimmunomodulation: special issue on stress and autoimmunity*. Basel: Karger; 2006.
564. Miller LE, Wessinghage D, Müller-Ladner U, Schölmerich J, Falk W, Kerner T, et al. In vitro superfusion method to study nerve-immune cell interactions in human synovial membrane in long-standing rheumatoid arthritis and osteoarthritis. *Ann N Y Acad Sci* 1999;**876**:266–75.
565. Straub RH, Cutolo M. Circadian rhythms in rheumatoid arthritis: implications for pathophysiology and therapeutic management. *Arthritis Rheum* 2007;**56**:399–408.
566. Matera L, Mori M, Galetto A. Effect of prolactin on the antigen presenting function of monocyte-derived dendritic cells. *Lupus* 2001;**10**:728–34.
567. Takeda S, Elefteriou F, Levasseur R, Liu X, Zhao L, Parker KL, et al. Leptin regulates bone formation via the sympathetic nervous system. *Cell* 2002;**111**:305–17.
568. Sanders VM, Kasprowicz DJ, Kohm AP, Swanson MA. Neurotransmitter receptors on lymphocytes and other lymphoid cells. In: Ader R, Felten DL, Cohen N, editors. *Psychneuroimmunology*. San Diego, CA: Academic Press; 2001. p. 161–96.
569. Heijnen CJ, Roupe van der Voort C, Wulffraat N, van der Net J, Kuis W, Kavelaars A, et al. Functional alpha 1-adrenergic receptors on leukocytes of patients with polyarticular juvenile rheumatoid arthritis. *J Neuroimmunol* 1996;**71**:223–6.
570. Heijnen CJ, Rouppe van der Voort C, Van de Pol M, Kavelaars A. Cytokines regulate alpha(1)-adrenergic receptor mRNA expression in human monocytic cells and endothelial cells. *J Neuroimmunol* 2002;**125**:66–72.
571. Pace TW, Hu F, Miller AH. Cytokine-effects on glucocorticoid receptor function: relevance to glucocorticoid resistance and the pathophysiology and treatment of major depression. *Brain Behav Immun* 2007;**21**:9–19.
572. Lang CH, Hong-Brown L, Frost RA. Cytokine inhibition of JAK-STAT signaling: a new mechanism of growth hormone resistance. *Pediatr Nephrol* 2005;**20**:306–12.
573. Kelley KW. From hormones to immunity: the physiology of immunology. *Brain Behav Immun* 2004;**18**:95–113.
574. Hotamisligil GS. Inflammatory pathways and insulin action. *Int J Obes Relat Metab Disord* 2003;**27**(Suppl. 3):S53–5.
575. Lombardi MS, Kavelaars A, Schedlowski M, Bijlsma JW, Okihara KL, Van de PM, et al. Decreased expression and activity of G-protein-coupled receptor kinases in peripheral blood mononuclear cells of patients with rheumatoid arthritis. *FASEB J* 1999;**13**:715–25.
576. Lombardi MS, Kavelaars A, Cobelens PM, Schmidt RE, Schedlowski M, Heijnen CJ. Adjuvant arthritis induces down-regulation of G protein-coupled receptor kinases in the immune system. *J Immunol* 2001;**166**:1635–40.
577. Baschant U, Tuckermann J. The role of the glucocorticoid receptor in inflammation and immunity. *J Steroid Biochem Mol Biol* 2010;**120**:69–75.
578. Sanders VM, Straub RH. Norepinephrine, the beta-adrenergic receptor, and immunity. *Brain Behav Immun* 2002;**16**:290–332.
579. Flierl MA, Rittirsch D, Nadeau BA, Sarma JV, Day DE, Lentsch AB, et al. Upregulation of phagocyte-derived catecholamines augments the acute inflammatory response. *PLoS One* 2009;**4**:e4414.
580. Spengler RN, Allen RM, Remick DG, Strieter RM, Kunkel SL. Stimulation of alpha-adrenergic receptor augments the production of macrophage-derived tumor necrosis factor. *J Immunol* 1990;**145**:1430–4.
581. Straub RH, Günzler C, Miller LE, Cutolo M, Schölmerich J, Schill S. Anti-inflammatory cooperativity of corticosteroids and norepinephrine in rheumatoid arthritis synovial tissue in vivo and in vitro. *FASEB J* 2002;**16**:993–1000.

582. Bevan JA. Some functional consequences of variation in adrenergic synaptic cleft width and in nerve density and distribution. *Fed Proc* 1977;**36**:2439–43.
583. Bevan JA, Su C. Variation of intra- and perisynaptic adrenergic transmitter concentrations with width of synaptic cleft in vascular tissue. *J Pharmacol Exp Ther* 1974;**190**:30–8.
584. Schaible HG, Straub RH. Function of the sympathetic supply in acute and chronic experimental joint inflammation. *Auton Neurosci* 2014;**182**:55–64.
585. Felten DL, Felten SY, Bellinger DL, Carlson SL, Ackerman KD, Madden KS, et al. Noradrenergic sympathetic neural interactions with the immune system: structure and function. *Immunol Rev* 1987;**100**:225–60.
586. Jesseph JM, Felten DL. Noradrenergic innervation of the gut-associated lymphoid tissues (GALT) in the rabbit [abstract]. *Anat Rec* 1984;**208**:81A.
587. Weihe E, Nohr D, Michel S, Muller S, Zentel HJ, Fink T, et al. Molecular anatomy of the neuro-immune connection. *Int J Neurosci* 1991;**59**:1–23.
588. Nance DM, Sanders VM. Autonomic innervation and regulation of the immune system (1987–2007). *Brain Behav Immun* 2007;**21**:736–45.
589. Niijima A, Hori T, Aou S, Oomura Y. The effects of interleukin-1 beta on the activity of adrenal, splenic and renal sympathetic nerves in the rat. *J Auton Nerv Syst* 1991;**36**:183–92.
590. Labrie F. Intracrinology. *Mol Cell Endocrinol* 1991;**78**:C113–8.
591. Cutolo M, Straub RH, Bijlsma JW. Neuroendocrine-immune interactions in synovitis. *Nat Clin Pract Rheumatol* 2007;**3**:627–34.
592. Lieberman GE, Lewis GP, Peters TJ. A membrane-bound enzyme in rabbit aorta capable of inhibiting adenosine-diphosphate-induced platelet aggregation. *Lancet* 1977;**2**:330–2.
593. Segel GB, Ryan DH, Lichtman MA. Ecto-nucleotide triphosphatase activity of human lymphocytes: studies of normal and CLL lymphocytes. *J Cell Physiol* 1985;**124**:424–32.
594. Pearson JD, Carleton JS, Gordon JL. Metabolism of adenine nucleotides by ectoenzymes of vascular endothelial and smooth-muscle cells in culture. *Biochem J* 1980;**190**:421–9.
595. Serra MC, Bazzoni F, Della BV, Greskowiak M, Rossi F. Activation of human neutrophils by substance P. Effect on oxidative metabolism, exocytosis, cytosolic Ca2+ concentration and inositol phosphate formation. *J Immunol* 1988;**141**:2118–24.
596. Piercey MF, Dobry PJ, Einspahr FJ, Schroeder LA, Masiques N. Use of substance P fragments to differentiate substance P receptors of different tissues. *Regul Pept* 1982;**3**:337–49.
597. Lundberg JM, Hemsen A, Larsson O, Rudehill A, Saria A, Fredholm BB. Neuropeptide Y receptor in pig spleen: binding characteristics, reduction of cyclic AMP formation and calcium antagonist inhibition of vasoconstriction. *Eur J Pharmacol* 1988;**145**:21–9.
598. Cox HM, Cuthbert AW. The effects of neuropeptide Y and its fragments upon basal and electrically stimulated ion secretion in rat jejunum mucosa. *Br J Pharmacol* 1990;**101**:247–52.
599. De Meester I, Durinx C, Bal G, Proost P, Struyf S, Goossens F, et al. Natural substrates of dipeptidyl peptidase IV. *Adv Exp Med Biol* 2000;**477**:67–87.
600. Blalock JE, Smith EM. Human leukocyte interferon: structural and biological relatedness to adrenocorticotropic hormone and endorphins. *Proc Natl Acad Sci U S A* 1980;**77**:5972–4.
601. Klein JR. The immune system as a regulator of thyroid hormone activity. *Exp Biol Med (Maywood)* 2006;**231**:229–36.
602. Kelley KW, Weigent DA, Kooijman R. Protein hormones and immunity. *Brain Behav Immun* 2007;**21**:384–92.
603. Josefsson E, Bergquist J, Ekman R, Tarkowski A. Catecholamines are synthesized by mouse lymphocytes and regulate function of these cells by induction of apoptosis. *Immunology* 1996;**88**:140–6.

604. Cosentino M, Fietta AM, Ferrari M, Rasini E, Bombelli R, Carcano E, et al. Human CD4+CD25+ regulatory T cells selectively express tyrosine hydroxylase and contain endogenous catecholamines subserving an autocrine/paracrine inhibitory functional loop. *Blood* 2007;**109**:632–42.
605. Flierl MA, Rittirsch D, Nadeau BA, Chen AJ, Sarma JV, Zetoune FS, et al. Phagocyte-derived catecholamines enhance acute inflammatory injury. *Nature* 2007;**449**:721–5.
606. Miller LE, Jüsten HP, Schölmerich J, Straub RH. The loss of sympathetic nerve fibers in the synovial tissue of patients with rheumatoid arthritis is accompanied by increased norepinephrine release from synovial macrophages. *FASEB J* 2000;**14**:2097–107.
607. Capellino S, Cosentino M, Wolff C, Schmidt M, Grifka J, Straub RH. Catecholamine-producing cells in the synovial tissue during arthritis: modulation of sympathetic neurotransmitters as new therapeutic target. *Ann Rheum Dis* 2010;**69**:1853–60.
608. Besedovsky H, Sorkin E, Felix D, Haas H. Hypothalamic changes during the immune response. *Eur J Immunol* 1977;**7**:323–5.
609. Besedovsky H, Sorkin E, Keller M, Muller J. Changes in blood hormone levels during the immune response. *Proc Soc Exp Biol Med* 1975;**150**:466–70.
610. Blalock JE. The immune system as a sensory organ. *J Immunol* 1984;**132**:1067–70.
611. Besedovsky HO, del Rey A. Immune-neuro-endocrine interactions. *Endocr Rev* 1996;**17**:64–102.
612. Watkins LR, Goehler LE, Relton JK, Tartaglia N, Silbert L, Martin D, et al. Blockade of interleukin-1 induced hyperthermia by subdiaphragmatic vagotomy: evidence for vagal mediation of immune-brain communication. *Neurosci Lett* 1995;**183**:27–31.
613. Bluthe RM, Walter V, Parnet P, Laye S, Lestage J, Verrier D, et al. Lipopolysaccharide induces sickness behaviour in rats by a vagal mediated mechanism. *C R Acad Sci III* 1994;**317**:499–503.
614. Romeo HE, Tio DL, Rahman SU, Chiappelli F, Taylor AN. The glossopharyngeal nerve as a novel pathway in immune-to-brain communication: relevance to neuroimmune surveillance of the oral cavity. *J Neuroimmunol* 2001;**115**:91–100.
615. Brenn D, Richter F, Schaible HG. Sensitization of unmyelinated sensory fibers of the joint nerve to mechanical stimuli by interleukin-6 in the rat: an inflammatory mechanism of joint pain. *Arthritis Rheum* 2007;**56**:351–9.
616. Boettger MK, Weber K, Grossmann D, Gajda M, Bauer R, Bar KJ, et al. Spinal tumor necrosis factor alpha neutralization reduces peripheral inflammation and hyperalgesia and suppresses autonomic responses in experimental arthritis: a role for spinal tumor necrosis factor alpha during induction and maintenance of peripheral inflammation. *Arthritis Rheum* 2010;**62**:1308–18.
617. Richter F, Natura G, Ebbinghaus M, von Banchet GS, Hensellek S, Konig C, et al. Interleukin-17 sensitizes joint nociceptors to mechanical stimuli and contributes to arthritic pain through neuronal interleukin-17 receptors in rodents. *Arthritis Rheum* 2012;**64**:4125–34.
618. Li Y, Ji A, Weihe E, Schafer MK. Cell-specific expression and lipopolysaccharide-induced regulation of tumor necrosis factor alpha (TNFalpha) and TNF receptors in rat dorsal root ganglion. *J Neurosci* 2004;**24**:9623–31.
619. Barajon I, Serrao G, Arnaboldi F, Opizzi E, Ripamonti G, Balsari A, et al. Toll-like receptors 3, 4, and 7 are expressed in the enteric nervous system and dorsal root ganglia. *J Histochem Cytochem* 2009;**57**:1013–23.
620. Franchimont D, Bouma G, Galon J, Wolkersdorfer GW, Haidan A, Chrousos GP, et al. Adrenal cortical activation in murine colitis. *Gastroenterology* 2000;**119**:1560–8.
621. Engstrom L, Rosen K, Angel A, Fyrberg A, Mackerlova L, Konsman JP, et al. Systemic immune challenge activates an intrinsically regulated local inflammatory circuit in the adrenal gland. *Endocrinology* 2008;**149**:1436–50.

622. Besedovsky HO, del Rey A, Sorkin E, Dinarello C. Immunoregulatory feedback between interleukin-1 and glucocorticoid hormones. *Science* 1986;**233**:652–4.
623. Munck A, Guyre PM. Glucocorticoid physiology, pharmacology and stress. *Adv Exp Med Biol* 1986;**196**:81–96.
624. Nadeau S, Rivest S. Glucocorticoids play a fundamental role in protecting the brain during innate immune response. *J Neurosci* 2003;**23**:5536–44.
625. Dhabhar FS, McEwen BS. Bidirectional effects of stress and glucocorticoid hormones on immune function: possible explanations for paradoxical observations. In: Ader R, Felten DL, Cohen N, editors. *Psychoneuroimmunology*. San Diego: Academic Press; 2001. p. 301–38.
626. Cruz-Topete D, Cidlowski JA. One hormone, two actions: anti- and pro-inflammatory effects of glucocorticoids. *Neuroimmunomodulation* 2014;**22**:20–32.
627. Mastorakos G, Chrousos GP, Weber JS. Recombinant interleukin-6 activates the hypothalamic-pituitary-adrenal axis in humans. *J Clin Endocrinol Metab* 1993;**77**:1690–4.
628. Spath-Schwalbe E, Hansen K, Schmidt F, Schrezenmeier H, Marshall L, Burger K, et al. Acute effects of recombinant human interleukin-6 on endocrine and central nervous sleep functions in healthy men. *J Clin Endocrinol Metab* 1998;**83**:1573–9.
629. Rettori V, Gimeno MF, Karara A, Gonzalez MC, McCann SM. Interleukin 1 alpha inhibits prostaglandin E2 release to suppress pulsatile release of luteinizing hormone but not follicle-stimulating hormone. *Proc Natl Acad Sci U S A* 1991;**88**:2763–7.
630. Ogilvie KM, Held HK, Roberts ME, Hales DB, Rivier C. The inhibitory effect of intracerebroventricularly injected interleukin 1beta on testosterone secretion in the rat: role of steroidogenic acute regulatory protein. *Biol Reprod* 1999;**60**:527–33.
631. Rivier C. Role of endotoxin and interleukin-1 in modulating ACTH, LH and sex steroid secretion. *Adv Exp Med Biol* 1990;**274**:295–301.
632. Turnbull AV, Rivier C. Inhibition of gonadotropin-induced testosterone secretion by the intracerebroventricular injection of interleukin-1 beta in the male rat. *Endocrinology* 1997;**138**:1008–13.
633. Herrmann M, Schölmerich J, Straub RH. Influence of cytokines and growth factors on distinct steroidogenic enzymes in vitro: a short tabular data collection. *Ann N Y Acad Sci* 2002;**966**:166–86.
634. Eskandari F, Webster JI, Sternberg EM. Neural immune pathways and their connection to inflammatory diseases. *Arthritis Res Ther* 2003;**5**:251–65.
635. Ehrhart-Bornstein M, Hinson JP, Bornstein SR, Scherbaum WA, Vinson GP. Intraadrenal interactions in the regulation of adrenocortical steroidogenesis. *Endocr Rev* 1998;**19**:101–43.
636. Elenkov IJ, Webster EL, Torpy DJ, Chrousos GP. Stress, corticotropin-releasing hormone, glucocorticoids, and the immune/inflammatory response: acute and chronic effects. *Ann N Y Acad Sci* 1999;**876**:1–11.
637. Kalantaridou S, Makrigiannakis A, Zoumakis E, Chrousos GP. Peripheral corticotropin-releasing hormone is produced in the immune and reproductive systems: actions, potential roles and clinical implications. *Front Biosci* 2007;**12**:572–80.
638. Ottaviani E, Franchini A, Genedani S. ACTH and its role in immune-neuroendocrine functions. A comparative study. *Curr Pharm Des* 1999;**5**:673–81.
639. Barnes PJ. Mechanisms and resistance in glucocorticoid control of inflammation. *J Steroid Biochem Mol Biol* 2010;**120**:76–85.
640. Rogatsky I, Ivashkiv LB. Glucocorticoid modulation of cytokine signaling. *Tissue Antigens* 2006;**68**:1–12.
641. Mittelstadt PR, Galon J, Franchimont D, O'Shea JJ, Ashwell JD. Glucocorticoid-inducible genes that regulate T-cell function. *Ernst Schering Res Found Workshop* 2002;**319–39**.

642. Buttgereit F, Scheffold A. Rapid glucocorticoid effects on immune cells. *Steroids* 2002;**67**:529–34.
643. Fimmel S, Zouboulis CC. Influence of physiological androgen levels on wound healing and immune status in men. *Aging Male* 2005;**8**:166–74.
644. Chen CC, Parker Jr CR. Adrenal androgens and the immune system. *Semin Reprod Med* 2004;**22**:369–77.
645. Cutolo M, Seriolo B, Villaggio B, Pizzorni C, Craviotto C, Sulli A. Androgens and estrogens modulate the immune and inflammatory responses in rheumatoid arthritis. *Ann N Y Acad Sci* 2002;**966**:131–42.
646. Olsen NJ, Kovacs WJ. Effects of androgens on T and B lymphocyte development. *Immunol Res* 2001;**23**:281–8.
647. Olsen NJ, Kovacs WJ. Gonadal steroids and immunity. *Endocr Rev* 1996;**17**:369–84.
648. Dimeloe S, Nanzer A, Ryanna K, Hawrylowicz C. Regulatory T cells, inflammation and the allergic response-The role of glucocorticoids and Vitamin D. *J Steroid Biochem Mol Biol* 2010;**120**:86–95.
649. Miller J, Gallo RL. Vitamin D and innate immunity. *Dermatol Ther* 2010;**23**:13–22.
650. Kamen DL, Tangpricha V. Vitamin D and molecular actions on the immune system: modulation of innate and autoimmunity. *J Mol Med* 2010;**88**:441–50.
651. Maruotti N, Cantatore FP. Vitamin D and the immune system. *J Rheumatol* 2010;**37**:491–5.
652. Szodoray P, Nakken B, Gaal J, Jonsson R, Szegedi A, Zold E, et al. The complex role of vitamin D in autoimmune diseases. *Scand J Immunol* 2008;**68**:261–9.
653. Cutolo M, Otsa K, Uprus M, Paolino S, Seriolo B. Vitamin D in rheumatoid arthritis. *Autoimmun Rev* 2007;**7**:59–64.
654. Mathieu C, van EE, Decallonne B, Guilietti A, Gysemans C, Bouillon R, et al. Vitamin D and 1,25-dihydroxyvitamin D3 as modulators in the immune system. *J Steroid Biochem Mol Biol* 2004;**89–90**:449–52.
655. Griffin MD, Xing N, Kumar R. Vitamin D and its analogs as regulators of immune activation and antigen presentation. *Annu Rev Nutr* 2003;**23**:117–45.
656. Hattori N. Expression, regulation and biological actions of growth hormone (GH) and ghrelin in the immune system. *Growth Horm IGF Res* 2009;**19**:187–97.
657. Redelman D, Welniak LA, Taub D, Murphy WJ. Neuroendocrine hormones such as growth hormone and prolactin are integral members of the immunological cytokine network. *Cell Immunol* 2008;**252**:111–21.
658. Meazza C, Pagani S, Travaglino P, Bozzola M. Effect of growth hormone (GH) on the immune system. *Pediatr Endocrinol Rev* 2004;**1**(Suppl 3):490–5.
659. Jeay S, Sonenshein GE, Postel-Vinay MC, Kelly PA, Baixeras E. Growth hormone can act as a cytokine controlling survival and proliferation of immune cells: new insights into signaling pathways. *Mol Cell Endocrinol* 2002;**188**:1–7.
660. Dorshkind K, Horseman ND. The roles of prolactin, growth hormone, insulin-like growth factor-I, and thyroid hormones in lymphocyte development and function: insights from genetic models of hormone and hormone receptor deficiency. *Endocr Rev* 2000;**21**:292–312.
661. Yu-Lee LY. Prolactin modulation of immune and inflammatory responses. *Recent Prog Horm Res* 2002;**57**:435–55.
662. Vera-Lastra O, Jara LJ, Espinoza LR. Prolactin and autoimmunity. *Autoimmun Rev* 2002;**1**:360–4.
663. De Bellis A, Bizzarro A, Pivonello R, Lombardi G, Bellastella A. Prolactin and autoimmunity. *Pituitary* 2005;**8**:25–30.

664. Szczepanik M. Melatonin and its influence on immune system. *J Physiol Pharmacol* 2007;**58**(Suppl 6):115–24.
665. Carrillo-Vico A, Reiter RJ, Lardone PJ, Herrera JL, Fernandez-Montesinos R, Guerrero JM, et al. The modulatory role of melatonin on immune responsiveness. *Curr Opin Investig Drugs* 2006;**7**:423–31.
666. Maestroni GJ. The immunotherapeutic potential of melatonin. *Expert Opin Investig Drugs* 2001;**10**:467–76.
667. Hotchkiss AK, Nelson RJ. Melatonin and immune function: hype or hypothesis? *Crit Rev Immunol* 2002;**22**:351–71.
668. Cutolo M, Maestroni GJ. The melatonin-cytokine connection in rheumatoid arthritis. *Ann Rheum Dis* 2005;**64**:1109–11.
669. Klein JR. Physiological relevance of thyroid stimulating hormone and thyroid stimulating hormone receptor in tissues other than the thyroid. *Autoimmunity* 2003;**36**:417–21.
670. Helderman JH, Reynolds TC, Strom TB. The insulin receptor as a universal marker of activated lymphocytes. *Eur J Immunol* 1978;**8**:589–95.
671. Garcia NW, Greives TJ, Zysling DA, French SS, Chester EM, Demas GE. Exogenous insulin enhances humoural immune responses in short-day, but not long-day, Siberian hamsters (Phodopus sungorus). *Proc Biol Sci* 2010;**277**:2211–8.
672. Wurm S, Neumeier M, Weigert J, Wanninger J, Gerl M, Gindner A, et al. Insulin induces monocytic CXCL8 secretion by the mitogenic signalling pathway. *Cytokine* 2008;**44**:185–90.
673. LaPensee CR, Hugo ER, Ben-Jonathan N. Insulin stimulates interleukin-6 expression and release in LS14 human adipocytes through multiple signaling pathways. *Endocrinology* 2008;**149**:5415–22.
674. Hill AF, Polvino WJ, Wilson DB. The significance of glucose, insulin and potassium for immunology and oncology: a new model of immunity. *J Immune Based Ther Vaccines* 2005;**3**:5–16.
675. Helderman JH. The insulin receptor on activated immunocompetent cells. *Exp Gerontol* 1993;**28**:323–7.
676. Snow EC. Insulin and growth hormone function as minor growth factors that potentiate lymphocyte activation. *J Immunol* 1985;**135**:776s–8s.
677. Hasko G, Linden J, Cronstein B, Pacher P. Adenosine receptors: therapeutic aspects for inflammatory and immune diseases. *Nat Rev Drug Discov* 2008;**7**:759–70.
678. Bours MJ, Swennen EL, Di VF, Cronstein BN, Dagnelie PC. Adenosine 5'-triphosphate and adenosine as endogenous signaling molecules in immunity and inflammation. *Pharmacol Ther* 2006;**112**:358–404.
679. Kohm AP, Sanders VM. Norepinephrine and beta 2-adrenergic receptor stimulation regulate CD4+ T and B lymphocyte function in vitro and in vivo. *Pharmacol Rev* 2001;**53**:487–525.
680. Straub RH, Härle P. Sympathetic neurotransmitters in joint inflammation. *Rheum Dis Clin North Am* 2005;**31**:43–59.
681. Bedoui S, Miyake S, Straub RH, von Hörsten S, Yamamura T. More sympathy for autoimmunity with neuropeptide Y? *Trends Immunol* 2004;**25**:508–12.
682. Elenkov IJ, Wilder RL, Chrousos GP, Vizi ES. The sympathetic nervous system—an integrative interface between two supersystems: the brain and the immune system. *Pharmacol Rev* 2000;**52**:595–638.
683. Hasko G, Cronstein BN. Adenosine: an endogenous regulator of innate immunity. *Trends Immunol* 2004;**25**:33–9.
684. Straub RH, Wiest R, Strauch UG, Härle P, Schölmerich J. The role of the sympathetic nervous system in intestinal inflammation. *Gut* 2006;**55**:1640–9.

685. Bellinger DL, Millar BA, Perez S, Carter J, Wood C, Thyagarajan S, et al. Sympathetic modulation of immunity: relevance to disease. *Cell Immunol* 2008;**252**:27–56.
686. Finley MJ, Happel CM, Kaminsky DE, Rogers TJ. Opioid and nociceptin receptors regulate cytokine and cytokine receptor expression. *Cell Immunol* 2008;**252**:146–54.
687. Stefano GB, Kream R. Endogenous opiates, opioids, and immune function: evolutionary brokerage of defensive behaviors. *Semin Cancer Biol* 2008;**18**:190–8.
688. Eisenstein TK, Rahim RT, Feng P, Thingalaya NK, Meissler JJ. Effects of opioid tolerance and withdrawal on the immune system. *J Neuroimmune Pharmacol* 2006;**1**:237–49.
689. Martin-Kleiner I, Balog T, Gabrilovac J. Signal transduction induced by opioids in immune cells: a review. *Neuroimmunomodulation* 2006;**13**:1–7.
690. Sharp BM. Multiple opioid receptors on immune cells modulate intracellular signaling. *Brain Behav Immun* 2006;**20**:9–14.
691. Smith EM. Opioid peptides in immune cells. *Adv Exp Med Biol* 2003;**521**:51–68.
692. Sacerdote P. Opioid-induced immunosuppression. *Curr Opin Support Palliat Care* 2008;**2**:14–8.
693. Jessop DS. Endomorphins as agents for the treatment of chronic inflammatory disease. *Bio-Drugs* 2006;**20**:161–6.
694. Rosas-Ballina M, Tracey KJ. Cholinergic control of inflammation. *J Intern Med* 2009;**265**:663–79.
695. Van Der Zanden EP, Boeckxstaens GE, de Jonge WJ. The vagus nerve as a modulator of intestinal inflammation. *Neurogastroenterol Motil* 2009;**21**:6–17.
696. Hosoi T, Nomura Y. Functional role of acetylcholine in the immune system. *Front Biosci* 2004;**9**:2414–9.
697. Kawashima K, Fujii T. Expression of non-neuronal acetylcholine in lymphocytes and its contribution to the regulation of immune function. *Front Biosci* 2004;**9**:2063–85.
698. Kawashima K, Fujii T. Extraneuronal cholinergic system in lymphocytes. *Pharmacol Ther* 2000;**86**:29–48.
699. Kawashima K, Fujii T. The lymphocytic cholinergic system and its contribution to the regulation of immune activity. *Life Sci* 2003;**74**:675–96.
700. Yadav M, Goetzl EJ. Vasoactive intestinal peptide-mediated Th17 differentiation: an expanding spectrum of vasoactive intestinal peptide effects in immunity and autoimmunity. *Ann N Y Acad Sci* 2008;**1144**:83–9.
701. Delgado M, Ganea D. Anti-inflammatory neuropeptides: a new class of endogenous immunoregulatory agents. *Brain Behav Immun* 2008;**22**:1146–51.
702. Gonzalez-Rey E, Anderson P, Delgado M. Emerging roles of vasoactive intestinal peptide: a new approach for autoimmune therapy. *Ann Rheum Dis* 2007;**66**(Suppl 3):iii70–6.
703. Leceta J, Gomariz RP, Martinez C, Carrion M, Arranz A, Juarranz Y. Vasoactive intestinal peptide regulates Th17 function in autoimmune inflammation. *Neuroimmunomodulation* 2007;**14**:134–8.
704. Pittman QJ. A neuro-endocrine-immune symphony. *J Neuroendocrinol* 2011;**23**:1296–7.
705. Russell JA, Walley KR. Vasopressin and its immune effects in septic shock. *J Innate Immun* 2010;**2**:446–60.
706. Jessop DS. Neuropeptides in the immune system: functional roles in health and disease. *Front Horm Res* 2002;**29**:50–68.
707. Tuluc F, Lai JP, Kilpatrick LE, Evans DL, Douglas SD. Neurokinin 1 receptor isoforms and the control of innate immunity. *Trends Immunol* 2009;**30**:271–6.
708. Zhang Y, Berger A, Milne CD, Paige CJ. Tachykinins in the immune system. *Curr Drug Targets* 2006;**7**:1011–20.

709. O'Connor TM, O'Connell J, O'Brien DI, Goode T, Bredin CP, Shanahan F. The role of substance P in inflammatory disease. *J Cell Physiol* 2004;**201**:167–80.
710. Kraneveld AD, Nijkamp FP. Tachykinins and neuro-immune interactions in asthma. *Int Immunopharmacol* 2001;**1**:1629–50.
711. Holzer P. Implications of tachykinins and calcitonin gene-related peptide in inflammatory bowel disease. *Digestion* 1998;**59**:269–83.
712. Berczi I, Chalmers IM, Nagy E, Warrington RJ. The immune effects of neuropeptides. *Baillieres Clin Rheumatol* 1996;**10**:227–57.
713. Torii H, Hosoi J, Asahina A, Granstein RD. Calcitonin gene-related peptide and Langerhans cell function. *J Investig Dermatol Symp Proc* 1997;**2**:82–6.
714. Saldanha C, Tougas G, Grace E. Evidence for anti-inflammatory effect of normal circulating plasma cortisol. *Clin Exp Rheumatol* 1986;**4**:365–6.
715. Schauenstein K, Fassler R, Dietrich H, Schwarz S, Kromer G, Wick G. Disturbed immune-endocrine communication in autoimmune disease. Lack of corticosterone response to immune signals in obese strain chickens with spontaneous autoimmune thyroiditis. *J Immunol* 1987;**139**:1830–3.
716. Sternberg EM, Hill JM, Chrousos GP, Kamilaris T, Listwak SJ, Gold PW, et al. Inflammatory mediator-induced hypothalamic-pituitary-adrenal axis activation is defective in streptococcal cell wall arthritis- susceptible Lewis rats. *Proc Natl Acad Sci U S A* 1989;**86**:2374–8.
717. Mason D, MacPhee I, Antoni F. The role of the neuroendocrine system in determining genetic susceptibility to experimental allergic encephalomyelitis in the rat. *Immunology* 1990;**70**:1–5.
718. Million M, Tache Y, Anton P. Susceptibility of Lewis and Fischer rats to stress-induced worsening of TNB-colitis: protective role of brain CRF. *Am J Physiol* 1999;**276**:G1027–36.
719. Gutierrez MA, Garcia ME, Rodriguez JA, Mardonez G, Jacobelli S, Rivero S. Hypothalamic-pituitary-adrenal axis function in patients with active rheumatoid arthritis: a controlled study using insulin hypoglycemia stress test and prolactin stimulation. *J Rheumatol* 1999;**26**:277–81.
720. Imrich R, Lukac J, Rovensky J, Radikova Z, Penesova A, Kvetnansky R, et al. Lower adrenocortical and adrenomedullary responses to hypoglycemia in premenopausal women with systemic sclerosis. *J Rheumatol* 2006;**33**:2235–41.
721. Rovensky J, Bakosova J, Koska J, Ksinantova L, Jezova D, Vigas M. Somatotropic, lactotropic and adrenocortical responses to insulin-induced hypoglycemia in patients with rheumatoid arthritis. *Ann N Y Acad Sci* 2002;**966**:263–70.
722. Dekkers JC, Geenen R, Godaert GL, Glaudemans KA, Lafeber FP, van Doornen LJ, et al. Experimentally challenged reactivity of the hypothalamic pituitary adrenal axis in patients with recently diagnosed rheumatoid arthritis. *J Rheumatol* 2001;**28**:1496–504.
723. Pool AJ, Whipp BJ, Skasick AJ, Alavi A, Bland JM, Axford JS. Serum cortisol reduction and abnormal prolactin and CD4+/CD8+ T-cell response as a result of controlled exercise in patients with rheumatoid arthritis and systemic lupus erythematosus despite unaltered muscle energetics. *Rheumatology (Oxford)* 2004;**43**:43–8.
724. Straub RH, Dhabhar FS, Bijlsma JW, Cutolo M. How psychological stress via hormones and nerve fibers may exacerbate rheumatoid arthritis. *Arthritis Rheum* 2005;**52**:16–26.
725. Baerwald CG, Panayi GS, Lanchbury JS. Corticotropin releasing hormone promoter region polymorphisms in rheumatoid arthritis. *J Rheumatol* 1997;**24**:215–6.
726. DeRijk RH. Single nucleotide polymorphisms related to HPA axis reactivity. *Neuroimmunomodulation* 2009;**16**:340–52.

727. Stark K, Straub RH, Blazickova S, Hengstenberg C, Rovensky J. Genetics in neuroendocrine immunology: implications for rheumatoid arthritis and osteoarthritis. *Ann N Y Acad Sci* 2010;**1193**:10–4.
728. Harbuz MS, Jessop DS. Is there a defect in cortisol production in rheumatoid arthritis? *Rheumatology (Oxford)* 1999;**38**:298–302.
729. Jessop DS, Harbuz MS. A defect in cortisol production in rheumatoid arthritis: why are we still looking? *Rheumatology (Oxford)* 2005;**44**:1097–100.
730. Warren RS, Starnes Jr HF, Gabrilove JL, Oettgen HF, Brennan MF. The acute metabolic effects of tumor necrosis factor administration in humans. *Arch Surg* 1987;**122**:1396–400.
731. Späth-Schwalbe E, Born J, Schrezenmeier H, Bornstein SR, Stromeyer P, Drechsler S, et al. Interleukin-6 stimulates the hypothalamus-pituitary-adrenocortical axis in man. *J Clin Endocrinol Metab* 1994;**79**:1212–4.
732. Spath-Schwalbe E, Porzsolt F, Digel W, Born J, Kloss B, Fehm HL. Elevated plasma cortisol levels during interferon-gamma treatment. *Immunopharmacology* 1989;**17**:141–5.
733. Roosth J, Pollard RB, Brown SL, Meyer III WJ. Cortisol stimulation by recombinant interferon-alpha 2. *J Neuroimmunol* 1986;**12**:311–6.
734. Gisslinger H, Svoboda T, Clodi M, Gilly B, Ludwig H, Havelec L, et al. Interferon-alpha stimulates the hypothalamic-pituitary-adrenal axis in vivo and in vitro. *Neuroendocrinology* 1993;**57**:489–95.
735. Jablons DM, Mule JJ, McIntosh JK, Sehgal PB, May LT, Huang CM, et al. IL-6/IFN-beta-2 as a circulating hormone. Induction by cytokine administration in humans. *J Immunol* 1989;**142**:1542–7.
736. Tsigos C, Papanicolaou DA, Defensor R, Mitsiadis CS, Kyrou I, Chrousos GP. Dose effects of recombinant human interleukin-6 on pituitary hormone secretion and energy expenditure. *Neuroendocrinology* 1997;**66**:54–62.
737. Straub RH, Paimela L, Peltomaa R, Schölmerich J, Leirisalo-Repo M. Inadequately low serum levels of steroid hormones in relation to IL-6 and TNF in untreated patients with early rheumatoid arthritis and reactive arthritis. *Arthritis Rheum* 2002;**46**:654–62.
738. Straub RH, Besedovsky HO. Integrated evolutionary, immunological, and neuroendocrine framework for the pathogenesis of chronic disabling inflammatory diseases. *FASEB J* 2003;**17**:2176–83.
739. Crofford LJ, Kalogeras KT, Mastorakos G, Magiakou MA, Wells J, Kanik KS, et al. Circadian relationships between interleukin (IL)-6 and hypothalamic- pituitary-adrenal axis hormones: failure of IL-6 to cause sustained hypercortisolism in patients with early untreated rheumatoid arthritis. *J Clin Endocrinol Metab* 1997;**82**:1279–83.
740. Jäättelä M, Ilvesmaki V, Voutilainen R, Stenman UH, Saksela E. Tumor necrosis factor as a potent inhibitor of adrenocorticotropin- induced cortisol production and steroidogenic P450 enzyme gene expression in cultured human fetal adrenal cells. *Endocrinology* 1991;**128**:623–9.
741. Straub RH, Konecna L, Hrach S, Rothe G, Kreutz M, Schölmerich J, et al. Serum dehydroepiandrosterone (DHEA) and DHEA sulfate are negatively correlated with serum interleukin-6 (IL-6), and DHEA inhibits IL-6 secretion from mononuclear cells in man in vitro: possible link between endocrinosenescence and immunosenescence. *J Clin Endocrinol Metab* 1998;**83**:2012–7.
742. Edwards C. Sixty years after Hench–corticosteroids and chronic inflammatory disease. *J Clin Endocrinol Metab* 2012;**97**:1443–51.
743. Kirwan JR. The effect of glucocorticoids on joint destruction in rheumatoid arthritis. The Arthritis and Rheumatism Council Low-Dose Glucocorticoid Study Group. *N Engl J Med* 1995;**333**:142–6.

744. Boers M, Verhoeven AC, Markusse HM, van de Laar MA, Westhovens R, van Denderen JC, et al. Randomised comparison of combined step-down prednisolone, methotrexate and sulphasalazine with sulphasalazine alone in early rheumatoid arthritis. *Lancet* 1997;**350**:309–18.
745. van Everdingen AA, Jacobs JW, Siewertsz Van Reesema DR, Bijlsma JW. Low-dose prednisone therapy for patients with early active rheumatoid arthritis: clinical efficacy, disease-modifying properties, and side effects: a randomized, double-blind, placebo-controlled clinical trial. *Ann Intern Med* 2002;**136**:1–12.
746. Wassenberg S, Rau R, Steinfeld P, Zeidler H. Very low-dose prednisolone in early rheumatoid arthritis retards radiographic progression over two years: a multicenter, double-blind, placebo-controlled trial. *Arthritis Rheum* 2005;**52**:3371–80.
747. Cooper MS, Stewart PM. Corticosteroid insufficiency in acutely ill patients. *N Engl J Med* 2003;**348**:727–34.
748. Straub RH, Härle P, Sarzi-Puttini P, Cutolo M. Tumor necrosis factor-neutralizing therapies improve altered hormone axes: an alternative mode of antiinflammatory action. *Arthritis Rheum* 2006;**54**:2039–46.
749. Straub RH, Härle P, Yamana S, Matsuda T, Takasugi K, Kishimoto T, et al. Anti-interleukin-6 receptor antibody therapy favors adrenal androgen secretion in patients with rheumatoid arthritis: a randomized, double-blind, placebo-controlled study. *Arthritis Rheum* 2006;**54**:1778–85.
750. Straub RH, Pongratz G, Cutolo M, Wijbrandts CA, Baeten D, Fleck M, et al. Increased cortisol relative to adrenocorticotropic hormone predicts improvement during anti-tumor necrosis factor therapy in rheumatoid arthritis. *Arthritis Rheum* 2008;**58**:976–84.
751. Wolff C, Krinner K, Schroeder JA, Straub RH. Inadequate corticosterone levels relative to arthritic inflammation are accompanied by altered mitochondria/cholesterol breakdown in adrenal cortex: a steroid-inhibiting role of IL-1beta in rats. *Ann Rheum Dis* 2014. http://dx.doi.org/10.1136/annrheumdis-2013-203885.
752. Ruperto N, Brunner HI, Quartier P, Constantin T, Wulffraat N, Horneff G, et al. Two randomized trials of canakinumab in systemic juvenile idiopathic arthritis. *N Engl J Med* 2012;**367**:2396–406.
753. Goldbach-Mansky R. Immunology in clinic review series; focus on autoinflammatory diseases: update on monogenic autoinflammatory diseases: the role of interleukin (IL)-1 and an emerging role for cytokines beyond IL-1. *Clin Exp Immunol* 2012;**167**:391–404.
754. Larsen CM, Faulenbach M, Vaag A, Volund A, Ehses JA, Seifert B, et al. Interleukin-1-receptor antagonist in type 2 diabetes mellitus. *N Engl J Med* 2007;**356**:1517–26.
755. Lockshin MD. Sex ratio and rheumatic disease. *Autoimmun Rev* 2002;**1**:162–7.
756. Masi AT. Incidence of rheumatoid arthritis: do the observed age-sex interaction patterns support a role of androgenic-anabolic steroid deficiency in its pathogenesis? *Br J Rheumatol* 1994;**33**:697–9.
757. Uhlig T, Kvien TK, Glennas A, Smedstad LM, Forre O. The incidence and severity of rheumatoid arthritis, results from a county register in Oslo, Norway. *J Rheumatol* 1998;**25**:1078–84.
758. Grimaldi CM, Michael DJ, Diamond B. Cutting edge: expansion and activation of a population of autoreactive marginal zone B cells in a model of estrogen-induced lupus. *J Immunol* 2001;**167**:1886–90.
759. Magliozzi R, Howell O, Vora A, Serafini B, Nicholas R, Puopolo M, et al. Meningeal B-cell follicles in secondary progressive multiple sclerosis associate with early onset of disease and severe cortical pathology. *Brain* 2007;**130**:1089–104.
760. Hassan J, Yanni G, Hegarty V, Feighery C, Bresnihan B, Whelan A. Increased numbers of CD5+ B cells and T cell receptor (TCR) gamma delta+ T cells are associated with younger age of onset in rheumatoid arthritis (RA). *Clin Exp Immunol* 1996;**103**:353–6.

761. Athreya BH, Rafferty JH, Sehgal GS, Lahita RG. Adenohypophyseal and sex hormones in pediatric rheumatic diseases. *J Rheumatol* 1993;**20**:725–30.
762. Barry NN, McGuire JL, van Vollenhoven RF. Dehydroepiandrosterone in systemic lupus erythematosus: relationship between dosage, serum levels, and clinical response. *J Rheumatol* 1998;**25**:2352–6.
763. Cutolo M, Balleari E, Giusti M, Monachesi M, Accardo S. Sex hormone status of male patients with rheumatoid arthritis: evidence of low serum concentrations of testosterone at baseline and after human chorionic gonadotropin stimulation. *Arthritis Rheum* 1988;**31**:1314–7.
764. de la Torre B, Fransson J, Scheynius A. Blood dehydroepiandrosterone sulphate (DHEAS) levels in pemphigoid/pemphigus and psoriasis. *Clin Exp Rheumatol* 1995;**13**:345–8.
765. Deighton CM, Watson MJ, Walker DJ. Sex hormones in postmenopausal HLA-identical rheumatoid arthritis discordant sibling pairs. *J Rheumatol* 1992;**19**:1663–7.
766. Dessein PH, Joffe BI, Stanwix AE, Moomal Z. Hyposecretion of the adrenal androgen dehydroepiandrosterone sulfate and its relation to clinical variables in inflammatory arthritis. *Arthritis Res* 2001;**3**:183–8.
767. Feher KG, Feher T. Plasma dehydroepiandrosterone, dehydroepiandrosterone sulphate and androsterone sulphate levels and their interaction with plasma proteins in rheumatoid arthritis. *Exp Clin Endocrinol* 1984;**84**:197–202.
768. Giltay EJ, van Schaardenburg D, Gooren LJ, Dijkmans BA. Dehydroepiandrosterone sulfate in patients with rheumatoid arthritis. *Ann N Y Acad Sci* 1999;**876**:152–4.
769. Gordon D, Beastall GH, Thomson JA, Sturrock RD. Androgenic status and sexual function in males with rheumatoid arthritis and ankylosing spondylitis. *Q J Med* 1986;**60**:671–9.
770. Hall GM, Perry LA, Spector TD. Depressed levels of dehydroepiandrosterone sulphate in postmenopausal women with rheumatoid arthritis but no relation with axial bone density. *Ann Rheum Dis* 1993;**52**:211–4.
771. Hedman M, Nilsson E, de la Torre B. Low blood and synovial fluid levels of sulpho-conjugated steroids in rheumatoid arthritis. *Clin Exp Rheumatol* 1992;**10**:25–30.
772. Kanik KS, Chrousos GP, Schumacher HR, Crane ML, Yarboro CH, Wilder RL. Adrenocorticotropin, glucocorticoid, and androgen secretion in patients with new onset synovitis/rheumatoid arthritis: relations with indices of inflammation. *J Clin Endocrinol Metab* 2000;**85**:1461–6.
773. Lahita RG, Bradlow HL, Ginzler E, Pang S, New M. Low plasma androgens in women with systemic lupus erythematosus. *Arthritis Rheum* 1987;**30**:241–8.
774. Masi AT, Josipovic DB, Jefferson WE. Low adrenal androgenic-anabolic steroids in women with rheumatoid arthritis (RA): gas–liquid chromatographic studies of RA patients and matched normal control women indicating decreased 11-deoxy-17-ketosteroid excretion. *Semin Arthritis Rheum* 1984;**14**:1–23.
775. Mateo L, Nolla JM, Bonnin MR, Navarro MA, Roig-Escofet D. Sex hormone status and bone mineral density in men with rheumatoid arthritis. *J Rheumatol* 1995;**22**:1455–60.
776. Mirone L, Altomonte L, D'Agostino P, Zoli A, Barini A, Magaro M. A study of serum androgen and cortisol levels in female patients with rheumatoid arthritis. Correlation with disease activity. *Clin Rheumatol* 1996;**15**:15–9.
777. Nilsson E, del la TB, Hedman M, Goobar J, Thorner A. Blood dehydroepiandrosterone sulphate (DHEAS) levels in polymyalgia rheumatica/giant cell arteritis and primary fibromyalgia. *Clin Exp Rheumatol* 1994;**12**:415–7.
778. Sambrook PN, Eisman JA, Champion GD, Pocock NA. Sex hormone status and osteoporosis in postmenopausal women with rheumatoid arthritis. *Arthritis Rheum* 1988;**31**:973–8.

779. Spector TD, Ollier W, Perry LA, Silman AJ, Thompson PW, Edwards A. Free and serum testosterone levels in 276 males: a comparative study of rheumatoid arthritis, ankylosing spondylitis and healthy controls. *Clin Rheumatol* 1989;**8**:37–41.
780. Straub RH, Zeuner M, Antoniou E, Schölmerich J, Lang B. Dehydroepiandrosterone sulfate is positively correlated with soluble interleukin 2 receptor and soluble intercellular adhesion molecule in systemic lupus erythematosus. *J Rheumatol* 1996;**23**:856–61.
781. Tengstrand B, Carlstrom K, Fellander-Tsai L, Hafstrom I. Abnormal levels of serum dehydroepiandrosterone, estrone, and estradiol in men with rheumatoid arthritis: high correlation between serum estradiol and current degree of inflammation. *J Rheumatol* 2003;**30**:2338–43.
782. Soares PM, Borba EF, Bonfa E, Hallak J, Correa AL, Silva CA. Gonad evaluation in male systemic lupus erythematosus. *Arthritis Rheum* 2007;**56**:2352–61.
783. Weidler C, Struharova S, Schmidt M, Ugele B, Schölmerich J, Straub RH. Tumor necrosis factor inhibits conversion of dehydroepiandrosterone sulfate (DHEAS) to DHEA in rheumatoid arthritis synovial cells: a prerequisite for local androgen deficiency. *Arthritis Rheum* 2005;**52**:1721–9.
784. Dulos J, van der Vleuten MA, Kavelaars A, Heijnen CJ, Boots AM. CYP7B expression and activity in fibroblast-like synoviocytes from patients with rheumatoid arthritis: regulation by proinflammatory cytokines. *Arthritis Rheum* 2005;**52**:770–8.
785. Schmidt M, Hartung R, Capellino S, Cutolo M, Pfeifer-Leeg A, Straub RH. Estrone/17beta-estradiol conversion to, and tumor necrosis factor inhibition by, estrogen metabolites in synovial cells of patients with rheumatoid arthritis and patients with osteoarthritis. *Arthritis Rheum* 2009;**60**:2913–22.
786. Hanna FS, Bell RJ, Cicuttini FM, Davison SL, Wluka AE, Davis SR. The relationship between endogenous testosterone, preandrogens, and sex hormone binding globulin and knee joint structure in women at midlife. *Semin Arthritis Rheum* 2007;**37**:56–62.
787. Stark K, Straub RH, Rovensky J, Blazickova S, Eiselt G, Schmidt M. *CYPB5A* polymorphism increases androgens and reduces risk of rheumatoid artzhritis in women. *Arthritis Res Ther* 2015, in press.
788. Castagnetta LA, Cutolo M, Granata OM, Di Falco M, Bellavia V, Carruba G. Endocrine endpoints in rheumatoid arthritis. *Ann N Y Acad Sci* 1999;**876**:180–91.
789. Schmidt M, Weidler C, Naumann H, Schölmerich J, Straub RH. Androgen conversion in osteoarthritis and rheumatoid arthritis synoviocytes—androstenedione and testosterone inhibit estrogen formation and favor production of more potent 5alpha-reduced androgens. *Arthritis Res Ther* 2005;**7**:R938–48.
790. Weidler C, Härle P, Schedel J, Schmidt M, Schölmerich J, Straub RH. Patients with rheumatoid arthritis and systemic lupus erythematosus have increased renal excretion of mitogenic estrogens in relation to endogenous antiestrogens. *J Rheumatol* 2004;**31**:489–94.
791. Masi AT, Aldag JC, Chatterton RT. Neuroendocrine immune perturbations in rheumatoid arthritis: causes, consequences, or confounders in the disease process? *J Rheumatol* 2003;**30**:2302–5.
792. Masi AT, Elmore KB, Rehman AA, Chatterton RT, Goertzen NJ, Aldag JC. Lower serum androstenedione levels in pre-rheumatoid arthritis versus normal control women: correlations with lower serum cortisol levels. *Autoimmune Dis* 2013. http://dx.doi.org/10.1155/2013/593493.
793. Pikwer M, Giwercman A, Bergstrom U, Nilsson JA, Jacobsson LT, Turesson C. Association between testosterone levels and risk of future rheumatoid arthritis in men: a population-based case–control study. *Ann Rheum Dis* 2014;**73**:573–9.
794. Heikkila R, Aho K, Heliovaara M, Knekt P, Reunanen A, Aromaa A, et al. Serum androgen-anabolic hormones and the risk of rheumatoid arthritis. *Ann Rheum Dis* 1998;**57**:281–5.

795. Karlson EW, Chibnik LB, McGrath M, Chang SC, Keenan BT, Costenbader KH, et al. A prospective study of androgen levels, hormone-related genes and risk of rheumatoid arthritis. *Arthritis Res Ther* 2009;**11**:R97.
796. Crofford LJ, Sano H, Karalis K, Friedman TC, Epps HR, Remmers EF, et al. Corticotropin-releasing hormone in synovial fluids and tissues of patients with rheumatoid arthritis and osteoarthritis. *J Immunol* 1993;**151**:1587–96.
797. Murphy EP, McEvoy A, Conneely OM, Bresnihan B, Fitzgerald O. Involvement of the nuclear orphan receptor NURR1 in the regulation of corticotropin-releasing hormone expression and actions in human inflammatory arthritis. *Arthritis Rheum* 2001;**44**:782–93.
798. Webster EL, Barrientos RM, Contoreggi C, Isaac MG, Ligier S, Gabry KE, et al. Corticotropin releasing hormone (CRH) antagonist attenuates adjuvant induced arthritis: role of CRH in peripheral inflammation. *J Rheumatol* 2002;**29**:1252–61.
799. Walker SE, Jacobson JD. Roles of prolactin and gonadotropin-releasing hormone in rheumatic diseases. *Rheum Dis Clin North Am* 2000;**26**:713–36.
800. Matera L. Endocrine, paracrine and autocrine actions of prolactin on immune cells. *Life Sci* 1996;**59**:599–614.
801. Leanos-Miranda A, Cardenas-Mondragon G. Serum free prolactin concentrations in patients with systemic lupus erythematosus are associated with lupus activity. *Rheumatology (Oxford)* 2006;**45**:97–101.
802. Fojtikova M, Tomasova SJ, Filkova M, Lacinova Z, Gatterova J, Pavelka K, et al. Elevated prolactin levels in patients with rheumatoid arthritis: association with disease activity and structural damage. *Clin Exp Rheumatol* 2010;**28**:849–54.
803. Lange T, Dimitrov S, Fehm HL, Westermann J, Born J. Shift of monocyte function toward cellular immunity during sleep. *Arch Intern Med* 2006;**166**:1695–700.
804. Cutolo M, Maestroni GJ, Otsa K, Aakre O, Villaggio B, Capellino S, et al. Circadian melatonin and cortisol levels in rheumatoid arthritis patients in winter time: a north and south Europe comparison. *Ann Rheum Dis* 2005;**64**:212–6.
805. Smolders J, Damoiseaux J, Menheere P, Hupperts R. Vitamin D as an immune modulator in multiple sclerosis, a review. *J Neuroimmunol* 2008;**194**:7–17.
806. De Vito P, Incerpi S, Pedersen JZ, Luly P, Davis FB, Davis PJ. Thyroid hormones as modulators of immune activities at the cellular level. *Thyroid* 2011;**21**:879–90.
807. Nishizawa Y, Fushiki S, Amakata Y. Thyroxine-induced production of superoxide anion by human alveolar neutrophils and macrophages: a possible mechanism for the exacerbation of bronchial asthma with the development of hyperthyroidism. *In Vivo* 1998;**12**:253–7.
808. Foster MP, Montecino-Rodriguez E, Dorshkind K. Proliferation of bone marrow pro-B cells is dependent on stimulation by the pituitary/thyroid axis. *J Immunol* 1999;**163**:5883–90.
809. Huang SA, Bianco AC. Reawakened interest in type III iodothyronine deiodinase in critical illness and injury. *Nat Clin Pract Endocrinol Metab* 2008;**4**:148–55.
810. Torpy DJ, Tsigos C, Lotsikas AJ, Defensor R, Chrousos GP, Papanicolaou DA. Acute and delayed effects of a single-dose injection of interleukin-6 on thyroid function in healthy humans. *Metabolism* 1998;**47**:1289–93.
811. Kumar K, Kole AK, Karmakar PS, Ghosh A. The spectrum of thyroid disorders in systemic lupus erythematosus. *Rheumatol Int* 2012;**32**:73–8.
812. Hashimoto H, Igarashi N, Yachie A, Miyawaki T, Hashimoto T, Sato T. The relationship between serum levels of interleukin-6 and thyroid hormone during the follow-up study in children with nonthyroidal illness: marked inverse correlation in Kawasaki and infectious disease. *Endocr J* 1996;**43**:31–8.

813. Vanderschueren-Lodeweyckx M, Eggermont E, Cornette C, Beckers C, Malvaux P, Eeckels R. Decreased serum thyroid hormone levels and increased TSH response to TRH in infants with coeliac disease. *Clin Endocrinol (Oxf)* 1977;**6**:361–7.
814. Farthing MJ, Rees LH, Edwards CR, Byfield PG, Himsworth RL, Dawson AM. Thyroid hormones and the regulation of thyroid function in men with coeliac disease. *Clin Endocrinol (Oxf)* 1982;**16**:525–35.
815. Hotz J, Goebell H, Hartmann I, Forster S, Hackenberg K, Tharandt L. Endocrinologic findings in Crohn's disease. *Schweiz Med Wochenschr* 1981;**111**:214–20.
816. Chong SK, Grossman A, Walker-Smith JA, Rees LH. Endocrine dysfunction in children with Crohn's disease. *J Pediatr Gastroenterol Nutr* 1984;**3**:529–34.
817. Kiessling WR, Pflughaupt KW, Haubitz I, Mertens HG. Thyroid function in multiple sclerosis. *Acta Neurol Scand* 1980;**62**:255–8.
818. Durelli L, Oggero A, Verdun E, Isoardo GL, Barbero P, Bergamasco B, et al. Thyroid function and anti-thyroid antibodies in MS patients screened for interferon treatment. A multicenter study. *J Neurol Sci* 2001;**193**:17–22.
819. Bowness P, Shotliff K, Middlemiss A, Myles AB. Prevalence of hypothyroidism in patients with polymyalgia rheumatica and giant cell arteritis. *Br J Rheumatol* 1991;**30**:349–51.
820. Weismann K, Verdich J, Howitz J, Hummer L. Normal function of the thyroid gland in PUVA-treated psoriatics. *Acta Derm Venereol* 1980;**60**:432–4.
821. Templ E, Koeller M, Riedl M, Wagner O, Graninger W, Luger A. Anterior pituitary function in patients with newly diagnosed rheumatoid arthritis. *Br J Rheumatol* 1996;**35**:350–6.
822. Wellby ML, Kennedy JA, Pile K, True BS, Barreau P. Serum interleukin-6 and thyroid hormones in rheumatoid arthritis. *Metabolism* 2001;**50**:463–7.
823. Ozgen AG, Keser G, Erdem N, Aksu K, Gumusdis G, Kabalak T, et al. Hypothalamus-hypophysis-thyroid axis, triiodothyronine and antithyroid antibodies in patients with primary and secondary Sjogren's syndrome. *Clin Rheumatol* 2001;**20**:44–8.
824. Gordon MB, Klein I, Dekker A, Rodnan GP, Medsger Jr TA. Thyroid disease in progressive systemic sclerosis: increased frequency of glandular fibrosis and hypothyroidism. *Ann Intern Med* 1981;**95**:431–5.
825. Molnar I, Czirjak L. Euthyroid sick syndrome and inhibitory effect of sera on the activity of thyroid 5'-deiodinase in systemic sclerosis. *Clin Exp Rheumatol* 2000;**18**:719–24.
826. Miller FW, Moore GF, Weintraub BD, Steinberg AD. Prevalence of thyroid disease and abnormal thyroid function test results in patients with systemic lupus erythematosus. *Arthritis Rheum* 1987;**30**:1124–31.
827. Volzke H, Krohn U, Wallaschofski H, Ludemann J, John U, Kerner W. The spectrum of thyroid disorders in adult type 1 diabetes mellitus. *Diabetes Metab Res Rev* 2007;**23**:227–33.
828. Umpierrez GE, Latif KA, Murphy MB, Lambeth HC, Stentz F, Bush A, et al. Thyroid dysfunction in patients with type 1 diabetes: a longitudinal study. *Diabetes Care* 2003;**26**:1181–5.
829. Raterman HG, Jamnitski A, Lems WF, Voskuyl AE, Dijkmans BA, Bos WH, et al. Improvement of thyroid function in hypothyroid patients with rheumatoid arthritis after 6 months of adalimumab treatment: a pilot study. *J Rheumatol* 2011;**38**:247–51.
830. Boelen A, Kwakkel J, Fliers E. Beyond low plasma T3: local thyroid hormone metabolism during inflammation and infection. *Endocr Rev* 2011;**32**:670–93.
831. Garcia-Leme J, Fortes ZB, Sannomiya P, Farsky SP. Insulin, glucocorticoids and the control of inflammatory responses. *Agents Actions Suppl* 1992;**36**:99–118.
832. Hyun E, Ramachandran R, Hollenberg MD, Vergnolle N. Mechanisms behind the anti-inflammatory actions of insulin. *Crit Rev Immunol* 2011;**31**:307–40.

833. Dandona P, Aljada A, Mohanty P. The anti-inflammatory and potential anti-atherogenic effect of insulin: a new paradigm. *Diabetologia* 2002;**45**:924–30.
834. Calder PC, Dimitriadis G, Newsholme P. Glucose metabolism in lymphoid and inflammatory cells and tissues. *Curr Opin Clin Nutr Metab Care* 2007;**10**:531–40.
835. Heemskerk VH, Daemen MA, Buurman WA. Insulin-like growth factor-1 (IGF-1) and growth hormone (GH) in immunity and inflammation. *Cytokine Growth Factor Rev* 1999;**10**:5–14.
836. Clark R. The somatogenic hormones and insulin-like growth factor-1: stimulators of lymphopoiesis and immune function. *Endocr Rev* 1997;**18**:157–79.
837. Svenson KL, Lundqvist G, Wide L, Hallgren R. Impaired glucose handling in active rheumatoid arthritis: relationship to the secretion of insulin and counter-regulatory hormones. *Metabolism* 1987;**36**:940–3.
838. Dessein PH, Joffe BI. Insulin resistance and impaired beta cell function in rheumatoid arthritis. *Arthritis Rheum* 2006;**54**:2765–75.
839. El Magadmi M, Ahmad Y, Turkie W, Yates AP, Sheikh N, Bernstein RM, et al. Hyperinsulinemia, insulin resistance, and circulating oxidized low density lipoprotein in women with systemic lupus erythematosus. *J Rheumatol* 2006;**33**:50–6.
840. Sarzi-Puttini P, Atzeni F, Schölmerich J, Cutolo M, Straub RH. Anti-TNF antibody therapy improves glucocorticoid- induced insulin-like growth factor-1 (IGF-1) resistance without influencing myoglobin and IGF-1 binding proteins 1 and 3. *Ann Rheum Dis* 2005;**65**:301–5.
841. Bennett AE, Silverman ED, Miller III JJ, Hintz RL. Insulin-like growth factors I and II in children with systemic onset juvenile arthritis. *J Rheumatol* 1988;**15**:655–8.
842. Johansson AG, Baylink DJ, af EE, Lindh E, Mohan S, Ljunghall S. Circulating levels of insulin-like growth factor-I and -II, and IGF-binding protein-3 in inflammation and after parathyroid hormone infusion. *Bone Miner* 1994;**24**:25–31.
843. Lemmey A, Maddison P, Breslin A, Cassar P, Hasso N, McCann R, et al. Association between insulin-like growth factor status and physical activity levels in rheumatoid arthritis. *J Rheumatol* 2001;**28**:29–34.
844. Andreassen M, Frystyk J, Faber J, Kristensen LO. GH activity and markers of inflammation: a crossover study in healthy volunteers treated with GH and a GH receptor antagonist. *Eur J Endocrinol* 2012;**166**:811–9.
845. Masternak MM, Bartke A. Growth hormone, inflammation and aging. *Pathobiol Aging Age Relat Dis* 2012;**2**:1–6 [ID 17293].
846. Denko CW, Malemud CJ. Role of the growth hormone/insulin-like growth factor-1 paracrine axis in rheumatic diseases. *Semin Arthritis Rheum* 2005;**35**:24–34.
847. Coari G, Di FM, Iagnocco A, Di Novi MR, Mauceri MT, Ciocci A. Intra-articular somatostatin 14 reduces synovial thickness in rheumatoid arthritis: an ultrasonographic study. *Int J Clin Pharmacol Res* 1995;**15**:27–32.
848. Paran D, Elkayam O, Mayo A, Paran H, Amit M, Yaron M, et al. A pilot study of a long acting somatostatin analogue for the treatment of refractory rheumatoid arthritis. *Ann Rheum Dis* 2001;**60**:888–91.
849. Koseoglu F, Koseoglu T. Long acting somatostatin analogue for the treatment of refractory RA. *Ann Rheum Dis* 2002;**61**:573–4.
850. Imhof AK, Gluck L, Gajda M, Lupp A, Brauer R, Schaible HG, et al. Differential antiinflammatory and antinociceptive effects of the somatostatin analogs octreotide and pasireotide in a mouse model of immune-mediated arthritis. *Arthritis Rheum* 2011;**63**:2352–62.
851. Pozzo AM, Kemp SF. Growth and growth hormone treatment in children with chronic diseases. *Endocrinol Metab Clin North Am* 2012;**41**:747–59.

852. Benigni A, Cassis P, Remuzzi G. Angiotensin II revisited: new roles in inflammation, immunology and aging. *EMBO Mol Med* 2010;**2**:247–57.
853. Crowley SD, Vasievich MP, Ruiz P, Gould SK, Parsons KK, Pazmino AK, et al. Glomerular type 1 angiotensin receptors augment kidney injury and inflammation in murine autoimmune nephritis. *J Clin Invest* 2009;**119**:943–53.
854. Dalbeth N, Edwards J, Fairchild S, Callan M, Hall FC. The non-thiol angiotensin-converting enzyme inhibitor quinapril suppresses inflammatory arthritis. *Rheumatology (Oxford)* 2005;**44**:24–31.
855. Sagawa K, Nagatani K, Komagata Y, Yamamoto K. Angiotensin receptor blockers suppress antigen-specific T cell responses and ameliorate collagen-induced arthritis in mice. *Arthritis Rheum* 2005;**52**:1920–8.
856. Cheng J, Ke Q, Jin Z, Wang H, Kocher O, Morgan JP, et al. Cytomegalovirus infection causes an increase of arterial blood pressure. *PLoS Pathog* 2009;**5**:e1000427.
857. Ohtani R, Ohashi Y, Muranaga K, Itoh N, Okamoto H. Changes in activity of the renin-angiotensin system of the rat by induction of acute inflammation. *Life Sci* 1989;**44**:237–41.
858. Doerschug KC, Delsing AS, Schmidt GA, Ashare A. Renin-angiotensin system activation correlates with microvascular dysfunction in a prospective cohort study of clinical sepsis. *Crit Care* 2010;**14**:R24.
859. Walsh DA, Catravas J, Wharton J. Angiotensin converting enzyme in human synovium: increased stromal [(125)I]351A binding in rheumatoid arthritis. *Ann Rheum Dis* 2000;**59**:125–31.
860. Samoriadova OS, Zharova EA, Masenko VP, Balabanova RM, Vil'chinskaia MI, Nasonov EL. The renin-angiotensin-aldosterone system and arterial hypertension in patients with rheumatoid arthritis. *Klin Med (Mosk)* 1991;**69**:69–71.
861. Shilkina NP, Stoliarova SA, Iunonin IE, Driazhenkova IV. Neurohumoral regulation of blood pressure in rheumatic patients. *Ter Arkh* 2009;**81**:37–41.
862. Otero M, Lago R, Gomez R, Dieguez C, Lago F, Gomez-Reino J, et al. Towards a pro-inflammatory and immunomodulatory emerging role of leptin. *Rheumatology (Oxford)* 2006;**45**:944–50.
863. Kirchgessner TG, Uysal KT, Wiesbrock SM, Marino MW, Hotamisligil GS. Tumor necrosis factor-alpha contributes to obesity-related hyperleptinemia by regulating leptin release from adipocytes. *J Clin Invest* 1997;**100**:2777–82.
864. Zumbach MS, Boehme MW, Wahl P, Stremmel W, Ziegler R, Nawroth PP. Tumor necrosis factor increases serum leptin levels in humans. *J Clin Endocrinol Metab* 1997;**82**:4080–2.
865. Faggioni R, Fantuzzi G, Fuller J, Dinarello CA, Feingold KR, Grunfeld C. IL-1 beta mediates leptin induction during inflammation. *Am J Physiol* 1998;**274**:R204–8.
866. Finck BN, Kelley KW, Dantzer R, Johnson RW. In vivo and in vitro evidence for the involvement of tumor necrosis factor-alpha in the induction of leptin by lipopolysaccharide. *Endocrinology* 1998;**139**:2278–83.
867. Finck BN, Johnson RW. Tumor necrosis factor (TNF)-alpha induces leptin production through the p55 TNF receptor. *Am J Physiol Regul Integr Comp Physiol* 2000;**278**:R537–43.
868. Zhang HH, Kumar S, Barnett AH, Eggo MC. Tumour necrosis factor-alpha exerts dual effects on human adipose leptin synthesis and release. *Mol Cell Endocrinol* 2000;**159**:79–88.
869. Santos-Alvarez J, Goberna R, Sanchez-Margalet V. Human leptin stimulates proliferation and activation of human circulating monocytes. *Cell Immunol* 1999;**194**:6–11.
870. Deng J, Liu Y, Yang M, Wang S, Zhang M, Wang X, et al. Leptin exacerbates collagen-induced arthritis via enhancement of Th17 cell response. *Arthritis Rheum* 2012;**64**:3564–73.
871. Wilson CA, Bekele G, Nicolson M, Ravussin E, Pratley RE. Relationship of the white blood cell count to body fat: role of leptin. *Br J Haematol* 1997;**99**:447–51.

872. Faggioni R, Feingold KR, Grunfeld C. Leptin regulation of the immune response and the immunodeficiency of malnutrition. *FASEB J* 2001;**15**:2565–71.
873. Bornstein SR, Licinio J, Tauchnitz R, Engelmann L, Negrao AB, Gold P, et al. Plasma leptin levels are increased in survivors of acute sepsis: associated loss of diurnal rhythm, in cortisol and leptin secretion. *J Clin Endocrinol Metab* 1998;**83**:280–3.
874. Anders HJ, Rihl M, Heufelder A, Loch O, Schattenkirchner M. Leptin serum levels are not correlated with disease activity in patients with rheumatoid arthritis. *Metabolism* 1999;**48**:745–8.
875. Garcia-Gonzalez A, Gonzalez-Lopez L, Valera-Gonzalez IC, Cardona-Munoz EG, Salazar-Paramo M, Gonzalez-Ortiz M, et al. Serum leptin levels in women with systemic lupus erythematosus. *Rheumatol Int* 2002;**22**:138–41.
876. Kimata H. Elevated serum leptin in AEDS. *Allergy* 2002;**57**:179.
877. Härle P, Pongratz G, Weidler C, Büttner R, Schölmerich J, Straub RH. Possible role of leptin in hypoandrogenicity in patients with systemic lupus erythematosus and rheumatoid arthritis. *Ann Rheum Dis* 2004;**63**:809–16.
878. Lago F, Dieguez C, Gomez-Reino J, Gualillo O. Adipokines as emerging mediators of immune response and inflammation. *Nat Clin Pract Rheumatol* 2007;**3**:716–24.
879. Ehling A, Schaffler A, Herfarth H, Tarner IH, Anders S, Distler O, et al. The potential of adiponectin in driving arthritis. *J Immunol* 2006;**176**:4468–78.
880. Neumann E, Frommer KW, Vasile M, Muller-Ladner U. Adipocytokines as driving forces in rheumatoid arthritis and related inflammatory diseases? *Arthritis Rheum* 2011;**63**:1159–69.
881. Herrada AA, Contreras FJ, Marini NP, Amador CA, Gonzalez PA, Cortes CM, et al. Aldosterone promotes autoimmune damage by enhancing Th17-mediated immunity. *J Immunol* 2010;**184**:191–202.
882. Stepien T, Lawnicka H, Komorowski J, Stepien H, Siejka A. Growth hormone-releasing hormone stimulates the secretion of interleukin 17 from human peripheral blood mononuclear cells in vitro. *Neuro Endocrinol Lett* 2010;**31**:852–6.
883. Liu M, Hu X, Wang Y, Peng F, Yang Y, Chen X, et al. Effect of high-dose methylprednisolone treatment on Th17 cells in patients with multiple sclerosis in relapse. *Acta Neurol Scand* 2009;**120**:235–41.
884. Miljkovic Z, Momcilovic M, Miljkovic D, Mostarica-Stojkovic M. Methylprednisolone inhibits IFN-gamma and IL-17 expression and production by cells infiltrating central nervous system in experimental autoimmune encephalomyelitis. *J Neuroinflammation* 2009;**6**:37–46.
885. Muls N, Jnaoui K, Dang HA, Wauters A, Van SJ, Sindic CJ, et al. Upregulation of IL-17, but not of IL-9, in circulating cells of CIS and relapsing MS patients. Impact of corticosteroid therapy on the cytokine network. *J Neuroimmunol* 2012;**243**:73–80.
886. Wang C, Dehghani B, Li Y, Kaler LJ, Vandenbark AA, Offner H. Oestrogen modulates experimental autoimmune encephalomyelitis and interleukin-17 production via programmed death 1. *Immunology* 2009;**126**:329–35.
887. Relloso M, Aragoneses-Fenoll L, Lasarte S, Bourgeois C, Romera G, Kuchler K, et al. Estradiol impairs the Th17 immune response against Candida albicans. *J Leukoc Biol* 2012;**91**:159–65.
888. Lelu K, Laffont S, Delpy L, Paulet PE, Perinat T, Tschanz SA, et al. Estrogen receptor alpha signaling in T lymphocytes is required for estradiol-mediated inhibition of Th1 and Th17 cell differentiation and protection against experimental autoimmune encephalomyelitis. *J Immunol* 2011;**187**:2386–93.
889. Plum SM, Park EJ, Strawn SJ, Moore EG, Sidor CF, Fogler WE. Disease modifying and antiangiogenic activity of 2-methoxyestradiol in a murine model of rheumatoid arthritis. *BMC Musculoskelet Disord* 2009;**10**(46):1–13.

890. Tyagi AM, Srivastava K, Mansoori MN, Trivedi R, Chattopadhyay N, Singh D. Estrogen deficiency induces the differentiation of IL-17 secreting Th17 cells: a new candidate in the pathogenesis of osteoporosis. *PLoS One* 2012;**7**:e44552.
891. Yates MA, Li Y, Chlebeck P, Proctor T, Vandenbark AA, Offner H. Progesterone treatment reduces disease severity and increases IL-10 in experimental autoimmune encephalomyelitis. *J Neuroimmunol* 2010;**220**:136–9.
892. Lee JH, Ulrich B, Cho J, Park J, Kim CH. Progesterone promotes differentiation of human cord blood fetal T cells into T regulatory cells but suppresses their differentiation into Th17 cells. *J Immunol* 2011;**187**:1778–87.
893. Tang J, Zhou R, Luger D, Zhu W, Silver PB, Grajewski RS, et al. Calcitriol suppresses antiretinal autoimmunity through inhibitory effects on the Th17 effector response. *J Immunol* 2009;**182**:4624–32.
894. Jeffery LE, Burke F, Mura M, Zheng Y, Qureshi OS, Hewison M, et al. 1,25-Dihydroxyvitamin D3 and IL-2 combine to inhibit T cell production of inflammatory cytokines and promote development of regulatory T cells expressing CTLA-4 and FoxP3. *J Immunol* 2009;**183**:5458–67.
895. Colin EM, Asmawidjaja PS, van Hamburg JP, Mus AM, van DM, Hazes JM, et al. 1,25-dihydroxyvitamin D3 modulates Th17 polarization and interleukin-22 expression by memory T cells from patients with early rheumatoid arthritis. *Arthritis Rheum* 2010;**62**:132–42.
896. Chang SH, Chung Y, Dong C. Vitamin D suppresses Th17 cytokine production by inducing C/EBP homologous protein (CHOP) expression. *J Biol Chem* 2010;**285**:38751–5.
897. Joshi S, Pantalena LC, Liu XK, Gaffen SL, Liu H, Rohowsky-Kochan C, et al. 1,25-Dihydroxyvitamin D(3) ameliorates Th17 autoimmunity via transcriptional modulation of interleukin-17A. *Mol Cell Biol* 2011;**31**:3653–69.
898. van Hamburg JP, Asmawidjaja PS, Davelaar N, Mus AM, Cornelissen F, van Leeuwen JP, et al. TNF blockade requires 1,25(OH)2D3 to control human Th17-mediated synovial inflammation. *Ann Rheum Dis* 2012;**71**:606–12.
899. Tian Y, Wang C, Ye Z, Xiao X, Kijlstra A, Yang P. Effect of 1,25-dihydroxyvitamin D3 on Th17 and Th1 response in patients with Behcet's disease. *Invest Ophthalmol Vis Sci* 2012;**53**:6434–41.
900. De Rosa V, Procaccini C, Cali G, Pirozzi G, Fontana S, Zappacosta S, et al. A key role of leptin in the control of regulatory T cell proliferation. *Immunity* 2007;**26**:241–55.
901. Tiittanen M, Huupponen JT, Knip M, Vaarala O. Insulin treatment in patients with type 1 diabetes induces upregulation of regulatory T-cell markers in peripheral blood mononuclear cells stimulated with insulin in vitro. *Diabetes* 2006;**55**:3446–54.
902. Taylor AW, Lee DJ. The alpha-melanocyte stimulating hormone induces conversion of effector T cells into treg cells. *J Transplant* 2011;1–7 [ID 246856].
903. Auriemma M, Brzoska T, Klenner L, Kupas V, Goerge T, Voskort M, et al. Alpha-MSH-stimulated tolerogenic dendritic cells induce functional regulatory T cells and ameliorate ongoing skin inflammation. *J Invest Dermatol* 2012;**132**:1814–24.
904. Polanczyk MJ, Carson BD, Subramanian S, Afentoulis M, Vandenbark AA, Ziegler SF, et al. Cutting edge: estrogen drives expansion of the CD4+CD25+ regulatory T cell compartment. *J Immunol* 2004;**173**:2227–30.
905. Prieto GA, Rosenstein Y. Oestradiol potentiates the suppressive function of human CD4 CD25 regulatory T cells by promoting their proliferation. *Immunology* 2006;**118**:58–65.
906. Tai P, Wang J, Jin H, Song X, Yan J, Kang Y, et al. Induction of regulatory T cells by physiological level estrogen. *J Cell Physiol* 2008;**214**:456–64.
907. Luo CY, Wang L, Sun C, Li DJ. Estrogen enhances the functions of CD4(+)CD25(+) Foxp3(+) regulatory T cells that suppress osteoclast differentiation and bone resorption in vitro. *Cell Mol Immunol* 2011;**8**:50–8.

908. Mao G, Wang J, Kang Y, Tai P, Wen J, Zou Q, et al. Progesterone increases systemic and local uterine proportions of CD4+CD25+ Treg cells during midterm pregnancy in mice. *Endocrinology* 2010;**151**:5477–88.
909. Lee JH, Lydon JP, Kim CH. Progesterone suppresses the mTOR pathway and promotes generation of induced regulatory T cells with increased stability. *Eur J Immunol* 2012;**42**:2683–96.
910. Fijak M, Schneider E, Klug J, Bhushan S, Hackstein H, Schuler G, et al. Testosterone replacement effectively inhibits the development of experimental autoimmune orchitis in rats: evidence for a direct role of testosterone on regulatory T cell expansion. *J Immunol* 2011;**186**:5162–72.
911. Chen X, Oppenheim JJ, Winkler-Pickett RT, Ortaldo JR, Howard OM. Glucocorticoid amplifies IL-2-dependent expansion of functional FoxP3(+)CD4(+)CD25(+) T regulatory cells in vivo and enhances their capacity to suppress EAE. *Eur J Immunol* 2006;**36**:2139–49.
912. Ling Y, Cao X, Yu Z, Ruan C. Circulating dendritic cells subsets and CD4+Foxp3+ regulatory T cells in adult patients with chronic ITP before and after treatment with high-dose dexamethasome. *Eur J Haematol* 2007;**79**:310–6.
913. Azab NA, Bassyouni IH, Emad Y, Abd El-Wahab GA, Hamdy G, Mashahit MA. CD4+CD25+ regulatory T cells (TREG) in systemic lupus erythematosus (SLE) patients: the possible influence of treatment with corticosteroids. *Clin Immunol* 2008;**127**:151–7.
914. Braitch M, Harikrishnan S, Robins RA, Nichols C, Fahey AJ, Showe L, et al. Glucocorticoids increase CD4CD25 cell percentage and Foxp3 expression in patients with multiple sclerosis. *Acta Neurol Scand* 2009;**119**:239–45.
915. Xie Y, Wu M, Song R, Ma J, Shi Y, Qin W, et al. A glucocorticoid amplifies IL-2-induced selective expansion of CD4(+)CD25(+)FOXP3(+) regulatory T cells in vivo and suppresses graft-versus-host disease after allogeneic lymphocyte transplantation. *Acta Biochim Biophys Sin* 2009;**41**:781–91.
916. Stary G, Klein I, Bauer W, Koszik F, Reininger B, Kohlhofer S, et al. Glucocorticosteroids modify Langerhans cells to produce TGF-beta and expand regulatory T cells. *J Immunol* 2011;**186**:103–12.
917. Daniel C, Sartory NA, Zahn N, Radeke HH, Stein JM. Immune modulatory treatment of trinitrobenzene sulfonic acid colitis with calcitriol is associated with a change of a T helper (Th) 1/Th17 to a Th2 and regulatory T cell profile. *J Pharmacol Exp Ther* 2008;**324**:23–33.
918. Baeke F, Korf H, Overbergh L, Verstuyf A, Thorrez L, Van LL, et al. The vitamin D analog, TX527, promotes a human CD4+CD25highCD127low regulatory T cell profile and induces a migratory signature specific for homing to sites of inflammation. *J Immunol* 2011;**186**:132–42.
919. van der Aar AM, Sibiryak DS, Bakdash G, van Capel TM, van der Kleij HP, Opstelten DJ, et al. Vitamin D3 targets epidermal and dermal dendritic cells for induction of distinct regulatory T cells. *J Allergy Clin Immunol* 2011;**127**:1532–40.
920. Kang SW, Kim SH, Lee N, Lee WW, Hwang KA, Shin MS, et al. 1,25-Dihyroxyvitamin D3 promotes FOXP3 expression via binding to vitamin D response elements in its conserved noncoding sequence region. *J Immunol* 2012;**188**:5276–82.
921. Straub RH. Neuronal regulation of inflammation & related pain mechanisms. In: Firestein GS, Budd RC, Gabriel SE, McInnes IB, O'Dell JR, editors. Kelley's textbook of rheumatology. Philadelphia, PA: Elsevier; 2011.
922. Stricker S. Untersuchungen über die Gefäßwurzel des Ischiadicus. *Ber Akad Wiss Wien* 1876;**3**:173–85.
923. Bayliss WM. On the origin from the spinal cord of the vaso-dilator fibres of the hind-limb, and on the nature of these fibres. *J Physiol* 1901;**26**:173–209.

924. Bruce AN. Über die Beziehung der sensiblen Nervenendigungen zum Entzündungsvorgang. *Arch Exp Pathol Pharmakol* 1910;**63**:424–33.
925. Breslauer F. Die Pathogenese des trophischen Gewebeschadens nach der Nervenverletzung. *Chir Deut Z* 1919;**150**:50–81.
926. Lewis T. Experiments relating to cutaneous hyperalgesia and its spread through somatic nerves. *Clin Sci* 1936;**2**:373–423.
927. Chapman LF, Ramos A, Goodell H, Wolff HG. Neurokinin—a polypeptide formed during neuronal activity in man. Observations on the axon reflex and antidromic dorsal root stimulation. *Trans Am Neurol Assoc* 1960;**85**:42–5.
928. Kelly M. The neurogenic factor in rheumatic inflammation. *Med J Aust* 1951;**1**:859–64.
929. Jancso N, Jancso-Gabor A, Szolcsanyi J. Direct evidence for neurogenic inflammation and its prevention by denervation and by pretreatment with capsaicin. *Br J Pharmacol Chemother* 1967;**31**:138–51.
930. Bellinger DL, Lorton D, Lubahn C, Felten DL. Innervation of lymphoid organs—association of nerves with cells of the immune system and their implications in disease. In: Ader R, Felten DL, Cohen N, editors. *Psychoneuroimmunology*. San Diego, CA: Academic Press; 2001. p. 55–111.
931. Straub RH. Complexity of the bi-directional neuroimmune junction in the spleen. *Trends Pharmacol Sci* 2004;**25**:640–6.
932. Baerwald C, Graefe C, von Wichert P, Krause A. Decreased density of beta-adrenergic receptors on peripheral blood mononuclear cells in patients with rheumatoid arthritis. *J Rheumatol* 1992;**19**:204–10.
933. Lorton D, Bellinger DL, Schaller JA, Shewmaker E, Osredkar T, Lubahn C. Altered sympathetic-to-immune cell signaling via beta(2)-adrenergic receptors in adjuvant arthritis. *Clin Dev Immunol* 2013;1–17 [ID 764395].
934. Glick EN. Asymmetrical rheumatoid arthritis after poliomyelitis. *Br Med J* 1967;**3**:26–8.
935. Jacqueline F. A case of evolutive polyarthritis with localisation controlateral to a hemiplegia. *Rev Rhum Mal Osteoartic* 1953;**20**:323–4.
936. Thompson M, Bywaters EGL. Unilateral rheumatoid arthritis following hemiplegia. *Ann Rheum Dis* 1962;**21**:370.
937. Bland JH, Eddy WM. Hemiplegia and rheumatoid hemiarthritis. *Arthritis Rheum* 1968;**11**: 72–80.
938. Garwolinska H. Effect of hemiplegia on the course of rheumatoid arthritis. *Reumatologia* 1972;**10**:259–61.
939. Velayos EE, Cohen BS. The effect of stroke on well-established rheumatoid arthritis. *Md State Med J* 1972;**21**:38–42.
940. Yaghmai I, Rooholamini SM, Faunce HF. Unilateral rheumatoid arthritis: protective effect of neurologic deficits. *AJR Am J Roentgenol* 1977;**128**:299–301.
941. Smith RD. Effect of hemiparesis on rheumatoid arthritis. *Arthritis Rheum* 1979;**22**:1419–20.
942. Carcassi A, Boschi S, Tundo G, Macri P. Unilateral rheumatoid arthritis. *Minerva Med* 1981;**72**:951–6.
943. Ueno Y, Sawada K, Imura H. Protective effect of neural lesion on rheumatoid arthritis. *Arthritis Rheum* 1983;**26**:118.
944. Hamilton S. Unilateral rheumatoid arthritis in hemiplegia. *J Can Assoc Radiol* 1983;**34**:49–50.
945. Nakamura K, Akizuki M, Kimura A, Chino N. A case of polyarthritis developed on the non-paralytic side in a hemiplegic patient. *Ryumachi* 1994;**34**:656–61.
946. Lapadula G, Iannone F, Zuccaro C, Covelli M, Grattagliano V, Pipitone V. Recovery of erosive rheumatoid arthritis after human immunodeficiency virus-1 infection and hemiplegia. *J Rheumatol* 1997;**24**:747–51.

947. Keyszer G, Langer T, Kornhuber M, Taute B, Horneff G. Neurovascular mechanisms as a possible cause of remission of rheumatoid arthritis in hemiparetic limbs. *Ann Rheum Dis* 2004;**63**:1349–51.
948. Dolan AL. Asymmetric rheumatoid vasculitis in a hemiplegic patient. *Ann Rheum Dis* 1995;**54**:532.
949. Glynn JJ, Clayton ML. Sparing effect of hemiplegia on tophaceous gout. *Ann Rheum Dis* 1976;**35**:534–5.
950. Sethi S, Sequeira W. Sparing effect of hemiplegia on scleroderma. *Ann Rheum Dis* 1990;**49**:999–1000.
951. Veale D, Farrell M, Fitzgerald O. Mechanism of joint sparing in a patient with unilateral psoriatic arthritis and a longstanding hemiplegia. *Br J Rheumatol* 1993;**32**:413–6.
952. Kane D, Lockhart JC, Balint PV, Mann C, Ferrell WR, McInnes IB. Protective effect of sensory denervation in inflammatory arthritis (evidence of regulatory neuroimmune pathways in the arthritic joint). *Ann Rheum Dis* 2005;**64**:325–7.
953. Bordin G, Atzeni F, Bettazzi L, Beyene NB, Carrabba M, Sarzi-Puttini P. Unilateral polymyalgia rheumatica with controlateral sympathetic dystrophy syndrome. A case of asymmetrical involvement due to pre-existing peripheral palsy. *Rheumatology (Oxford)* 2006;**45**:1578–80.
954. Tarkowski E, Naver H, Wallin BG, Blomstrand C, Tarkowski A. Lateralization of T-lymphocyte responses in patients with stroke. Effect of sympathetic dysfunction? *Stroke* 1995;**26**:57–62.
955. Lee JC, Salonen DC, Inman RD. Unilateral hemochromatosis arthropathy on a neurogenic basis. *J Rheumatol* 1997;**24**:2476–8.
956. Benzing T, Brandes R, Sellin L, Schermer B, Lecker S, Walz G, et al. Upregulation of RGS7 may contribute to TNF-induced changes of central nervous function. *Nat Med* 1999;**5**:913–8.
957. Millan MJ. The induction of pain: an integrative review. *Prog Neurobiol* 1999;**57**:1–164.
958. Basbaum AI, Bautista DM, Scherrer G, Julius D. Cellular and molecular mechanisms of pain. *Cell* 2009;**139**:267–84.
959. Schaible HG, Del Rosso A, Matucci-Cerinic M. Neurogenic aspects of inflammation. *Rheum Dis Clin North Am* 2005;**31**:77–101.
960. Sluka KA, Westlund-High KN. Neurologic regulation of inflammation. In: Firestein GS, Budd RC, Harris Jr ED, McInnes IB, Ruddy S, Sergent JS, editors. Kelley's textbook of rheumatology. Philadelphia, PA: Saunders/Elsevier; 2008. p. 411–9.
961. Carolan EJ, Casale TB. Effects of neuropeptides on neutrophil migration through noncellular and endothelial barriers. *J Allergy Clin Immunol* 1993;**92**:589–98.
962. Saban MR, Saban R, Bjorling D, Haak-Frendscho M. Involvement of leukotrienes, TNF-alpha, and the LFA-1/ICAM-1 interaction in substance P-induced granulocyte infiltration. *J Leukoc Biol* 1997;**61**:445–51.
963. Hood VC, Cruwys SC, Urban L, Kidd BL. Differential role of neurokinin receptors in human lymphocyte and monocyte chemotaxis. *Regul Pept* 2000;**96**:17–21.
964. Westlund KN, Sun YC, Sluka KA, Dougherty PM, Sorkin LS, Willis WD. Neural changes in acute arthritis in monkeys. II. Increased glutamate immunoreactivity in the medial articular nerve. *Brain Res Brain Res Rev* 1992;**17**:15–27.
965. Xu XJ, Hokfelt T, Wiesenfeld-Hallin Z. Galanin and spinal pain mechanisms: where do we stand in 2008? *Cell Mol Life Sci* 2008;**65**:1813–9.
966. Trejter M, Brelinska R, Warchol JB, Butowska W, Neri G, Rebuffat P, et al. Effects of galanin on proliferation and apoptosis of immature rat thymocytes. *Int J Mol Med* 2002;**10**:183–6.
967. Su Y, Ganea D, Peng X, Jonakait GM. Galanin down-regulates microglial tumor necrosis factor-alpha production by a post-transcriptional mechanism. *J Neuroimmunol* 2003;**134**:52–60.

968. Treede RD, Kenshalo DR, Gracely RH, Jones AK. The cortical representation of pain. *Pain* 1999;**79**:105–11.
969. Segond von Banchet G, Petrow PK, Brauer R, Schaible HG. Monoarticular antigen-induced arthritis leads to pronounced bilateral upregulation of the expression of neurokinin 1 and bradykinin 2 receptors in dorsal root ganglion neurons of rats. *Arthritis Res* 2000;**2**:424–7.
970. Hu P, McLachlan EM. Macrophage and lymphocyte invasion of dorsal root ganglia after peripheral nerve lesions in the rat. *Neuroscience* 2002;**112**:23–38.
971. Amaya F, Oh-hashi K, Naruse Y, Iijima N, Ueda M, Shimosato G, et al. Local inflammation increases vanilloid receptor 1 expression within distinct subgroups of DRG neurons. *Brain Res* 2003;**963**:190–6.
972. Hensellek S, Brell P, Schaible HG, Brauer R, Segond von Banchet G. The cytokine TNFalpha increases the proportion of DRG neurones expressing the TRPV1 receptor via the TNFR1 receptor and ERK activation. *Mol Cell Neurosci* 2007;**36**:381–91.
973. Boettger MK, Fischer N, Gajda M, Brauer R, Schaible HG. Experimental arthritis causes tumor necrosis factor-alpha-dependent infiltration of macrophages into rat dorsal root ganglia which correlates with pain-related behavior. *Pain* 2009;**145**:151–9.
974. Schaible HG, Schmelz M, Tegeder I. Pathophysiology and treatment of pain in joint disease. *Adv Drug Deliv Rev* 2006;**58**:323–42.
975. Sorkin LS, Westlund KN, Sluka KA, Dougherty PM, Willis WD. Neural changes in acute arthritis in monkeys. IV. Time-course of amino acid release into the lumbar dorsal horn. *Brain Res Brain Res Rev* 1992;**17**:39–50.
976. Schaible HG, Ebersberger A, von Banchet GS. Mechanisms of pain in arthritis. *Ann N Y Acad Sci* 2002;**966**:343–54.
977. Schmelz M, Michael K, Weidner C, Schmidt R, Torebjork HE, Handwerker HO. Which nerve fibers mediate the axon reflex flare in human skin? *Neuroreport* 2000;**11**:645–8.
978. Boettger MK, Hensellek S, Richter F, Gajda M, Stockigt R, von Banchet GS, et al. Antinociceptive effects of tumor necrosis factor alpha neutralization in a rat model of antigen-induced arthritis: evidence of a neuronal target. *Arthritis Rheum* 2008;**58**:2368–78.
979. Kangrga I, Randic M. Tachykinins and calcitonin gene-related peptide enhance release of endogenous glutamate and aspartate from the rat spinal dorsal horn slice. *J Neurosci* 1990;**10**:2026–38.
980. Levine JD, Dardick SJ, Basbaum AI, Scipio E. Reflex neurogenic inflammation. I. Contribution of the peripheral nervous system to spatially remote inflammatory responses that follow injury. *J Neurosci* 1985;**5**:1380–6.
981. Merry P, Kidd BL, Mapp PI, Stevens CR, Morris CJ, Blake DR. Mechanisms of persistent synovitis. *Scand J Rheumatol Suppl* 1988;**76**:85–93.
982. Christianson CA, Corr M, Firestein GS, Mobargha A, Yaksh TL, Svensson CI. Characterization of the acute and persistent pain state present in K/BxN serum transfer arthritis. *Pain* 2010;**151**:394–403.
983. Rees H, Sluka KA, Westlund KN, Willis WD. The role of glutamate and GABA receptors in the generation of dorsal root reflexes by acute arthritis in the anaesthetized rat. *J Physiol* 1995;**484**:437–45.
984. Sluka KA, Westlund KN. An experimental arthritis model in rats: the effects of NMDA and non-NMDA antagonists on aspartate and glutamate release in the dorsal horn. *Neurosci Lett* 1993;**149**:99–102.
985. Neugebauer V, Lucke T, Schaible HG. N-methyl-D-aspartate (NMDA) and non-NMDA receptor antagonists block the hyperexcitability of dorsal horn neurons during development of acute arthritis in rat's knee joint. *J Neurophysiol* 1993;**70**:1365–77.

986. Dirig DM, Isakson PC, Yaksh TL. Effect of COX-1 and COX-2 inhibition on induction and maintenance of carrageenan-evoked thermal hyperalgesia in rats. *J Pharmacol Exp Ther* 1998;**285**:1031–8.
987. Sluka KA, Willis WD, Westlund KN. Joint inflammation and hyperalgesia are reduced by spinal bicuculline. *Neuroreport* 1993;**5**:109–12.
988. Watkins LR, Maier SF. Beyond neurons: evidence that immune and glial cells contribute to pathological pain states. *Physiol Rev* 2002;**82**:981–1011.
989. Hains LE, Loram LC, Weiseler JL, Frank MG, Bloss EB, Sholar P, et al. Pain intensity and duration can be enhanced by prior challenge: initial evidence suggestive of a role of microglial priming. *J Pain* 2010;**11**:1004–14.
990. Bao L, Zhu Y, Elhassan AM, Wu Q, Xiao B, Zhu J, et al. Adjuvant-induced arthritis: IL-1 beta, IL-6 and TNF-alpha are up-regulated in the spinal cord. *Neuroreport* 2001;**12**:3905–8.
991. de Mos M, Laferriere A, Millecamps M, Pilkington M, Sturkenboom MC, Huygen FJ, et al. Role of NFkappaB in an animal model of complex regional pain syndrome-type I (CRPS-I). *J Pain* 2009;**10**:1161–9.
992. Sluka KA, Rees H, Chen PS, Tsuruoka M, Willis WD. Inhibitors of G-proteins and protein kinases reduce the sensitization to mechanical stimulation and the desensitization to heat of spinothalamic tract neurons induced by intradermal injection of capsaicin in the primate. *Exp Brain Res* 1997;**115**:15–24.
993. Khasar SG, Lin YH, Martin A, Dadgar J, McMahon T, Wang D, et al. A novel nociceptor signaling pathway revealed in protein kinase C epsilon mutant mice. *Neuron* 1999;**24**:253–60.
994. Lin Q, Peng YB, Willis WD. Possible role of protein kinase C in the sensitization of primate spinothalamic tract neurons. *J Neurosci* 1996;**16**:3026–34.
995. Gao YJ, Ji RR. Targeting astrocyte signaling for chronic pain. *Neurotherapeutics* 2010;**7**:482–93.
996. Dominguez E, Mauborgne A, Mallet J, Desclaux M, Pohl M. SOCS3-mediated blockade of JAK/STAT3 signaling pathway reveals its major contribution to spinal cord neuroinflammation and mechanical allodynia after peripheral nerve injury. *J Neurosci* 2010;**30**:5754–66.
997. Boyle DL, Jones TL, Hammaker D, Svensson CI, Rosengren S, Albani S, et al. Regulation of peripheral inflammation by spinal p38 MAP kinase in rats. *PLoS Med* 2006;**3**:e338.
998. Inoue K, Tsuda M, Koizumi S. ATP- and adenosine-mediated signaling in the central nervous system: chronic pain and microglia: involvement of the ATP receptor P2X4. *J Pharmacol Sci* 2004;**94**:112–4.
999. Sorkin LS, Boyle DL, Hammaker D, Herman DS, Vail E, Firestein GS. MKK3, an upstream activator of p38, contributes to formalin phase 2 and late allodynia in mice. *Neuroscience* 2009;**162**:462–71.
1000. Katsura H, Obata K, Mizushima T, Sakurai J, Kobayashi K, Yamanaka H, et al. Activation of Src-family kinases in spinal microglia contributes to mechanical hypersensitivity after nerve injury. *J Neurosci* 2006;**26**:8680–90.
1001. Vanegas H, Schaible HG. Prostaglandins and cyclooxygenases [correction of cycloxygenases] in the spinal cord. *Prog Neurobiol* 2001;**64**:327–63.
1002. Boyle DL, Moore J, Yang L, Sorkin LS, Firestein GS. Spinal adenosine receptor activation inhibits inflammation and joint destruction in rat adjuvant-induced arthritis. *Arthritis Rheum* 2002;**46**:3076–82.
1003. Sorkin LS, Maruyama K, Boyle DL, Yang L, Marsala M, Firestein GS. Spinal adenosine agonist reduces c-fos and astrocyte activation in dorsal horn of rats with adjuvant-induced arthritis. *Neurosci Lett* 2003;**340**:119–22.
1004. Morioka N, Tanabe H, Inoue A, Dohi T, Nakata Y. Noradrenaline reduces the ATP-stimulated phosphorylation of p38 MAP kinase via beta-adrenergic receptors-cAMP-protein kinase A-dependent mechanism in cultured rat spinal microglia. *Neurochem Int* 2009;**55**:226–34.

1005. Gogas KR, Cho HJ, Botchkina GI, Levine JD, Basbaum AI. Inhibition of noxious stimulus-evoked pain behaviors and neuronal fos-like immunoreactivity in the spinal cord of the rat by supraspinal morphine. *Pain* 1996;**65**:9–15.
1006. Qian L, Tan KS, Wei SJ, Wu HM, Xu Z, Wilson B, et al. Microglia-mediated neurotoxicity is inhibited by morphine through an opioid receptor-independent reduction of NADPH oxidase activity. *J Immunol* 2007;**179**:1198–209.
1007. Laye S, Bluthe RM, Kent S, Combe C, Medina C, Parnet P, et al. Subdiaphragmatic vagotomy blocks induction of IL-1 beta mRNA in mice brain in response to peripheral LPS. *Am J Physiol* 1995;**268**:R1327–31.
1008. Goshen I, Yirmiya R. Interleukin-1 (IL-1): a central regulator of stress responses. *Front Neuroendocrinol* 2009;**30**:30–45.
1009. Chen Y, Michaelis M, Jänig W, Devor M. Adrenoreceptor subtype mediating sympathetic-sensory coupling in injured sensory neurons. *J Neurophysiol* 1996;**76**:3721–30.
1010. Gonzales R, Goldyne ME, Taiwo YO, Levine JD. Production of hyperalgesic prostaglandins by sympathetic postganglionic neurons. *J Neurochem* 1989;**53**:1595–8.
1011. Goldstein RS, Bruchfeld A, Yang L, Qureshi AR, Gallowitsch-Puerta M, Patel NB, et al. Cholinergic anti-inflammatory pathway activity and High Mobility Group Box-1 (HMGB1) serum levels in patients with rheumatoid arthritis. *Mol Med* 2007;**13**:210–5.
1012. Dhabhar FS, McEwen BS. Bi-directional effects of stress on immune function: possible explanations for salubrious as well as harmful effects. In: Ader R, editor. *Psychoneuroimmunology*. Burlington, San Diego, London: Elsevier; 2007. p. 723–60.
1013. Miao FJ, Jänig W, Levine J. Role of sympathetic postganglionic neurons in synovial plasma extravasation induced by bradykinin. *J Neurophysiol* 1996;**75**:715–24.
1014. Spiegel A, Shivtiel S, Kalinkovich A, Ludin A, Netzer N, Goichberg P, et al. Catecholaminergic neurotransmitters regulate migration and repopulation of immature human CD34+ cells through Wnt signaling. *Nat Immunol* 2007;**8**:1123–31.
1015. Speidl WS, Toller WG, Kaun C, Weiss TW, Pfaffenberger S, Kastl SP, et al. Catecholamines potentiate LPS-induced expression of MMP-1 and MMP-9 in human monocytes and in the human monocytic cell line U937: possible implications for peri-operative plaque instability. *FASEB J* 2004;**18**:603–5.
1016. Straub RH, Mayer M, Kreutz M, Leeb S, Schölmerich J, Falk W. Neurotransmitters of the sympathetic nerve terminal are powerful chemoattractants for monocytes. *J Leukoc Biol* 2000;**67**:553–8.
1017. Levine JD, Khasar SG, Green PG. Neurogenic inflammation and arthritis. *Ann N Y Acad Sci* 2006;**1069**:155–67.
1018. Jänig W. Vagal afferent neurons and pain. In: Basbaum AI, Bushnell MC, editors. *Science of pain*. San Diego, CA: Academic Press; 2009. p. 245–52.
1019. Borovikova LV, Ivanova S, Zhang M, Yang H, Botchkina GI, Watkins LR, et al. Vagus nerve stimulation attenuates the systemic inflammatory response to endotoxin. *Nature* 2000;**405**:458–62.
1020. Rosas-Ballina M, Ochani M, Parrish WR, Ochani K, Harris YT, Huston JM, et al. Splenic nerve is required for cholinergic antiinflammatory pathway control of TNF in endotoxemia. *Proc Natl Acad Sci U S A* 2008;**105**:11008–13.
1021. Saeed RW, Varma S, Peng-Nemeroff T, Sherry B, Balakhaneh D, Huston J, et al. Cholinergic stimulation blocks endothelial cell activation and leukocyte recruitment during inflammation. *J Exp Med* 2005;**201**:1113–23.
1022. Waldburger JM, Boyle DL, Edgar M, Sorkin LS, Levine YA, Pavlov VA, et al. Spinal p38 MAP kinase regulates peripheral cholinergic outflow. *Arthritis Rheum* 2008;**58**:2919–21.

1023. Smith CH, Barker JN, Morris RW, MacDonald DM, Lee TH. Neuropeptides induce rapid expression of endothelial cell adhesion molecules and elicit granulocytic infiltration in human skin. *J Immunol* 1993;**151**:3274–82.
1024. Chalothorn D, Zhang H, Clayton JA, Thomas SA, Faber JE. Catecholamines augment collateral vessel growth and angiogenesis in hindlimb ischemia. *Am J Physiol Heart Circ Physiol* 2005;**289**:H947–59.
1025. Ruff MR, Wahl SM, Pert CB. Substance P receptor-mediated chemotaxis of human monocytes. *Peptides* 1985;**6**(Suppl 2):107–11.
1026. Numao T, Agrawal DK. Neuropeptides modulate human eosinophil chemotaxis. *J Immunol* 1992;**149**:3309–15.
1027. Serra MC, Calzetti F, Ceska M, Cassatella MA. Effect of substance P on superoxide anion and IL-8 production by human PMNL. *Immunology* 1994;**82**:63–9.
1028. Kavelaars A, van de Pol M, Zijlstra J, Heijnen CJ. Beta 2-adrenergic activation enhances interleukin-8 production by human monocytes. *J Neuroimmunol* 1997;**77**:211–6.
1029. Xu J, Xu F, Barrett E. Metalloelastase in lungs and alveolar macrophages is modulated by extracellular substance P in mice. *Am J Physiol Lung Cell Mol Physiol* 2008;**295**:L162–70.
1030. Straub RH. *Tables of molecular and functional neuroendocrine immune interactions*. Eching, Germany: Biozol; 2000.
1031. Sowa NA, Taylor-Blake B, Zylka MJ. Ecto-5'-nucleotidase (CD73) inhibits nociception by hydrolyzing AMP to adenosine in nociceptive circuits. *J Neurosci* 2010;**30**:2235–44.
1032. Montesinos MC, Takedachi M, Thompson LF, Wilder TF, Fernandez P, Cronstein BN. The antiinflammatory mechanism of methotrexate depends on extracellular conversion of adenine nucleotides to adenosine by ecto-5'-nucleotidase: findings in a study of ecto-5'-nucleotidase gene-deficient mice. *Arthritis Rheum* 2007;**56**:1440–5.
1033. Zernecke A, Bidzhekov K, Ozuyaman B, Fraemohs L, Liehn EA, Luscher-Firzlaff JM, et al. CD73/ecto-5'-nucleotidase protects against vascular inflammation and neointima formation. *Circulation* 2006;**113**:2120–7.
1034. Ernst PB, Garrison JC, Thompson LF. Much ado about adenosine: adenosine synthesis and function in regulatory T cell biology. *J Immunol* 2010;**185**:1993–8.
1035. Aloe L, Tuveri MA, Carcassi U, Levi-Montalcini R. Nerve growth factor in the synovial fluid of patients with chronic arthritis. *Arthritis Rheum* 1992;**35**:351–5.
1036. Aloe L, Probert L, Kollias G, Bracci-Laudiero L, Spillantini MG, Levi-Montalcini R. The synovium of transgenic arthritic mice expressing human tumor necrosis factor contains a high level of nerve growth factor. *Growth Factors* 1993;**9**:149–55.
1037. Miller LE, Weidler C, Falk W, Angele P, Schaumburger J, Schölmerich J, et al. Increased prevalence of semaphorin 3C, a repellent of sympathetic nerve fibers, in the synovial tissue of patients with rheumatoid arthritis. *Arthritis Rheum* 2004;**50**:1156–63.
1038. Fassold A, Falk W, Anders S, Hirsch T, Mirsky VM, Straub RH. Soluble neuropilin-2, a nerve repellent receptor, is increased in rheumatoid arthritis synovium and aggravates sympathetic fiber repulsion and arthritis. *Arthritis Rheum* 2009;**60**:2892–901.
1039. Straub RH. Autoimmune disease and innervation. *Brain Behav Immun* 2007;**21**:528–34.
1040. Reynolds ML, Fitzgerald M. Long-term sensory hyperinnervation following neonatal skin wounds. *J Comp Neurol* 1995;**358**:487–98.
1041. Lorton D, Lubahn C, Schaller J, Bellinger D. Noradrenergic (NA) nerves in spleens from rats with adjuvant arthritis (AA) undergo an injury and sprouting responses that parallels changes in nerve growth factor (NGF)—positive cells and tissue LEV. *Brain Behav Immun* 2003;**17**:186–7.
1042. Straub RH, Rauch L, Fassold A, Lowin T, Pongratz G. Neuronally released sympathetic neurotransmitters stimulate splenic interferon-gamma secretion from T cells in early type II collagen-induced arthritis. *Arthritis Rheum* 2008;**58**:3450–60.

1043. Mei Q, Mundinger TO, Lernmark A, Taborsky Jr GJ. Early, selective, and marked loss of sympathetic nerves from the islets of BioBreeder diabetic rats. *Diabetes* 2002;**51**:2997–3002.
1044. Lorton D, Lubahn C, Sweeney S, Major A, Lindquist CA, Schaller J, et al. Differences in the injury/sprouting response of splenic noradrenergic nerves in Lewis rats with adjuvant-induced arthritis compared with rats treated with 6-hydroxydopamine. *Brain Behav Immun* 2009;**23**:276–85.
1045. Allen JM, Iggulden HL, McHale NG. Beta-adrenergic inhibition of bovine mesenteric lymphatics. *J Physiol* 1986;**374**:401–11.
1046. McHale NG, Allen JM, Iggulden HL. Mechanism of alpha-adrenergic excitation in bovine lymphatic smooth muscle. *Am J Physiol* 1987;**252**:H873–8.
1047. Maestroni GJ. Dendritic cell migration controlled by alpha 1b-adrenergic receptors. *J Immunol* 2000;**165**:6743–7.
1048. Kaneider NC, Kaser A, Dunzendorfer S, Tilg H, Patsch JR, Wiedermann CJ. Neurokinin-1 receptor interacts with PrP(106–126)-induced dendritic cell migration and maturation. *J Neuroimmunol* 2005;**158**:153–8.
1049. Marriott I, Bost KL. Expression of authentic substance P receptors in murine and human dendritic cells. *J Neuroimmunol* 2001;**114**:131–41.
1050. Maestroni GJ. Short exposure of maturing, bone marrow-derived dendritic cells to norepinephrine: impact on kinetics of cytokine production and Th development. *J Neuroimmunol* 2002;**129**:106–14.
1051. Straub RH, Rauch L, Rauh L, Pongratz G. Sympathetic inhibition of IL-6, IFN-gamma, and KC/CXCL1 and sympathetic stimulation of TGF-beta in spleen of early arthritic mice. *Brain Behav Immun* 2011;**25**:1708–15.
1052. Straub RH, Linde HJ, Männel DN, Schölmerich J, Falk W. A bacteria-induced switch of sympathetic effector mechanisms augments local inhibition of TNF-alpha and IL-6 secretion in the spleen. *FASEB J* 2000;**14**:1380–8.
1053. Levine JD, Clark R, Devor M, Helms C, Moskowitz MA, Basbaum AI. Intraneuronal substance P contributes to the severity of experimental arthritis. *Science* 1984;**226**:547–9.
1054. Mikami N, Watanabe K, Hashimoto N, Miyagi Y, Sueda K, Fukada S, et al. Calcitonin gene-related peptide enhances experimental autoimmune encephalomyelitis by promoting Th17-cell functions. *Int Immunol* 2012;**24**:681–91.
1055. Ebbinghaus M, Gajda M, Boettger MK, Schaible HG, Brauer R. The anti-inflammatory effects of sympathectomy in murine antigen-induced arthritis are associated with a reduction of Th1 and Th17 responses. *Ann Rheum Dis* 2012;**71**:253–61.
1056. Cunin P, Caillon A, Corvaisier M, Garo E, Scotet M, Blanchard S, et al. The tachykinins substance P and hemokinin-1 favor the generation of human memory Th17 cells by inducing IL-1beta, IL-23, and TNF-like 1A expression by monocytes. *J Immunol* 2011;**186**:4175–82.
1057. Barros PO, Ferreira TB, Vieira MM, Almeida CR, Araujo-Lima CF, Silva-Filho RG, et al. Substance P enhances Th17 phenotype in individuals with generalized anxiety disorder: an event resistant to glucocorticoid inhibition. *J Clin Immunol* 2011;**31**:51–9.
1058. Manni M, Granstein RD, Maestroni G. Beta2-adrenergic agonists bias TLR-2 and NOD2 activated dendritic cells towards inducing an IL-17 immune response. *Cytokine* 2011;**55**:380–6.
1059. Prado C, Contreras F, Gonzalez H, Diaz P, Elgueta D, Barrientos M, et al. Stimulation of dopamine receptor D5 expressed on dendritic cells potentiates Th17-mediated immunity. *J Immunol* 2012;**188**:3062–70.
1060. Nakano K, Yamaoka K, Hanami K, Saito K, Sasaguri Y, Yanagihara N, et al. Dopamine induces IL-6-dependent IL-17 production via D1-like receptor on CD4 naive T cells and D1-like receptor antagonist SCH-23390 inhibits cartilage destruction in a human rheumatoid arthritis/SCID mouse chimera model. *J Immunol* 2011;**186**:3745–52.

1061. Kim BJ, Jones HP. Epinephrine-primed murine bone marrow-derived dendritic cells facilitate production of IL-17A and IL-4 but not IFN-gamma by CD4+ T cells. *Brain Behav Immun* 2010;**24**:1126–36.

1062. Härle P, Pongratz G, Albrecht J, Tarner IH, Straub RH. An early sympathetic nervous system influence exacerbates collagen-induced arthritis via CD4+CD25+ cells. *Arthritis Rheum* 2008;**58**:2347–55.

1063. Bosmann M, Meta F, Ruemmler R, Haggadone MD, Sarma JV, Zetoune FS, et al. Regulation of IL-17 Family Members by Adrenal Hormones During Experimental Sepsis in Mice. *Am J Pathol* 2013;**182**:1124–30.

1064. Li N, Mu L, Wang J, Zhang J, Xie X, Kong Q, et al. Activation of the adenosine A2A receptor attenuates experimental autoimmune myasthenia gravis severity. *Eur J Immunol* 2012;**42**:1140–51.

1065. Nakagome K, Imamura M, Okada H, Kawahata K, Inoue T, Hashimoto K, et al. Dopamine D1-like receptor antagonist attenuates Th17-mediated immune response and ovalbumin antigen-induced neutrophilic airway inflammation. *J Immunol* 2011;**186**:5975–82.

1066. Kipnis J, Cardon M, Avidan H, Lewitus GM, Mordechay S, Rolls A, et al. Dopamine, through the extracellular signal-regulated kinase pathway, downregulates CD4+CD25+ regulatory T-cell activity: implications for neurodegeneration. *J Neurosci* 2004;**24**:6133–43.

1067. Bhowmick S, Singh A, Flavell RA, Clark RB, O'Rourke J, Cone RE. The sympathetic nervous system modulates CD4(+)FoxP3(+) regulatory T cells via a TGF-beta-dependent mechanism. *J Leukoc Biol* 2009;**86**:1275–83.

1068. Seiffert K, Hosoi J, Torii H, Ozawa H, Ding W, Campton K, et al. Catecholamines inhibit the antigen-presenting capability of epidermal Langerhans cells. *J Immunol* 2002;**168**:6128–35.

1069. Frohman EM, Vayuvegula B, Gupta S, van den Noort S. Norepinephrine inhibits gamma-interferon-induced major histocompatibility class II (Ia) antigen expression on cultured astrocytes via beta-2-adrenergic signal transduction mechanisms. *Proc Natl Acad Sci U S A* 1988;**85**:1292–6.

1070. Loughlin AJ, Woodroofe MN, Cuzner ML. Modulation of interferon-gamma-induced major histocompatibility complex class II and Fc receptor expression on isolated microglia by transforming growth factor-beta 1, interleukin-4, noradrenaline and glucocorticoids. *Immunology* 1993;**79**:125–30.

1071. Seiffert K, Granstein RD. Neuroendocrine regulation of skin dendritic cells. *Ann N Y Acad Sci* 2006;**1088**:195–206.

1072. Spengler RN, Chensue SW, Giacherio DA, Blenk N, Kunkel SL. Endogenous norepinephrine regulates tumor necrosis factor-alpha production from macrophages in vitro. *J Immunol* 1994;**152**:3024–31.

1073. Sanders VM, Munson AE. Norepinephrine and the antibody response. *Pharmacol Rev* 1985;**37**:229–48.

1074. Edgar VA, Silberman DM, Cremaschi GA, Zieher LM, Genaro AM. Altered lymphocyte catecholamine reactivity in mice subjected to chronic mild stress. *Biochem Pharmacol* 2003;**65**:15–23.

1075. Kohm AP, Sanders VM. Norepinephrine: a messenger from the brain to the immune system. *Immunol Today* 2000;**21**:539–42.

1076. Weihe E, Nohr D, Millan MJ, Stein C, Muller S, Gramsch C, et al. Peptide neuroanatomy of adjuvant-induced arthritic inflammation in rat. *Agents Actions* 1988;**25**:255–9.

1077. McDougall JJ, Bray RC, Sharkey KA. Morphological and immunohistochemical examination of nerves in normal and injured collateral ligaments of rat, rabbit, and human knee joints. *Anat Rec* 1997;**248**:29–39.

1078. Imai S, Tokunaga Y, Konttinen YT, Maeda T, Hukuda S, Santavirta S. Ultrastructure of the synovial sensory peptidergic fibers is distinctively altered in different phases of adjuvant induced arthritis in rats: ultramorphological characterization combined with morphometric and immunohistochemical study for substance P, calcitonin gene related peptide, and protein gene product 9.5. *J Rheumatol* 1997;**24**:2177–87.

1079. Reinert A, Kaske A, Mense S. Inflammation-induced increase in the density of neuropeptide-immunoreactive nerve endings in rat skeletal muscle. *Exp Brain Res* 1998;**121**:174–80.

1080. Forsgren S, Hockerfelt U, Norrgard O, Henriksson R, Franzen L. Pronounced substance P innervation in irradiation-induced enteropathy–a study on human colon. *Regul Pept* 2000;**88**:1–13.

1081. Feher E, Altdorfer K, Bagameri G, Feher J. Neuroimmune interactions in experimental colitis. An immunoelectron microscopic study. *Neuroimmunomodulation* 2001;**9**:247–55.

1082. Lorton D, Lubahn C, Lindquist CA, Schaller J, Washington C, Bellinger DL. Changes in the density and distribution of sympathetic nerves in spleens from Lewis rats with adjuvant-induced arthritis suggest that an injury and sprouting response occurs. *J Comp Neurol* 2005;**489**:260–73.

1083. Kakurai M, Monteforte R, Suto H, Tsai M, Nakae S, Galli SJ. Mast cell-derived tumor necrosis factor can promote nerve fiber elongation in the skin during contact hypersensitivity in mice. *Am J Pathol* 2006;**169**:1713–21.

1084. Yamaoka J, Di ZH, Sun W, Kawana S. Changes in cutaneous sensory nerve fibers induced by skin-scratching in mice. *J Dermatol Sci* 2007;**46**:41–51.

1085. Watanabe N, Horie S, Spina D, Michael GJ, Page CP, Priestley JV. Immunohistochemical localization of transient receptor potential vanilloid subtype 1 in the trachea of ovalbumin-sensitized Guinea pigs. *Int Arch Allergy Immunol* 2008;**146**(Suppl 1):28–32.

1086. Skobowiat C, Gonkowski S, Calka J. Phenotyping of sympathetic chain ganglia (SChG) neurons in porcine colitis. *J Vet Med Sci* 2010;**72**:1269–74.

1087. Pernthaler H, Pfurtscheller G, Klima G, Plattner R, Schmid T, Kofler M, et al. Regeneration of sympathetic activities in small bowel transplants. *Eur Surg Res* 1993;**25**:316–20.

1088. Koistinaho J, Wadhwani KC, Balbo A, Rapoport SI. Regeneration of perivascular adrenergic innervation in rat tibial nerve after nerve crush. *Acta Neuropathol* 1991;**81**:486–90.

1089. Lorton D, Hewitt D, Bellinger DL, Felten SY, Felten DL. Noradrenergic reinnervation of the rat spleen following chemical sympathectomy with 6-hydroxydopamine: pattern and time course of reinnervation. *Brain Behav Immun* 1990;**4**:198–222.

1090. Buma P, Elmans L, van den Berg WB, Schrama LH. Neurovascular plasticity in the knee joint of an arthritic mouse model. *Anat Rec* 2000;**260**:51–61.

1091. Kiecolt-Glaser JK, Marucha PT, Malarkey WB, Mercado AM, Glaser R. Slowing of wound healing by psychological stress. *Lancet* 1995;**346**:1194–6.

1092. Eijkelkamp N, Engeland CG, Gajendrareddy PK, Marucha PT. Restraint stress impairs early wound healing in mice via alpha-adrenergic but not beta-adrenergic receptors. *Brain Behav Immun* 2007;**21**:409–12.

1093. Kishimoto S. The regeneration of substance P-containing nerve fibers in the process of burn wound healing in the guinea pig skin. *J Invest Dermatol* 1984;**83**:219–23.

1094. Senapati A, Anand P, McGregor GP, Ghatei MA, Thompson RP, Bloom SR. Depletion of neuropeptides during wound healing in rat skin. *Neurosci Lett* 1986;**71**:101–5.

1095. Dunnick CA, Gibran NS, Heimbach DM. Substance P has a role in neurogenic mediation of human burn wound healing. *J Burn Care Rehabil* 1996;**17**:390–6.

1096. Khalil Z, Helme R. Sensory peptides as neuromodulators of wound healing in aged rats. *J Gerontol A Biol Sci Med Sci* 1996;**51**:B354–61.

1097. Nakamura M, Kawahara M, Morishige N, Chikama T, Nakata K, Nishida T. Promotion of corneal epithelial wound healing in diabetic rats by the combination of a substance P-derived peptide (FGLM-NH2) and insulin-like growth factor-1. *Diabetologia* 2003;**46**:839–42.
1098. Delgado AV, McManus AT, Chambers JP. Exogenous administration of Substance P enhances wound healing in a novel skin-injury model. *Exp Biol Med* 2005;**230**:271–80.
1099. Felderbauer P, Bulut K, Hoeck K, Deters S, Schmidt WE, Hoffmann P. Substance P induces intestinal wound healing via fibroblasts–evidence for a TGF-beta-dependent effect. *Int J Colorectal Dis* 2007;**22**:1475–80.
1100. Muangman P, Tamura RN, Muffley LA, Isik FF, Scott JR, Xie C, et al. Substance P enhances wound closure in nitric oxide synthase knockout mice. *J Surg Res* 2009;**153**:201–9.
1101. Kishimoto S, Maruo M, Ohse C, Yasuno H, Kimura H, Nagai T, et al. The regeneration of the sympathetic catecholaminergic nerve fibers in the process of burn wound healing in guinea pigs. *J Invest Dermatol* 1982;**79**:141–6.
1102. Donaldson DJ, Mahan JT. Influence of catecholamines on epidermal cell migration during wound closure in adult newts. *Comp Biochem Physiol C* 1984;**78**:267–70.
1103. Perez E, Lopez-Briones LG, Gallar J, Belmonte C. Effects of chronic sympathetic stimulation on corneal wound healing. *Invest Ophthalmol Vis Sci* 1987;**28**:221–4.
1104. Gosain A, Muthu K, Gamelli RL, DiPietro LA. Norepinephrine suppresses wound macrophage phagocytic efficiency through alpha- and beta-adrenoreceptor dependent pathways. *Surgery* 2007;**142**:170–9.
1105. Gosain A, Gamelli RL, DiPietro LA. Norepinephrine-mediated suppression of phagocytosis by wound neutrophils. *J Surg Res* 2009;**152**:311–8.
1106. Gosain A, Jones SB, Shankar R, Gamelli RL, DiPietro LA. Norepinephrine modulates the inflammatory and proliferative phases of wound healing. *J Trauma* 2006;**60**:736–44.
1107. Souza BR, Santos JS, Costa AM. Blockade of beta1- and beta2-adrenoceptors delays wound contraction and re-epithelialization in rats. *Clin Exp Pharmacol Physiol* 2006;**33**:421–30.
1108. Romana-Souza B, Santos JS, Monte-Alto-Costa A. Beta-1 and beta-2, but not alpha-1 and alpha-2, adrenoceptor blockade delays rat cutaneous wound healing. *Wound Repair Regen* 2009;**17**:230–9.
1109. Jones MA, Marfurt CF. Sympathetic stimulation of corneal epithelial proliferation in wounded and nonwounded rat eyes. *Invest Ophthalmol Vis Sci* 1996;**37**:2535–47.
1110. Montesinos MC, Gadangi P, Longaker M, Sung J, Levine J, Nilsen D, et al. Wound healing is accelerated by agonists of adenosine A2 (G alpha s-linked) receptors. *J Exp Med* 1997;**186**:1615–20.
1111. Feoktistov I, Biaggioni I, Cronstein BN. Adenosine receptors in wound healing, fibrosis and angiogenesis. *Handb Exp Pharmacol* 2009;383–97.
1112. Lotz M, Vaughan JH, Carson DA. Effect of neuropeptides on production of inflammatory cytokines by human monocytes. *Science* 1988;**241**:1218–21.
1113. Lorton D, Lubahn C, Engan C, Schaller J, Felten DL, Bellinger DL. Local application of capsaicin into the draining lymph nodes attenuates expression of adjuvant-induced arthritis. *Neuroimmunomodulation* 2000;**7**:115–25.
1114. Levine JD, Coderre TJ, Helms C, Basbaum AI. Beta 2-adrenergic mechanisms in experimental arthritis. *Proc Natl Acad Sci U S A* 1988;**85**:4553–6.
1115. Lorton D, Lubahn C, Klein N, Schaller J, Bellinger DL. Dual role for noradrenergic innervation of lymphoid tissue and arthritic joints in adjuvant-induced arthritis. *Brain Behav Immun* 1999;**13**:315–34.
1116. Härle P, Mobius D, Carr DJ, Schölmerich J, Straub RH. An opposing time-dependent immune-modulating effect of the sympathetic nervous system conferred by altering the

cytokine profile in the local lymph nodes and spleen of mice with type II collagen-induced arthritis. *Arthritis Rheum* 2005;**52**:1305–13.

1117. Dhabhar FS, McEwen BS. Enhancing versus suppressive effects of stress hormones on skin immune function. *Proc Natl Acad Sci U S A* 1999;**96**:1059–64.
1118. Straub RH, Grum F, Strauch UG, Capellino S, Bataille F, Bleich A, et al. Anti-inflammatory role of sympathetic nerves in chronic intestinal inflammation. *Gut* 2008;**57**:911–21.
1119. Del Rey A, Wolff C, Wildmann J, Randolf A, Hahnel A, Besedovsky HO, et al. Disrupted joint-immune-brain communication during experimental arthritis. *Arthritis Rheum* 2008;**58**:3090–9.
1120. Miller LE, Grifka J, Schölmerich J, Straub RH. Norepinephrine from synovial tyrosine hydroxylase positive cells is a strong indicator of synovial inflammation in rheumatoid arthritis. *J Rheumatol* 2002;**29**:427–35.
1121. Capellino S, Weber K, Gelder M, Härle P, Straub RH. First appearance and location of catecholaminergic cells during experimental arthritis and elimination by chemical sympathectomy. *Arthritis Rheum* 2012;**64**:1110–8.
1122. Jenei-Lanzl Z, Capellino S, Kees F, Fleck M, Lowin T, Straub RH. Anti-inflammatory effects of cell-based therapy with tyrosine hydroxylase-positive catecholaminergic cells in experimental arthritis. *Ann Rheum Dis* 2015;**74**:444–51.
1123. Wang H, Yu M, Ochani M, Amella CA, Tanovic M, Susarla S, et al. Nicotinic acetylcholine receptor alpha7 subunit is an essential regulator of inflammation. *Nature* 2003;**421**:384–8.
1124. Westman M, Saha S, Morshed M, Lampa J. Lack of acetylcholine nicotine alpha 7 receptor suppresses development of collagen-induced arthritis and adaptive immunity. *Clin Exp Immunol* 2010;**162**:62–7.
1125. van Maanen MA, Stoof SP, Larosa GJ, Vervoordeldonk MJ, Tak PP. Role of the cholinergic nervous system in rheumatoid arthritis: aggravation of arthritis in nicotinic acetylcholine receptor alpha7 subunit gene knockout mice. *Ann Rheum Dis* 2010;**69**:1717–23.
1126. van Maanen MA, Lebre MC, van der PT, Larosa GJ, Elbaum D, Vervoordeldonk MJ, et al. Stimulation of nicotinic acetylcholine receptors attenuates collagen-induced arthritis in mice. *Arthritis Rheum* 2009;**60**:114–22.
1127. Waldburger JM, Boyle DL, Pavlov VA, Tracey KJ, Firestein GS. Acetylcholine regulation of synoviocyte cytokine expression by the alpha7 nicotinic receptor. *Arthritis Rheum* 2008;**58**:3439–49.
1128. Bruchfeld A, Goldstein RS, Chavan S, Patel NB, Rosas-Ballina M, Kohn N, et al. Whole blood cytokine attenuation by cholinergic agonists ex vivo and relationship to vagus nerve activity in rheumatoid arthritis. *J Intern Med* 2010;**268**:94–101.
1129. Westman M, Engstrom M, Catrina AI, Lampa J. Cell specific synovial expression of nicotinic alpha 7 acetylcholine receptor in rheumatoid arthritis and psoriatic arthritis. *Scand J Immunol* 2009;**70**:136–40.
1130. Grimsholm O, Rantapaa-Dahlqvist S, Dalen T, Forsgren S. Unexpected finding of a marked non-neuronal cholinergic system in human knee joint synovial tissue. *Neurosci Lett* 2008;**442**:128–33.
1131. Tilan J, Kitlinska J. Sympathetic neurotransmitters and tumor angiogenesis-link between stress and cancer progression. *J Oncol* 2010;1–6 [ID 539706].
1132. Capellino S, Falk W, Straub RH. Reserpine as a new therapeutical agent in arthritis. *Arthritis Rheum* 2009;**58**:S730.
1133. Capellino S, Cosentino M, Luini A, Bombelli R, Lowin T, Cutolo M, et al. Increased expression of dopamine receptors in synovial fibroblasts from patients with rheumatoid arthritis: inhibitory effects of dopamine on interleukin-8 and interleukin-6. *Arthritis Rheumatol* 2014;**66**:2685–93.

1134. Ribatti D, Conconi MT, Nussdorfer GG. Nonclassic endogenous novel [corrected] regulators of angiogenesis. *Pharmacol Rev* 2007;**59**:185–205.

1135. Lai KB, Sanderson JE, Yu CM. Suppression of collagen production in norepinephrine stimulated cardiac fibroblasts culture: differential effect of alpha and beta-adrenoreceptor antagonism. *Cardiovasc Drugs Ther* 2009;**23**:271–80.

1136. Teeters JC, Erami C, Zhang H, Faber JE. Systemic alpha 1A-adrenoceptor antagonist inhibits neointimal growth after balloon injury of rat carotid artery. *Am J Physiol Heart Circ Physiol* 2003;**284**:H385–92.

1137. Zhang H, Facemire CS, Banes AJ, Faber JE. Different alpha-adrenoceptors mediate migration of vascular smooth muscle cells and adventitial fibroblasts in vitro. *Am J Physiol Heart Circ Physiol* 2002;**282**:H2364–70.

1138. Zhang H, Faber JE. Trophic effect of norepinephrine on arterial intima-media and adventitia is augmented by injury and mediated by different alpha1-adrenoceptor subtypes. *Circ Res* 2001;**89**:815–22.

1139. Aranguiz-Urroz P, Canales J, Copaja M, Troncoso R, Vicencio JM, Carrillo C, et al. Beta(2)-adrenergic receptor regulates cardiac fibroblast autophagy and collagen degradation. *Biochim Biophys Acta* 2011;**1812**:23–31.

1140. Lai KB, Sanderson JE, Yu CM. High dose norepinephrine-induced apoptosis in cultured rat cardiac fibroblast. *Int J Cardiol* 2009;**136**:33–9.

1141. Banfi C, Cavalca V, Veglia F, Brioschi M, Barcella S, Mussoni L, et al. Neurohormonal activation is associated with increased levels of plasma matrix metalloproteinase-2 in human heart failure. *Eur Heart J* 2005;**26**:481–8.

1142. Briest W, Rassler B, Deten A, Leicht M, Morwinski R, Neichel D, et al. Norepinephrine-induced interleukin-6 increase in rat hearts: differential signal transduction in myocytes and non-myocytes. *Pflugers Arch* 2003;**446**:437–46.

1143. Leicht M, Briest W, Zimmer HG. Regulation of norepinephrine-induced proliferation in cardiac fibroblasts by interleukin-6 and p42/p44 mitogen activated protein kinase. *Mol Cell Biochem* 2003;**243**:65–72.

1144. Bürger A, Benicke M, Deten A, Zimmer HG. Catecholamines stimulate interleukin-6 synthesis in rat cardiac fibroblasts. *Am J Physiol Heart Circ Physiol* 2001;**281**:H14–21.

1145. Raap T, Justen HP, Miller LE, Cutolo M, Schölmerich J, Straub RH. Neurotransmitter modulation of interleukin 6 (IL-6) and IL-8 secretion of synovial fibroblasts in patients with rheumatoid arthritis compared to osteoarthritis. *J Rheumatol* 2000;**27**:2558–65.

1146. Kimball ES, Fisher MC. Potentiation of IL-1-induced BALB/3T3 fibroblast proliferation by neuropeptides. *J Immunol* 1988;**141**:4203–8.

1147. Ziche M, Morbidelli L, Pacini M, Dolara P, Maggi CA. NK1-receptors mediate the proliferative response of human fibroblasts to tachykinins. *Br J Pharmacol* 1990;**100**:11–4.

1148. Kähler CM, Sitte BA, Reinisch N, Wiedermann CJ. Stimulation of the chemotactic migration of human fibroblasts by substance P. *Eur J Pharmacol* 1993;**249**:281–6.

1149. Harrison NK, Dawes KE, Kwon OJ, Barnes PJ, Laurent GJ, Chung KF. Effects of neuropeptides on human lung fibroblast proliferation and chemotaxis. *Am J Physiol* 1995;**268**:L278–83.

1150. Sakuta H, Inaba K, Muramatsu S. Calcitonin gene-related peptide enhances cytokine-induced IL-6 production by fibroblasts. *Cell Immunol* 1995;**165**:20–5.

1151. Kaminski DA, Randall TD. Adaptive immunity and adipose tissue biology. *Trends Immunol* 2010;**31**:384–90.

1152. Straub RH, Cutolo M, Buttgereit F, Pongratz G. Energy regulation and neuroendocrine-immune control in chronic inflammatory diseases. *J Intern Med* 2010;**267**:543–60.

1153. Bartness TJ, Song CK. Thematic review series: adipocyte biology. Sympathetic and sensory innervation of white adipose tissue. *J Lipid Res* 2007;**48**:1655–72.
1154. Gross K, Karagiannides I, Thomou T, Koon HW, Bowe C, Kim H, et al. Substance P promotes expansion of human mesenteric preadipocytes through proliferative and antiapoptotic pathways. *Am J Physiol Gastrointest Liver Physiol* 2009;**296**:G1012–9.
1155. Melnyk A, Himms-Hagen J. Resistance to aging-associated obesity in capsaicin-desensitized rats one year after treatment. *Obes Res* 1995;**3**:337–44.
1156. Cherruau M, Morvan FO, Schirar A, Saffar JL. Chemical sympathectomy-induced changes in TH-, VIP-, and CGRP-immunoreactive fibers in the rat mandible periosteum: influence on bone resorption. *J Cell Physiol* 2003;**194**:341–8.
1157. Aitken SJ, Landao-Bassonga E, Ralston SH, Idris AI. Beta2-adrenoreceptor ligands regulate osteoclast differentiation in vitro by direct and indirect mechanisms. *Arch Biochem Biophys* 2009;**482**:96–103.
1158. Elefteriou F. Regulation of bone remodeling by the central and peripheral nervous system. *Arch Biochem Biophys* 2008;**473**:231–6.
1159. Suzuki A, Palmer G, Bonjour JP, Caverzasio J. Catecholamines stimulate the proliferation and alkaline phosphatase activity of MC3T3-E1 osteoblast-like cells. *Bone* 1998;**23**:197–203.
1160. Huang HH, Brennan TC, Muir MM, Mason RS. Functional alpha1- and beta2-adrenergic receptors in human osteoblasts. *J Cell Physiol* 2009;**220**:267–75.
1161. Lerner UH, Persson E. Osteotropic effects by the neuropeptides calcitonin gene-related peptide, substance P and vasoactive intestinal peptide. *J Musculoskelet Neuronal Interact* 2008;**8**:154–65.
1162. Naot D, Cornish J. The role of peptides and receptors of the calcitonin family in the regulation of bone metabolism. *Bone* 2008;**43**:813–8.
1163. Kojima T, Yamaguchi M, Kasai K. Substance P stimulates release of RANKL via COX-2 expression in human dental pulp cells. *Inflamm Res* 2006;**55**:78–84.
1164. Wang L, Zhao R, Shi X, Wei T, Halloran BP, Clark DJ, et al. Substance P stimulates bone marrow stromal cell osteogenic activity, osteoclast differentiation, and resorption activity in vitro. *Bone* 2009;**45**:309–20.
1165. Leden I, Eriksson A, Lilja B, Sturfelt G, Sundkvist G. Autonomic nerve function in rheumatoid arthritis of varying severity. *Scand J Rheumatol* 1983;**12**:166–70.
1166. Kuis W, Jong-de Vos v, Sinnema G, Kavelaars A, Prakken B, Helders PM, et al. The autonomic nervous system and the immune system in juvenile rheumatoid arthritis. *Brain Behav Immun* 1996;**10**:387–98.
1167. Perry F, Heller PH, Kamiya J, Levine JD. Altered autonomic function in patients with arthritis or with chronic myofascial pain. *Pain* 1989;**39**:77–84.
1168. Dekkers JC, Geenen R, Godaert GL, Bijlsma JW, van Doornen LJ. Elevated sympathetic nervous system activity in patients with recently diagnosed rheumatoid arthritis with active disease. *Clin Exp Rheumatol* 2004;**22**:63–70.
1169. Glück T, Oertel M, Reber T, Zietz B, Schölmerich J, Straub RH. Altered function of the hypothalamic stress axes in patients with moderately active systemic lupus erythematosus. I. The hypothalamus-autonomic nervous system axis. *J Rheumatol* 2000;**27**:903–10.
1170. Snow MH, Mikuls TR. Rheumatoid arthritis and cardiovascular disease: the role of systemic inflammation and evolving strategies of prevention. *Curr Opin Rheumatol* 2005;**17**:234–41.
1171. Oikarinen J, Hamalainen L, Oikarinen A. Modulation of glucocorticoid receptor activity by cyclic nucleotides and its implications on the regulation of human skin fibroblast growth and protein synthesis. *Biochim Biophys Acta* 1984;**799**:158–65.

1172. Schmidt P, Holsboer F, Spengler D. Beta(2)-adrenergic receptors potentiate glucocorticoid receptor transactivation via G protein betagamma-subunits and the phosphoinositide 3-kinase pathway. *Mol Endocrinol* 2001;**15**:553–64.

1173. Straub RH, Herfarth H, Falk W, Andus T, Schölmerich J. Uncoupling of the sympathetic nervous system and the hypothalamic-pituitary-adrenal axis in inflammatory bowel disease? *J Neuroimmunol* 2002;**126**:116–25.

1174. Tracey KJ. Physiology and immunology of the cholinergic antiinflammatory pathway. *J Clin Invest* 2007;**117**:289–96.

1175. Pereira da Silva JA, Carmo-Fonseca M. Peptide containing nerves in human synovium: immunohistochemical evidence for decreased innervation in rheumatoid arthritis. *J Rheumatol* 1990;**17**:1592–9.

1176. Mapp PI, Walsh DA, Garrett NE, Kidd BL, Cruwys SC, Polak JM, et al. Effect of three animal models of inflammation on nerve fibres in the synovium. *Ann Rheum Dis* 1994;**53**:240–6.

1177. Weidler C, Holzer C, Harbuz M, Hofbauer R, Angele P, Schölmerich J, et al. Low density of sympathetic nerve fibres and increased density of brain derived neurotrophic factor positive cells in RA synovium. *Ann Rheum Dis* 2005;**64**:13–20.

1178. Nissalo S, Hietanen J, Malmstrom M, Hukkanen M, Polak J, Konttinen YT. Disorder-specific changes in innervation in oral lichen planus and lichenoid reactions. *J Oral Pathol Med* 2000;**29**:361–9.

1179. Koeck FX, Bobrik V, Fassold A, Grifka J, Kessler S, Straub RH. Marked loss of sympathetic nerve fibers in chronic Charcot foot of diabetic origin compared to ankle joint osteoarthritis. *J Orthop Res* 2009;**27**:736–41.

1180. Ferrero S, Haas S, Remorgida V, Camerini G, Fulcheri E, Ragni N, et al. Loss of sympathetic nerve fibers in intestinal endometriosis. *Fertil Steril* 2010;**94**:2817–9.

1181. Arnold J, Barcena de Arellano ML, Ruster C, Vercellino GF, Chiantera V, Schneider A, et al. Imbalance between sympathetic and sensory innervation in peritoneal endometriosis. *Brain Behav Immun* 2012;**26**:132–41.

1182. Haas S, Capellino S, Phan NQ, Bohm M, Luger TA, Straub RH, et al. Low density of sympathetic nerve fibers relative to substance P-positive nerve fibers in lesional skin of chronic pruritus and prurigo nodularis. *J Dermatol Sci* 2010;**58**:193–7.

1183. Sipos G, Sipos P, Altdorfer K, Pongor E, Feher E. Correlation and immunolocalization of substance P nerve fibers and activated immune cells in human chronic gastritis. *Anat Rec (Hoboken)* 2008;**291**:1140–8.

1184. Matthews PJ, Aziz Q, Facer P, Davis JB, Thompson DG, Anand P. Increased capsaicin receptor TRPV1 nerve fibres in the inflamed human oesophagus. *Eur J Gastroenterol Hepatol* 2004;**16**:897–902.

1185. Naukkarinen A, Nickoloff BJ, Farber EM. Quantification of cutaneous sensory nerves and their substance P content in psoriasis. *J Invest Dermatol* 1989;**92**:126–9.

1186. Dirmeier M, Capellino S, Schubert T, Angele P, Anders S, Straub RH. Lower density of synovial nerve fibres positive for calcitonin gene-related peptide relative to substance P in rheumatoid arthritis but not in osteoarthritis. *Rheumatology (Oxford)* 2008;**47**:36–40.

1187. Dick WC, Jubb R, Buchanan WW, Williamson J, Whaley K, Porter BB. Studies on the sympathetic control of normal and diseased synovial blood vessels: the effect of alpha and beta receptor stimulation and inhibition, monitored by the 133xenon clearance technique. *Clin Sci* 1971;**40**:197–209.

1188. McDougall JJ. Abrogation of alpha-adrenergic vasoactivity in chronically inflamed rat knee joints. *Am J Physiol Regul Integr Comp Physiol* 2001;**281**:R821–7.

1189. Kavelaars A. Regulated expression of alpha-1 adrenergic receptors in the immune system. *Brain Behav Immun* 2002;**16**:799–807.

1190. Mishima K, Otani H, Tanabe T, Kawasaki H, Oshiro A, Saito N, et al. Molecular mechanisms for alpha2-adrenoceptor-mediated regulation of synoviocyte populations. *Jpn J Pharmacol* 2001;**85**:214–26.

1191. Wahle M, Krause A, Ulrichs T, Jonas D, von Wichert P, Burmester GR, et al. Disease activity related catecholamine response of lymphocytes from patients with rheumatoid arthritis. *Ann N Y Acad Sci* 1999;**876**:287–96.

1192. Meinel T, Pongratz G, Rauch L, Straub RH. Neuronal alpha1/2-adrenergic stimulation of IFN-gamma, IL-6, and CXCL-1 in murine spleen in late experimental arthritis. *Brain Behav Immun* 2013;**33**:80–9.

1193. Fortier LA, Nixon AJ. Distributional changes in substance P nociceptive fiber patterns in naturally osteoarthritic articulations. *J Rheumatol* 1997;**24**:524–30.

1194. Inoue H, Shimoyama Y, Hirabayashi K, Kajigaya H, Yamamoto S, Oda H, et al. Production of neuropeptide substance P by synovial fibroblasts from patients with rheumatoid arthritis and osteoarthritis. *Neurosci Lett* 2001;**303**:149–52.

1195. Cosentino M, Zaffaroni M, Ferrari M, Marino F, Bombelli R, Rasini E, et al. Interferon-gamma and interferon-beta affect endogenous catecholamines in human peripheral blood mononuclear cells: implications for multiple sclerosis. *J Neuroimmunol* 2005;**162**:112–21.

1196. Busch-Dienstfertig M, Stein C. Opioid receptors and opioid peptide-producing leukocytes in inflammatory pain–basic and therapeutic aspects. *Brain Behav Immun* 2010;**24**:683–94.

1197. Barnes PJ. Glucocorticoids. *Chem Immunol Allergy* 2014;**100**:311–6.

1198. Gruol DJ, Campbell NF, Bourgeois S. Cyclic AMP-dependent protein kinase promotes glucocorticoid receptor function. *J Biol Chem* 1986;**261**:4909–14.

1199. Nakada MT, Stadel JM, Poksay KS, Crooke ST. Glucocorticoid regulation of beta-adrenergic receptors in 3T3-L1 preadipocytes. *Mol Pharmacol* 1987;**31**:377–84.

1200. Dong Y, Aronsson M, Gustafsson JA, Okret S. The mechanism of cAMP-induced glucocorticoid receptor expression. Correlation to cellular glucocorticoid response. *J Biol Chem* 1989;**264**:13679–83.

1201. DiBattista JA, Martel-Pelletier J, Cloutier JM, Pelletier JP. Modulation of glucocorticoid receptor expression in human articular chondrocytes by cAMP and prostaglandins. *J Rheumatol Suppl* 1991;**27**:102–5.

1202. Korn SH, Wouters EF, Wesseling G, Arends JW, Thunnissen FB. Interaction between glucocorticoids and beta2-agonists: alpha and beta glucocorticoid-receptor mRNA expression in human bronchial epithelial cells. *Biochem Pharmacol* 1998;**56**:1561–9.

1203. Eickelberg O, Roth M, Lorx R, Bruce V, Rudiger J, Johnson M, et al. Ligand-independent activation of the glucocorticoid receptor by beta2-adrenergic receptor agonists in primary human lung fibroblasts and vascular smooth muscle cells. *J Biol Chem* 1999;**274**:1005–10.

1204. Stein C, Schafer M, Machelska H. Attacking pain at its source: new perspectives on opioids. *Nat Med* 2003;**9**:1003–8.

1205. Li Z, Proud D, Zhang C, Wiehler S, McDougall JJ. Chronic arthritis down-regulates peripheral mu-opioid receptor expression with concomitant loss of endomorphin 1 antinociception. *Arthritis Rheum* 2005;**52**:3210–9.

1206. Shen H, Aeschlimann A, Reisch N, Gay RE, Simmen BR, Michel BA, et al. Kappa and delta opioid receptors are expressed but down-regulated in fibroblast-like synoviocytes of patients with rheumatoid arthritis and osteoarthritis. *Arthritis Rheum* 2005;**52**:1402–10.

1207. Straub RH, Wolff C, Fassold A, Hofbauer R, Chover-Gonzalez A, Richards LJ, et al. Antiinflammatory role of endomorphins in osteoarthritis, rheumatoid arthritis, and adjuvant-induced polyarthritis. *Arthritis Rheum* 2008;**58**:456–66.

1208. Stein C, Hassan AH, Przewlocki R, Gramsch C, Peter K, Herz A. Opioids from immunocytes interact with receptors on sensory nerves to inhibit nociception in inflammation. *Proc Natl Acad Sci U S A* 1990;**87**:5935–9.
1209. Stein A, Yassouridis A, Szopko C, Helmke K, Stein C. Intraarticular morphine versus dexamethasone in chronic arthritis. *Pain* 1999;**83**:525–32.
1210. Delgado M, Toscano MG, Benabdellah K, Cobo M, O'Valle F, Gonzalez-Rey E, et al. In vivo delivery of lentiviral vectors expressing vasoactive intestinal peptide complementary DNA as gene therapy for collagen-induced arthritis. *Arthritis Rheum* 2008;**58**:1026–37.
1211. Delgado M, Abad C, Martinez C, Leceta J, Gomariz RP. Vasoactive intestinal peptide prevents experimental arthritis by downregulating both autoimmune and inflammatory components of the disease. *Nat Med* 2001;**7**:563–8.
1212. Juarranz MG, Santiago B, Torroba M, Gutierrez-Canas I, Palao G, Galindo M, et al. Vasoactive intestinal peptide modulates proinflammatory mediator synthesis in osteoarthritic and rheumatoid synovial cells. *Rheumatology (Oxford)* 2004;**43**:416–22.
1213. Juarranz Y, Gutierrez-Canas I, Santiago B, Carrion M, Pablos JL, Gomariz RP. Differential expression of vasoactive intestinal peptide and its functional receptors in human osteoarthritic and rheumatoid synovial fibroblasts. *Arthritis Rheum* 2008;**58**:1086–95.
1214. Lundberg JM, Anggard A, Pernow J, Hokfelt T. Neuropeptide Y-, substance P- and VIP-immunoreactive nerves in cat spleen in relation to autonomic vascular and volume control. *Cell Tissue Res* 1985;**239**:9–18.
1215. Fried G, Terenius L, Brodin E, Efendic S, Dockray G, Fahrenkrug J, et al. Neuropeptide Y, enkephalin and noradrenaline coexist in sympathetic neurons innervating the bovine spleen. *Cell Tissue Res* 1986;**243**:495–508.
1216. Nast A, Malysheva O, Krause A, Wahle M, Baerwald CG. Intracellular calcium responses to cholinergic stimulation of lymphocytes from healthy donors and patients with rheumatoid arthritis. *Rheumatol Int* 2009;**29**:497–502.
1217. Waldburger JM, Firestein GS. Regulation of peripheral inflammation by the central nervous system. *Curr Rheumatol Rep* 2010;**12**:370–8.
1218. van Maanen MA, Stoof SP, Van Der Zanden EP, de Jonge WJ, Janssen RA, Fischer DF, et al. The alpha7 nicotinic acetylcholine receptor on fibroblast-like synoviocytes and in synovial tissue from rheumatoid arthritis patients: a possible role for a key neurotransmitter in synovial inflammation. *Arthritis Rheum* 2009;**60**:1272–81.
1219. Koopman FA, Stoof SP, Straub RH, van Maanen MA, Vervoordeldonk MJ, Tak PP. Restoring the balance of the autonomic nervous system as an innovative approach to the treatment of rheumatoid arthritis. *Mol Med* 2011;**17**:937–48.
1220. Carlens C, Brandt L, Klareskog L, Lampa J, Askling J. The inflammatory reflex and risk for rheumatoid arthritis: a case–control study of human vagotomy. *Ann Rheum Dis* 2007;**66**:414–6.
1221. Melville H. *Moby Dick or the whale*. Chicago: Encyclopaedia Britannica; 1952.
1222. Rassow J, Hauser K, Netzker R, Deutzmann R. *Biochemistry*. Stuttgart: Georg Thieme; 2008.
1223. Kimber NE, Ross JJ, Mason SL, Speedy DB. Energy balance during an ironman triathlon in male and female triathletes. *Int J Sport Nutr Exerc Metab* 2002;**12**:47–62.
1224. Rolfe DF, Brown GC. Cellular energy utilization and molecular origin of standard metabolic rate in mammals. *Physiol Rev* 1997;**77**:731–58.
1225. Blaxter K. *Energy metabolism in animals and man*. Cambridge, New York, New Rochelle, Melbourne, Sydney: Cambridge University Press; 1989.

1226. Iemitsu M, Itoh M, Fujimoto T, Tashiro M, Nagatomi R, Ohmori H, et al. Whole-body energy mapping under physical exercise using positron emission tomography. *Med Sci Sports Exerc* 2000;**32**:2067–70.

1227. Pabst R, Trepel F. 72-Hour perfusion of the isolated spleen at normothermia. *Res Exp Med (Berl)* 1974;**164**:247–57.

1228. Calder PC. Fuel utilization by cells of the immune system. *Proc Nutr Soc* 1995;**54**:65–82.

1229. Maciver NJ, Jacobs SR, Wieman HL, Wofford JA, Coloff JL, Rathmell JC. Glucose metabolism in lymphocytes is a regulated process with significant effects on immune cell function and survival. *J Leukoc Biol* 2008;**84**:949–57.

1230. Frauwirth KA, Thompson CB. Regulation of T lymphocyte metabolism. *J Immunol* 2004;**172**:4661–5.

1231. Straub RH. Interaction of the endocrine system with inflammation: a function of energy and volume regulation. *Arthritis Res Ther* 2014;**16**:203–17.

1232. Lippert H. *Lehrbuch Anatomie*. Munich: Elsevier; 2003.

1233. Princiotta MF, Finzi D, Qian SB, Gibbs J, Schuchmann S, Buttgereit F, et al. Quantitating protein synthesis, degradation, and endogenous antigen processing. *Immunity* 2003;**18**:343–54.

1234. Buttgereit F, Brand MD. A hierarchy of ATP-consuming processes in mammalian cells. *Biochem J* 1995;**312**:163–7.

1235. Schmid D, Burmester GR, Tripmacher R, Kuhnke A, Buttgereit F. Bioenergetics of human peripheral blood mononuclear cell metabolism in quiescent, activated, and glucocorticoid-treated states. *Biosci Rep* 2000;**20**:289–302.

1236. Kuhnke A, Burmester GR, Krauss S, Buttgereit F. Bioenergetics of immune cells to assess rheumatic disease activity and efficacy of glucocorticoid treatment. *Ann Rheum Dis* 2003;**62**:133–9.

1237. Kunsch K. *Der Mensch in Zahlen [The human being in numbers]*. Gustav Fischer: Suttgart Jena Lübeck Ulm; 1997.

1238. Ghesquière B, Wong BW, Kuchnio A, Carmeliet P. Metabolism of stromal and immune cells in health and disease. *Nature* 2014;**511**:167–76.

1239. Michal G. *Biochemical pathways*. Heidelberg Berlin: Spektrum Akademischer Verlag; 1999.

1240. Schedlowski M, Hosch W, Oberbeck R, Benschop RJ, Jacobs R, Raab HR, et al. Catecholamines modulate human NK cell circulation and function via spleen-independent beta 2-adrenergic mechanisms. *J Immunol* 1996;**156**:93–9.

1241. Born J, Lange T, Hansen K, Molle M, Fehm HL. Effects of sleep and circadian rhythm on human circulating immune cells. *J Immunol* 1997;**158**:4454–64.

1242. Merl V, Peters A, Oltmanns KM, Kern W, Hubold C, Hallschmid M, et al. Preserved circadian rhythm of serum insulin concentration at low plasma glucose during fasting in lean and overweight humans. *Metabolism* 2004;**53**:1449–53.

1243. Wildenhoff KE, Johansen JP, Karstoft H, Yde H, Sorensen NS. Diurnal variations in the concentrations of blood acetoacetate and 3-hydroxybutyrate. The ketone body peak around midnight and its relationship to free fatty acids, glycerol, insulin, growth hormone and glucose in serum and plasma. *Acta Med Scand* 1974;**195**:25–8.

1244. Fraser G, Trinder J, Colrain IM, Montgomery I. Effect of sleep and circadian cycle on sleep period energy expenditure. *J Appl Physiol (1985)* 1989;**66**:830–6.

1245. Ravussin E, Lillioja S, Anderson TE, Christin L, Bogardus C. Determinants of 24-hour energy expenditure in man. Methods and results using a respiratory chamber. *J Clin Invest* 1986;**78**:1568–78.

1246. Boyle PJ, Scott JC, Krentz AJ, Nagy RJ, Comstock E, Hoffman C. Diminished brain glucose metabolism is a significant determinant for falling rates of systemic glucose utilization during sleep in normal humans. *J Clin Invest* 1994;**93**:529–35.
1247. Buttgereit F, Burmester GR, Brand MD. Bioenergetics of immune functions: fundamental and therapeutic aspects. *Immunol Today* 2000;**21**:192–9.
1248. Higami Y, Barger JL, Page GP, Allison DB, Smith SR, Prolla TA, et al. Energy restriction lowers the expression of genes linked to inflammation, the cytoskeleton, the extracellular matrix, and angiogenesis in mouse adipose tissue. *J Nutr* 2006;**136**:343–52.
1249. Okano K, Tsukazaki T, Ohtsuru A, Namba H, Osaki M, Iwasaki K, et al. Parathyroid hormone-related peptide in synovial fluid and disease activity of rheumatoid arthritis. *Br J Rheumatol* 1996;**35**:1056–62.
1250. Del Rey A, Roggero E, Randolf A, Mahuad C, McCann S, Rettori V, et al. IL-1 resets glucose homeostasis at central levels. *Proc Natl Acad Sci U S A* 2006;**103**:16039–44.
1251. Pacheco-Lopez G, Bermudez-Rattoni F. Brain-immune interactions and the neural basis of disease-avoidant ingestive behaviour. *Philos Trans R Soc Lond B Biol Sci* 2011;**366**:3389–405.
1252. Dantzer R, Kelley KW. Twenty years of research on cytokine-induced sickness behavior. *Brain Behav Immun* 2007;**21**:153–60.
1253. Konsman JP, Parnet P, Dantzer R. Cytokine-induced sickness behaviour: mechanisms and implications. *Trends Neurosci* 2002;**25**:154–9.
1254. Carter ME, Soden ME, Zweifel LS, Palmiter RD. Genetic identification of a neural circuit that suppresses appetite. *Nature* 2013;**503**:111–4.
1255. Straub RH. Evolutionary medicine and chronic inflammatory state – known and new concepts in pathophysiology. *J Mol Med* 2012;**90**:523–34.
1256. Meyer-Hermann M. A mathematical model for the germinal center morphology and affinity maturation. *J Theor Biol* 2002;**216**:273–300.
1257. Aletaha D, Neogi T, Silman AJ, Funovits J, Felson DT, Bingham III CO, et al. Rheumatoid arthritis classification criteria: an American College of Rheumatology/European League Against Rheumatism collaborative initiative. *Arthritis Rheum* 2010;**62**:2569–81.
1258. Davies JW, Lamke LO, Liljedahl SO. A guide to the rate of non-renal water loss from patients with burns. *Br J Plast Surg* 1974;**27**:325–9.
1259. Jacob M, Chappell D, Hofmann-Kiefer K, Conzen P, Peter K, Rehm M. Determinants of insensible fluid loss. Perspiration, protein shift and endothelial glycocalyx. *Anaesthesist* 2007;**56**:747–58.
1260. Berg JM, Tymoczko JL, Stryer L. *Biochemistry*. New York: W.H. Freeman; 2002.
1261. Tredget EE, Yu YM. The metabolic effects of thermal injury. *World J Surg* 1992;**16**:68–79.
1262. Bader M, Ganten D. Regulation of renin: new evidence from cultured cells and genetically modified mice. *J Mol Med (Berl)* 2000;**78**:130–9.
1263. Hattangady N, Olala L, Bollag WB, Rainey WE. Acute and chronic regulation of aldosterone production. *Mol Cell Endocrinol* 2012;**350**:151–62.
1264. Selye H. The general adaptation syndrome and the diseases of adaptation. *Am J Med* 1951;**10**:549–55.
1265. Dorffel Y, Latsch C, Stuhlmuller B, Schreiber S, Scholze S, Burmester GR, et al. Preactivated peripheral blood monocytes in patients with essential hypertension. *Hypertension* 1999;**34**:113–7.
1266. Phillips MI, Kagiyama S. Angiotensin II as a pro-inflammatory mediator. *Curr Opin Investig Drugs* 2002;**3**:569–77.
1267. Boos CJ, Lip GY. Is hypertension an inflammatory process? *Curr Pharm Des* 2006;**12**:1623–35.
1268. Anderson W, Mackay IR. *Intolerant bodies*. Baltimore: Johns Hopkins University Press; 2014.

1269. Nesse RM. How is Darwinian medicine useful? *West J Med* 2001;**174**:358–60.
1270. Mayr E. The idea of teleology. *J Hist Ideas* 1992;**53**:117–35.
1271. Cortez D, Marin R, Toledo-Flores D, Froidevaux L, Liechti A, Waters PD, et al. Origins and functional evolution of Y chromosomes across mammals. *Nature* 2014;**508**:488–93.
1272. Rast JP, Buckley KM. Lamprey immunity is far from primitive. *Proc Natl Acad Sci U S A* 2013;**110**:5746–7.
1273. Ritchie M. Neuroanatomy and physiology of the avian hypothalamic/pituitary axis: clinical aspects. *Vet Clin North Am Exot Anim Pract* 2014;**17**:13–22.
1274. Rai S, Szeitz A, Roberts BW, Christie Q, Didier W, Eom J, et al. A putative corticosteroid hormone in Pacific lamprey, Entosphenus tridentatus. *Gen Comp Endocrinol* 2014;**10**.
1275. Yang DL, Yang Y, He Z. Roles of plant hormones and their interplay in rice immunity. *Mol Plant* 2013;**6**:675–85.
1276. Funakoshi K, Nakano M. The sympathetic nervous system of anamniotes. *Brain Behav Evol* 2007;**69**:105–13.
1277. Connor KL, Colabroy KL, Gerratana B. A heme peroxidase with a functional role as an L-tyrosine hydroxylase in the biosynthesis of anthramycin. *Biochemistry* 2011;**50**:8926–36.
1278. Lyte M. Microbial endocrinology: host-microbiota neuroendocrine interactions influencing brain and behavior. *Gut Microbes* 2014;**5**:381–9.
1279. Dantzer R, O'Connor JC, Freund GG, Johnson RW, Kelley KW. From inflammation to sickness and depression: when the immune system subjugates the brain. *Nat Rev Neurosci* 2008;**9**:46–56.
1280. Hoglund E, Sorensen C, Bakke MJ, Nilsson GE, Overli O. Attenuation of stress-induced anorexia in brown trout (Salmo trutta) by pre-treatment with dietary l-tryptophan. *Br J Nutr* 2007;**97**:786–9.
1281. Clippinger TL, Bennett RA, Johnson CM, Vliet KA, Deem SL, Oros J, et al. Morbidity and mortality associated with a new mycoplasma species from captive American alligators (Alligator mississippiensis). *J Zoo Wildl Med* 2000;**31**:303–14.
1282. Diamond J. *The world until yesterday: what we can learn from traditional societies*. New York: Viking Penguin; 2012.
1283. Baur M, Ziegler G. *Die Odyssee des Menschen—Es begann in Afrika*. München: Econ Ullstein List Verlag; 2001.
1284. Caspari R, Lee SH. Older age becomes common late in human evolution. *Proc Natl Acad Sci U S A* 2004;**101**:10895–900.
1285. Braun J, van der Heijde D, Pincus T. Novel anti-rheumatic therapies challenge old views on ankylsoing spondylitis and other spondylarthropathies. *Clin Exp Rheumatol* 2002;**20**:S1–2.
1286. Ringsdal VS, Andreasen JJ. Ankylosing spondylitis–experience with a self administered questionnaire: an analytical study. *Ann Rheum Dis* 1989;**48**:924–7.
1287. Kurtzke JF, Page WF, Murphy FM, Norman Jr JE. Epidemiology of multiple sclerosis in US veterans. 4. Age at onset. *Neuroepidemiology* 1992;**11**:226–35.
1288. Bell DA, Rigby R, Stiller CR, Clark WF, Harth M, Ebers G. HLA antigens in systemic lupus erythematosus: relationship to disease severity, age at onset, and sex. *J Rheumatol* 1984;**11**:475–9.
1289. Monsen U. Inflammatory bowel disease. An epidemiological and genetic study. *Acta Chir Scand Suppl* 1990;**559**:1–42.
1290. Smith JB, Tulloch JE, Meyer LJ, Zone JJ. The incidence and prevalence of dermatitis herpetiformis in Utah. *Arch Dermatol* 1992;**128**:1608–10.
1291. Suarez-Almazor ME, Soskolne CL, Saunders LD, Russell AS. Outcome in rheumatoid arthritis. A 1985 inception cohort study. *J Rheumatol* 1994;**21**:1438–46.

1292. Hietanen J, Salo OP. Pemphigus: an epidemiological study of patients treated in Finnish hospitals between 1969 and 1978. *Acta Derm Venereol* 1982;**62**:491–6.

1293. Brenner S, Wohl Y. A survey of sex differences in 249 pemphigus patients and possible explanations. *Skinmed* 2007;**6**:163–5.

1294. Selmi C, Mayo MJ, Bach N, Ishibashi H, Invernizzi P, Gish RG, et al. Primary biliary cirrhosis in monozygotic and dizygotic twins: genetics, epigenetics, and environment. *Gastroenterology* 2004;**127**:485–92.

1295. Garcia-Carrasco M, Siso A, Ramos-Casals M, Rosas J, de la RG, Gil V, et al. Raynaud's phenomenon in primary Sjogren's syndrome. Prevalence and clinical characteristics in a series of 320 patients. *J Rheumatol* 2002;**29**:726–30.

1296. Watson ME, Newman RJ, Payne AM, Abdelrahim M, Francis GL. The effect of macrophage conditioned media on Leydig cell function. *Ann Clin Lab Sci* 1994;**24**:84–95.

1297. Gruschwitz MS, Brezinschek R, Brezinschek HP. Cytokine levels in the seminal plasma of infertile males. *J Androl* 1996;**17**:158–63.

1298. Gerard N, Caillaud M, Martoriati A, Goudet G, Lalmanach AC. The interleukin-1 system and female reproduction. *J Endocrinol* 2004;**180**:203–12.

1299. Adashi EY, Resnick CE, Croft CS, Payne DW. Tumor necrosis factor alpha inhibits gonadotropin hormonal action in nontransformed ovarian granulosa cells. A modulatory noncytotoxic property. *J Biol Chem* 1989;**264**:11591–7.

1300. Fray MD, Mann GE, Bleach EC, Knight PG, Clarke MC, Charleston B. Modulation of sex hormone secretion in cows by acute infection with bovine viral diarrhoea virus. *Reproduction* 2002;**123**:281–9.

1301. Xiao E, Ferin M. Stress-related disturbances of the menstrual cycle. *Ann Med* 1997;**29**:215–9.

1302. Hulter BM, Lundberg PO. Sexual function in women with advanced multiple sclerosis. *J Neurol Neurosurg Psychiatry* 1995;**59**:83–6.

1303. Gonzalez-Crespo MR, Gomez-Reino JJ, Merino R, Ciruelo E, Gomez-Reino FJ, Muley R, et al. Menstrual disorders in girls with systemic lupus erythematosus treated with cyclophosphamide. *Br J Rheumatol* 1995;**34**:737–41.

1304. Pasoto SG, Mendonca BB, Bonfa E. Menstrual disturbances in patients with systemic lupus erythematosus without alkylating therapy: clinical, hormonal and therapeutic associations. *Lupus* 2002;**11**:175–80.

1305. Silva CA, Hilario MO, Febronio MV, Oliveira SK, Terreri MT, Sacchetti SB, et al. Risk factors for amenorrhea in juvenile systemic lupus erythematosus (JSLE): a Brazilian multicentre cohort study. *Lupus* 2007;**16**:531–6.

1306. Ostensen M, Almberg K, Koksvik HS. Sex, reproduction, and gynecological disease in young adults with a history of juvenile chronic arthritis. *J Rheumatol* 2000;**27**:1783–7.

1307. Tsigos C, Papanicolaou DA, Kyrou I, Raptis SA, Chrousos GP. Dose-dependent effects of recombinant human interleukin-6 on the pituitary-testicular axis. *J Interferon Cytokine Res* 1999;**19**:1271–6.

1308. Singh AK, Mahlios J, Mignot E. Genetic association, seasonal infections and autoimmune basis of narcolepsy. *J Autoimmun* 2013;**43**:26–31.

1309. Simmons D. Prevalence and age of onset of type 1 diabetes in adult Asians in the Coventry Diabetes Study. *Diabet Med* 1990;**7**:238–41.

1310. Croker BP, Dawson DV, Sanfilippo F. IgA nephropathy. Correlation of clinical and histologic features. *Lab Invest* 1983;**48**:19–24.

1311. Sklar CA, Qazi R, David R. Juvenile autoimmune thyroiditis. Hormonal status at presentation and after long-term follow-up. *Am J Dis Child* 1986;**140**:877–80.

1312. Cooper GS, Miller FW, Pandey JP. The role of genetic factors in autoimmune disease: implications for environmental research. *Environ Health Perspect* 1999;**107**(Suppl. 5):693–700.

1313. Svendsen AJ, Holm NV, Kyvik K, Petersen PH, Junker P. Relative importance of genetic effects in rheumatoid arthritis: historical cohort study of Danish nationwide twin population. *BMJ* 2002;**324**:264–6.
1314. Silman AJ, MacGregor AJ, Thomson W, Holligan S, Carthy D, Farhan A, et al. Twin concordance rates for rheumatoid arthritis: results from a nationwide study. *Br J Rheumatol* 1993;**32**:903–7.
1315. Ringold DA, Nicoloff JT, Kesler M, Davis H, Hamilton A, Mack T. Further evidence for a strong genetic influence on the development of autoimmune thyroid disease: the California twin study. *Thyroid* 2002;**12**:647–53.
1316. Deapen D, Escalante A, Weinrib L, Horwitz D, Bachman B, Roy-Burman P, et al. A revised estimate of twin concordance in systemic lupus erythematosus. *Arthritis Rheum* 1992;**35**:311–8.
1317. Duffy DL, Spelman LS, Martin NG. Psoriasis in Australian twins. *J Am Acad Dermatol* 1993;**29**:428–34.
1318. Jarvinen P. Occurrence of ankylosing spondylitis in a nationwide series of twins. *Arthritis Rheum* 1995;**38**:381–3.
1319. Mathews JD, Whittingham S, Hooper BM, Mackay IR, Stenhouse NS. Association of autoantibodies with smoking, cardiovascular morbidity, and death in the Busselton population. *Lancet* 1973;**2**:754–8.
1320. Hernandez AM, Liang MH, Willett WC, Stampfer MJ, Colditz GA, Rosner B, et al. Reproductive factors, smoking, and the risk for rheumatoid arthritis. *Epidemiology* 1990;**1**:285–91.
1321. Heliovaara M, Aho K, Aromaa A, Knekt P, Reunanen A. Smoking and risk of rheumatoid arthritis. *J Rheumatol* 1993;**20**:1830–5.
1322. Voigt LF, Koepsell TD, Nelson JL, Dugowson CE, Daling JR. Smoking, obesity, alcohol consumption, and the risk of rheumatoid arthritis. *Epidemiology* 1994;**5**:525–32.
1323. Silman AJ, Newman J, MacGregor AJ. Cigarette smoking increases the risk of rheumatoid arthritis. Results from a nationwide study of disease-discordant twins. *Arthritis Rheum* 1996;**39**:732–5.
1324. Stolt P, Bengtsson C, Nordmark B, Lindblad S, Lundberg I, Klareskog L, et al. Quantification of the influence of cigarette smoking on rheumatoid arthritis: results from a population based case–control study, using incident cases. *Ann Rheum Dis* 2003;**62**:835–41.
1325. Klareskog L, Stolt P, Lundberg K, Kallberg H, Bengtsson C, Grunewald J, et al. A new model for an etiology of rheumatoid arthritis: smoking may trigger HLA-DR (shared epitope)-restricted immune reactions to autoantigens modified by citrullination. *Arthritis Rheum* 2006;**54**:38–46.
1326. Kallberg H, Jacobsen S, Bengtsson C, Pedersen M, Padyukov L, Garred P, et al. Alcohol consumption is associated with decreased risk of rheumatoid arthritis: results from two Scandinavian case–control studies. *Ann Rheum Dis* 2009;**68**:222–7.
1327. Rosell M, Wesley AM, Rydin K, Klareskog L, Alfredsson L. Dietary fish and fish oil and the risk of rheumatoid arthritis. *Epidemiology* 2009;**20**:896–901.
1328. Parks CG, Conrad K, Cooper GS. Occupational exposure to crystalline silica and autoimmune disease. *Environ Health Perspect* 1999;**107**(Suppl. 5):793–802.
1329. Khuder SA, Peshimam AZ, Agraharam S. Environmental risk factors for rheumatoid arthritis. *Rev Environ Health* 2002;**17**:307–15.
1330. Videm V, Cortes A, Thomas R, Brown MA. Current smoking is associated with incident ankylosing spondylitis—the HUNT population-based Norwegian Health Study. *J Rheumatol* 2014;**41**:2041–8.
1331. Hernan MA, Olek MJ, Ascherio A. Cigarette smoking and incidence of multiple sclerosis. *Am J Epidemiol* 2001;**154**:69–74.

1332. Tobin MV, Logan RF, Langman MJ, McConnell RB, Gilmore IT. Cigarette smoking and inflammatory bowel disease. *Gastroenterology* 1987;**93**:316–21.
1333. Ghaussy NO, Sibbitt Jr WL, Qualls CR. Cigarette smoking, alcohol consumption, and the risk of systemic lupus erythematosus: a case–control study. *J Rheumatol* 2001;**28**:2449–53.
1334. Hardy CJ, Palmer BP, Muir KR, Sutton AJ, Powell RJ. Smoking history, alcohol consumption, and systemic lupus erythematosus: a case–control study. *Ann Rheum Dis* 1998;**57**:451–5.
1335. Carlsson S, Midthjell K, Grill V. Smoking is associated with an increased risk of type 2 diabetes but a decreased risk of autoimmune diabetes in adults: an 11-year follow-up of incidence of diabetes in the Nord-Trondelag study. *Diabetologia* 2004;**47**:1953–6.
1336. Brenner S, Tur E, Shapiro J, Ruocco V, D'Avino M, Ruocco E, et al. Pemphigus vulgaris: environmental factors. Occupational, behavioral, medical, and qualitative food frequency questionnaire. *Int J Dermatol* 2001;**40**:562–9.
1337. Williams GC. Pleiotropy, natural selection, and the evolution of senescence. *Evolution* 1957;**11**:398–411.
1338. Cramer DW, Xu H, Harlow BL. Does "incessant" ovulation increase risk for early menopause? *Am J Obstet Gynecol* 1995;**172**:568–73.
1339. Chang SH, Kim CS, Lee KS, Kim H, Yim SV, Lim YJ, et al. Premenopausal factors influencing premature ovarian failure and early menopause. *Maturitas* 2007;**58**:19–30.
1340. Straub RH. Neuroendocrine immunology: new pathogenetic aspects and clinical application. *Z Rheumatol* 2011;**70**:767–74.
1341. LaFleur C, Granados J, Vargas-Alarcon G, Ruiz-Morales J, Villarreal-Garza C, Higuera L, et al. HLA-DR antigen frequencies in Mexican patients with dengue virus infection: HLA-DR4 as a possible genetic resistance factor for dengue hemorrhagic fever. *Hum Immunol* 2002;**63**:1039–44.
1342. McKiernan SM, Hagan R, Curry M, McDonald GS, Kelly A, Nolan N, et al. Distinct MHC class I and II alleles are associated with hepatitis C viral clearance, originating from a single source. *Hepatology* 2004;**40**:108–14.
1343. Goulder PJ, Watkins DI. Impact of MHC class I diversity on immune control of immunodeficiency virus replication. *Nat Rev Immunol* 2008;**8**:619–30.
1344. Pertovaara M, Raitala A, Juonala M, Kahonen M, Lehtimaki T, Viikari JS, et al. Autoimmunity and atherosclerosis: functional polymorphism of PTPN22 is associated with phenotypes related to the risk of atherosclerosis. The Cardiovascular Risk in Young Finns Study. *Clin Exp Immunol* 2007;**147**:265–9.
1345. Thio CL, Mosbruger TL, Kaslow RA, Karp CL, Strathdee SA, Vlahov D, et al. Cytotoxic T-lymphocyte antigen 4 gene and recovery from hepatitis B virus infection. *J Virol* 2004;**78**:11258–62.
1346. Raitala A, Karjalainen J, Oja SS, Kosunen TU, Hurme M. Helicobacter pylori-induced indoleamine 2,3-dioxygenase activity in vivo is regulated by TGFB1 and CTLA4 polymorphisms. *Mol Immunol* 2007;**44**:1011–4.
1347. Gu CC, Hunt SC, Kardia S, Turner ST, Chakravarti A, Schork N, et al. An investigation of genome-wide associations of hypertension with microsatellite markers in the family blood pressure program (FBPP). *Hum Genet* 2007;**121**:577–90.
1348. Ammendola M, Bottini N, Pietropolli A, Saccucci P, Gloria-Bottini F. Association between PTPN22 and endometriosis. *Fertil Steril* 2008;**89**:993–4.
1349. Roldan MB, White C, Witchel SF. Association of the GAA1013–>GAG polymorphism of the insulin-like growth factor-1 receptor (IGF1R) gene with premature pubarche. *Fertil Steril* 2007;**88**:410–7.
1350. Morgan AW, Keyte VH, Babbage SJ, Robinson JI, Ponchel F, Barrett JH, et al. FcgammaRIIIA-158V and rheumatoid arthritis: a confirmation study. *Rheumatology (Oxford)* 2003;**42**:528–33.

1351. van de Winkel JG, Anderson CL. Biology of human immunoglobulin G Fc receptors. *J Leukoc Biol* 1991;**49**:511–24.
1352. Rekand T, Langeland N, Aarli JA, Vedeler CA. Fcgamma receptor IIIA polymorphism as a risk factor for acute poliomyelitis. *J Infect Dis* 2002;**186**:1840–3.
1353. Naylor R, Richardson SJ, McAllan BM. Boom and bust: a review of the physiology of the marsupial genus Antechinus. *J Comp Physiol B* 2008;**178**:545–62.
1354. Meyer-Hermann ME, Maini PK. Cutting edge: back to "one-way" germinal centers. *J Immunol* 2005;**174**:2489–93.
1355. Hitze B, Hubold C, Schlichting K, van DR, Lehnert H, Entringer S, et al. How the selfish brain organizes its supply and demand. *Front Neuroenerg* 2010;**2**:7–17.
1356. Boyer D, Walsh PD. Modelling the mobility of living organisms in heterogeneous landscapes: does memory improve foraging success? *Philos Transact A Math Phys Eng Sci* 2010;**368**:5645–59.
1357. Nairne JS, Pandeirada JN. Adaptive memory: ancestral priorities and the mnemonic value of survival processing. *Cogn Psychol* 2010;**61**:1–22.
1358. Fall T, Ingelsson E. Genome-wide association studies of obesity and metabolic syndrome. *Mol Cell Endocrinol* 2014;**382**:740–57.
1359. Spalding KL, Arner E, Westermark PO, Bernard S, Buchholz BA, Bergmann O, et al. Dynamics of fat cell turnover in humans. *Nature* 2008;**453**:783–7.
1360. Kuzawa CW. Adipose tissue in human infancy and childhood: an evolutionary perspective. *Am J Phys Anthropol* 1998;**41**(Suppl. 27):177–209.
1361. Hales CN, Barker DJ. The thrifty phenotype hypothesis. *Br Med Bull* 2001;**60**:5–20.
1362. Hales CN, Barker DJ. Type 2 (non-insulin-dependent) diabetes mellitus: the thrifty phenotype hypothesis. *Diabetologia* 1992;**35**:595–601.
1363. Raison CL, Capuron L, Miller AH. Cytokines sing the blues: inflammation and the pathogenesis of depression. *Trends Immunol* 2006;**27**:24–31.
1364. Dickens C, Creed F. The burden of depression in patients with rheumatoid arthritis. *Rheumatology (Oxford)* 2001;**40**:1327–30.
1365. Wolfe F, Michaud K. Fatigue, rheumatoid arthritis, and anti-tumor necrosis factor therapy: an investigation in 24,831 patients. *J Rheumatol* 2004;**31**:2115–20.
1366. Rupp I, Boshuizen HC, Jacobi CE, Dinant HJ, van den Bos GA. Impact of fatigue on health-related quality of life in rheumatoid arthritis. *Arthritis Rheum* 2004;**51**:578–85.
1367. Kozora E, Ellison MC, West S. Depression, fatigue, and pain in systemic lupus erythematosus (SLE): relationship to the American College of Rheumatology SLE neuropsychological battery. *Arthritis Rheum* 2006;**55**:628–35.
1368. Nery FG, Borba EF, Hatch JP, Soares JC, Bonfa E, Neto FL. Major depressive disorder and disease activity in systemic lupus erythematosus. *Compr Psychiatry* 2007;**48**:14–9.
1369. Palkonyai E, Kolarz G, Kopp M, Bogye G, Temesvari P, Palkonyay L, et al. Depressive symptoms in early rheumatoid arthritis: a comparative longitudinal study. *Clin Rheumatol* 2007;**26**:753–8.
1370. Zakeri Z, Shakiba M, Narouie B, Mladkova N, Ghasemi-Rad M, Khosravi A. Prevalence of depression and depressive symptoms in patients with systemic lupus erythematosus: iranian experience. *Rheumatol Int* 2012;**32**:1179–87.
1371. Basu N, Murray AD, Jones GT, Reid DM, Macfarlane GJ, Waiter GD. Neural correlates of fatigue in granulomatosis with polyangiitis: a functional magnetic resonance imaging study. *Rheumatology (Oxford)* 2014;**53**:2080–7.
1372. Dantzer R, Wollman EE, Yirmiya R. *Brain behavior and immunity: special issue of on cytokines and depression*. San Diego: Academic Press; 2002.

1373. Reichenberg A, Yirmiya R, Schuld A, Kraus T, Haack M, Morag A, et al. Cytokine-associated emotional and cognitive disturbances in humans. *Arch Gen Psychiatry* 2001;**58**:445–52.

1374. Wegner A, Elsenbruch S, Maluck J, Grigoleit JS, Engler H, Jager M, et al. Inflammation-induced hyperalgesia: effects of timing, dosage, and negative affect on somatic pain sensitivity in human experimental endotoxemia. *Brain Behav Immun* 2014;**41**: 46–54.

1375. Musselman DL, Lawson DH, Gumnick JF, Manatunga AK, Penna S, Goodkin RS, et al. Paroxetine for the prevention of depression induced by high-dose interferon alfa. *N Engl J Med* 2001;**344**:961–6.

1376. Moreland LW, Genovese MC, Sato R, Singh A. Effect of etanercept on fatigue in patients with recent or established rheumatoid arthritis. *Arthritis Rheum* 2006;**55**:287–93.

1377. Wells G, Li T, Maxwell L, Maclean R, Tugwell P. Responsiveness of patient reported outcomes including fatigue, sleep quality, activity limitation, and quality of life following treatment with abatacept for rheumatoid arthritis. *Ann Rheum Dis* 2008;**67**:260–5.

1378. Minnock P, Kirwan J, Veale D, Fitzgerald O, Bresnihan B. Fatigue is an independent outcome measure and is sensitive to change in patients with psoriatic arthritis. *Clin Exp Rheumatol* 2010;**28**:401–4.

1379. Mease PJ. Certolizumab pegol in the treatment of rheumatoid arthritis: a comprehensive review of its clinical efficacy and safety. *Rheumatology (Oxford)* 2011;**50**:261–70.

1380. Norheim KB, Jonsson G, Omdal R. Biological mechanisms of chronic fatigue. *Rheumatology (Oxford)* 2011;**50**:1009–18.

1381. Lee YC, Lu B, Edwards RR, Wasan AD, Nassikas NJ, Clauw DJ, et al. The role of sleep problems in central pain processing in rheumatoid arthritis. *Arthritis Rheum* 2013;**65**:59–68.

1382. Wolfe F, Hawley DJ, Wilson K. The prevalence and meaning of fatigue in rheumatic disease. *J Rheumatol* 1996;**23**:1407–17.

1383. van Oers ML, Bossema ER, Thoolen BJ, Hartkamp A, Dekkers JC, Godaert GL, et al. Variability of fatigue during the day in patients with primary Sjogren's syndrome, systemic lupus erythematosus, and rheumatoid arthritis. *Clin Exp Rheumatol* 2010;**28**:715–21.

1384. Iaboni A, Ibanez D, Gladman DD, Urowitz MB, Moldofsky H. Fatigue in systemic lupus erythematosus: contributions of disordered sleep, sleepiness, and depression. *J Rheumatol* 2006;**33**:2453–7.

1385. Cote I, Trojan DA, Kaminska M, Cardoso M, Benedetti A, Weiss D, et al. Impact of sleep disorder treatment on fatigue in multiple sclerosis. *Mult Scler* 2013;**19**:480–9.

1386. Strober BE, Sobell JM, Duffin KC, Bao Y, Guerin A, Yang H, et al. Sleep quality and other patient-reported outcomes improve after patients with psoriasis with suboptimal response to other systemic therapies are switched to adalimumab: results from PROGRESS, an open-label Phase IIIB trial. *Br J Dermatol* 2012;**167**:1374–81.

1387. Jones SD, Koh WH, Steiner A, Garrett SL, Calin A. Fatigue in ankylosing spondylitis: its prevalence and relationship to disease activity, sleep, and other factors. *J Rheumatol* 1996;**23**:487–90.

1388. Omachi TA. Measures of sleep in rheumatologic diseases: Epworth Sleepiness Scale (ESS), Functional Outcome of Sleep Questionnaire (FOSQ), Insomnia Severity Index (ISI), and Pittsburgh Sleep Quality Index (PSQI). *Arthritis Care Res (Hoboken)* 2011;**63**(Suppl. 11):S287–96.

1389. Wells GA, Li T, Kirwan JR, Peterson J, Aletaha D, Boers M, et al. Assessing quality of sleep in patients with rheumatoid arthritis. *J Rheumatol* 2009;**36**:2077–86.

1390. Moldofsky H. Rheumatic manifestations of sleep disorders. *Curr Opin Rheumatol* 2010;**22**:59–63.

1391. Valencia-Flores M, Resendiz M, Castano VA, Santiago V, Campos RM, Sandino S, et al. Objective and subjective sleep disturbances in patients with systemic lupus erythematosus. *Arthritis Rheum* 1999;**42**:2189–93.

1392. Costa DD, Bernatsky S, Dritsa M, Clarke AE, Dasgupta K, Keshani A, et al. Determinants of sleep quality in women with systemic lupus erythematosus. *Arthritis Rheum* 2005;**53**:272–8.
1393. Gudbjornsson B, Broman JE, Hetta J, Hallgren R. Sleep disturbances in patients with primary Sjogren's syndrome. *Br J Rheumatol* 1993;**32**:1072–6.
1394. Milette K, Hudson M, Korner A, Baron M, Thombs BD. Sleep disturbances in systemic sclerosis: evidence for the role of gastrointestinal symptoms, pain and pruritus. *Rheumatology (Oxford)* 2013;**52**:1715–20.
1395. Rudwaleit M, Gooch K, Michel B, Herold M, Thorner A, Wong R, et al. Adalimumab improves sleep and sleep quality in patients with active ankylosing spondylitis. *J Rheumatol* 2011;**38**:79–86.
1396. Deodhar A, Braun J, Inman RD, Mack M, Parasuraman S, Buchanan J, et al. Golimumab reduces sleep disturbance in patients with active ankylosing spondylitis: results from a randomized, placebo-controlled trial. *Arthritis Care Res (Hoboken)* 2010;**62**:1266–71.
1397. Hultgren S, Broman JE, Gudbjornsson B, Hetta J, Lindqvist U. Sleep disturbances in outpatients with ankylosing spondylitisa questionnaire study with gender implications. *Scand J Rheumatol* 2000;**29**:365–9.
1398. Tascilar NF, Tekin NS, Ankarali H, Sezer T, Atik L, Emre U, et al. Sleep disorders in Behcet's disease, and their relationship with fatigue and quality of life. *J Sleep Res* 2012;**21**:281–8.
1399. Ranjbaran Z, Keefer L, Farhadi A, Stepanski E, Sedghi S, Keshavarzian A. Impact of sleep disturbances in inflammatory bowel disease. *J Gastroenterol Hepatol* 2007;**22**:1748–53.
1400. Keefer L, Stepanski EJ, Ranjbaran Z, Benson LM, Keshavarzian A. An initial report of sleep disturbance in inactive inflammatory bowel disease. *J Clin Sleep Med* 2006;**2**:409–16.
1401. Neau JP, Paquereau J, Auche V, Mathis S, Godeneche G, Ciron J, et al. Sleep disorders and multiple sclerosis: a clinical and polysomnography study. *Eur Neurol* 2012;**68**:8–15.
1402. Clark CM, Fleming JA, Li D, Oger J, Klonoff H, Paty D. Sleep disturbance, depression, and lesion site in patients with multiple sclerosis. *Arch Neurol* 1992;**49**:641–3.
1403. Callis Duffin K, Wong B, Horn EJ, Krueger GG. Psoriatic arthritis is a strong predictor of sleep interference in patients with psoriasis. *J Am Acad Dermatol* 2009;**60**:604–8.
1404. Brown RE, Basheer R, McKenna JT, Strecker RE, McCarley RW. Control of sleep and wakefulness. *Physiol Rev* 2012;**92**:1087–187.
1405. Steiger A. Sleep and the hypothalamo-pituitary-adrenocortical system. *Sleep Med Rev* 2002;**6**:125–38.
1406. Dresler M, Spoormaker VI, Beitinger P, Czisch M, Kimura M, Steiger A, et al. Neuroscience-driven discovery and development of sleep therapeutics. *Pharmacol Ther* 2014;**141**:300–34.
1407. Antonijevic I. HPA axis and sleep: identifying subtypes of major depression. *Stress* 2008;**11**:15–27.
1408. Buckley TM, Schatzberg AF. On the interactions of the hypothalamic-pituitary-adrenal (HPA) axis and sleep: normal HPA axis activity and circadian rhythm, exemplary sleep disorders. *J Clin Endocrinol Metab* 2005;**90**:3106–14.
1409. Holsboer F, von BU, Steiger A. Effects of intravenous corticotropin-releasing hormone upon sleep-related growth hormone surge and sleep EEG in man. *Neuroendocrinology* 1988;**48**:32–8.
1410. Born J, DeKloet ER, Wenz H, Kern W, Fehm HL. Gluco- and antimineralocorticoid effects on human sleep: a role of central corticosteroid receptors. *Am J Physiol* 1991;**260**:E183–8.
1411. Born J, Spath-Schwalbe E, Schwakenhofer H, Kern W, Fehm HL. Influences of corticotropin-releasing hormone, adrenocorticotropin, and cortisol on sleep in normal man. *J Clin Endocrinol Metab* 1989;**68**:904–11.
1412. Fehm HL, Spath-Schwalbe E, Pietrowsky R, Kern W, Born J. Entrainment of nocturnal pituitary-adrenocortical activity to sleep processes in man–a hypothesis. *Exp Clin Endocrinol* 1993;**101**:267–76.

1413. Besedovsky L, Born J, Lange T. Blockade of mineralocorticoid receptors enhances naive T-helper cell counts during early sleep in humans. *Brain Behav Immun* 2012;**26**:1116–21.
1414. Lorton D, Lubahn CL, Estus C, Millar BA, Carter JL, Wood CA, et al. Bidirectional communication between the brain and the immune system: implications for physiological sleep and disorders with disrupted sleep. *Neuroimmunomodulation* 2006;**13**:357–74.
1415. Imeri L, Opp MR. How (and why) the immune system makes us sleep. *Nat Rev Neurosci* 2009;**10**:199–210.
1416. Pollmächer T, Schreiber W, Gudewill S, Vedder H, Fassbender K, Wiedemann K, et al. Influence of endotoxin on nocturnal sleep in humans. *Am J Physiol* 1993;**264**:R1077–83.
1417. Raison CL, Rye DB, Woolwine BJ, Vogt GJ, Bautista BM, Spivey JR, et al. Chronic interferon-alpha administration disrupts sleep continuity and depth in patients with hepatitis C: association with fatigue, motor slowing, and increased evening cortisol. *Biol Psychiatry* 2010;**68**:942–9.
1418. Opp MR. Sleep and psychoneuroimmunology. *Immunol Allergy Clin North Am* 2009;**29**:295–307.
1419. Cohen O, Reichenberg A, Perry C, Ginzberg D, Pollmacher T, Soreq H, et al. Endotoxin-induced changes in human working and declarative memory associate with cleavage of plasma "readthrough" acetylcholinesterase. *J Mol Neurosci* 2003;**21**:199–212.
1420. Taylor-Gjevre RM, Gjevre JA, Nair BV, Skomro RP, Lim HJ. Improved sleep efficiency after anti-tumor necrosis factor alpha therapy in rheumatoid arthritis patients. *Ther Adv Musculoskelet Dis* 2011;**3**:227–33.
1421. Fragiadaki K, Tektonidou MG, Konsta M, Chrousos GP, Sfikakis PP. Sleep disturbances and interleukin 6 receptor inhibition in rheumatoid arthritis. *J Rheumatol* 2012;**39**:60–2.
1422. Wells G, Li T, Tugwell P. Investigation into the impact of abatacept on sleep quality in patients with rheumatoid arthritis, and the validity of the MOS-Sleep questionnaire Sleep Disturbance Scale. *Ann Rheum Dis* 2010;**69**:1768–73.
1423. Genovese MC, Schiff M, Luggen M, Becker JC, Aranda R, Teng J, et al. Efficacy and safety of the selective co-stimulation modulator abatacept following 2 years of treatment in patients with rheumatoid arthritis and an inadequate response to anti-tumour necrosis factor therapy. *Ann Rheum Dis* 2008;**67**:547–54.
1424. Vgontzas AN, Zoumakis E, Lin HM, Bixler EO, Trakada G, Chrousos GP. Marked decrease in sleepiness in patients with sleep apnea by etanercept, a tumor necrosis factor-alpha antagonist. *J Clin Endocrinol Metab* 2004;**89**:4409–13.
1425. Buckley TM, Schatzberg AF. On the interactions of the hypothalamic-pituitary-adrenal (HPA) axis and sleep: normal HPA axis activity and circadian rhythm, exemplary sleep disorders. *J Clin Endocrinol Metab* 2005;**90**:3106–14.
1426. Hart BL. Biological basis of the behavior of sick animals. *Neurosci Biobehav Rev* 1988;**12**:123–37.
1427. Rennie KL, Hughes J, Lang R, Jebb SA. Nutritional management of rheumatoid arthritis: a review of the evidence. *J Hum Nutr Diet* 2003;**16**:97–109.
1428. van de Laar MA, Nieuwenhuis JM, Former-Boon M, Hulsing J, van der Korst JK. Nutritional habits of patients suffering from seropositive rheumatoid arthritis: a screening of 93 Dutch patients. *Clin Rheumatol* 1990;**9**:483–8.
1429. Hansen GV, Nielsen L, Kluger E, Thysen M, Emmertsen H, Stengaard-Pedersen K, et al. Nutritional status of Danish rheumatoid arthritis patients and effects of a diet adjusted in energy intake, fish-meal, and antioxidants. *Scand J Rheumatol* 1996;**25**:325–30.
1430. Lundberg AC, Akesson A, Akesson B. Dietary intake and nutritional status in patients with systemic sclerosis. *Ann Rheum Dis* 1992;**51**:1143–8.

1431. Payne A. Nutrition and diet in the clinical management of multiple sclerosis. *J Hum Nutr Diet* 2001;**14**:349–57.
1432. Bacon MC, White PH, Raiten DJ, Craft N, Margolis S, Levander OA, et al. Nutritional status and growth in juvenile rheumatoid arthritis. *Semin Arthritis Rheum* 1990;**20**:97–106.
1433. Roubenoff R, Roubenoff RA, Cannon JG, Kehayias JJ, Zhuang H, Wson-Hughes B, et al. Rheumatoid cachexia: cytokine-driven hypermetabolism accompanying reduced body cell mass in chronic inflammation. *J Clin Invest* 1994;**93**:2379–86.
1434. Yamauchi T, Sato H. Nutritional status, activity pattern, and dietary intake among the Baka hunter-gatherers in the village camps in cameroon. *Afr Study Mongr* 2000;**21**:67–82.
1435. Tsigos C, Stefanaki C, Lambrou GI, Boschiero D, Chrousos GP. Stress and inflammatory biomarkers and symptoms are associated with bio-impedance measures. *Eur J Clin Invest* 2015;**45**:126–34.
1436. Montalcini T, Romeo S, Ferro Y, Migliaccio V, Gazzaruso C, Pujia A. Osteoporosis in chronic inflammatory disease: the role of malnutrition. *Endocrine* 2013;**43**:59–64.
1437. Liefmann R. Endocrine imbalance in rheumatoid arthritis and rheumatoid spondylitis; hyperglycemia unresponsiveness, insulin resistance, increased gluconeogenesis and mesenchymal tissue degeneration; preliminary report. *Acta Med Scand* 1949;**136**:226–32.
1438. Roubenoff R, Roubenoff RA, Ward LM, Stevens MB. Catabolic effects of high-dose corticosteroids persist despite therapeutic benefit in rheumatoid arthritis. *Am J Clin Nutr* 1990;**52**:1113–7.
1439. Walsmith J, Roubenoff R. Cachexia in rheumatoid arthritis. *Int J Cardiol* 2002;**85**:89–99.
1440. Owen OE, Reichard Jr GA, Patel MS, Boden G. Energy metabolism in feasting and fasting. *Adv Exp Med Biol* 1979;**111**:169–88.
1441. Marcora SM, Chester KR, Mittal G, Lemmey AB, Maddison PJ. Randomized phase 2 trial of anti-tumor necrosis factor therapy for cachexia in patients with early rheumatoid arthritis. *Am J Clin Nutr* 2006;**84**:1463–72.
1442. Fong Y, Moldawer LL, Marano M, Wei H, Barber A, Manogue K, et al. Cachectin/TNF or IL-1 alpha induces cachexia with redistribution of body proteins. *Am J Physiol* 1989;**256**:R659–65.
1443. Kaufmann J, Kielstein V, Kilian S, Stein G, Hein G. Relation between body mass index and radiological progression in patients with rheumatoid arthritis. *J Rheumatol* 2003;**30**:2350–5.
1444. Escalante A, Haas RW, del RI. Paradoxical effect of body mass index on survival in rheumatoid arthritis: role of comorbidity and systemic inflammation. *Arch Intern Med* 2005;**165**:1624–9.
1445. van der Helm-van Mil AH, van der Kooij SM, Allaart CF, Toes RE, Huizinga TW. A high body mass index has a protective effect on the amount of joint destruction in small joints in early rheumatoid arthritis. *Ann Rheum Dis* 2008;**67**:769–74.
1446. Westhoff G, Rau R, Zink A. Radiographic joint damage in early rheumatoid arthritis is highly dependent on body mass index. *Arthritis Rheum* 2007;**56**:3575–82.
1447. Joslin EP. The treatment of diabetes mellitus. *Can Med Assoc J* 1916;**6**:673–84.
1448. Pemberton R, Foster GL. Studies on arthritis in the army based on four hundred cases (iii). studies on the nitrogen, urea, carbon dioxid combining power, calcium, total fat and cholesterol of the fasting blood, renal function, blood sugar and sugar tolerance. *Arch Intern Med* 1920;**25**:243–82.
1449. Rabinowitch IM. The influence of infection upon the reaction of the diabetic to insulin treatment. *Can Med Assoc J* 1924;**14**:481–2.
1450. Root HF. Insulin resistance and bronze diabetes. *N Engl J Med* 1929;**201**:201–6.
1451. Moller DE, Flier JS. Insulin resistance–mechanisms, syndromes, and implications. *N Engl J Med* 1991;**325**:938–48.

1452. Gregor MF, Hotamisligil GS. Inflammatory mechanisms in obesity. *Annu Rev Immunol* 2011;**29**:415–45.
1453. Hotamisligil GS, Erbay E. Nutrient sensing and inflammation in metabolic diseases. *Nat Rev Immunol* 2008;**8**:923–34.
1454. Schenk S, Saberi M, Olefsky JM. Insulin sensitivity: modulation by nutrients and inflammation. *J Clin Invest* 2008;**118**:2992–3002.
1455. Goldfine AB, Fonseca V, Shoelson SE. Therapeutic approaches to target inflammation in type 2 diabetes. *Clin Chem* 2011;**57**:162–7.
1456. Dallman MF. Stress-induced obesity and the emotional nervous system. *Trends Endocrinol Metab* 2010;**21**:159–65.
1457. Brunner EJ, Chandola T, Marmot MG. Prospective effect of job strain on general and central obesity in the Whitehall II Study. *Am J Epidemiol* 2007;**165**:828–37.
1458. Block JP, He Y, Zaslavsky AM, Ding L, Ayanian JZ. Psychosocial stress and change in weight among US adults. *Am J Epidemiol* 2009;**170**:181–92.
1459. Korkeila M, Kaprio J, Rissanen A, Koshenvuo M, Sorensen TI. Predictors of major weight gain in adult Finns: stress, life satisfaction and personality traits. *Int J Obes Relat Metab Disord* 1998;**22**:949–57.
1460. Serlachius A, Hamer M, Wardle J. Stress and weight change in university students in the United Kingdom. *Physiol Behav* 2007;**92**:548–53.
1461. Straub RH. Systemic disease sequelae in chronic inflammatory diseases and chronic psychological stress: comparison and pathophysiological model. *Ann N Y Acad Sci* 2014;**1318**:7–17.
1462. Himsworth HP. Diabetes mellitus: Its differentiation into insulin-sensitive and insulin-insensitive types. *Lancet* 1936;**227**:127–30.
1463. Thomsen V. Das Trauma und der Kohlenhydratstoffwechsel. *Acta Med Scand* 1936;**90**:918–25.
1464. Graham G. A review of the causes of diabetes mellitus. *Br Med J* 1940;**2**:479–82.
1465. Arendt EC, Pattee CJ. Studies on obesity. I. The insulin-glucose tolerance curve. *J Clin Endocrinol Metab* 1956;**16**:367–74.
1466. Collins J. Insulin resistance in schizophrenia. *Med J Aust* 1957;**44**:467–70.
1467. Yalow RS, Berson SA. Immunoassay of endogenous plasma insulin in man. *J Clin Invest* 1960;**39**:1157–75.
1468. Randle PJ, Garland PB, Hales CN, Newsholme EA. The glucose fatty-acid cycle. Its role in insulin sensitivity and the metabolic disturbances of diabetes mellitus. *Lancet* 1963;**1**:785–9.
1469. van Praag HM, Leijnse B. Depression, glucose tolerance, peripheral glucose uptake and their alterations under the influence of anti-depressive drugs of the hydrazine type. *Psychopharmacologia* 1965;**8**:67–78.
1470. Butterfield WJH, Wichelow MJ. Peripheral glucose metabolism in control subjects and diabetic patients during glucose, glucose-insulin and insulin sensitivity tests. *Diabetologia* 1965;**1**:43–53.
1471. Shen SW, Reaven GM, Farquhar JW. Comparison of impedance to insulin-mediated glucose uptake in normal subjects and in subjects with latent diabetes. *J Clin Invest* 1970;**49**:2151–60.
1472. DeFronzo RA, Tobin JD, Andres R. Glucose clamp technique: a method for quantifying insulin secretion and resistance. *Am J Physiol* 1979;**237**:E214–23.
1473. Wolfe RR. Substrate utilization/insulin resistance in sepsis/trauma. *Baillieres Clin Endocrinol Metab* 1997;**11**:645–57.
1474. Kasuga M, Zick Y, Blithe DL, Crettaz M, Kahn CR. Insulin stimulates tyrosine phosphorylation of the insulin receptor in a cell-free system. *Nature* 1982;**298**:667–9.

1475. Ciaraldi TP, Kolterman OG, Scarlett JA, Kao M, Olefsky JM. Role of glucose transport in the postreceptor defect of non-insulin-dependent diabetes mellitus. *Diabetes* 1982;**31**:1016–22.
1476. Grunberger G, Zick Y, Gorden P. Defect in phosphorylation of insulin receptors in cells from an insulin-resistant patient with normal insulin binding. *Science* 1984;**223**:932–4.
1477. Garvey WT, Olefsky JM, Marshall S. Insulin induces progressive insulin resistance in cultured rat adipocytes. Sequential effects at receptor and multiple postreceptor sites. *Diabetes* 1986;**35**:258–67.
1478. Krieger DR, Landsberg L. Mechanisms in obesity-related hypertension: role of insulin and catecholamines. *Am J Hypertens* 1988;**1**:84–90.
1479. DeFronzo RA. Lilly lecture 1987. The triumvirate: beta-cell, muscle, liver. A collusion responsible for NIDDM. *Diabetes* 1988;**37**:667–87.
1480. Reaven GM. Banting lecture 1988. Role of insulin resistance in human disease. *Diabetes* 1988;**37**:1595–607.
1481. Uchida I, Asoh T, Shirasaka C, Tsuji H. Effect of epidural analgesia on postoperative insulin resistance as evaluated by insulin clamp technique. *Br J Surg* 1988;**75**:557–62.
1482. Greisen J, Juhl CB, Grofte T, Vilstrup H, Jensen TS, Schmitz O. Acute pain induces insulin resistance in humans. *Anesthesiology* 2001;**95**:578–84.
1483. Feingold KR, Grunfeld C. Role of cytokines in inducing hyperlipidemia. *Diabetes* 1992;**41**(Suppl. 2):97–101.
1484. Moberg E, Kollind M, Lins PE, Adamson U. Acute mental stress impairs insulin sensitivity in IDDM patients. *Diabetologia* 1994;**37**:247–51.
1485. Keltikangas-Jarvinen L, Ravaja N, Raikkonen K, Lyytinen H. Insulin resistance syndrome and autonomically mediated physiological responses to experimentally induced mental stress in adolescent boys. *Metabolism* 1996;**45**:614–21.
1486. Björntorp P. Neuroendocrine perturbations as a cause of insulin resistance. *Diabetes Metab Res Rev* 1999;**15**:427–41.
1487. Chrousos GP. The role of stress and the hypothalamic-pituitary-adrenal axis in the pathogenesis of the metabolic syndrome: neuro-endocrine and target tissue-related causes. *Int J Obes Relat Metab Disord* 2000;**24**(Suppl. 2):S50–5.
1488. Seematter G, Guenat E, Schneiter P, Cayeux C, Jequier E, Tappy L. Effects of mental stress on insulin-mediated glucose metabolism and energy expenditure in lean and obese women. *Am J Physiol Endocrinol Metab* 2000;**279**:E799–805.
1489. Tso TK, Huang HY, Chang CK, Liao YJ, Huang WN. Clinical evaluation of insulin resistance and beta-cell function by the homeostasis model assessment in patients with systemic lupus erythematosus. *Clin Rheumatol* 2004;**23**:416–20.
1490. Kiortsis DN, Mavridis AK, Vasakos S, Nikas SN, Drosos AA. Effects of infliximab treatment on insulin resistance in patients with rheumatoid arthritis and ankylosing spondylitis. *Ann Rheum Dis* 2005;**64**:765–6.
1491. Stagakis I, Bertsias G, Karvounaris S, Kavousanaki M, Virla D, Raptopoulou A, et al. Anti-tumor necrosis factor therapy improves insulin resistance, beta cell function and insulin signaling in active rheumatoid arthritis patients with high insulin resistance. *Arthritis Res Ther* 2012;**14**:R141.
1492. Fleischman A, Shoelson SE, Bernier R, Goldfine AB. Salsalate improves glycemia and inflammatory parameters in obese young adults. *Diabetes Care* 2008;**31**:289–94.
1493. Goldfine AB, Fonseca V, Jablonski KA, Pyle L, Staten MA, Shoelson SE. The effects of salsalate on glycemic control in patients with type 2 diabetes: a randomized trial. *Ann Intern Med* 2010;**152**:346–57.

1494. Schultz O, Oberhauser F, Saech J, Rubbert-Roth A, Hahn M, Krone W, et al. Effects of inhibition of interleukin-6 signalling on insulin sensitivity and lipoprotein (a) levels in human subjects with rheumatoid diseases. *PLoS One* 2010;**5**:e14328.
1495. Gonzalez-Gay MA, De Matias JM, Gonzalez-Juanatey C, Garcia-Porrua C, Sanchez-Andrade A, Martin J, et al. Anti-tumor necrosis factor-alpha blockade improves insulin resistance in patients with rheumatoid arthritis. *Clin Exp Rheumatol* 2006;**24**:83–6.
1496. Kiortsis DN, Mavridis AK, Filippatos TD, Vasakos S, Nikas SN, Drosos AA. Effects of infliximab treatment on lipoprotein profile in patients with rheumatoid arthritis and ankylosing spondylitis. *J Rheumatol* 2006;**33**:921–3.
1497. Dubreuil M, Rho YH, Man A, Zhu Y, Zhang Y, Love TJ, et al. Diabetes incidence in psoriatic arthritis, psoriasis and rheumatoid arthritis: a UK population-based cohort study. *Rheumatology (Oxford)* 2014;**53**:346–52.
1498. Landsberg L. Role of the sympathetic adrenal system in the pathogenesis of the insulin resistance syndrome. *Ann N Y Acad Sci* 1999;**892**:84–90.
1499. Landsberg L, Aronne LJ, Beilin LJ, Burke V, Igel LI, Lloyd-Jones D, et al. Obesity-related hypertension: pathogenesis, cardiovascular risk, and treatment: a position paper of The Obesity Society and the American Society of Hypertension. *J Clin Hypertens (Greenwich)* 2013;**15**:14–33.
1500. Chrousos GP, Tsigos C. *Annals of the New York Academy of Science: stress, obesity, and metabolic syndrome*. Malden, MA: John Wiley & Sons, Inc.; 2006.
1501. Myers Jr MG, Olson DP. Central nervous system control of metabolism. *Nature* 2012;**491**:357–63.
1502. Kaaja R, Kujala S, Manhem K, Katzman P, Kibarskis A, Antikainen R, et al. Effects of sympatholytic therapy on insulin sensitivity indices in hypertensive postmenopausal women. *Int J Clin Pharmacol Ther* 2007;**45**:394–401.
1503. Mahfoud F, Schlaich M, Kindermann I, Ukena C, Cremers B, Brandt MC, et al. Effect of renal sympathetic denervation on glucose metabolism in patients with resistant hypertension: a pilot study. *Circulation* 2011;**123**:1940–6.
1504. Bergman RN, Ider YZ, Bowden CR, Cobelli C. Quantitative estimation of insulin sensitivity. *Am J Physiol* 1979;**236**:E667–77.
1505. Borai A, Livingstone C, Kaddam I, Ferns G. Selection of the appropriate method for the assessment of insulin resistance. *BMC Med Res Methodol* 2011;**11**:158–67.
1506. Abdul-Ghani MA, Matsuda M, Balas B, DeFronzo RA. Muscle and liver insulin resistance indexes derived from the oral glucose tolerance test. *Diabetes Care* 2007;**30**:89–94.
1507. Syed Ikmal SI, Zaman Huri H, Vethakkan SR, Wan Ahmad WA. Potential biomarkers of insulin resistance and atherosclerosis in type 2 diabetes mellitus patients with coronary artery disease. *Int J Endocrinol* 2013;**2013**:1–11. http://dx.doi.org/10.1155/2013/698567.
1508. Montague CT, Farooqi IS, Whitehead JP, Soos MA, Rau H, Wareham NJ, et al. Congenital leptin deficiency is associated with severe early-onset obesity in humans. *Nature* 1997;**387**:903–8.
1509. Vaisse C, Clement K, Guy-Grand B, Froguel P. A frameshift mutation in human MC4R is associated with a dominant form of obesity. *Nat Genet* 1998;**20**:113–4.
1510. Peters A, Langemann D. Build-ups in the supply chain of the brain: on the neuroenergetic cause of obesity and type 2 diabetes mellitus. *Front Neuroenerg* 2009;**1**:2–12.
1511. Jauch-Chara K, Oltmanns KM. Obesity—a neuropsychological disease? Systematic review and neuropsychological model. *Prog Neurobiol* 2014;**114**:84–101.
1512. Keen-Rhinehart E, Ondek K, Schneider JE. Neuroendocrine regulation of appetitive ingestive behavior. *Front Neurosci* 2013;**7**:213.

1513. Osborn O, Olefsky JM. The cellular and signaling networks linking the immune system and metabolism in disease. *Nat Med* 2012;**18**:363–74.
1514. Shoelson SE, Herrero L, Naaz A. Obesity, inflammation, and insulin resistance. *Gastroenterology* 2007;**132**:2169–80.
1515. Nakae J, Oki M, Cao Y. The FoxO transcription factors and metabolic regulation. *FEBS Lett* 2008;**582**:54–67.
1516. Glass CK, Olefsky JM. Inflammation and lipid signaling in the etiology of insulin resistance. *Cell Metab* 2012;**15**:635–45.
1517. Ley RE, Turnbaugh PJ, Klein S, Gordon JI. Microbial ecology: human gut microbes associated with obesity. *Nature* 2006;**444**:1022–3.
1518. Johnson AM, Olefsky JM. The origins and drivers of insulin resistance. *Cell* 2013;**152**:673–84.
1519. Jin C, Henao-Mejia J, Flavell RA. Innate immune receptors: key regulators of metabolic disease progression. *Cell Metab* 2013;**17**:873–82.
1520. Neel JV. Diabetes mellitus: a "thrifty" genotype rendered detrimental by "progress"? *Am J Hum Genet* 1962;**14**:353–62.
1521. Neel JV. The "thrifty genotype" in 1998. *Nutr Rev* 1999;**57**:S2–9.
1522. Reaven GM. Hypothesis: muscle insulin resistance is the ("not-so") thrifty genotype. *Diabetologia* 1998;**41**:482–4.
1523. Levitan RD, Wendland B. Novel "thrifty" models of increased eating behaviour. *Curr Psychiatry Rep* 2013;**15**:408.
1524. Cahill Jr GF. Human evolution and insulin-dependent (IDD) and non-insulin dependent diabetes (NIDD). *Metabolism* 1979;**28**:389–93.
1525. Sebert S, Sharkey D, Budge H, Symonds ME. The early programming of metabolic health: is epigenetic setting the missing link? *Am J Clin Nutr* 2011;**94**:1953S–8S.
1526. Roseboom TJ, Watson ED. The next generation of disease risk: are the effects of prenatal nutrition transmitted across generations? Evidence from animal and human studies. *Placenta* 2012;**33**(Suppl. 2):e40–4.
1527. Gluckman PD, Hanson MA. The developmental origins of the metabolic syndrome. *Trends Endocrinol Metab* 2004;**15**:183–7.
1528. Fernandez-Real JM, Ricart W. Insulin resistance and inflammation in an evolutionary perspective: the contribution of cytokine genotype/phenotype to thriftiness. *Diabetologia* 1999;**42**:1367–74.
1529. Hotamisligil GS. Inflammation and metabolic disorders. *Nature* 2006;**444**:860–7.
1530. Kitano H, Oda K, Kimura T, Matsuoka Y, Csete M, Doyle J, et al. Metabolic syndrome and robustness tradeoffs. *Diabetes* 2004;**53**(Suppl. 3):S6–15.
1531. Schwartz MW, Niswender KD. Adiposity signaling and biological defense against weight gain: absence of protection or central hormone resistance? *J Clin Endocrinol Metab* 2004;**89**:5889–97.
1532. Taubes G. *Good calories, bad calories—challenging the conventional wisdom on diet, weight control, and disease*. New York: Knopf; 2007.
1533. Kuipers RS, Luxwolda MF, jck-Brouwer DA, Eaton SB, Crawford MA, Cordain L, et al. Estimated macronutrient and fatty acid intakes from an East African Paleolithic diet. *Br J Nutr* 2010;**104**:1666–87.
1534. DeFronzo RA. Banting Lecture. From the triumvirate to the ominous octet: a new paradigm for the treatment of type 2 diabetes mellitus. *Diabetes* 2009;**58**:773–95.
1535. Pharmazeutika Geigy. *Wissenschaftliche Tabellen*. Wehr: Ciba-Geigy; 1973.
1536. Peters A, Schweiger U, Pellerin L, Hubold C, Oltmanns KM, Conrad M, et al. The selfish brain: competition for energy resources. *Neurosci Biobehav Rev* 2004;**28**:143–80.

1537. Quine S, Lyle D, Pierce J. Stressors experienced by relatives of patients in an innovative rehabilitation program. *Health Soc Work* 1993;**18**:114–22.
1538. McAlonan GM, Lee AM, Cheung V, Cheung C, Tsang KW, Sham PC, et al. Immediate and sustained psychological impact of an emerging infectious disease outbreak on health care workers. *Can J Psychiatry* 2007;**52**:241–7.
1539. Zunhammer M, Eberle H, Eichhammer P, Busch V. Somatic symptoms evoked by exam stress in university students: the role of alexithymia, neuroticism, anxiety and depression. *PLoS One* 2013;**8**:e84911.
1540. Borella P, Bargellini A, Rovesti S, Pinelli M, Vivoli R, Solfrini V, et al. Emotional stability, anxiety, and natural killer activity under examination stress. *Psychoneuroendocrinology* 1999;**24**:613–27.
1541. Aggarwal B, Liao M, Christian A, Mosca L. Influence of caregiving on lifestyle and psychosocial risk factors among family members of patients hospitalized with cardiovascular disease. *J Gen Intern Med* 2009;**24**:93–8.
1542. Fredman L, Doros G, Cauley JA, Hillier TA, Hochberg MC. Caregiving, metabolic syndrome indicators, and 1-year decline in walking speed: results of Caregiver-SOF. *J Gerontol A Biol Sci Med Sci* 2010;**65**:565–72.
1543. von Känel R, Mausbach BT, Dimsdale JE, Mills PJ, Patterson TL, Ancoli-Israel S, et al. Cardiometabolic effects in caregivers of nursing home placement and death of their spouse with Alzheimer's disease. *J Am Geriatr Soc* 2011;**59**:2037–44.
1544. Reeves KW, Bacon K, Fredman L. Caregiving associated with selected cancer risk behaviors and screening utilization among women: cross-sectional results of the 2009 BRFSS. *BMC Public Health* 2012;**12**:685.
1545. Capistrant BD, Berkman LF, Glymour MM. Does duration of spousal caregiving affect risk of depression onset? Evidence from the health and retirement study. *Am J Geriatr Psychiatry* 2014;**22**:766–70.
1546. Kiecolt-Glaser JK, Preacher KJ, MacCallum RC, Atkinson C, Malarkey WB, Glaser R. Chronic stress and age-related increases in the proinflammatory cytokine IL-6. *Proc Natl Acad Sci U S A* 2003;**100**:9090–5.
1547. Agardh EE, Ahlbom A, Andersson T, Efendic S, Grill V, Hallqvist J, et al. Work stress and low sense of coherence is associated with type 2 diabetes in middle-aged Swedish women. *Diabetes Care* 2003;**26**:719–24.
1548. Esquirol Y, Bongard V, Mabile L, Jonnier B, Soulat JM, Perret B. Shift work and metabolic syndrome: respective impacts of job strain, physical activity, and dietary rhythms. *Chronobiol Int* 2009;**26**:544–59.
1549. Edwards EM, Stuver SO, Heeren TC, Fredman L. Job strain and incident metabolic syndrome over 5 years of follow-up: the coronary artery risk development in young adults study. *J Occup Environ Med* 2012;**54**:1447–52.
1550. Mullington JM, Haack M, Toth M, Serrador JM, Meier-Ewert HK. Cardiovascular, inflammatory, and metabolic consequences of sleep deprivation. *Prog Cardiovasc Dis* 2009;**51**:294–302.
1551. Kivimaki M, Virtanen M, Elovainio M, Kouvonen A, Vaananen A, Vahtera J. Work stress in the etiology of coronary heart disease–a meta-analysis. *Scand J Work Environ Health* 2006;**32**:431–42.
1552. Björntorp P. Thrifty genes and human obesity. Are we chasing ghosts? *Lancet* 2001;**358**:1006–8.
1553. Castagnetta LA, Carruba G, Granata OM, Stefano R, Miele M, Schmidt M, et al. Increased estrogen formation and estrogen to androgen ratio in the synovial fluid of patients with rheumatoid arthritis. *J Rheumatol* 2003;**30**:2597–605.

1554. Pongratz G, Straub RH. Role of peripheral nerve fibres in acute and chronic inflammation in arthritis. *Nat Rev Rheumatol* 2013;**9**:117–26.
1555. Lutgendorf SK, Garand L, Buckwalter KC, Reimer TT, Hong SY, Lubaroff DM. Life stress, mood disturbance, and elevated interleukin-6 in healthy older women. *J Gerontol A Biol Sci Med Sci* 1999;**54**:M434–9.
1556. Sjögren E, Leanderson P, Kristenson M, Ernerudh J. Interleukin-6 levels in relation to psychosocial factors: studies on serum, saliva, and in vitro production by blood mononuclear cells. *Brain Behav Immun* 2006;**20**:270–8.
1557. Müller N, Riedel M, Scheppach C, Brandstätter B, Sokullu S, Krampe K, et al. Beneficial antipsychotic effects of celecoxib add-on therapy compared to risperidone alone in schizophrenia. *Am J Psychiatry* 2002;**159**:1029–34.
1558. Rosenblat JD, Cha DS, Mansur RB, McIntyre RS. Inflamed moods: a review of the interactions between inflammation and mood disorders. *Prog Neuropsychopharmacol Biol Psychiatry* 2014;**53**:23–34.
1559. Müller N. The role of anti-inflammatory treatment in psychiatric disorders. *Psychiatr Danub* 2013;**25**:292–8.
1560. Adiels M, Olofsson SO, Taskinen MR, Boren J. Overproduction of very low-density lipoproteins is the hallmark of the dyslipidemia in the metabolic syndrome. *Arterioscler Thromb Vasc Biol* 2008;**28**:1225–36.
1561. Rossner S, Lofmark C. Dyslipoproteinaemia in patients with active, chronic polyarthritis. A study on serum lipoproteins and triglyceride clearance (intravenous fat tolerance test). *Atherosclerosis* 1977;**28**:41–52.
1562. Lakatos J, Harsagyi A. Serum total, HDL, LDL cholesterol, and triglyceride levels in patients with rheumatoid arthritis. *Clin Biochem* 1988;**21**:93–6.
1563. Ilowite NT, Samuel P, Ginzler E, Jacobson MS. Dyslipoproteinemia in pediatric systemic lupus erythematosus. *Arthritis Rheum* 1988;**31**:859–63.
1564. Ettinger Jr WH, Hazzard WR. Elevated apolipoprotein-B levels in corticosteroid-treated patients with systemic lupus erythematosus. *J Clin Endocrinol Metab* 1988;**67**:425–8.
1565. Borba EF, Bonfa E. Dyslipoproteinemias in systemic lupus erythematosus: influence of disease, activity, and anticardiolipin antibodies. *Lupus* 1997;**6**:533–9.
1566. Jones SM, Harris CP, Lloyd J, Stirling CA, Reckless JP, McHugh NJ. Lipoproteins and their subfractions in psoriatic arthritis: identification of an atherogenic profile with active joint disease. *Ann Rheum Dis* 2000;**59**:904–9.
1567. Hahn BH, Grossman J, Chen W, McMahon M. The pathogenesis of atherosclerosis in autoimmune rheumatic diseases: roles of inflammation and dyslipidemia. *J Autoimmun* 2007;**28**:69–75.
1568. Grunfeld C, Feingold KR. Tumor necrosis factor, cytokines, and the hyperlipidemia of infection. *Trends Endocrinol Metab* 1991;**2**:213–9.
1569. Sherman ML, Spriggs DR, Arthur KA, Imamura K, Frei III E, Kufe DW. Recombinant human tumor necrosis factor administered as a five-day continuous infusion in cancer patients: phase I toxicity and effects on lipid metabolism. *J Clin Oncol* 1988;**6**:344–50.
1570. Khovidhunkit W, Kim MS, Memon RA, Shigenaga JK, Moser AH, Feingold KR, et al. Effects of infection and inflammation on lipid and lipoprotein metabolism: mechanisms and consequences to the host. *J Lipid Res* 2004;**45**:1169–96.
1571. Pond CM, Mattacks CA. Interactions between adipose tissue around lymph nodes and lymphoid cells in vitro. *J Lipid Res* 1995;**36**:2219–31.
1572. Pond CM. Paracrine relationships between adipose and lymphoid tissues: implications for the mechanism of HIV-associated adipose redistribution syndrome. *Trends Immunol* 2003;**24**:13–8.

1573. Straub RH, Lowin T, Klatt S, Wolff C, Rauch L. Increased density of sympathetic nerve fibers in metabolically activated fat tissue surrounding human synovium and mouse lymph nodes in arthritis. *Arthritis Rheum* 2011;**63**:3234–42.
1574. Miller AH, Pariante CM, Pearce BD. Effects of cytokines on glucocorticoid receptor expression and function. Glucocorticoid resistance and relevance to depression. *Adv Exp Med Biol* 1999;**461**:107–16.
1575. Webster JC, Oakley RH, Jewell CM, Cidlowski JA. Proinflammatory cytokines regulate human glucocorticoid receptor gene expression and lead to the accumulation of the dominant negative beta isoform: a mechanism for the generation of glucocorticoid resistance. *Proc Natl Acad Sci U S A* 2001;**98**:6865–70.
1576. Masi AT, Aldag JC, Jacobs JW. Rheumatoid arthritis: neuroendocrine immune integrated physiopathogenetic perspectives and therapy. *Rheum Dis Clin North Am* 2005;**31**:131–60.
1577. Djurhuus CB, Gravholt CH, Nielsen S, Pedersen SB, Moller N, Schmitz O. Additive effects of cortisol and growth hormone on regional and systemic lipolysis in humans. *Am J Physiol Endocrinol Metab* 2004;**286**:E488–94.
1578. Pedersen SB, Kristensen K, Hermann PA, Katzenellenbogen JA, Richelsen B. Estrogen controls lipolysis by up-regulating alpha2A-adrenergic receptors directly in human adipose tissue through the estrogen receptor alpha. Implications for the female fat distribution. *J Clin Endocrinol Metab* 2004;**89**:1869–78.
1579. Elbers JM, Asscheman H, Seidell JC, Gooren LJ. Effects of sex steroid hormones on regional fat depots as assessed by magnetic resonance imaging in transsexuals. *Am J Physiol* 1999;**276**:E317–25.
1580. Rivier C, Vale W. In the rat, interleukin-1 alpha acts at the level of the brain and the gonads to interfere with gonadotropin and sex steroid secretion. *Endocrinology* 1989;**124**:2105–9.
1581. El Maghraoui A, Tellal S, Chaouir S, Lebbar K, Bezza A, Nouijai A, et al. Bone turnover markers, anterior pituitary and gonadal hormones, and bone mass evaluation using quantitative computed tomography in ankylosing spondylitis. *Clin Rheumatol* 2005;**24**:346–51.
1582. Suehiro RM, Borba EF, Bonfa E, Okay TS, Cocuzza M, Soares PM, et al. Testicular Sertoli cell function in male systemic lupus erythematosus. *Rheumatology (Oxford)* 2008;**47**:1692–7.
1583. Johnson EO, Moutsopoulos HM. Neuroendocrine manifestations in Sjogren's syndrome. Relation to the neurobiology of stress. *Ann N Y Acad Sci* 2000;**917**:797–808.
1584. Tengstrand B, Carlstrom K, Hafstrom I. Bioavailable testosterone in men with rheumatoid arthritis-high frequency of hypogonadism. *Rheumatology (Oxford)* 2002;**41**:285–9.
1585. Silva CA, Deen ME, Febronio MV, Oliveira SK, Terreri MT, Sacchetti SB, et al. Hormone profile in juvenile systemic lupus erythematosus with previous or current amenorrhea. *Rheumatol Int* 2011;**31**:1037–43.
1586. Koller MD, Templ E, Riedl M, Clodi M, Wagner O, Smolen JS, et al. Pituitary function in patients with newly diagnosed untreated systemic lupus erythematosus. *Ann Rheum Dis* 2004;**63**:1677–80.
1587. Villiger PM, Caliezi G, Cottin V, Forger F, Senn A, Ostensen M. Effects of TNF antagonists on sperm characteristics in patients with spondyloarthritis. *Ann Rheum Dis* 2010;**69**:1842–4.
1588. Richter JG, Becker A, Specker C, Schneider M. Hypogonadism in Wegener's granulomatosis. *Scand J Rheumatol* 2008;**37**:365–9.
1589. Aikawa NE, Sallum AM, Leal MM, Bonfa E, Pereira RM, Silva CA. Menstrual and hormonal alterations in juvenile dermatomyositis. *Clin Exp Rheumatol* 2010;**28**:571–5.
1590. Wallenius M, Skomsvoll JF, Irgens LM, Salvesen KA, Nordvag BY, Koldingsnes W, et al. Fertility in women with chronic inflammatory arthritides. *Rheumatology (Oxford)* 2011;**50**:1162–7.

1591. Nakajima A, Sendo W, Tsutsumino M, Koseki Y, Ichikawa N, Akama H, et al. Acute sympathetic hyperfunction in overlapping syndromes of systemic lupus erythematosus and polymyositis. *J Rheumatol* 1998;**25**:1638–41.
1592. Härle P, Straub RH, Wiest R, Mayer A, Schölmerich J, Atzeni F, et al. Increase of sympathetic outflow measured by neuropeptide Y and decrease of the hypothalamic-pituitary-adrenal axis tone in patients with systemic lupus erythematosus and rheumatoid arthritis: another example of uncoupling of response systems. *Ann Rheum Dis* 2006;**65**:51–6.
1593. Dekkers JC, Geenen R, Godaert GL, Bijlsma JW, Doornen LJP. Sympathetic and parasympathetic nervous system activity at night in patients with recently diagnosed rheumatoid arthritis. In: Dekkers JC, editor. Psychophysiological responsiveness in recently diagnosed patients with rheumatoid arthritis (Thesis). Dordrecht: Dekkers; 2003. p. 55–74.
1594. Scherrer U, Sartori C. Insulin as a vascular and sympathoexcitatory hormone: implications for blood pressure regulation, insulin sensitivity, and cardiovascular morbidity. *Circulation* 1997;**96**:4104–13.
1595. Levine JD, Dardick SJ, Roizen MF, Helms C, Basbaum AI. Contribution of sensory afferents and sympathetic efferents to joint injury in experimental arthritis. *J Neurosci* 1986;**6**:3423–9.
1596. Straub RH, del Rey A, Besedovsky HO. Emerging concepts for the pathogenesis of chronic disabling inflammatory diseases: neuroendocrine-immune interactions and evolutionary biology. In: Ader R, editor. Psychoneuroimmunology. San Diego, CA: Elsevier/Academic Press; 2007. p. 217–32.
1597. Straub RH, Cutolo M, Zietz B, Schölmerich J. The process of aging changes the interplay of the immune, endocrine and nervous systems. *Mech Ageing Dev* 2001;**122**:1591–611.
1598. Yasuda M, Yasuda D, Tomooka K, Nobunaga M. Plasma concentration of human atrial natriuretic hormone in patients with connective tissue diseases. *Clin Rheumatol* 1993;**12**:231–5.
1599. Straub RH, Hall C, Kramer BK, Elbracht R, Palitzsch KD, Lang B, et al. Atrial natriuretic factor and digoxin-like immunoreactive factor in diabetic patients: their interrelation and the influence of the autonomic nervous system. *J Clin Endocrinol Metab* 1996;**81**:3385–9.
1600. Peters MJ, Welsh P, McInnes IB, Wolbink G, Dijkmans BA, Sattar N, et al. Tumour necrosis factor blockade reduces circulating N-terminal pro-brain natriuretic peptide levels in patients with active rheumatoid arthritis: results from a prospective cohort study. *Ann Rheum Dis* 2010;**69**:1281–5.
1601. Provan S, Angel K, Semb AG, Atar D, Kvien TK. NT-proBNP predicts mortality in patients with rheumatoid arthritis: results from 10-year follow-up of the EURIDISS study. *Ann Rheum Dis* 2010;**69**:1946–50.
1602. Szekely M. The vagus nerve in thermoregulation and energy metabolism. *Auton Neurosci* 2000;**85**:26–38.
1603. Thorens B, Larsen PJ. Gut-derived signaling molecules and vagal afferents in the control of glucose and energy homeostasis. *Curr Opin Clin Nutr Metab Care* 2004;**7**:471–8.
1604. Mussa BM, Verberne AJ. The dorsal motor nucleus of the vagus and regulation of pancreatic secretory function. *Exp Physiol* 2013;**98**:25–37.
1605. Burcelin R, Dolci W, Thorens B. Portal glucose infusion in the mouse induces hypoglycemia: evidence that the hepatoportal glucose sensor stimulates glucose utilization. *Diabetes* 2000;**49**:1635–42.
1606. Moore MC, Satake S, Baranowski B, Hsieh PS, Neal DW, Cherrington AD. Effect of hepatic denervation on peripheral insulin sensitivity in conscious dogs. *Am J Physiol Endocrinol Metab* 2002;**282**:E286–96.
1607. Vatamaniuk MZ, Horyn OV, Vatamaniuk OK, Doliba NM. Acetylcholine affects rat liver metabolism via type 3 muscarinic receptors in hepatocytes. *Life Sci* 2003;**72**:1871–82.

1608. Xue C, Aspelund G, Sritharan KC, Wang JP, Slezak LA, Andersen DK. Isolated hepatic cholinergic denervation impairs glucose and glycogen metabolism. *J Surg Res* 2000;**90**:19–25.
1609. Inoue S, Nagase H, Satoh S, Saito M, Egawa M, Tanaka K, et al. Role of the efferent and afferent vagus nerve in the development of ventromedial hypothalamic (VMH) obesity. *Brain Res Bull* 1991;**27**:511–5.
1610. Straub RH, Zeuner M, Lock G, Rath H, Hein R, Schölmerich J, et al. Autonomic and sensorimotor neuropathy in patients with systemic lupus erythematosus and systemic sclerosis. *J Rheumatol* 1996;**23**:87–92.
1611. Toussirot E, Bahjaoui-Bouhaddi M, Poncet JC, Cappelle S, Henriet MT, Wendling D, et al. Abnormal autonomic cardiovascular control in ankylosing spondylitis. *Ann Rheum Dis* 1999;**58**:481–7.
1612. Borman P, Gokoglu F, Kocaoglu S, Yorgancioglu ZR. The autonomic dysfunction in patients with ankylosing spondylitis: a clinical and electrophysiological study. *Clin Rheumatol* 2008;**27**:1267–73.
1613. Syngle A, Verma I, Garg N, Krishan P. Autonomic dysfunction in psoriatic arthritis. *Clin Rheumatol* 2013;**32**:1059–64.
1614. Sloan RP, McCreath H, Tracey KJ, Sidney S, Liu K, Seeman T. RR interval variability is inversely related to inflammatory markers: the CARDIA study. *Mol Med* 2007;**13**:178–84.
1615. Paton JF, Boscan P, Pickering AE, Nalivaiko E. The yin and yang of cardiac autonomic control: vago-sympathetic interactions revisited. *Brain Res Brain Res Rev* 2005;**49**:555–65.
1616. Weiss G, Schett G. Anaemia in inflammatory rheumatic diseases. *Nat Rev Rheumatol* 2013;**9**:205–15.
1617. Bertero MT, Caligaris-Cappio F. Anemia of chronic disorders in systemic autoimmune diseases. *Haematologica* 1997;**82**:375–81.
1618. Kötter I, Wacker A, Koch S, Henes J, Richter C, Engel A, et al. Anakinra in patients with treatment-resistant adult-onset Still's disease: four case reports with serial cytokine measurements and a review of the literature. *Semin Arthritis Rheum* 2007;**37**:189–97.
1619. Nicolas G, Bennoun M, Devaux I, Beaumont C, Grandchamp B, Kahn A, et al. Lack of hepcidin gene expression and severe tissue iron overload in upstream stimulatory factor 2 (USF2) knockout mice. *Proc Natl Acad Sci U S A* 2001;**98**:8780–5.
1620. Beard JL. Iron biology in immune function, muscle metabolism and neuronal functioning. *J Nutr* 2001;**131**:568S–79S.
1621. Schaible UE, Kaufmann SH. Iron and microbial infection. *Nat Rev Microbiol* 2004;**2**:946–53.
1622. Marx JJ. Iron and infection: competition between host and microbes for a precious element. *Best Pract Res Clin Haematol* 2002;**15**:411–26.
1623. Theurl I, Fritsche G, Ludwiczek S, Garimorth K, Bellmann-Weiler R, Weiss G. The macrophage: a cellular factory at the interphase between iron and immunity for the control of infections. *Biometals* 2005;**18**:359–67.
1624. Hibbert JM, Creary MS, Gee BE, Buchanan ID, Quarshie A, Hsu LL. Erythropoiesis and myocardial energy requirements contribute to the hypermetabolism of childhood sickle cell anemia. *J Pediatr Gastroenterol Nutr* 2006;**43**:680–7.
1625. Fernandez-Real JM, Lopez-Bermejo A, Ricart W. Cross-talk between iron metabolism and diabetes. *Diabetes* 2002;**51**:2348–54.
1626. Wasserman DH, Lavina H, Lickley A, Vranic M. Effect of hematocrit reduction on hormonal and metabolic responses to exercise. *J Appl Physiol* 1985;**58**:1257–62.
1627. Kopic S, Geibel JP. Gastric acid, calcium absorption, and their impact on bone health. *Physiol Rev* 2013;**93**:189–268.
1628. Feske S. Calcium signalling in lymphocyte activation and disease. *Nat Rev Immunol* 2007;**7**:690–702.

1629. Stapleton FB, Hanissian AS, Miller LA. Hypercalciuria in children with juvenile rheumatoid arthritis: association with hematuria. *J Pediatr* 1985;**107**:235–9.
1630. Sakalli H, Arslan D, Yucel AE. The effect of oral and parenteral vitamin D supplementation in the elderly: a prospective, double-blinded, randomized, placebo-controlled study. *Rheumatol Int* 2012;**32**:2279–83.
1631. Udy AA, Putt MT, Shanmugathasan S, Roberts JA, Lipman J. Augmented renal clearance in the Intensive Care Unit: an illustrative case series. *Int J Antimicrob Agents* 2010;**35**:606–8.
1632. Fabrizi F, Martin P, Dixit V, Messa P. Hepatitis C virus infection and kidney disease: a meta-analysis. *Clin J Am Soc Nephrol* 2012;**7**:549–57.
1633. Pacifici R. Osteoimmunology and its implications for transplantation. *Am J Transplant* 2013;**13**:2245–54.
1634. Braun T, Schett G. Pathways for bone loss in inflammatory disease. *Curr Osteoporos Rep* 2012;**10**:101–8.
1635. Redlich K, Smolen JS. Inflammatory bone loss: pathogenesis and therapeutic intervention. *Nat Rev Drug Discov* 2012;**11**:234–50.
1636. Geusens P, Lems WF. Osteoimmunology and osteoporosis. *Arthritis Res Ther* 2011;**13**:242.
1637. Mundy GR. Osteoporosis and inflammation. *Nutr Rev* 2007;**65**:S147–51.
1638. Schett G, Hayer S, Zwerina J, Redlich K, Smolen JS. Mechanisms of Disease: the link between RANKL and arthritic bone disease. *Nat Clin Pract Rheumatol* 2005;**1**:47–54.
1639. Biber J, Hernando N, Forster I, Murer H. Regulation of phosphate transport in proximal tubules. *Pflugers Arch* 2009;**458**:39–52.
1640. Murer H, Lotscher M, Kaissling B, Levi M, Kempson SA, Biber J. Renal brush border membrane Na/Pi-cotransport: molecular aspects in PTH-dependent and dietary regulation. *Kidney Int* 1996;**49**:1769–73.
1641. Geerse DA, Bindels AJ, Kuiper MA, Roos AN, Spronk PE, Schultz MJ. Treatment of hypophosphatemia in the intensive care unit: a review. *Crit Care* 2010;**14**:R147.
1642. Marinella MA. Refeeding syndrome and hypophosphatemia. *J Intensive Care Med* 2005;**20**:155–9.
1643. Sievanen H. Immobilization and bone structure in humans. *Arch Biochem Biophys* 2010;**503**:146–52.
1644. Uhthoff HK, Jaworski ZF. Bone loss in response to long-term immobilisation. *J Bone Joint Surg (Br)* 1978;**60-B**:420–9.
1645. Tsakalakos N, Magiasis B, Tsekoura M, Lyritis G. The effect of short-term calcitonin administration on biochemical bone markers in patients with acute immobilization following hip fracture. *Osteoporos Int* 1993;**3**:337–40.
1646. Yusuf MB, Ikem IC, Oginni LM, Akinyoola AL, Badmus TA, Idowu AA, et al. Comparison of serum and urinary calcium profile of immobilized and ambulant trauma patients. *Bone* 2013;**57**:361–6.
1647. Rivier C. Neuroendocrine effects of cytokines in the rat. *Rev Neurosci* 1993;**4**:223–37.
1648. Petzke F, Heppner C, Mbulamberi D, Winkelmann W, Chrousos GP, Allolio B, et al. Hypogonadism in Rhodesian sleeping sickness: evidence for acute and chronic dysfunction of the hypothalamic-pituitary-gonadal axis. *Fertil Steril* 1996;**65**:68–75.
1649. Reincke M, Arlt W, Heppner C, Petzke F, Chrousos GP, Allolio B. Neuroendocrine dysfunction in African trypanosomiasis. The role of cytokines. *Ann N Y Acad Sci* 1998;**840**:809–21.

1650. Oktenli C, Doganci L, Ozgurtas T, Araz RE, Tanyuksel M, Musabak U, et al. Transient hypogonadotrophic hypogonadism in males with acute toxoplasmosis: suppressive effect of interleukin-1 beta on the secretion of GnRH. *Hum Reprod* 2004;**19**:859–66.

1651. Kalyani RR, Gavini S, Dobs AS. Male hypogonadism in systemic disease. *Endocrinol Metab Clin North Am* 2007;**36**:333–48.

1652. Amling M, Takeda S, Karsenty G. A neuro(endocrine) regulation of bone remodeling. *Bioessays* 2000;**22**:970–5.

1653. Takeda S, Karsenty G. Molecular bases of the sympathetic regulation of bone mass. *Bone* 2008;**42**:837–40.

1654. Schnedl C, Pieber TR, Amrein K. Vitamin D intervention trials in critical illness. *Inflamm Allergy Drug Targets* 2013;**12**:282–7.

1655. Sauneuf B, Brunet J, Lucidarme O, du Cheyron D. Prevalence and risk factors of vitamin D deficiency in critically ill patients. *Inflamm Allergy Drug Targets* 2013;**12**:223–9.

1656. Cutolo M, Pizzorni C, Sulli A. Vitamin D endocrine system involvement in autoimmune rheumatic diseases. *Autoimmun Rev* 2011;**11**:84–7.

1657. Pinheiro da SF, Zampieri FG, Barbeiro HV, Filho FT, Goulart AC, Jorgetti V, et al. Decreased parathyroid hormone levels despite persistent hypocalcemia in patients with kidney failure recovering from septic shock. *Endocr Metab Immune Disord Drug Targets* 2013;**13**:135–42.

1658. Lind L, Carlstedt F, Rastad J, Stiernstrom H, Stridsberg M, Ljunggren O, et al. Hypocalcemia and parathyroid hormone secretion in critically ill patients. *Crit Care Med* 2000;**28**:93–9.

1659. Toribio RE, Kohn CW, Hardy J, Rosol TJ. Alterations in serum parathyroid hormone and electrolyte concentrations and urinary excretion of electrolytes in horses with induced endotoxemia. *J Vet Intern Med* 2005;**19**:223–31.

1660. Prosnitz AR, Leonard MB, Shults J, Zemel BS, Hollis BW, Denson LA, et al. Changes in vitamin D and parathyroid hormone metabolism in incident pediatric Crohn's disease. *Inflamm Bowel Dis* 2013;**19**:45–53.

1661. Sainaghi PP, Bellan M, Antonini G, Bellomo G, Pirisi M. Unsuppressed parathyroid hormone in patients with autoimmune/inflammatory rheumatic diseases: implications for vitamin D supplementation. *Rheumatology (Oxford)* 2011;**50**:2290–6.

1662. Lange U, Teichmann J, Strunk J, Muller-Ladner U, Schmidt KL. Association of 1.25 vitamin D3 deficiency, disease activity and low bone mass in ankylosing spondylitis. *Osteoporos Int* 2005;**16**:1999–2004.

1663. Jensen T, Hansen M, Madsen JC, Kollerup G, Stoltenberg M, Florescu A, et al. Serum levels of parathyroid hormone and markers of bone metabolism in patients with rheumatoid arthritis. Relationship to disease activity and glucocorticoid treatment. *Scand J Clin Lab Invest* 2001;**61**:491–501.

1664. af Ekenstam E, Benson L, Hallgren R, Wide L, Ljunghall S. Impaired secretion of parathyroid hormone in patients with rheumatoid arthritis: relationship to inflammatory activity. *Clin Endocrinol (Oxf)* 1990;**32**:323–8.

1665. Oxlund H, Ortoft G, Thomsen JS, Danielsen CC, Ejersted C, Andreassen TT. The anabolic effect of PTH on bone is attenuated by simultaneous glucocorticoid treatment. *Bone* 2006;**39**:244–52.

1666. Deans C, Wigmore S, Paterson-Brown S, Black J, Ross J, Fearon KC. Serum parathyroid hormone-related peptide is associated with systemic inflammation and adverse prognosis in gastroesophageal carcinoma. *Cancer* 2005;**103**:1810–8.

1667. Walsh NC, Crotti TN, Goldring SR, Gravallese EM. Rheumatic diseases: the effects of inflammation on bone. *Immunol Rev* 2005;**208**:228–51.

1668. Fierer J, Burton DW, Haghighi P, Deftos LJ. Hypercalcemia in disseminated coccidioidomycosis: expression of parathyroid hormone-related peptide is characteristic of granulomatous inflammation. *Clin Infect Dis* 2012;**55**:e61–6.

1669. Funk JL, Moser AH, Grunfeld C, Feingold KR. Parathyroid hormone-related protein is induced in the adult liver during endotoxemia and stimulates the hepatic acute phase response. *Endocrinology* 1997;**138**:2665–73.

1670. Funk JL, Chen J, Downey KJ, Davee SM, Stafford G. Blockade of parathyroid hormone-related protein prevents joint destruction and granuloma formation in streptococcal cell wall-induced arthritis. *Arthritis Rheum* 2003;**48**:1721–31.

1671. Zdobnov E, GroupOrthoDB. *Database of orthologous groups*; 2014. http://www.orthodb.org.

1672. Parrinello N, Vizzini A, Arizza V, Salerno G, Parrinello D, Cammarata M, et al. Enhanced expression of a cloned and sequenced Ciona intestinalis TNFalpha-like (CiTNF alpha) gene during the LPS-induced inflammatory response. *Cell Tissue Res* 2008;**334**:305–17.

1673. Kasahara M, Suzuki T, Pasquier LD. On the origins of the adaptive immune system: novel insights from invertebrates and cold-blooded vertebrates. *Trends Immunol* 2004;**25**:105–11.

1674. Sherwood NM, Tello JA, Roch GJ. Neuroendocrinology of protochordates: insights from Ciona genomics. *Comp Biochem Physiol A Mol Integr Physiol* 2006;**144**:254–71.

1675. Kiechl S, Wittmann J, Giaccari A, Knoflach M, Willeit P, Bozec A, et al. Blockade of receptor activator of nuclear factor-kappaB (RANKL) signaling improves hepatic insulin resistance and prevents development of diabetes mellitus. *Nat Med* 2013;**19**:358–63.

1676. Keller JJ, Kang JH, Lin HC. Association between osteoporosis and psoriasis: results from the Longitudinal Health Insurance Database in Taiwan. *Osteoporos Int* 2013;**24**:1835–41.

1677. Sambrook PN, Geusens P. The epidemiology of osteoporosis and fractures in ankylosing spondylitis. *Ther Adv Musculoskelet Dis* 2012;**4**:287–92.

1678. Bultink IE, Vis M, van der Horst-Bruinsma IE, Lems WF. Inflammatory rheumatic disorders and bone. *Curr Rheumatol Rep* 2012;**14**:224–30.

1679. Bultink IE. Osteoporosis and fractures in systemic lupus erythematosus. *Arthritis Care Res (Hoboken)* 2012;**64**:2–8.

1680. Kampman MT, Eriksen EF, Holmoy T. Multiple sclerosis, a cause of secondary osteoporosis? What is the evidence and what are the clinical implications? *Acta Neurol Scand Suppl* 2011;**44–9**.

1681. Wohl Y, Dreiher J, Cohen AD. Pemphigus and osteoporosis: a case–control study. *Arch Dermatol* 2010;**146**:1126–31.

1682. Ali T, Lam D, Bronze MS, Humphrey MB. Osteoporosis in inflammatory bowel disease. *Am J Med* 2009;**122**:599–604.

1683. Rooney T, Scherzer R, Shigenaga JK, Graf J, Imboden JB, Grunfeld C. Levels of plasma fibrinogen are elevated in well-controlled rheumatoid arthritis. *Rheumatology (Oxford)* 2011;**50**:1458–65.

1684. Hoppe B, Dorner T. Coagulation and the fibrin network in rheumatic disease: a role beyond haemostasis. *Nat Rev Rheumatol* 2012;**8**:738–46.

1685. van den Oever IA, Sattar N, Nurmohamed MT. Thromboembolic and cardiovascular risk in rheumatoid arthritis: role of the haemostatic system. *Ann Rheum Dis* 2014;**73**:954–7.

1686. Stadnicki A. Involvement of coagulation and hemostasis in inflammatory bowel diseases. *Curr Vasc Pharmacol* 2012;**10**:659–69.

1687. Marzano AV, Tedeschi A, Spinelli D, Fanoni D, Crosti C, Cugno M. Coagulation activation in autoimmune bullous diseases. *Clin Exp Immunol* 2009;**158**:31–6.

1688. Aranow C, Ginzler EM. Epidemiology of cardiovascular disease in systemic lupus erythematosus. *Lupus* 2000;**9**:166–9.

1689. Bruce IN, Gladman DD, Urowitz MB. Premature atherosclerosis in systemic lupus erythematosus. *Rheum Dis Clin North Am* 2000;**26**:257–78.

1690. Cohen Tervaert JW. Cardiovascular disease due to accelerated atherosclerosis in systemic vasculitides. *Best Pract Res Clin Rheumatol* 2013;**27**:33–44.

1691. del Rincon ID, Williams K, Stern MP, Freeman GL, Escalante A. High incidence of cardiovascular events in a rheumatoid arthritis cohort not explained by traditional cardiac risk factors. *Arthritis Rheum* 2001;**44**:2737–45.

1692. Doria A, Sherer Y, Meroni PL, Shoenfeld Y. Inflammation and accelerated atherosclerosis: basic mechanisms. *Rheum Dis Clin North Am* 2005;**31**:355–62.

1693. Doria A, Shoenfeld Y, Wu R, Gambari PF, Puato M, Ghirardello A, et al. Risk factors for subclinical atherosclerosis in a prospective cohort of patients with systemic lupus erythematosus. *Ann Rheum Dis* 2003;**62**:1071–7.

1694. Farhey Y, Hess EV. Accelerated atherosclerosis and coronary disease in SLE. *Lupus* 1997;**6**:572–7.

1695. Fukumoto S, Tsumagari T, Kinjo M, Tanaka K. Coronary atherosclerosis in patients with systemic lupus erythematosus at autopsy. *Acta Pathol Jpn* 1987;**37**:1–9.

1696. Haskard DO. Accelerated atherosclerosis in inflammatory rheumatic diseases. *Scand J Rheumatol* 2004;**33**:281–92.

1697. Ilowite NT. Premature atherosclerosis in systemic lupus erythematosus. *J Rheumatol* 2000;**27**(Suppl. 58):15–9.

1698. Jonsson SW, Backman C, Johnson O, Karp K, Lundstrom E, Sundqvist KG, et al. Increased prevalence of atherosclerosis in patients with medium term rheumatoid arthritis. *J Rheumatol* 2001;**28**:2597–602.

1699. Joris I, Majno G. Atherosclerosis and inflammation. *Adv Exp Med Biol* 1978;**104**:227–43.

1700. La Montagna G, Cacciapuoti F, Buono R, Manzella D, Mennillo GA, Arciello A, et al. Insulin resistance is an independent risk factor for atherosclerosis in rheumatoid arthritis. *Diab Vasc Dis Res* 2007;**4**:130–5.

1701. Manzi S, Wasko MC. Inflammation-mediated rheumatic diseases and atherosclerosis. *Ann Rheum Dis* 2000;**59**:321–5.

1702. Nikpour M, Urowitz MB, Gladman DD. Premature atherosclerosis in systemic lupus erythematosus. *Rheum Dis Clin North Am* 2005;**31**:329–54.

1703. Park YB, Ahn CW, Choi HK, Lee SH, In BH, Lee HC, et al. Atherosclerosis in rheumatoid arthritis: morphologic evidence obtained by carotid ultrasound. *Arthritis Rheum* 2002;**46**:1714–9.

1704. Ramonda R, Lo NA, Modesti V, Nalotto L, Musacchio E, Iaccarino L, et al. Atherosclerosis in psoriatic arthritis. *Autoimmun Rev* 2011;**10**:773–8.

1705. Stojan G, Petri M. Atherosclerosis in systemic lupus erythematosus. *J Cardiovasc Pharmacol* 2013;**62**:255–62.

1706. Szekanecz Z, Koch AE. Vascular involvement in rheumatic diseases: 'vascular rheumatology'. *Arthritis Res Ther* 2008;**10**:224.

1707. Urowitz M, Gladman D, Bruce I. Atherosclerosis and systemic lupus erythematosus. *Curr Rheumatol Rep* 2000;**2**:19–23.

1708. Van Doornum S, McColl G, Wicks IP. Accelerated atherosclerosis: an extraarticular feature of rheumatoid arthritis? *Arthritis Rheum* 2002;**46**:862–73.

1709. Zinger H, Sherer Y, Shoenfeld Y. Atherosclerosis in autoimmune rheumatic diseases-mechanisms and clinical findings. *Clin Rev Allergy Immunol* 2009;**37**:20–8.

1710. Manly DA, Boles J, Mackman N. Role of tissue factor in venous thrombosis. *Annu Rev Physiol* 2011;**73**:515–25.

1711. ten Cate H. Tissue factor-driven thrombin generation and inflammation in atherosclerosis. *Thromb Res* 2012;**129**(Suppl. 2):S38–40.
1712. Bode M, Mackman N. Regulation of tissue factor gene expression in monocytes and endothelial cells: thromboxane A as a new player. *Vasc Pharmacol* 2014;**62**:57–62.
1713. Ingegnoli F, Fantini F, Griffini S, Soldi A, Meroni PL, Cugno M. Anti-tumor necrosis factor alpha therapy normalizes fibrinolysis impairment in patients with active rheumatoid arthritis. *Clin Exp Rheumatol* 2010;**28**:254–7.
1714. Dulai R, Perry M, Twycross-Lewis R, Morrissey D, Atzeni F, Greenwald S. The effect of tumor necrosis factor-alpha antagonists on arterial stiffness in rheumatoid arthritis: a literature review. *Semin Arthritis Rheum* 2012;**42**:1–8.
1715. Esmon CT. Interactions between the innate immune and blood coagulation systems. *Trends Immunol* 2004;**25**:536–42.
1716. Weiss C, Bierhaus A, Kinscherf R, Hack V, Luther T, Nawroth PP, et al. Tissue factor-dependent pathway is not involved in exercise-induced formation of thrombin and fibrin. *J Appl Physiol (1985)* 2002;**92**:211–8.
1717. Bartsch P, Welsch B, Albert M, Friedmann B, Levi M, Kruithof EK. Balanced activation of coagulation and fibrinolysis after a 2-h triathlon. *Med Sci Sports Exerc* 1995;**27**:1465–70.
1718. Herren T, Bartsch P, Haeberli A, Straub PW. Increased thrombin-antithrombin III complexes after 1 h of physical exercise. *J Appl Physiol (1985)* 1992;**73**:2499–504.
1719. Mustonen P, Lepantalo M, Lassila R. Physical exertion induces thrombin formation and fibrin degradation in patients with peripheral atherosclerosis. *Arterioscler Thromb Vasc Biol* 1998;**18**:244–9.
1720. Kestin AS, Ellis PA, Barnard MR, Errichetti A, Rosner BA, Michelson AD. Effect of strenuous exercise on platelet activation state and reactivity. *Circulation* 1993;**88**:1502–11.
1721. Zhu GJ, Abbadini M, Donati MB, Mussoni L. Tissue-type plasminogen activator release in response to epinephrine in perfused rat hindlegs. *Am J Physiol* 1989;**256**:H404–10.
1722. Mustonen P, Lassila R. Epinephrine augments platelet recruitment to immobilized collagen in flowing blood–evidence for a von Willebrand factor-mediated mechanism. *Thromb Haemost* 1996;**75**:175–81.
1723. Goto S, Ikeda Y, Murata M, Handa M, Takahashi E, Yoshioka A, et al. Epinephrine augments von Willebrand factor-dependent shear-induced platelet aggregation. *Circulation* 1992;**86**:1859–63.
1724. Dünser MW, Hasibeder WR. Sympathetic overstimulation during critical illness: adverse effects of adrenergic stress. *J Intensive Care Med* 2009;**24**:293–316.
1725. Frideman M, Rosenman RH, Carroll V. Changes in the serum cholesterol and blood clotting time in men subjected to cyclic variation of occupational stress. *Circulation* 1958;**17**:852–61.
1726. Wright RJ, Newby DE, Stirling D, Ludlam CA, Macdonald IA, Frier BM. Effects of acute insulin-induced hypoglycemia on indices of inflammation: putative mechanism for aggravating vascular disease in diabetes. *Diabetes Care* 2010;**33**:1591–7.
1727. Strother SV, Bull JM, Branham SA. Activation of coagulation during therapeutic whole body hyperthermia. *Thromb Res* 1986;**43**:353–60.
1728. Shibolet S, Fisher S, Gilat T, Bank H, Heller H. Fibrinolysis and hemorrhages in fatal heatstroke. *N Engl J Med* 1962;**266**:169–73.
1729. He M, He X, Xie Q, Chen F, He S. Angiotensin II induces the expression of tissue factor and its mechanism in human monocytes. *Thromb Res* 2006;**117**:579–90.
1730. Celi A, Cianchetti S, Dell'Omo G, Pedrinelli R. Angiotensin II, tissue factor and the thrombotic paradox of hypertension. *Expert Rev Cardiovasc Ther* 2010;**8**:1723–9.

1731. Grant PJ, Davies JA, Tate GM, Boothby M, Prentice CR. Effects of physiological concentrations of vasopressin on haemostatic function in man. *Clin Sci (Lond)* 1985;**69**:471–6.

1732. Tomasiak M, Stelmach H, Rusak T, Ciborowski M, Radziwon P. Vasopressin acts on platelets to generate procoagulant activity. *Blood Coagul Fibrinolysis* 2008;**19**:615–24.

1733. Wun T, Paglieroni T, Lachant NA. Physiologic concentrations of arginine vasopressin activate human platelets in vitro. *Br J Haematol* 1996;**92**:968–72.

1734. Harbuz MS, Rees RG, Eckland D, Jessop DS, Brewerton D, Lightman SL. Paradoxical responses of hypothalamic corticotropin-releasing factor (CRF) messenger ribonucleic acid (mRNA) and CRF-41 peptide and adenohypophysial proopiomelanocortin mRNA during chronic inflammatory stress. *Endocrinology* 1992;**130**:1394–400.

1735. Harbuz MS, Jessop DS, Chowdrey HS, Blackwell JM, Larsen PJ, Lightman SL. Evidence for altered control of hypothalamic CRF in immune-mediated diseases. *Ann N Y Acad Sci* 1995;**771**:449–58.

1736. Ortega A, de Prada MT Perez, Mateos-Caceres PJ, Ramos MP, Gonzalez-Armengol JJ, Gonzalez Del Castillo JM, et al. Effect of parathyroid-hormone-related protein on human platelet activation. *Clin Sci (Lond)* 2007;**113**:319–27.

1737. Doolittle RF. *Stanching the flow: the evolution of vertebrate blood clotting*. Mill Valley, CA: University Science Books; 2012.

1738. Pannell D, Brisebois R, Talbot M, Trottier V, Clement J, Garraway N, et al. Causes of death in Canadian Forces members deployed to Afghanistan and implications on tactical combat casualty care provision. *J Trauma* 2011;**71**:S401–7.

1739. Moore RY, Eichler VB. Loss of a circadian adrenal corticosterone rhythm following suprachiasmatic lesions in the rat. *Brain Res* 1972;**42**:201–6.

1740. Buttgereit F, Doering G, Schaeffler A, Witte S, Sierakowski S, Gromnica-Ihle E, et al. Efficacy of modified-release versus standard prednisone to reduce duration of morning stiffness of the joints in rheumatoid arthritis (CAPRA-1): a double-blind, randomised controlled trial. *Lancet* 2008;**371**:205–14.

1741. Spies CM, Cutolo M, Straub RH, Burmester GR, Buttgereit F. More night than day–circadian rhythms in polymyalgia rheumatica and ankylosing spondylitis. *J Rheumatol* 2010;**37**:894–9.

1742. Gupta MA, Gupta AK. Sleep-wake disorders and dermatology. *Clin Dermatol* 2013;**31**:118–26.

1743. Fox AW, Davis RL. Migraine chronobiology. *Headache* 1998;**38**:436–41.

1744. Marsh III EE, Biller J, Adams Jr HP, Marler JR, Hulbert JR, Love BB, et al. Circadian variation in onset of acute ischemic stroke. *Arch Neurol* 1990;**47**:1178–80.

1745. Ebata T, Aizawa H, Kamide R, Niimura M. The characteristics of nocturnal scratching in adults with atopic dermatitis. *Br J Dermatol* 1999;**141**:82–6.

1746. Smolensky MH, Lemmer B, Reinberg AE. Chronobiology and chronotherapy of allergic rhinitis and bronchial asthma. *Adv Drug Deliv Rev* 2007;**59**:852–82.

1747. Doria A, Iaccarino L, Arienti S, Ghirardello A, Zampieri S, Rampudda ME, et al. Th2 immune deviation induced by pregnancy: the two faces of autoimmune rheumatic diseases. *Reprod Toxicol* 2006;**22**:234–41.

1748. Clowse ME, Magder LS, Witter F, Petri M. The impact of increased lupus activity on obstetric outcomes. *Arthritis Rheum* 2005;**52**:514–21.

1749. Rahman FZ, Rahman J, Al-Suleiman SA, Rahman MS. Pregnancy outcome in lupus nephropathy. *Arch Gynecol Obstet* 2005;**271**:222–6.

1750. Doria A, Cutolo M, Ghirardello A, Zampieri S, Vescovi F, Sulli A, et al. Steroid hormones and disease activity during pregnancy in systemic lupus erythematosus. *Arthritis Rheum* 2002;**47**:202–9.

1751. Chao TC, Phuangsab A, Van Alten PJ, Walter RJ. Steroid sex hormones and macrophage function: regulation of chemiluminescence and phagocytosis. *Am J Reprod Immunol* 1996;**35**:106–13.
1752. Piccinni MP, Giudizi MG, Biagiotti R, Beloni L, Giannarini L, Sampognaro S, et al. Progesterone favors the development of human T helper cells producing Th2-type cytokines and promotes both IL-4 production and membrane CD30 expression in established Th1 cell clones. *J Immunol* 1995;**155**:128–33.
1753. Montes MJ, Tortosa CG, Borja C, Abadia AC, Gonzalez-Gomez F, Ruiz C, et al. Constitutive secretion of interleukin-6 by human decidual stromal cells in culture. Regulatory effect of progesterone. *Am J Reprod Immunol* 1995;**34**:188–94.
1754. Szekeres-Bartho J, Wegmann TG. A progesterone-dependent immunomodulatory protein alters the Th1/Th2 balance. *J Reprod Immunol* 1996;**31**:81–95.
1755. Le N, Yousefi S, Vaziri N, Carandang G, Ocariz J, Cesario T. The effect of beta-estradiol, progesterone and testosterone on the production of human leukocyte derived interferons. *J Biol Regul Homeost Agents* 1988;**2**:199–204.
1756. Szekeres-Bartho J, Kinsky R, Chaouat G. The effect of a progesterone-induced immunologic blocking factor on NK- mediated resorption. *Am J Reprod Immunol* 1990;**24**:105–7.
1757. Mantovani G, Maccio A, Esu S, Lai P, Santona MC, Massa E, et al. Medroxyprogesterone acetate reduces the in vitro production of cytokines and serotonin involved in anorexia/cachexia and emesis by peripheral blood mononuclear cells of cancer patients. *Eur J Cancer* 1997;**33**:602–7.
1758. Correale J, Arias M, Gilmore W. Steroid hormone regulation of cytokine secretion by proteolipid protein- specific CD4+ T cell clones isolated from multiple sclerosis patients and normal control subjects. *J Immunol* 1998;**161**:3365–74.
1759. Hench PS. The reversibility of certain rheumatic and nonrheumatic conditions by the use of cortisone or of the pituitary adrenocotropic hormone. *Ann Intern Med* 1952;**36**:1–38.
1760. Russell AS, Johnston C, Chew C, Maksymowych WP. Evidence for reduced Th1 function in normal pregnancy: a hypothesis for the remission of rheumatoid arthritis. *J Rheumatol* 1997;**24**:1045–50.
1761. Ostensen M, Husby G. Pregnancy and rheumatic disease. A review of recent studies in rheumatoid arthritis and ankylosing spondylitis. *Klin Wochenschr* 1984;**62**:891–5.
1762. Tchorzewski H, Krasomski G, Biesiada L, Glowacka E, Banasik M, Lewkowicz P. IL-12, IL-6 and IFN-gamma production by lymphocytes of pregnant women with rheumatoid arthritis remission during pregnancy. *Mediat Inflamm* 2000;**9**:289–93.
1763. Elenkov IJ, Wilder RL, Bakalov VK, Link AA, Dimitrov MA, Fisher S, et al. IL-12, TNF-alpha, and hormonal changes during late pregnancy and early postpartum: implications for autoimmune disease activity during these times. *J Clin Endocrinol Metab* 2001;**86**:4933–8.
1764. Munoz-Valle JF, Vazquez-Del Mercado M, Garcia-Iglesias T, Orozco-Barocio G, Bernard-Medina G, Martinez-Bonilla G, et al. T(H)1/T(H)2 cytokine profile, metalloprotease-9 activity and hormonal status in pregnant rheumatoid arthritis and systemic lupus erythematosus patients. *Clin Exp Immunol* 2003;**131**:377–84.
1765. Ostensen M, Sicher P, Forger F, Villiger PM. Activation markers of peripheral blood mononuclear cells in late pregnancy and after delivery: a pilot study. *Ann Rheum Dis* 2005;**64**:318–20.
1766. Confavreux C, Hutchinson M, Hours MM, Cortinovis-Tourniaire P, Moreau T. Rate of pregnancy-related relapse in multiple sclerosis. Pregnancy in Multiple Sclerosis Group. *N Engl J Med* 1998;**339**:285–91.
1767. Stagnaro-Green A. Postpartum thyroiditis. *Best Pract Res Clin Endocrinol Metab* 2004;**18**:303–16.

1768. Ilnyckyji A, Blanchard JF, Rawsthorne P, Bernstein CN. Perianal Crohn's disease and pregnancy: role of the mode of delivery. *Am J Gastroenterol* 1999;**94**:3274–8.

1769. Ostensen M, Motta M. Therapy insight: the use of antirheumatic drugs during nursing. *Nat Clin Pract Rheumatol* 2007;**3**:400–6.

1770. Baer AN, Witter FR, Petri M. Lupus and pregnancy. *Obstet Gynecol Surv* 2011;**66**:639–53.

1771. Ostensen M. The effect of pregnancy on ankylosing spondylitis, psoriatic arthritis, and juvenile rheumatoid arthritis. *Am J Reprod Immunol* 1992;**28**:235–7.

1772. Houtchens M. Multiple sclerosis and pregnancy. *Clin Obstet Gynecol* 2013;**56**:342–9.

1773. Daneshpazhooh M, Chams-Davatchi C, Valikhani M, Aghabagheri A, Mortazavizadeh SM, Barzegari M, et al. Pemphigus and pregnancy: a 23-year experience. *Indian J Dermatol Venereol Leprol* 2011;**77**:534–6323.

1774. Weetman AP. Immunity, thyroid function and pregnancy: molecular mechanisms. *Nat Rev Endocrinol* 2010;**6**:311–8.

1775. Schramm C, Herkel J, Beuers U, Kanzler S, Galle PR, Lohse AW. Pregnancy in autoimmune hepatitis: outcome and risk factors. *Am J Gastroenterol* 2006;**101**:556–60.

1776. Parks CG, D'Aloisio AA, DeRoo LA, Huiber K, Rider LG, Miller FW, et al. Childhood socioeconomic factors and perinatal characteristics influence development of rheumatoid arthritis in adulthood. *Ann Rheum Dis* 2013;**72**:350–6.

1777. Spitzer C, Wegert S, Wollenhaupt J, Wingenfeld K, Barnow S, Grabe HJ. Gender-specific association between childhood trauma and rheumatoid arthritis: a case–control study. *J Psychosom Res* 2013;**74**:296–300.

1778. Dube SR, Fairweather D, Pearson WS, Felitti VJ, Anda RF, Croft JB. Cumulative childhood stress and autoimmune diseases in adults. *Psychosom Med* 2009;**71**:243–50.

1779. Herrmann M, Schölmerich J, Straub RH. Stress and rheumatic diseases. *Rheum Dis Clin North Am* 2000;**26**:737–63.

1780. Nielsen NM, Pedersen BV, Stenager E, Koch-Henriksen N, Frisch M. Stressful life-events in childhood and risk of multiple sclerosis: a Danish nationwide cohort study. *Mult Scler* 2014;**20**:1609–15.

1781. O'Donovan A, Cohen BE, Seal KH, Bertenthal D, Margaretten M, Nishimi K, et al. Elevated risk for autoimmune disorders in Iraq and Afghanistan veterans with posttraumatic stress disorder. *Biol Psychiatry* 2014;**77**:365–74. http://dx.doi.org/10.1016/j.biopsych.2014.06.015 Epub ahead of print.

1782. Chen Y, Huang JZ, Qiang Y, Wang J, Han MM. Investigation of stressful life events in patients with systemic sclerosis. *J Zhejiang Univ Sci B* 2008;**9**:853–6.

1783. Dhabhar FS. Effects of stress on immune function: the good, the bad, and the beautiful. *Immunol Res* 2014;**58**:193–210.

1784. Brown RF, Tennant CC, Sharrock M, Hodgkinson S, Dunn SM, Pollard JD. Relationship between stress and relapse in multiple sclerosis: Part II. Direct and indirect relationships. *Mult Scler* 2006;**12**:465–75.

1785. Morell-Dubois S, Carpentier O, Cottencin O, Queyrel V, Hachulla E, Hatron PY, et al. Stressful life events and pemphigus. *Dermatology* 2008;**216**:104–8.

1786. Hunter HJ, Griffiths CE, Kleyn CE. Does psychosocial stress play a role in the exacerbation of psoriasis? *Br J Dermatol* 2013;**169**:965–74.

1787. Burns MN, Nawacki E, Kwasny MJ, Pelletier D, Mohr DC. Do positive or negative stressful events predict the development of new brain lesions in people with multiple sclerosis? *Psychol Med* 2014;**44**:349–59.

1788. Steptoe A, Hamer M, Chida Y. The effects of acute psychological stress on circulating inflammatory factors in humans: a review and meta-analysis. *Brain Behav Immun* 2007;**21**:901–12.
1789. Bierhaus A, Wolf J, Andrassy M, Rohleder N, Humpert PM, Petrov D, et al. A mechanism converting psychosocial stress into mononuclear cell activation. *Proc Natl Acad Sci U S A* 2003;**100**:1920–5.
1790. Glaser R, Kiecolt-Glaser JK. Stress-induced immune dysfunction: implications for health. *Nat Rev Immunol* 2005;**5**:243–51.
1791. de Brouwer SJ, Van MH, Stormink C, Kraaimaat FW, Joosten I, Radstake TR, et al. Immune responses to stress in rheumatoid arthritis and psoriasis. *Rheumatology (Oxford)* 2014;**53**:1844–8.
1792. Vereecke L, Beyaert R, van LG. van LG. The ubiquitin-editing enzyme A20 (TNFAIP3) is a central regulator of immunopathology. *Trends Immunol* 2009;**30**:383–91.
1793. Straub RH. Rheumatoid arthritis: Stress in RA: a trigger of proinflammatory pathways? *Nat Rev Rheumatol* 2014;**10**:516–8.
1794. Imrich R, Rovensky J, Malis F, Zlnay M, Killinger Z, Kvetnansky R, et al. Low levels of dehydroepiandrosterone sulphate in plasma, and reduced sympathoadrenal response to hypoglycaemia in premenopausal women with rheumatoid arthritis. *Ann Rheum Dis* 2005;**64**:202–6.
1795. Straub RH, Pongratz G, Hirvonen H, Pohjolainen T, Mikkelsson M, Leirisalo-Repo M. Acute cold stress in rheumatoid arthritis inadequately activates stress responses and induces an increase of interleukin 6. *Ann Rheum Dis* 2009;**68**:572–8.
1796. Gouin JP, Hantsoo L, Kiecolt-Glaser JK. Immune dysregulation and chronic stress among older adults: a review. *Neuroimmunomodulation* 2008;**15**:251–9.
1797. Ader R. *Psychoneuroimmunology*. San Diego, CA: Elsevier—Academic Press; 2007.
1798. Hirano D, Nagashima M, Ogawa R, Yoshino S. Serum levels of interleukin 6 and stress related substances indicate mental stress condition in patients with rheumatoid arthritis. *J Rheumatol* 2001;**28**:490–5.
1799. Roupe van der Voort C, Heijnen CJ, Wulffraat N, Kuis W, Kavelaars A. Stress induces increases in IL-6 production by leucocytes of patients with the chronic inflammatory disease juvenile rheumatoid arthritis: a putative role for alpha(1)-adrenergic receptors. *J Neuroimmunol* 2000;**110**:223–9.
1800. Straub RH, Kalden JR. Stress of different types increases the proinflammatory load in rheumatoid arthritis. *Arthritis Res Ther* 2009;**11**:114.
1801. Kittner JM, Jacobs R, Pawlak CR, Heijnen CJ, Schedlowski M, Schmidt RE. Adrenaline-induced immunological changes are altered in patients with rheumatoid arthritis. *Rheumatology (Oxford)* 2002;**41**:1031–9.
1802. Motivala SJ, Khanna D, FitzGerald J, Irwin MR. Stress activation of cellular markers of inflammation in rheumatoid arthritis: protective effects of tumor necrosis factor alpha antagonists. *Arthritis Rheum* 2008;**58**:376–83.
1803. Franceschi C, Bonafe M, Valensin S, Olivieri F, De LM, Ottaviani E, et al. Inflamm-aging. An evolutionary perspective on immunosenescence. *Ann N Y Acad Sci* 2000;**908**:244–54.
1804. Reina-San-Martin B, Cosson A, Minoprio P. Lymphocyte polyclonal activation: a pitfall for vaccine design against infectious agents. *Parasitol Today* 2000;**16**:62–7.
1805. Wedderburn LR, Patel A, Varsani H, Woo P. The developing human immune system: T-cell receptor repertoire of children and young adults shows a wide discrepancy in the frequency of persistent oligoclonal T-cell expansions. *Immunology* 2001;**102**:301–9.

1806. Giachino C, Granziero L, Modena V, Maiocco V, Lomater C, Fantini F, et al. Clonal expansions of V delta 1+ and V delta 2+ cells increase with age and limit the repertoire of human gamma delta T cells. *Eur J Immunol* 1994;**24**:1914–8.
1807. Franceschi C, Valensin S, Fagnoni F, Barbi C, Bonafe M. Biomarkers of immunosenescence within an evolutionary perspective: the challenge of heterogeneity and the role of antigenic load. *Exp Gerontol* 1999;**34**:911–21.
1808. Schwab R, Szabo P, Manavalan JS, Weksler ME, Posnett DN, Pannetier C, et al. Expanded CD4+ and CD8+ T cell clones in elderly humans. *J Immunol* 1997;**158**:4493–9.
1809. Globerson A, Effros RB. Ageing of lymphocytes and lymphocytes in the aged. *Immunol Today* 2000;**21**:515–21.
1810. Vaziri H, Schachter F, Uchida I, Wei L, Zhu X, Effros R, et al. Loss of telomeric DNA during aging of normal and trisomy 21 human lymphocytes. *Am J Hum Genet* 1993;**52**:661–7.
1811. Pawelec G. T-cell immunity in the aging human. *Haematologica* 2014;**99**:795–7.
1812. Douek DC, McFarland RD, Keiser PH, Gage EA, Massey JM, Haynes BF, et al. Changes in thymic function with age and during the treatment of HIV infection. *Nature* 1998;**396**:690–5.
1813. Weyand CM, Yang Z, Goronzy JJ. T-cell aging in rheumatoid arthritis. *Curr Opin Rheumatol* 2014;**26**:93–100.
1814. Abo T, Watanabe H, Sato K, Iiai T, Moroda T, Takeda K, et al. Extrathymic T cells stand at an intermediate phylogenetic position between natural killer cells and thymus-derived T cells. *Nat Immun* 1995;**14**:173–87.
1815. Tarazona R, DelaRosa O, Alonso C, Ostos B, Espejo J, Pena J, et al. Increased expression of NK cell markers on T lymphocytes in aging and chronic activation of the immune system reflects the accumulation of effector/senescent T cells. *Mech Ageing Dev* 2000;**121**:77–88.
1816. Bruunsgaard H, Skinhoj P, Pedersen AN, Schroll M, Pedersen BK. Ageing, tumour necrosis factor-alpha (TNF-alpha) and atherosclerosis. *Clin Exp Immunol* 2000;**121**:255–60.
1817. Bruunsgaard H, Pedersen BK. Age-related inflammatory cytokines and disease. *Immunol Allergy Clin North Am* 2003;**23**:15–39.
1818. Fagiolo U, Cossarizza A, Scala E, Fanales-Belasio E, Ortolani C, Cozzi E, et al. Increased cytokine production in mononuclear cells of healthy elderly people. *Eur J Immunol* 1993;**23**:2375–8.
1819. Inadera H, Egashira K, Takemoto M, Ouchi Y, Matsushima K. Increase in circulating levels of monocyte chemoattractant protein-1 with aging. *J Interferon Cytokine Res* 1999;**19**:1179–82.
1820. Satoh T, Brown LM, Blattner WA, Maloney EM, Kurman CC, Nelson DL, et al. Serum neopterin, beta2-microglobulin, soluble interleukin-2 receptors, and immunoglobulin levels in healthy adolescents. *Clin Immunol Immunopathol* 1998;**88**:176–82.
1821. Fong YM, Marano MA, Moldawer LL, Wei H, Calvano SE, Kenney JS, et al. The acute splanchnic and peripheral tissue metabolic response to endotoxin in humans. *J Clin Invest* 1990;**85**:1896–904.
1822. Meseguer V, Alpizar YA, Luis E, Tajada S, Denlinger B, Fajardo O, et al. TRPA1 channels mediate acute neurogenic inflammation and pain produced by bacterial endotoxins. *Nat Commun* 2014;**5**(3125):1–14.
1823. Hutchinson MR, Buijs M, Tuke J, Kwok YH, Gentgall M, Williams D, et al. Low-dose endotoxin potentiates capsaicin-induced pain in man: evidence for a pain neuroimmune connection. *Brain Behav Immun* 2013;**30**:3–11.
1824. de Goeij M, van Eijk LT, Vanelderen P, Wilder-Smith OH, Vissers KC, van der Hoeven JG, et al. Systemic inflammation decreases pain threshold in humans in vivo. *PLoS One* 2013;**8**:e84159.
1825. Quan N. In-depth conversation: spectrum and kinetics of neuroimmune afferent pathways. *Brain Behav Immun* 2014;**40**:1–8.

1826. Turnbull AV, Pitossi FJ, Lebrun JJ, Lee S, Meltzer JC, Nance DM, et al. Inhibition of tumor necrosis factor-alpha action within the CNS markedly reduces the plasma adrenocorticotropin response to peripheral local inflammation in rats. *J Neurosci* 1997;**17**:3262–73.
1827. Pollard LC, Choy EH, Gonzalez J, Khoshaba B, Scott DL. Fatigue in rheumatoid arthritis reflects pain, not disease activity. *Rheumatology (Oxford)* 2006;**45**:885–9.
1828. Kirkwood TB, Austad SN. Why do we age? *Nature* 2000;**408**:233–8.
1829. Lamberts SW, van den Beld AW, van der Lely AJ. The endocrinology of aging. *Science* 1997;**278**:419–24.
1830. Orentreich N, Brind JL, Rizer RL, Vogelman JH. Age changes and sex differences in serum dehydroepiandrosterone sulfate concentrations throughout adulthood. *J Clin Endocrinol Metab* 1984;**59**:551–5.
1831. Perry HM. The endocrinology of aging. *Clin Chem* 1999;**45**:1369–76.
1832. Corpas E, Harman SM, Blackman MR. Human growth hormone and human aging. *Endocr Rev* 1993;**14**:20–39.
1833. MacLaughlin J, Holick MF. Aging decreases the capacity of human skin to produce vitamin D3. *J Clin Invest* 1985;**76**:1536–8.
1834. Orwoll ES, Meier DE. Alterations in calcium, vitamin D, and parathyroid hormone physiology in normal men with aging: relationship to the development of senile osteopenia. *J Clin Endocrinol Metab* 1986;**63**:1262–9.
1835. Tsai KS, Heath H, Kumar R, Riggs BL. Impaired vitamin D metabolism with aging in women. Possible role in pathogenesis of senile osteoporosis. *J Clin Invest* 1984;**73**:1668–72.
1836. Straub RH, Thies U, Jeron A, Palitzsch KD, Schölmerich J. Valid parameters for investigation of the pupillary light reflex in normal and diabetic subjects shown by factor analysis and partial correlation. *Diabetologia* 1994;**37**:414–9.
1837. Wieling W, van Brederode JF, de Rijk LG, Borst C, Dunning AJ. Reflex control of heart rate in normal subjects in relation to age: a data base for cardiac vagal neuropathy. *Diabetologia* 1982;**22**:163–6.
1838. Iwase S, Mano T, Watanabe T, Saito M, Kobayashi F. Age-related changes of sympathetic outflow to muscles in humans. *J Gerontol* 1991;**46**:M1–5.
1839. Jones PP, Davy KP, Alexander S, Seals DR. Age-related increase in muscle sympathetic nerve activity is associated with abdominal adiposity. *Am J Physiol* 1997;**272**:E976–80.
1840. Vanhoutte PM. Aging and vascular responsiveness. *J Cardiovasc Pharmacol* 1988;**12**(Suppl. 8):S11–9.
1841. Yamada Y, Miyajima E, Tochikubo O, Matsukawa T, Ishii M. Age-related changes in muscle sympathetic nerve activity in essential hypertension. *Hypertension* 1989;**13**:870–7.
1842. Palmer GJ, Ziegler MG, Lake CR. Response of norepinephrine and blood pressure to stress increases with age. *J Gerontol* 1978;**33**:482–7.
1843. Esler M, Skews H, Leonard P, Jackman G, Bobik A, Korner P. Age-dependence of noradrenaline kinetics in normal subjects. *Clin Sci* 1981;**60**:217–9.
1844. Hetland ML, Eldrup E, Bratholm P, Christensen NJ. The relationship between age and venous plasma concentrations of noradrenaline, catecholamine metabolites, DOPA and neuropeptide Y-like immunoreactivity in normal human subjects. *Scand J Clin Lab Invest* 1991;**51**:219–24.
1845. Kerckhoffs DA, Blaak EE, Van Baak MA, Saris WH. Effect of aging on beta-adrenergically mediated thermogenesis in men. *Am J Physiol* 1998;**274**:E1075–9.
1846. Mazzeo RS, Rajkumar C, Jennings G, Esler M. Norepinephrine spillover at rest and during submaximal exercise in young and old subjects. *J Appl Physiol* 1997;**82**:1869–74.
1847. Bellinger DL, Madden KS, Lorton D, Thyagarajan S, Felten DL. Age-related alterations in neural-immune interactions and neural strategies in immunosenescence. In: Ader R, Felten DL, Cohen N, editors. Psychoneuroimmunology. San Diego: Academic Press; 2001. p. 241–86.

1848. Nakata T, Nakajima K, Yamashina S, Yamada T, Momose M, Kasama S, et al. A pooled analysis of multicenter cohort studies of (123)I-MIBG imaging of sympathetic innervation for assessment of long-term prognosis in heart failure. *JACC Cardiovasc Imaging* 2013;**6**:772–84.
1849. Sakata K, Iida K, Mochizuki N, Ito M, Nakaya Y. Physiological changes in human cardiac sympathetic innervation and activity assessed by (123)I-metaiodobenzylguanidine (MIGB) imaging. *Circ J* 2009;**73**:310–5.
1850. Strogatz S. *SYNC—how order emerges from chaos in the universe, nature, and daily life.* New York: Hyperion; 2003.
1851. Hammers CM, Chen J, Lin C, Kacir S, Siegel DL, Payne AS, et al. Persistence of anti-desmoglein 3 IgG B-cell clones in pemphigus patients over years. *J Invest Dermatol* 2015;**135**:742–9.
1852. Sanchez RR, Pauli ML, Neuhaus IM, Yu SS, Arron ST, Harris HW, et al. Memory regulatory T cells reside in human skin. *J Clin Invest* 2014;**124**:1027–36.
1853. McDermott MF, Aksentijevich I, Galon J, McDermott EM, Ogunkolade BW, Centola M, et al. Germline mutations in the extracellular domains of the 55 kDa TNF receptor, TNFR1, define a family of dominantly inherited autoinflammatory syndromes. *Cell* 1999;**97**:133–44.
1854. Martinon F, Aksentijevich I. New players driving inflammation in monogenic autoinflammatory diseases. *Nat Rev Rheumatol* 2015;**11**:11–20.
1855. Frazer IH. Autoimmunity and persistent viral infection: two sides of the same coin? *J Autoimmun* 2008;**31**:216–8.
1856. Croia C, Serafini B, Bombardieri M, Kelly S, Humby F, Severa M, et al. Epstein-Barr virus persistence and infection of autoreactive plasma cells in synovial lymphoid structures in rheumatoid arthritis. *Ann Rheum Dis* 2013;**72**:1559–68.
1857. Cavalcante P, Serafini B, Rosicarelli B, Maggi L, Barberis M, Antozzi C, et al. Epstein-Barr virus persistence and reactivation in myasthenia gravis thymus. *Ann Neurol* 2010;**67**:726–38.
1858. Seksik P, Sokol H, Lepage P, Vasquez N, Manichanh C, Mangin I, et al. The role of bacteria in onset and perpetuation of inflammatory bowel disease. *Aliment Pharmacol Ther* 2006;**24**(Suppl. 3):11–8.
1859. Germain RN. Maintaining system homeostasis: the third law of Newtonian immunology. *Nat Immunol* 2012;**13**:902–6.
1860. Buckley CD, Pilling D, Lord JM, Akbar AN, Scheel-Toellner D, Salmon M. Fibroblasts regulate the switch from acute resolving to chronic persistent inflammation. *Trends Immunol* 2001;**22**:199–204.
1861. Ospelt C, Reedquist KA, Gay S, Tak PP. Inflammatory memories: is epigenetics the missing link to persistent stromal cell activation in rheumatoid arthritis? *Autoimmun Rev* 2011;**10**:519–24.
1862. Ospelt C, Gay S. The role of resident synovial cells in destructive arthritis. *Best Pract Res Clin Rheumatol* 2008;**22**:239–52.
1863. Misu T, Hoftberger R, Fujihara K, Wimmer I, Takai Y, Nishiyama S, et al. Presence of six different lesion types suggests diverse mechanisms of tissue injury in neuromyelitis optica. *Acta Neuropathol* 2013;**125**:815–27.
1864. Adams KM, Nelson JL. Microchimerism: an investigative frontier in autoimmunity and transplantation. *JAMA* 2004;**291**:1127–31.
1865. Prigogine I. Time, structure, and fluctuations: nobel lecture, 8 December 1977. *Science* 1978;**201**:777–85.
1866. Jantsch E. *The self-organizing universe—scientific and human implications of the emerging paradigm of evolution.* Oxford: Pergamon Press; 1980.

1867. Eigen M, Schuster P, Schuster P. *The hypercycle, a principle of natural self-organization.* Berlin: Springer-Verlag; 1978.

1868. Meyer-Hermann M, Figge MT, Straub RH. Mathematical modeling of the circadian rhythm of key neuroendocrine-immune system players in rheumatoid arthritis: a systems biology approach. *Arthritis Rheum* 2009;**60**:2585–94.

1869. Hall JM, Couse JF, Korach KS. The multifaceted mechanisms of estradiol and estrogen receptor signaling. *J Biol Chem* 2001;**276**:36869–72.

1870. Wiest R, Moleda L, Zietz B, Hellerbrand C, Schölmerich J, Straub RH. Uncoupling of sympathetic nervous system and hypothalamic-pituitary-adrenal axis in cirrhosis. *J Gastroenterol Hepatol* 2008;**23**:1901–8.

1871. Weidler C, Holzer C, Harbuz M, Hofbauer R, Angele P, Schölmerich J, et al. Low density of sympathetic nerve fibres and increased density of brain derived neurotrophic factor positive cells in RA synovium. *Ann Rheum Dis* 2005;**64**:13–20.

1872. Haas S, Straub RH. Disruption of rhythms of molecular clocks in primary synovial fibroblasts of patients with osteoarthritis and rheumatoid arthritis, role of IL-1beta/TNF. *Arthritis Res Ther* 2012;**14**:R122.

1873. Straub RH, Miller LE, Schölmerich J, Zietz B. Cytokines and hormones as possible links between endocrinosenescence and immunosenescence. *J Neuroimmunol* 2000;**109**:10–5.

1874. Wolff C, Wildmann J, Randolf A, Basedovsky HO, del Rey A, Straub RH. Mimicking disruption of the brain – immune system – joint communication results in expression of collagen type II-induced arthritis in non-susceptible PVG rats [abstract]. *Arthritis Rheum* 2011;**63**(Suppl):S442.

1875. DIAbetes Genetics Replication and Meta-analysis (DIAGRAM) Consortium, Mahajan A., Go M.J., Zhang W., Below J.E., Gaulton K.J., et al. Genome-wide trans-ancestry meta-analysis provides insight into the genetic architecture of type 2 diabetes susceptibility. *Nat Genet* 2014;**46**:234–44.

Index

Note: Page numbers followed by *f* indicate figures and *t* indicate tables.

B

C

D

E

F

I

J

K

L

M

P

Q

R

S

T

U

V

Printed in the United States
By Bookmasters